Programming in C++
for Engineering and Science

Programming in C++
for Engineering and Science

Larry Nyhoff

CRC Press
Taylor & Francis Group
Boca Raton London New York

CRC Press is an imprint of the
Taylor & Francis Group, an **informa** business

A CHAPMAN & HALL BOOK

CRC Press
Taylor & Francis Group
6000 Broken Sound Parkway NW, Suite 300
Boca Raton, FL 33487-2742

© 2013 by Taylor & Francis Group, LLC
CRC Press is an imprint of Taylor & Francis Group, an Informa business

No claim to original U.S. Government works

Printed in the United States of America on acid-free paper
Version Date: 20120711

International Standard Book Number: 978-1-4398-2534-1 (Paperback)

Library of Congress Cataloging-in-Publication Data

Nyhoff, Larry R.
    Programming in C++ for engineering and science / Larry Nyhoff.
       p. cm.
    Includes index.
    ISBN 978-1-4398-2534-1 (pbk.)
    1. Engineering--Data processing. 2. Science--Data processing. 3. C++ (Computer program language) I. Title.

TA345.N94 2012
005.13'3--dc23
                                                                 2012010215

Visit the Taylor & Francis Web site at
http://www.taylorandfrancis.com

and the CRC Press Web site at
http://www.crcpress.com

Contents

Preface

C++ is a general-purpose programming language that has both high-level and low-level language features. Bjarne Stroustrup developed it in 1979 at Bell Labs as a series of enhancements to the C programming language, which, although developed for system programming, has been used increasingly in engineering and scientific applications.

Because the first enhancement was the addition of classes, the resulting language was originally named "C with Classes," but was renamed C++ in 1983.

Bjarne Stroustrup

Along with overcoming some of the dangers and disadvantages of C, these and subsequent enhancements have resulted in a very powerful language in which very efficient programs can be written and developed using the object-oriented paradigm. A programming language standard for C++ (ISO/IEC148821998) was adopted in 1998 and revised in 2003 and is the basis for this text.

BACKGROUND AND CONTENT

This text grew out of many years of teaching courses in computing, including programming courses intended for students majoring in engineering and science. Although the Fortran language was first used, these courses are now taught using C++. However, most C++ textbooks are written for the general college student and thus include examples and some content that is not aimed at or especially relevant to science and engineering students.

In this text, nearly all of the examples and exercises involve engineering and scientific applications, including the following (and many more):

- Temperature conversion

- Radioactive decay

- Einstein's equation

- Pollution indexes

- Digital circuits

- Root finding, integration, differential equations

- Internet addresses

- A-C circuits

- Simulation

- Quality control

- Street networks

- Environmental data analysis

- Searching a chemistry database

- Oceanographic data analysis

- Electrical networks

- Coordinate transformations

- Data encryption

- Beam deflection

- Weather data analysis

- Oceanographic data analysis

Some examples are described and solved in detail, while for others the presentation in the text outlines the solution and the complete development is available on the text's website maintained by the author:

http://cs.calvin.edu/books/c++/engr-sci

This text also focuses on those features of C++ that are most important in engineering and science applications, with other features described in optional sections, appendices, or on the text's website. This makes it useable in a variety of courses ranging from a regular full-credit course to one with reduced credit such as a two-credit course that the author has taught many times, where the class lectures are supplemented by lab exercises—tutorial in nature—in which the students develop a program to solve some problem using the new language features presented in class.

PRESENTATION

The basic approach of the text is a *spiral* approach that revisits topics in increasingly more detail. For example, the basic C++ operations used to build expressions are presented first, and then predefined functions provided in C++ libraries are added. Once students have experience with functions, they learn how to define their own simple functions and then more complicated ones. Later they learn how to incorporate these into libraries of their own, thus extending the C++ language with custom-designed libraries.

Learning how to develop a program from scratch, however, can be a difficult and challenging task for novice programmers. A methodology used in this text for designing programs to solve problems, developed over years of teaching C++ to computer science, engineering, and science students and coauthoring texts in C++, is called *object-centered design* (OCD):

- Identify the objects in the problem that need to be processed.

- Identify the operations needed to do this processing.

- Develop an algorithm for this processing.

- Implement these objects, operations, and algorithm in a program.

- Test, correct, and revise the program.

Although this approach cannot technically be called *object-oriented design* (OOD), it does focus on the objects and operations on these objects in a problem. As new language constructs are learned, they are incorporated into the design process. For example, simple types of objects are used in early chapters, but Chapter 7 introduces students to some of the standard classes provided in C++ for processing more complex objects—those that have multiple attributes. In subsequent chapters, more classes are introduced and explained, and students gain more practice in using them and understanding the structure of a class. Once they have a good understanding of these predefined standard classes, in Chapter 14 they learn how to build their own classes to model objects, thus extending the C++ language to include a new custom-built type.

IMPORTANT FEATURES

- Standard C++ is used throughout.

- A "use it first—build it later" approach is used for key concepts such as functions (use predefined functions first, build functions later) and classes (use predefined classes first, build classes later). Various other topics are similarly introduced early and used, and are expanded later—a spiral kind of approach.

- The very powerful and useful Standard Template Library (STL) is introduced and some of the important class templates (e.g., vector) and function templates (e.g., sort()) are presented in detail.

- C++'s language features that are not provided in C are noted.

- Engineering and science examples, including numeric techniques, are emphasized.

- Programs for some examples are developed in detail; for others, the design of a program is outlined and a complete development is available on the text's website.

- Object-centered design (OCD) helps students develop programs to solve problems.

- Proper techniques of design and style are emphasized and used throughout.

- Test-yourself questions (with answers supplied) provide a quick check of understanding of the material being studied.

- Chapter summaries highlight key terms, important points, design and style suggestions, and common programming pitfalls.

- Each chapter has a carefully selected set of programming projects of varying degrees of difficulty that make use of the topics presented in that chapter. Solutions of selected projects are available on an instructor's website and can be used for in-class presentations.

PLANNED SUPPLEMENTARY MATERIALS

- A lab manual (perhaps online) containing laboratory exercises and projects coordinated with the text

- A website (http://cs.calvin.edu/books/c++/engr-sci) for the text containing

  - Source code for the programs in the text

  - Expanded presentations and source code for some examples

  - Links to important sites that correspond to items in the text

  - Corrections, additions, reference materials, and other supplementary materials

- A website for instructors containing

  - PowerPoint slides to use in class presentations

  - Solutions to exercises

- Other instructional materials and links to relevant items of interest

Acknowledgments

I EXPRESS MY SPECIAL APPRECIATION to Alan Apt, whose friendship extends over many years and who encouraged me to write this text; to Randi Cohen, David Tumarkin, Suzanne Lassandro, and Jennifer Ahringer, who managed all the details involved in getting it into production; and to Yong Bakos, for his technical review of the manuscript. And, of course, I pay homage to my wife, Shar, and to our children and grandchildren—Jeff, Rebecca, Megan, and Sara; Jim; Greg, Julie, Joshua, Derek, and Isabelle; Tom, Joan, Abigail, Micah, Lucas, Gabriel, Eden, and Josiah—for their love and understanding when my busyness restricted the time I could spend with them. Above all, I give thanks to God for the opportunity and ability to prepare this text.

About the Author

AFTER GRADUATING FROM CALVIN College in 1960 with a degree in mathematics, Larry Nyhoff went on to earn a master's degree in mathematics from the University of Michigan in 1961, and then returned to Calvin in 1963 to teach. After earning his PhD from Michigan State University in 1969, he settled in for an anticipated lifelong career as a mathematics professor and coauthored his first textbook, *Essentials of College Mathematics* (Holt, Rinehart, Winston, Inc.), in 1969.

However, as students began clamoring for computing courses in the '70s, Professor Nyhoff volunteered to help develop a curriculum and coauthored several manuals for the BASIC, FORTRAN, and COBOL programming languages. Following graduate work in computer science at Western Michigan University from 1981–1983, he made the transition from mathematics to computing and became a professor in the newly formed Computer Science Department.

A long stint of textbook writing soon commenced, beginning with a coauthored FORTRAN 77 programming text that was published by Macmillan in 1983. This was then followed by a Pascal programming text, which went through three editions and became a top seller. Over 25 other books followed, covering FORTRAN 90, Turbo Pascal, Modula-2, and Java, and including three editions of a very popular C++ text and an introductory text in data structures using C++. Several of these texts are still used world-wide and some have been translated into other languages, including Spanish, Chinese, and Greek.

A year before his retirement in 2003, after 41 years of full-time teaching, Professor Nyhoff was awarded the Presidential Award for Exemplary Teaching, Calvin College's highest faculty honor. Since retirement, he has continued instructing part-time, teaching sections of "Applied C++," a two-credit course required of all engineering students and also taken by several science students. This textbook is the result of preliminary versions used in that course over several semesters.

Introduction to Computing

CONTENTS

I wish these calculations had been executed by steam.

CHARLES BABBAGE

One machine can do the work of fifty ordinary men. No machine can do the work of one extraordinary man.

ELBERT HUBBARD

Where a computer like the ENIAC is equipped with 18,000 vacuum tubes and weighs 30 tons, computers in the future may have only 1000 vacuum tubes and weigh only 1-1/2 tons.

POPULAR MECHANICS (MARCH 1949)

640K ought to be enough for anyone.

BILL GATES (1981)

So IBM has equipped all XTs with what it considers to be the minimum gear for a serious personal computer. Now the 10-megabyte disk and the 128K of memory are naturals for a serious machine.

PETER NORTON (1983)

THE MODERN ELECTRONIC COMPUTER is one of the most important products of the twentieth century. It is an essential tool in many areas, including business, industry, government, science, and education; indeed, it has touched nearly every aspect of our lives. The impact of the twentieth-century information revolution brought about by the development of high-speed computing systems has been nearly as widespread as the impact of the nineteenth-century industrial revolution. In this chapter we begin with some background by describing computing systems, their main components, and how information is stored in them.

Early computers were very difficult to program. In fact, programming some of the earliest computers consisted of designing and building circuits to carry out the computations required to solve each new problem. Later, computer instructions could be coded in a language that the machine could understand. But these codes were very cryptic, and programming was therefore very tedious and error prone. Computers would not have gained widespread use if it had not been for the development of high-level programming languages that made it possible to enter instructions using an English-like syntax.

Fortran, C, C++, Java, and Python are some of the languages that are used extensively in engineering and scientific applications. This text will focus on C++ but will also describe some properties of its parent language, C, noting features that these two languages have in common, as well as their differences.

1.1 COMPUTING SYSTEMS

Four important concepts have shaped the history of computing:

1. The mechanization of arithmetic

2. The stored program

3. The graphical user interface

4. The computer network

This section briefly describes a few of the important events and devices that have implemented these concepts. Additional information can be found on the website for this book described in the preface.

1.1.1 Machines to Do Arithmetic

One of the earliest "personal calculators" was the *abacus* (Figure 1.1a), with movable beads strung on rods to count and to do calculations. Although its exact origin is unknown, the abacus was used by the Chinese perhaps 3000 to 4000 years ago and is still used today throughout Asia. Early merchants used the abacus in trading transactions. The ancient British stone monument *Stonehenge* (Figure 1.1b), located near Salisbury, England, was built between 1900 and 1600 BC and, evidently, was used to predict the changes of the seasons. In the twelfth century, a Persian teacher of mathematics in Baghdad, *Muhammad ibn-Musa al-Khowarizm*, developed some of the first step-by-step procedures for doing computations. The word *algorithm,* used for such procedures, is derived from his name.

(a)

(b)

(c)

FIGURE 1.1 (a) Abacus. (Image courtesy of the Computer History Museum.) (b) Stonehenge. (c) Slide rule.

The English mathematician William Oughtred invented a circular *slide rule* in the early 1600s, and more modern ones (Figure 1.1c) were used by engineers and scientists through the 1950s and into the 1960s to do rapid approximate computations.

In 1642, the young French mathematician *Blaise Pascal* invented one of the first mechanical adding machines to help his father with calculating taxes. This Pascaline (Figure 1.2a) was a *digital* calculator because it represented numerical information as discrete digits, as opposed to a graduated scale like that used in analog instruments of measurement such as slide rules and nondigital thermometers. Each digit was represented by a gear that had 10 different positions (a ten-state device) so that it could "count" from 0 through 9 and, upon reaching 10, would reset to 0 and advance the gear in the next column so as to represent the action of "carrying" to the next digit. In 1673, the German mathematician *Gottfried Wilhelm von Leibniz* invented an improved mechanical calculator (Figure 1.2b) that also used a system of gears and dials to do calculations. However, it was more reliable and accurate than the Pascaline and could perform all four of the basic arithmetic operations of addition, subtraction, multiplication, and division. A number of other mechanical calculators followed that further refined Pascal's and Leibniz's designs, and by the end of the nineteenth century, these calculators had become important tools in science, business, and commerce.

1.1.2 The Stored Program Concept

The fundamental idea that distinguishes computers from calculators is the concept of a *stored program* that controls the computation. A program is a sequence of instructions that the computer follows to solve some problem. An income tax form is a good analogy. Although a calculator can be a useful tool in the process, computing taxes involves much more than arithmetic. To produce the correct result, one must execute the form's precise sequence of steps of writing numbers down (storage), looking numbers up (retrieval), and computation to produce the correct result.

The stored program concept also gives the computer its amazing versatility. Unlike most other machines, which are engineered to mechanize a single task, a computer can be programmed to perform many different tasks. Although its hardware is designed for a very specific task—the mechanization of arithmetic—computer software programs enable the computer to perform a wide variety of tasks, from navigational control of the space shuttle to word processing to musical composition.

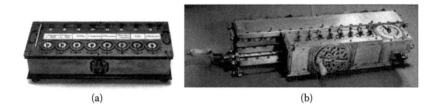

(a) (b)

FIGURE 1.2 (a) Pascaline. (b) Leibnitz's calculator. (Images courtesy of the Computer History Museum.)

The *Jacquard loom* (Figure 1.3a), invented in 1801 by the Frenchman Joseph Marie Jacquard, is an early example of a stored program automatically controlling a hardware device. Holes punched in metal cards directed the action of this loom: a hole punched in one of the cards would enable its corresponding thread to come through and be incorporated into the weave at a given point in the process; the absence of a hole would exclude an undesired thread. To change to a different weaving pattern, the operator of this loom would simply switch to another set of cards. Jacquard's loom is thus one of the first examples of a programmable machine, and many later computers would make similar use of punched cards.

(a)　　　　　　　　　　(b)

(c)

FIGURE 1.3   (a) Jacquard Loom. (Image courtesy of the Computer History Museum.) (b) Charles Babbage. (c) Difference Engine.

The English mathematician *Charles Babbage* (1792–1871) (Figure 1.3b) combined the two fundamental concepts of mechanized calculation and stored program control. In 1822, supported by the British government, he began work on a machine that he called the *Difference Engine* (Figure 1.3c). Comprised of a system of gears, the Difference Engine was designed to compute polynomials for preparing mathematical tables.

Babbage abandoned this effort and began the design of a much more sophisticated machine that he called his *Analytical Engine* (Figure 1.4a). It was to have over 50,000 components, and its operation was to be far more versatile and fully automatic, controlled by programs stored on punched cards, an idea based on Jacquard's earlier work. Although this machine was not built during his lifetime, it is an important part of the history of computing because many of the concepts of its design are used in modern computers. For this reason, Babbage is sometimes called the "Father of Computing." *Ada Augusta* (Figure 1.4b), Lord Byron's daughter, was one of the few people other than Babbage who understood the Analytical Engine's design. This enabled her to develop "programs" for the machine, and for this reason she is sometimes called "the first programmer." In the 1980s, the programming language Ada was named in her honor.

(a)

(b)

FIGURE 1.4   (a) Analytical Engine. (b) Ada Augusta.

During the next 100 years, the major significant event was the invention by *Herman Hollerith* of an electric tabulating machine (Figure 1.5a) that could tally census statistics stored on punched cards. This was noteworthy because the U.S. Census Bureau feared it would not be possible to complete the 1890 census before the next one was to be taken, but Hollerith's machine enabled it to be completed in 2-1/2 years. The Hollerith Tabulating Company later merged with other companies to form the International Business Machines (IBM) Corporation in 1924.

The development of electromechanical computing devices continued at a rapid pace for the next few decades. These included the "Z" machines, developed by the German engineer *Konrad Zuse* in the 1930s, which used *binary* arithmetic instead of decimal so that two-state devices could be used instead of ten-state devices. Some of his later machines replaced

(a)

(b)

FIGURE 1.5 (a) Hollerith's tabulating machine. (b) Harvard Mark I. (Images courtesy of the Computer History Museum.)

mechanical relays with vacuum tubes. Zuse also designed a high-level programming language called Plankalkül. World War II also spurred the development of computing devices, including the Collosus computers developed by Alan Turing and a British team to break codes generated by Germany's Enigma machine. The best-known computer built before 1945 was probably the Harvard Mark I (Figure 1.5b). Like Zuse's "Z" machines, it was driven by electromechanical relay technology. Repeating much of the work of Babbage, Howard Aiken and others at IBM constructed this large, automatic, general-purpose, electromechanical calculator, sponsored by the U.S. Navy and intended to compute mathematical and navigational tables.

In 1944, *Grace Murray Hopper* (1907–1992) began work as a *coder*—what we today would call a *programmer*—for the Mark I. Later, while working on its successor, the Mark II, she found one of the first computer "**bugs**"—an actual bug stuck in one of the thousands of relays.[1] To this day, efforts to find the cause of errors in programs are still referred to as "debugging." In the late 1950s, "Grandma COBOL," as she has affectionately been called, developed the FLOW-MATIC language, which was the basis for COBOL (COmmon Business-Oriented Language), a widely-used programming language for business applications.

John Atanasoff and *Clifford Berry* developed the first fully electronic binary computer (Figure 1.6a), the *ABC* (Atanasoff-Berry Computer), at Iowa State University during 1937–1942. It introduced the ideas of binary arithmetic, regenerative memory, and logic circuits. Unfortunately, because the ABC was never patented and others failed at the time to see its utility, it took three decades before Atanasoff and Berry received recognition for this remarkable technology. Until then, the Electronic Numerical Integrator and Computer, better known as the *ENIAC* (Figure 1.6b), bore the title of the first fully electronic computer. The designers, *J. Presper Eckert* and *John Mauchly*, began work on it in 1943 at the Moore School of Engineering at the University of Pennsylvania. When it was completed in 1946, this 30-ton machine had 18,000 vacuum tubes, 70,000 resistors, and 5 million soldered joints, and consumed 160 kilowatts of electrical power. Stories are told of how the lights in Philadelphia dimmed when the ENIAC was operating. This extremely large machine could multiply numbers approximately 1000 times faster than the Mark I, but it was quite limited in its applications and was used primarily by the Army Ordnance Department to calculate firing tables and trajectories for various types of artillery shells. Eckert and Mauchly later left the University of Pennsylvania to form the Eckert-Mauchly Computer Corporation, which built the *UNIVAC* (Universal Automatic Computer). Started in 1946 and completed in 1951, it was the first commercially available computer designed for both scientific and business applications. The UNIVAC achieved instant fame partly due to its correct (albeit not believed) prediction on national television of the election of President Eisenhower in the 1952 U.S. presidential election, based on 5% of the returns.

The instructions that controlled the ENIAC's operation were entered into the machine by rewiring some of the computer's circuits. This complicated process was very time-consuming, sometimes taking a number of people several days; during this time, the computer was idle.

---

[1] This bug has been preserved in the National Museum of American History of the Smithsonian Institution.

(a)

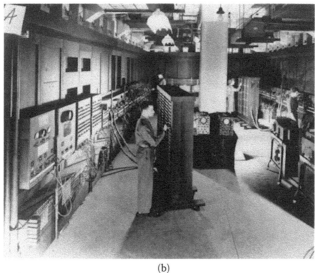

(b)

FIGURE 1.6   (a) The ABC. (b) ENIAC. (U.S. Army photo. Image courtesy of the Computer History Museum.)

In other early computers, the instructions were stored outside the machine on punched cards or some other medium, and were transferred into the machine one at a time for interpretation and execution.

It must be pointed out, however, that although men had built the machine, it was women who learned how to make it work to solve mathematical problems that would have taken hours by hand (Figure 1.7). And there were thousands of women doing similar work all across the United States. A documentary called *Top Secret Rosies: The Female Computers of World War II* that debuted in 2010 acknowledges their important work.[2]

---

[2] http://www.topsecretrosies.com/Top_Secret_Rosies/Home.html

FIGURE 1.7 Men built the ENIAC, but women made it work. (U.S. Army photo.)

In 1945, Princeton mathematician *John von Neumann* wrote *First Draft of a Report on the EDVAC (Electronic Discrete Variable Automatic Computer)* in which he described a scheme that required program instructions to be stored internally before execution. This led to his being credited as the inventor of the *stored-program concept*. The architectural design he described is still known as the *von Neumann architecture*. The advantage of executing instructions from a computer's memory rather than directly from a mechanical input device is that it eliminates time that the computer must spend waiting for instructions. Instructions can be processed more rapidly and, more importantly, they can be modified by the computer itself while computations are taking place. The introduction of this scheme to computer architecture was crucial to the development of general-purpose computers.

The actual physical components used in constructing a computer system are its **hardware**. Several generations of computers can be identified by the type of hardware used. The ENIAC and UNIVAC are examples of **first-generation** computers, which are characterized by their extensive use of vacuum tubes. Advances in electronics brought changes in computing systems, and in 1958 IBM introduced the first of the **second-generation** computers, the IBM 7090. These computers were built between 1959 and 1965 and used transistors in place of vacuum tubes. Consequently, these computers were smaller, required less power, generated far less heat, and were more reliable than their predecessors. They were also less expensive, as illustrated by the introduction of the first **minicomputer** in 1963, the PDP-8, which sold for $18,000, in contrast with earlier computers whose six-digit price tags limited their sales to large companies. The **third-generation** computers that followed used integrated circuits and introduced new techniques for better system utilization, such as multiprogramming and time sharing. The IBM System/360 introduced in 1964 is commonly accepted as the first of this generation of computers. Computers from the 1980s on, called **fourth-generation** computers, use very large-scale integrated circuits (VLSI) on silicon chips and other microelectronic advances to shrink their size and cost still more while enlarging their capabilities.

The first chip was the 4004 chip (Figure 1.8) designed by Intel's *Ted Hoff*, giving birth to the microprocessor, which marked the beginning of the fourth generation of computers.

FIGURE 1.8   Intel 4004 chip (1971).

This, along with the first use of an 8-inch floppy disk at IBM, ushered in the era of the personal computer. Robert Noyce, one of the cofounders of the Intel Corporation, contrasted microcomputers with the ENIAC as follows:

> An individual integrated circuit on a chip perhaps a quarter of an inch square now can embrace more electronic elements than the most complex piece of electronic equipment that could be built in 1950. Today's microcomputer, at a cost of perhaps $300, has more computing capacity than the first electronic computer, ENIAC. It is twenty times faster, has a larger memory, consumes the power of a light bulb rather than that of a locomotive, occupies 1/30,000 the volume and costs 1/10,000 as much. It is available by mail order or at your local hobby shop.

1.1.3 System Software

The stored-program concept was a significant improvement over manual programming methods, but early computers still were difficult to use because of the complex coding schemes required for representing programs and data. Consequently, in addition to improved hardware, computer manufacturers began to develop collections of programs known as **system software**, which make computers easier to use. One of the more important advances in this area was the development of **operating systems**, which allocate storage for programs and data and carry out many other supervisory functions. They also act as an interface between the user and the machine, interpreting commands given by the user from the keyboard, by a mouse click, or by a spoken command, and then directing the appropriate system software and hardware to carry them out. Two important early operating systems are *Unix* (1971) and *MS-DOS* (1981). Unix was developed in 1971 by *Ken Thompson* and *Dennis Ritchie* at AT&T's Bell Laboratories and is the only operating system that has been implemented on computers ranging from microcomputers to supercomputers. The most popular operating system for personal computers for many years was MS-DOS, developed in 1981 by *Bill Gates*, founder of the Microsoft Corporation. More recently, **graphical user interfaces (GUIs)**, such as MIT's X Window System for UNIX-based machines, Microsoft's Windows for personal computers, and Apple's Macintosh interface, were devised to provide a simpler and more intuitive interface between humans and computers.

As noted in the introduction to this chapter, one of the most important advances in system software was the development of **high-level languages**, which allow users to write programs in a language similar to natural language. A program written in a high-level language is known as a **source program**. For most high-level languages, the instructions that make up a source program must be translated into **machine language**, that is, the language used directly by a particular computer for all its calculations and processing. This machine-language program is called an **object program**. The programs that translate source programs into object programs are called **compilers**.

This summary of the history of computing has dealt mainly with the first two important concepts that have shaped the history of computers: the mechanization of arithmetic and the stored-program concept. Looking back, we marvel at the advances in technology that have, in little more than a half century, led from ENIAC to today's wide array of computer systems, ranging from smart phones, tablet PCs, and laptops to powerful desktop machines, to supercomputers capable of performing billions of operations each second, and to massively parallel computers that use thousands of microprocessors working together in parallel to solve large problems. Someone once noted that if progress in the automotive industry had been as rapid as in computer technology since 1960, today's automobile would have an engine that is less than 0.1 inch in length, would get 120,000 miles to a gallon of gas, would have a top speed of 240,000 miles per hour, and would cost $4.

1.1.4 The Graphical User Interface

The third key concept that has produced revolutionary change in the evolution of the computer is the graphical user interface (GUI). A user interface is the portion of a software program that responds to commands from the user. User interfaces have evolved greatly in the past two decades, in direct correlation to equally dramatic changes in the typical computer user.

In the early 1980s, the personal computer burst onto the scene. However, at the outset, the personal computer did not suit the average person very well. The explosion in the amount of commercially available application software spared computer users the task of learning to program in order to compose their own software; for example, the mere availability of the Lotus 1-2-3 spreadsheet software was enough to convince many to buy a PC. Even so, using a computer still required learning many precise and cryptic commands, if not outright programming skills.

In the early 1980s, the Apple Corporation decided to take steps to remedy this situation. The Apple II, like its new competitor, the IBM PC, employed a command-line interface, requiring users to learn difficult commands. In the late 1970s, Steve Jobs visited Xerox's Palo Alto Research Center (PARC) and viewed several technologies that amazed him: the laser printer, Ethernet, and the graphical user interface. It was the last of these that excited Jobs the most, for it offered the prospect of software that computer users could understand almost intuitively. In a 1995 interview he said, "I remember within 10 minutes of seeing the graphical user interface stuff, just knowing that every computer would work this way some day."

Drawing upon child development theories, Xerox PARC had developed the graphical user interface for a prototype computer called the Alto developed in 1973. The Alto featured a new device that had been dubbed a "mouse" by its inventor, PARC research scientist *Douglas Engelbart*. The mouse allowed the user to operate the computer by pointing to icons and selecting options from menus. At the time, however, the cost of the hardware that the Alto required made it unfeasible to market, and the brilliant concept went unused. Steve Jobs saw, however, that the same remarkable change in the computer hardware market that had made the personal computer feasible also made the graphical user interface a reasonable possibility. In 1984, in a famous commercial first run during halftime of the Super Bowl, Apple introduced the first GUI personal computer to the world: the Macintosh. In 1985, Microsoft responded with a competing product, the Windows operating system, but until Windows version 3.0 was released in 1990, Macintosh reigned unchallenged in the world of GUI microcomputing. Researchers at the Massachusetts Institute of Technology also brought GUI to the UNIX platform with the release of the X Window system in 1984.

The graphical user interface has made computers easy to use and has produced many new computer users. At the same time, it has greatly changed the character of computing: computers are now expected to be "user friendly." The personal computer, especially, must indeed be "personal" for the average person and not just for computer programmers.

1.1.5 Networks

The computer network is a fourth key concept that has greatly influenced the nature of modern computing. Defined simply, a computer network consists of two or more computers that have been connected in order to exchange resources. This could be hardware resources such as processing power, storage, or access to a printer; software resources such as a data file or access to a computer program; or messages between humans such as electronic mail or multimedia World Wide Web pages.

As computers became smaller, cheaper, more common, more versatile, and easier to use, computer use rose, and with it the number of computer users. Thus, computers had to be shared. In the early 1960s, timesharing was introduced, in which several persons make simultaneous use of a single computer called a *host* by way of a collection of terminals, each of which consists of a keyboard for input and either a printer or a monitor to display output. With a modem (short for "modulator/demodulator," because it both modulates binary digits into sounds that can travel over a phone line and, at the other end, demodulates such sounds back into bits), such a terminal connection could be over long distances.

Users, however, began to wish for the ability for one host computer to communicate with another. For example, transferring files from one host to another typically meant transporting tapes from one location to the other. In the late 1960s, the Department of Defense began exploring the development of a computer network by which its research centers at various universities could share their computer resources with each other. In 1969, the ARPANET began by connecting research center computers, enabling them to share software and data and to perform another kind of exchange that surprised everyone in terms of its popularity: electronic mail. Hosts were added to the ARPANET backbone in the 1970s, 1980s, and 1990s at an exponential rate, producing a global digital infrastructure that came to be known as the Internet.

Likewise, with the introduction of microcomputers in the late 1970s and early 1980s, users began to desire the ability for PCs to share resources. The invention of Ethernet network hardware and such network operating systems as Novell NetWare produced the Local Area Network, or LAN, enabling PC users to share printers and other peripherals, disk storage, software programs, and more. Microsoft also included networking capability as a major feature of its Windows NT operating system.

The growth of computer connectivity has continued at a surprising rate. Computers have become common, and they are used in isolation less and less. With the advent of affordable and widely available Internet Service Providers (ISPs) and WiFi, computer users can now connect to the growing global digital infrastructure almost anywhere.

1.1.6 A Brief History of C++

To simplify the task of transferring the Unix operating system to other computers, Ken Thompson began to search for a high-level language in which to rewrite Unix. None of the languages in existence at the time were appropriate; therefore, in 1970, Thompson began designing a new language called B. By 1972, it had become apparent that B was not adequate for implementing Unix. At that time, Dennis Ritchie, also at Bell Labs, designed a successor language to B that he called C, and approximately 90% of Unix was rewritten in C.

By the late 1970s, a new approach to programming appeared on the scene—**object-oriented programming (OOP)**—that emphasized the modeling of objects through classes and inheritance. A research group at Xerox PARC created the first truly object-oriented language, named Smalltalk-80. Another Bell Labs researcher, Bjarne Stroustrup, began the work of extending C with object-oriented features. In 1983, the redesigned and extended programming language C With Classes was introduced with the new name C++.

In the years that followed, as computer manufacturers developed C and C++ compilers for their machines, some added extensions and variations that were specific to their particular computers. As a consequence, programs written for one machine might not be usable on a different machine without modification. To remedy these problems, a standard for C++ was developed so that programs written in C++ are **portable**, which means they can be processed on several different machines with little or no alteration.

1.2 COMPUTER ORGANIZATION

The basic design of the Analytical Engine corresponded remarkably to that of modern computers in that it involved the four primary operations of a computer system: processing, storage, input, and output. It included a mill for carrying out the arithmetic computations according to a sequence of instructions (like the central processing unit in modern machines); the store was the machine's memory for storing up to one thousand 50-digit numbers and intermediate results; input was to be by means of punched cards; output was to be printed; and other components were designed for the transfer of information between components. When completed, it would have been as large as a locomotive, powered by steam, and able to calculate to six decimal places of accuracy very rapidly and print out results, all of which was to be controlled by a stored program!

The design of Babbage's Analytical Engine as a system of several separate components, each with its own particular function, was incorporated in many later computers and is, in fact, a common feature of most modern computers. In this section we briefly describe the major components of a modern computing system and how program instructions and data are stored and processed. A more complete description of computer architecture can be found on the website for this text described in the Preface.

1.2.1 Computing Systems

Most present-day computers exhibit a structure that is often referred to as the **von Neumann architecture** after Hungarian mathematician John von Neumann, whose pioneering work in the stored program concept and whose theories defined many key features of the modern computer. According to the von Neumann architecture (see Figure 1.9), the heart of the computing system is its **central processing unit (CPU)**. The CPU controls the operation of the entire system, performs the arithmetic and logic operations, and stores and retrieves instructions and data. Every task that a computer performs ultimately comes down to

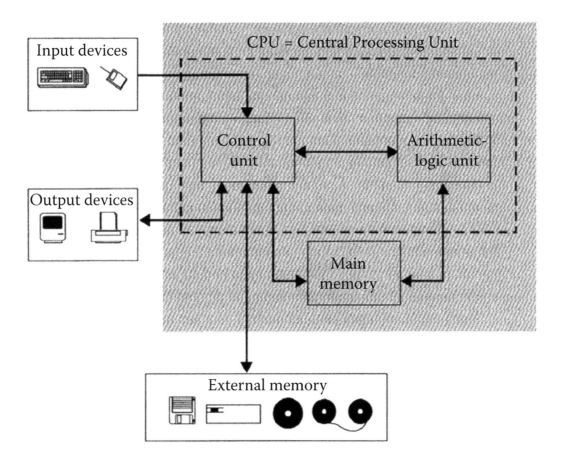

FIGURE 1.9   Major components of a computing system.

instructions and data that can be operated upon by the CPU. The instructions and data are stored in a high-speed memory unit, and the control unit fetches these instructions from memory, decodes them, and directs the system to execute the operations indicated by the instructions. Those operations that are arithmetical or logical in nature are carried out using the circuits of the **arithmetic-logic unit (ALU)** of the CPU. These operations of the CPU are known as *processing*.

In contrast to the one-instruction-at-a-time operation by the CPU in the von Neumann architecture, parallel processing computers improve performance by employing two or more CPUs. The world's fastest supercomputers employ thousands of CPU chips and for this reason are termed massively parallel processing computers. Parallel computing, however, requires a very different programming strategy in order to make use of the power of systems with thousands of processors.

1.2.2 Storage

The **memory** unit of a computer system serves several purposes. **Main memory** is also known as **internal**, **primary**, or **random access memory (RAM)**, and its main function is to store the instructions and data of the programs being executed. Most modern computers also have a smaller amount of high-speed memory called **cache memory** that is usually on the same chip as the CPU. It is used to speed up execution by storing a set of recent or current instructions being executed so they need not be fetched from main memory. Also, as part of the CPU's processing, it may need to temporarily write down (store) a number and read (retrieve) it later. The CPU can use main memory in this manner, but there is also a set of special high-speed memory locations within the CPU called **registers**. Values that are stored in registers can typically be accessed thousands of times faster than values that are stored in RAM.

One problem with RAM and registers is that they are volatile; that is, if the power to the computing system is shut off (either intentionally or accidentally), values that are stored in these memory components are lost. To provide long-term storage of software programs and data, most computing systems also have components that are called **secondary**, **external**, or **auxiliary storage**. Common forms of this type of storage include magnetic media such as hard disks and optical media such as CD-ROM and DVD, which make use of laser technology to store and retrieve information. These devices are nonvolatile, in that they provide long-term storage for large collections of data, even if power is lost. However, the time required to access data that is stored on such devices can be thousands of times greater than the access time for data stored in RAM.

Both main memory and secondary storage are collections of *two-state devices*. There are only two possible digits, 0 and 1, in the binary number system. Thus, if one of the states of a two-state device is interpreted as 0 and the other as 1, then a two-state device can be said to be a 1-bit device, because it is capable of representing a single **binary digit** (or "**bit**"). Such two-state devices are organized into groups called **bytes**, with one byte consisting of eight bits.

To indicate larger amounts of storage, some of the prefixes of the metric system are used—for example, *kilo*. However, there is an important difference. The metric system is convenient precisely because it is a decimal system, based on powers of 10, but modern computers are binary computers, based on powers of two. Thus, in computing, the prefix kilo usually is not used for 1000 but, rather, is equal to 2^{10} or 1024. Thus, a **kilobyte (KB)** is 1024 bytes, not 1000 bytes; one **megabyte (MB)** is 1024 KB or 1,048,576 bytes, not 1 million bytes; and one **gigabyte (GB)** is 1024 MB or 1,073,741,824 bytes, not 1 billion bytes.

Bytes are typically grouped together into **words**. The number of bits in a word is equal to the number of bits in a CPU data register. The word size thus varies on different computers, but common word sizes are 16 bits (= 2 bytes), 32 bits (= 4 bytes), and 64 bits (= 8 bytes). Associated with each word or byte is an address that can be used to directly access that word or byte. This makes possible *random access* (*direct access*): the ability to store information in a specific memory location and then to directly retrieve it later from that same location. The details of how various types of data are represented in a binary form and stored are described in Chapter 3.

1.2.3 Input and Output

For instructions and data to be processed by a computer's CPU, they must be **digitized**—that is, they must be encoded in binary form and transmitted to the CPU. This is the main function of **input devices**. The keyboard is the most common input device, followed by such pointing devices as the mouse, trackball, and joystick. Similarly, scanners convert and input graphics as binary information, and audio and video capture boards can encode and input sounds and video.

Once a CPU has completed a process, in order for that binary result to be meaningful to a human, it needs to be converted to another form. This is the main function of **output devices**. Two of the more common types of output device are monitors and printers. However, the varieties of output that can be generated by a computer are of a growing and surprising variety. Computers can output information as graphics, sound, video, and motion (in the case of robotics).

The communication between the CPU and input and output devices often happens by way of a **port**, a point of connection between the computer system's internal components and its **peripherals** (external components). Some ports, such as a monitor port, are designed for a single and specific use. Others, such as parallel and serial ports, are more flexible and can accommodate a variety of types of peripherals. Ports in turn connect to the computer system's **bus**, a kind of highway running through the computer system. By way of the bus, the computer system's components can send instructions and data to and from the CPU and memory.

1.2.4 Operating Systems

In order for a computer to be a general-purpose computer, it must first load a system software program called an **operating system (OS)**. In very general terms, this software program performs two main functions:

1. It serves as an interface between the computer user(s) and the system hardware.

2. It serves as an environment in which other software programs can run.

The OS and the computer system hardware together comprise a **platform** upon which additional functionality can be built. Some operating systems can run only on a single type of hardware. For example, the DOS and Windows operating systems run only on PC hardware. In contrast, the UNIX operating system will operate on several types of computer hardware.

1.2.5 Programming

Program instructions for the CPU must be stored in memory. They must be instructions that the machine can execute, and they must be expressed in a form that the machine can understand—that is, they must be written in the **machine language** for that machine. These instructions consist of two parts: (1) a numeric **opcode**, which represents a basic machine operation, such as load, multiply, add, and store; and (2) the address of the **operand**. Like all information stored in memory, these instructions must be represented in a binary form.

As an example, suppose that values have been stored in three memory locations with addresses 1024, 1025, and 1026, and that we want to multiply the first two values, add the third, and store the result in a fourth memory location, 1027. To perform this computation, the following instructions must be executed:

1. Fetch the contents of memory location 1024, and load it into a register in the ALU.

2. Fetch the contents of memory location 1025, and compute the product of this value and the value in the register.

3. Fetch the contents of memory location 1026, and add this value to the value in the register.

4. Store the contents of the register in memory location 1027.

If the opcodes for load, store, add, and multiply are 16, 17, 35, and 36, respectively, these four instructions might be written in machine language as follows:[3]

```
1.   00010000000000000000010000000000
2.   00100100000000000000010000000001
3.   00100011000000000000010000000010
4.   00010001000000000000010000000011
```
$$\underbrace{}_{\text{opcode}}\underbrace{}_{\text{operand}}$$

These instructions can then be stored in four (consecutive) memory locations. When the program is executed, the control unit will fetch each of these instructions, decode it

---

[3] In binary notation, the opcodes 16, 17, 35, and 36 are 10000, 10001, 100011, and 100100, respectively, and the addresses 1024, 1025, 1026, and 1027 are 10000000000, 10000000001, 10000000010, and 10000000011, respectively. See the text's website for more information about nondecimal number systems, including methods for converting base-10 numbers to base-2 (binary) numbers.

to determine the operation and the address of the operand, fetch the operand, and then perform the required operation, using the ALU if necessary.

Programming in the machine language of an early computer was obviously a very difficult and time-consuming task in which errors were common. Only later did it become possible to write programs in **assembly language**, which uses mnemonics (names) in place of numeric opcodes and variable names in place of numeric addresses. For example, the preceding sequence of instructions might be written in assembly language as

```
1. LOAD a, ACC
2. MULT b, ACC
3. ADD c, ACC
4. STOR ACC, x
```

An assembler, which is part of the system software, translates such assembly language instructions into machine language.

```
LOAD a, ACC                              00010000000000000000010000000000
MULT b, ACC      ━━━▶  Assembler  ━━▶    00100100000000000000010000000001
ADD c, ACC                               00100011000000000000010000000010
STORE ACC, x                             00010001000000000000010000000011
```

Today, most programs are written in **high-level languages** such as C++ and Java. Such programs are known as **source programs**. The instructions that make up a source program must be translated into machine language before they can be executed. For some languages (e.g., C++), a **compiler** that translates the source program into an **object program** carries this out. For example, for the preceding problem, a programmer might write the C++ statement

```
x = a * b + c;
```

which instructs the computer to multiply the values of a and b, add the value of c, and assign the value to x. A C++ compiler would translate this statement into a sequence of machine language instructions like those considered earlier.

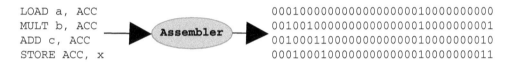

```
                                         00010000000000000000010000000000
                                         00100100000000000000010000000001
x = a * b + c;  ━━━▶  Compiler  ━━▶      00100011000000000000010000000010
                                         00010001000000000000010000000011
```

For a complete program like those in the chapters that follow, the compiler will convert each C++ statement into machine language. A **linker** will then be used to connect items such as input/output libraries that are defined outside of the resulting object file with their definitions to produce an **executable program**, which can then be loaded into memory and executed by the computer to generate the output produced by the program.

EXERCISES

1. Match each item in the first column with the associated item in the second column.

_____ peripheral devices   A.   high-speed memory used by the CPU

_____ bit   B.   central processing unit

_____ byte   C.   1024

_____ megabyte   D.   terminals, scanners, printers

_____ object program   E.   binary digit

_____ source program   F.   group of binary digits

_____ CPU   G.   1024 K bytes

_____ K   H.   written in machine language

_____ RAM   I.   written in high-level language

_____ cache   J.   language translator

Briefly define each of the terms in Exercises 2–16.

2. ALU

3. CPU

4. peripheral devices

5. bit

6. byte

7. word

8. K

9. megabyte

10. source program

11. object program

12. machine language

13. assembly language

14. compiler

15. assembler

16. interpreter

CHAPTER 2

Programming and Problem Solving— Software Engineering

If we really understand the problem, the answer will come out of it, because the answer is not separate from the problem.

JIDDU KRISHNAMURTI

People always get what they ask for; the only trouble is that they never know, until they get it, what it actually is that they have asked for.

ALDOUS HUXLEY

It's the only job I can think of where I get to be both an engineer and an artist. There's an incredible, rigorous, technical element to it, which I like because you have to do very precise thinking. On the other hand, it has a wildly creative side where the boundaries of imagination are the only real limitation.

ANDY HERTZFELD

2.1 A SNEAK PEAK AT C++

A program is a collection of statements written in a programming language. In the same way that *grammar rules* dictate how to construct English sentences, there are C++ grammar rules that govern how C++ statements are formed and combined into more complex statements and into programs. Much of this text is devoted to learning these rules, and in this section we take a first look at a few of these in a simple C++ program.

It is traditional to use as a first example a program like the one in Example 2.1 that displays a greeting. The user is prompted to enter his or her first name and then a greeting is output. We will use this program to illustrate the basic structure of C++ programs.

Example 2.1 Greeting a User

```
/* Program that greets the user.
   Written by John Doe for CS 104, Assignment 1, Feb. 2, 2012

   Input:  the name of the user
   Output: a personalized greeting
-----------------------------------------------------------------*/

#include <iostream>        // cin, cout, <<, >>
#include <string>          // string
using namespace std;

int main()
{
   cout << "What is your first name? ";
   string firstName;
   cin >> firstName;

   cout << "\nWelcome to CS 104, " << firstName <<"!\n";
}
```

SAMPLE RUN:

```
What is your first name? Sharlene    ⟵——— user input
Welcome to CS 104, Sharlene!
```

The first line of the program begins with the pair of characters /* and the seventh line ends with the pair */. In a C++ program, anything contained between these character pairs is a **comment**. This multiline comment in these opening lines of the program is **opening documentation** that gives information about the program such as what it does, who wrote it, when it was written (or last updated), and what is input to and output by the program. The dashes in the sixth line are optional and are used in the examples of this text as a border to set this opening documentation off from the program statements that follow.

The two lines that follow begin with #include and are called **compiler directives.** The first one instructs the compiler to add to the program the items in the *library* iostream that are needed to perform input and output; it will appear in all of our C++ programs. The second directive adds the items in the library string that are needed to process character strings. The // following each directive indicates that what follows to the end of the line is a comment. Here these comments indicate which items from the libraries are being used.

The next line using namespace std; will be present in nearly all of our programs. It informs the compiler that we want these to be the standard libraries from the **namespace** named std.[1] Without it, we would have to qualify each library item (such as cout) with the prefix std::; for example:

```
std::cout << "What is your first name? ";
```

However, this soon becomes annoying because the standard library identifiers such as cin and cout are used so frequently.

The rest of the program has the form

```
int main()
{
    A list of C++ statements
}
```

This is actually a function named main and is called the **main function** of the program. The C++ keyword int preceding the word main specifies the return type of the function and indicates that it will return an integer value to the operating system. Normal termination is indicated by returning zero; nonzero return values indicate abnormal termination. Some programmers use (and some older compilers may require) a **return statement** return 0; as the last statement in the program.

Execution of this program will begin with the first statement enclosed between the curly braces ({ and }) in this main function and proceed through the statements that follow it. Note that each statement must end with a semicolon.

In the program in Example 2.1, the << operator in the first statement will output a message to the screen (cout) that prompts the user to enter her or his first name:

```
cout << "What is your first name? ";
```

The next statement

```
string firstName;
```

---

[1] In 1997, the C++ ANSI standard gave new names to the standard libraries (e.g., iostream in place of iostream.h) and stored these names and others in containers called namespaces. The ANSI standard identifiers are stored in the namespace std. With non-ANSI-compliant compilers, it may be necessary to use the older library names (e.g., iostream.h instead of iostream, and math.h instead of cmath) and remove the using namespace std; line.

declares that the *variable* firstName will store a character string; and the statement

```
cin >> firstName;
```

uses the >> operator to read the character string entered by the user from the keyboard (cin) and stores it in variable firstName. The next statement

```
cout << "\nWelcome to CS 104, " << firstName << "!\n";
```

then displays on the screen a personalized greeting consisting of

1. a special character (\n) that causes an advance to a new line followed by the string

    Welcome to CS 104,

2. the character string that is stored in firstName

3. the character ! followed by the new-line character

2.2 PROGRAMMING AND PROBLEM SOLVING—AN OVERVIEW

A computer **program** is a sequence of instructions that must be followed to solve some problem, and the main reason that people learn programming is so that they can use the computer as a problem-solving tool. At least four steps or stages can be identified in the program-development process:

1. **Design:** Analyze the problem and design a solution, which results in an *algorithm* to solve the problem. This is usually the most difficult part of the development process because it basically requires that the programmer knows how to go about solving the problem.

2. **Coding:** Translate the design plan into the syntax of a high-level language such as C++ to produce a *program.*

3. **Testing, Execution, and Debugging:** Repeatedly test the program, removing errors (called *bugs*) until one is confident that it solves the problem.

4. **Maintenance:** Over time, the program is updated and modified, as necessary, to meet the changing needs of its users.

In this section these steps are illustrated with an example that is quite simple so that the main ideas are emphasized at each stage without getting lost in a maze of details.

2.2.1 Problem: Temperature Conversion

A marine biologist is conducting research on microorganisms in the Great Lakes. One part of this study involves the effect of sudden changes in water temperature. The reading she just recorded was 17.35°C, but some of the formulas she uses to analyze data require that the temperatures be in Fahrenheit. She would like a program she can use to convert a Celsius temperature to Fahrenheit.

2.2.2 Program Design

Problems to be solved are usually expressed in a natural language such as English and often are stated imprecisely, making it necessary to analyze the problem and formulate it more precisely. For the preceding problem, this is quite easy:

Given a temperature reading in Celsius, compute the equivalent Fahrenheit temperature.

For many problems, however, this may be considerably more difficult, because the initial descriptions may be quite vague and imprecise, perhaps because the people who pose the problems do not understand them well nor how to solve them nor what the computer's capabilities and limitations are.

We will call the approach used in this text to design software solutions to problems **object-centered design** (**OCD**) because it focuses on objects that are given in the problem (the *input*); objects that make up the solution of the problem (the *output*); and other objects that may be needed to obtain the solution.[2] In its simplest form, it consists of the following stages:

1. *Behavior:* Describe how you want the program to behave.

2. *Objects:* Identify the real-world objects in this description and categorize them.

3. *Operations:* Identify the operations needed to solve the problem.

4. *Algorithm:* Arrange these objects and operations in an order that solves the problem.

2.2.2.1 Behavior

We begin by writing out what we want our program to do (i.e., how we want it to behave). Because the remainder of our design depends on this step, we try to make it as precise as possible:

Behavior: The program should display a prompt for the Celsius temperature on the screen and should then read this Celsius temperature from the keyboard. It should then compute the corresponding Fahrenheit temperature and display it on the screen.

Note that we have generalized the problem to convert an arbitrary Celsius temperature and not just 17.35°C. Such **generalization** is an important aspect of analyzing a problem because programs should be sufficiently flexible to solve not only the given specific problem, but also any related problem of the same kind with little, if any, modification required.

---

[2] This is not the same as *object-oriented design*, which has a specific meaning in computing and will be described in the last chapters of this text. To avoid confusion, we will refer to objects (i.e., things) in a problem's description as *real-world objects* or as *problem objects* and use the term *software objects* for those things used to represent real-world objects in a programming language. The C++ standard uses the term *entities* for these software objects.

2.2.2.2 Objects

Once we have decided exactly what the program should do, we are ready for the next step of identifying the objects in the problem. One simple approach is to identify the noun phrases in our behavioral description, ignoring nouns like user and program:

Behavior: The program should display a <u>prompt for the Celsius temperature</u> on the <u>screen</u> and should then read this <u>Celsius temperature</u> from the <u>keyboard</u>. It should then compute the corresponding <u>Fahrenheit temperature</u> and display it on the screen.

This gives us the following list of objects:

> *Problem's Objects:*
> prompt for the Celsius temperature
> screen
> Celsius temperature
> keyboard
> Fahrenheit temperature

They must be represented in a programming language by *software objects*, which in most programming languages must have a specified *type* that tells what kind of values they can have. Some of them will have values that vary from one execution of the program to the next and/or during execution; they are called **variables** and must have *names*. Those whose values remain constant may or may not be named. In our example, we can classify our objects as follows:

	Software Objects		
Problem Objects	**Type**	**Kind**	**Name**
prompt for Celsius temperature	text string	constant	none
screen	output device	variable	`cout`
Celsius temperature	real number	variable	*celsius*
keyboard	input device	variable	`cin`
Fahrenheit temperature	real number	variable	*fahrenheit*

We will not name the first software object in our problem, the prompt, because its value does not change during execution of the program and is unlikely to change in the future. In a C++ program, such text string constants are enclosed within double quotes.

The predefined name `cout` in C++ refers to the output screen (or window). Because the contents of the screen will change during program execution, it is considered to be a variable.

We have chosen the name *celsius* for the third software object, the Celsius temperature. Its value will vary from one execution to the next because the user enters that value from the keyboard. Numeric objects that can store real values (i.e., numbers with decimal points) are represented by the type `double` (or `float`) in C++.

The predefined name `cin` in C++ refers to the keyboard. Because values will be entered from it during program execution, it too is considered to be a variable.

Finally, we have chosen the name *fahrenheit* for our last software object, the Fahrenheit temperature. It is a variable, because its value will be computed using the value of *celsius*.

2.2.2.3 Operations

Now that we have identified and classified the objects in our program, we can proceed to the next step, which is to identify the operations needed to solve the problem. For the objects we identified the nouns in our behavioral description; for the operations we can begin by identifying the verbs that describe actions of the program:

Behavior: The program should <u>display</u> a prompt for the Celsius temperature on the screen and should then <u>read</u> this Celsius temperature from the keyboard. It should then <u>compute</u> the corresponding Fahrenheit temperature and display it on the screen.

Using the objects we identified earlier, we can describe these operations as follows:

> *Problem's Operations:*
> Output a prompt for the Celsius temperature to `cout`
> Input a real value from `cin` and store it in *celsius*
> Compute *fahrenheit*
> Output the value of *fahrenheit* to `cout`

C++ provides the operator `<<` that we can use to output the prompt for the Celsius temperature and the value of *fahrenheit* to `cout` (i.e, the screen). Similarly, we can use C++'s operator `>>` to input a real value from `cin` (i.e., the keyboard) and store it in *celsius*.

Computing the value of *fahrenheit* requires some additional work. Using the formula

$$fahrenheit = 1.8 \times celsius + 32.0$$

means that we need to expand our list of operations:

> *Problem's Operations:*
> Output a prompt for the Celsius temperature to `cout`
> Input a real value from `cin` and store it in *celsius*
> Compute *fahrenheit*:
>   Multiply real values: 1.8 and *celsius*
>   Add real values: the preceding product and 32.0
> Output the value of *fahrenheit* to `cout`

In most languages, including C++, real values can be multiplied using the `*` operator and can be added using the `+` operator. Thus, C++ provides all of the operations needed to solve our problem. However, our formula for converting Celsius to Fahrenheit adds two new objects to our list.

| | Software Objects | | |
Problem Objects	Type	Kind	Name
prompt for Celsius temperature	text string	constant	none
screen	output device	variable	`cout`
Celsius temperature	real number	variable	*celsius*
keyboard	input device	variable	`cin`
Fahrenheit temperature	real number	variable	*fahrenheit*
1.8	real number	constant	none
32.0	real number	constant	none

2.2.2.4 Algorithm

Once all of the objects and operations have been identified, we are ready to arrange them into an algorithm. If the preceding steps have been done correctly, this is usually straightforward:

Algorithm:

1. Output a prompt for the Celsius temperature to `cout`.

2. Input a real value from `cin` and store it in *celsius*.

3. Compute *fahrenheit* = 1.8 * *celsius* + 32.0.

4. Output *fahrenheit* to `cout`.

The language used in this algorithm is sometimes called *pseudocode*, because it isn't written in any particular programming language, but it does bear some similarity to a program's code. The algorithm serves as a kind of blueprint for the program that comes next.

2.2.3 Coding in C++

Once we have designed an algorithm for our problem, we are ready to translate it into a high-level language such as C++. Beginning programmers may find it easiest to do this a step at a time:

- First create a **program stub** that contains
  - Opening documentation
  - Compiler directives that add items in libraries needed for some of the objects and operations
  - An empty main function

- Convert each step of the algorithm into code. If it uses a software object that hasn't already been declared, add a *declaration statement* that specifies the object's type and name.

For our example, we might begin as follows:

```
/* This program converts a Celsius temperature to Fahrenheit.

   Written by John Doe for CS 104 -- Project #1 -- Feb. 3, 2012

    Input:  the Celsius temperature
    Output: the corresponding Fahrenheit temperature
 -----------------------------------------------------------------*/

#include <iostream>      // cin, cout, <<, >>
using namespace std;


int main()
{
}
```

The program begins with opening documentation, as described in the preceding section. This is followed by a `#include` directive for the `iostream` library that provides the objects (`cin` and `cout`) and operations (`<<` and `>>`) listed in the comment that follows this directive. The C++ statements that implement our algorithm will be placed between the curly braces (`{` and `}`) that follow `int main()`. Because this program stub is a complete program, we could compile it to see that the syntax is correct. We could even execute it if we want to, although nothing will be produced.

Now we move on to converting the steps of the algorithm into code. If you are new to programming, you may find it helpful to actually include the steps of the algorithm (or summaries of them) between the curly braces `{` and `}` of `main()` as comments:

```
. . .
int main()
{
   // 1. Output a prompt for the Celsius temperature to cout.
   // 2. Input a real value from cin and store it in celsius.
   // 3. Compute fahrenheit = 1.8 * celsius + 32.
   // 4. Output fahrenheit to cout.
}
```

The double slashes (`//`) inform the compiler that what follows on this line is a comment.

Now we translate each step of the algorithm into C++ code, a line at a time. Teaching you the features of the C++ that enable you to do this will be the purpose of the chapters that follow; for now, we will just demonstrate this translation with our example.

The first line of the algorithm can be translated into the C++ statement

```
cout << "Enter the Celsius temperature: ";
```

and added to the main function:

```
. . .
int main()
{
   // 1. Output a prompt for the Celsius temperature to cout.
   cout << "Enter the Celsius temperature: ";

   // 2. Input a real value from cin and store it in celsius.
   // 3. Compute fahrenheit = 1.8 * celsius + 32.
   // 4. Output fahrenheit to cout.
}
```

We might now compile this program to see if there are any syntax errors, and if not, execute it to see how our prompt appears on the screen:

```
Enter the Celsius temperature:
```

The cursor on the screen will be positioned at the end of this line and is where the temperature value will be entered. If the program hadn't compiled—for example, if we forgot one of the double quotes (") or the semicolon at the end of the statement—we would make the correction and recompile and re-execute the corrected program. Similarly, if we don't really like the output produced—for example, if we decide we'd prefer the prompt Please enter the temperature in Celsius:—we could make the change, recompile, and re-execute.

Once we have an acceptable first statement, we can go on to the next step of the algorithm and write C++ statements for it:

```
. . .
int main()
{
   // 1. Output a prompt for the Celsius temperature to cout.
   cout << "Enter the Celsius temperature: ";

   // 2. Input a real value from cin and store it in celsius.
   double celsius;
   cin >> celsius

   // 3. Compute fahrenheit = 1.8 * celsius + 32.
   // 4. Output fahrenheit to cout.
}
```

Again, we can compile and execute this new version of the program.

Eventually, we will arrive at a complete program like that shown in Example 2.2:

Example 2.2   Temperature Converter

```
/* This program converts a Celsius temperature to Fahrenheit.

   Written by John Doe for CS 104 -- Project #1 -- Feb. 3, 2012

   Input:  the Celsius temperature
   Output: the corresponding Fahrenheit temperature
------------------------------------------------------------------*/

#include <iostream> // cin, cout, <<, >>
using namespace std;

int main()
{
  // 1. Output a prompt for the Celsius temperature to cout.
  cout << "Enter the Celsius temperature: ";

  // 2. Input a real value from cin and store it in celsius.
  double celsius;
  cin >> celsius;

  // 3. Compute fahrenheit = 1.8 * celsius + 32.
  double fahrenheit = 1.8 * celsius + 32;

  // 4. Output fahrenheit to cout.
  cout << celsius << " degrees Celsius is equivalent to "
       << fahrenheit << " degrees Fahrenheit.\n";
}
```

A software program called a *text editor* can be used to enter this program into a computer's memory, and later to correct any errors that are found. This text editor is commonly built into such programming environments such as Visual C++, but it may also be a standalone editor such as emacs in Unix systems.

2.2.4 Testing, Execution, and Debugging

There are a number of errors can be introduced into a program, including

- **syntax errors** that arise when some grammar rule of the programming language is violated;
- **run-time errors** that occur during program execution; and
- **logic errors** in the design of the algorithm on which the program is based.

Finding and fixing such errors is known as *debugging* the program.

The **compiler** will locate syntax errors when it attempts to translate the C++ program into the machine language of a given computer and will generate error messages that explain the (apparent) problem. For example, if we forgot to type the semicolon at the end of the line

```
double celsius;
```

in the program in Example 2.2, and entered

```
double celsius
```

instead, the compiler might display an error message like the following:

```
(19) error: missing ';' before identifier 'cin'
```

A different compiler might display a less precise diagnostic for the same error, such as

```
In function 'int main()':
(19) error: expected initializer before 'cin'
```

The compiler displays the number of the line it was processing when it detected that something was wrong, which is the line following the line containing the error. Learning to understand error messages that your compiler generates is an important skill.

The second kind of errors, called **run-time errors**, cannot be detected until the program is executed. They include such things as dividing by zero in an arithmetic expression, computing the square root of a negative number, and generating some value outside a given range. Once the cause of the error is determined, the offending statements or expressions must be replaced with correct ones, and the modified program must be recompiled and re-executed. For example, the program in Example 2.2 contains the statement

```
double fahrenheit = 1.8 * celsius + 32.0;
```

But suppose we misread the formula for the temperature conversion and typed a + in place of the first * operator:

```
double fahrenheit = 1.8 + celsius + 32.0;
```

Because this is a valid C++ statement, the compiler will not detect the error. The program will compile and execute, but it will produce incorrect values because the formula used to compute the Fahrenheit temperature is not correct.

To check for logic errors, a program must be run using sample data and the output checked for correctness. This **testing** of a program should be done several times using a variety of inputs that test the various parts of the program. If *any* combination of inputs produces incorrect output, the program contains a logic error.

Once this has been determined, finding the error is one of the more difficult aspects of programming. It may be necessary to trace the execution step by step until the point at which a computed value differs from an expected value is located. To simplify this tracing, most implementations of C++ provide an integrated *debugger* that allows a programmer to actually execute a program one line at a time, observing the effect(s) of each line's execution on the values produced. Once the error has been located, the text editor can be used to correct it.

A program should be tested with several different kinds of data (positive values, negative values, small values, large values, etc.) until one is reasonably confident of its correctness. However, it is almost never possible to test a program with every possible set of test data, so errors may turn up months—even years—later. As programs grow in size and complexity, the difficulty of testing them increases. No matter how much testing is done, more could always be done. It is never finished but only stopped, and there is no guarantee that all the errors in a program have been found and corrected. *Testing can only show the presence of errors, not their absence. It cannot prove that a program is correct; it can only show that it is incorrect.*

The effect of errors in a program written for a programming assignment is usually not serious. Perhaps the student loses a few points on that assignment or may be lucky and the grader doesn't even notice the error. For real-world problems, however, instead of a course grade, much more may be at stake: money, jobs, and even lives. Here are a few examples selected from a plethora of software horror stories:

- In September, 1999, the Mars Climate Orbiter crashed into the planet instead of reaching a safe orbit. A report by a NASA investigation board stated that the main reason for the loss of the spacecraft was a failure to convert measurements of rocket thrusts from English units to metric units in a section of ground-based navigation-related mission software.

- In June, 1996, an unmanned Ariane 5 rocket, developed by the European Space Agency at a cost of $7 billion, exploded 37 seconds after liftoff on its maiden flight. A report by a board of inquiry identified the cause of the failure as a complete loss of guidance and attitude information due to specification and design errors in the inertial reference system software. More specifically, a run-time error occurred when a 64-bit floating-point number was converted to a 16-bit integer.

- In March of 1991, DSC Communications shipped a software upgrade to its Bell customers for a product used in high-capacity telephone call routing and switching systems. During the summer, major telephone outages occurred in these systems in California, the District of Columbia, Maryland, Virginia, West Virginia, and Pennsylvania. These were caused by an error introduced into the signaling software when three lines of code in the several million lines of code were changed and the company felt it was unnecessary to retest the program.

- On February 25, 1991, during the Gulf War, a Patriot missile defense system at Dharan, Saudi Arabia, failed to track and intercept an incoming Scud missile. This missile hit an American Army barracks, killing 28 soldiers and injuring 98 others. An error in the guidance software produced an inaccurate calculation of the time since system start-up due to accumulated roundoff errors that result from inexact binary representations of real numbers. And this time calculation was a key factor in determining the exact location of the incoming missile. The sad epilogue is that corrected software arrived in Dharan on February 26, the next day.

These are but a few examples of program errors that are more than just a nuisance and can lead to very serious and even tragic results. In such cases, careful software design, coding, and extensive and thorough testing are mandatory. In safety-critical situations where errors cannot be tolerated, relying on the results of test runs may not be sufficient because *testing can show only the presence of errors, not their absence.* It may be necessary to give a deductive proof that the program is correct and that it will always produce the correct results (assuming no system malfunction).

2.2.5 Maintenance

In contrast to student programs that are often run once or twice and then discarded, real-world programs may represent a significant investment of a company's resources and be used for many years, during which time new features or enhancements may be added to upgrade the program. To illustrate, users of the program in Example 2.2 might find it more useful to be able to choose between Celsius-to-Fahrenheit and Fahrenheit-to-Celsius conversions and perhaps to add the Kelvin scale also. Because a program that offers these alternatives uses more advanced programming features, we will defer it until the next chapter.

CHAPTER SUMMARY

Key Terms

`#include`	debugger
algorithm	debugging
behavior	design
C++	generalization
`cin`	input (>>)
class	library
coding	logic error
comment (/*, */, //)	main function
compiler	maintenance
compiler directive	namespace
`cout`	object-centered design (OCD)
curly braces ({ and })	object-oriented design

objects

opening documentation

operations

output (<<)

program

software engineering

std

syntax error

testing

TEST YOURSELF

1. In a C++ program, anything contained between /* and */ is a _____.

2. Execution of a C++ program begins with the first statement enclosed between _____ in the _____ function.

3. Name the four stages of the software life cycle.

4. List the four steps in object-centered design.

5. The _____ in a problem can be identified by finding the nouns in the behavioral description of the problem.

6. The _____ in a problem can be identified by finding the verbs in the behavioral description of the problem.

7. Objects whose values will change are called _____.

8. The screen has the predefined name _____ in C++.

9. The keyboard has the predefined name _____ in C++.

10. _____ is the output operator in C++ and _____ is the input operator.

11. Finding the errors in a program is called _____.

12. What are three types of errors that can occur in developing a program?

EXERCISES

For each of the following problems, give a precise description of how a program to solve that problem must behave. Then describe the objects and operations needed to solve the problem and design an algorithm for it.

1. Calculate and display the perimeter and the area of a square with a given side. (The perimeter of a square where the length of each side is s is $4s$ and the area is s^2.)

2. Calculate and display the diameter, circumference, and the area of a circle with a given radius. (The diameter is twice the radius. For radius r, the circumference is $2\pi r$ and the area is πr^2 where π is the mathematical constant pi whose value is approximately 3.14159.)

3. Three resistors are arranged in parallel in the following circuit:

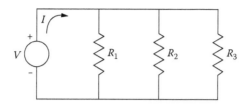

For given values of R_1, R_2, and R_3, calculate and display the combined resistance

$$\frac{1}{\dfrac{1}{R_1}+\dfrac{1}{R_2}+\dfrac{1}{R_3}}$$

4. The half-life of polonium is 140 days, which means that because of radioactive decay, the amount of polonium that remains after 140 days is one-half of the original amount. For a given initial amount of polonium and number of days, calculate the amount remaining after that many days.

PROGRAMMING PROBLEMS

1. Enter and execute the following C++ program on your computer system:

```
/* This program adds the values of variables x and y.

   Output: The value x + y
------------------------------------------------------------*/

#include <iostream>
using namespace std;

int main()
{
   int x = 214,    // the first value
       y = 2057,   // the second value
       sum = x + y;
   // output the resulting value
   cout << "\nThe sum of " << x << " and " << y
        << " is " << sum << endl;
}
```

2. Make the following changes in the program in Problem 1 and execute the modified program:

   a. Change 214 to 1723 in the statement that gives x a value.

   b. Change the variable names x and y to alpha and beta throughout.

c. Add the comment

```
// find their sum
```

following the declaration of sum.

d. Change the variable declarations to

```
int alpha = 214,                // the first value
    beta = 2057,                // the second value
    difference = alpha - beta,  // find their difference
    sum = alpha + beta;         // find their sum
```

and add the following statement before the output statement:

```
cout << "\nThe difference of " << alpha << " and "
     << beta << " is " << difference << endl;
```

3. Using the program in this chapter as a guide, write a C++ program to solve the problem in Exercise 1.

4. Using the program in this chapter as a guide, write a C++ program to solve the problem in Exercise 2.

5. Using the program in this chapter as a guide, write a C++ program to solve the problem in Exercise 3.

Types in C++

CONTENTS

There are three types of people in this world: Those who can count, and those who can't.

SEEN ON A BUMPER STICKER

Kindly enter them in your notebook.
And, in order to refer to them conveniently, let's call them A, B, and Z.

LEWIS CARROLL, "WHAT THE TORTOISE SAID TO ACHILLES"

For, contrary to the unreasoned opinion of the ignorant, the choice of a system of numeration is a mere matter of convention.

BLAISE PASCAL

IN THIS CHAPTER, WE look at some of the data types that are provided in C++, focusing on the simplest ones, called *fundamental types*. We also show how values of these types are represented in such a way that they can be stored in memory.

3.1 INTRODUCTORY EXAMPLE: RADIOACTIVE DECAY

3.1.1 Problem

A scientist at Uranium University is conducting research with the radioactive element polonium. The half-life of polonium is 140 days, which means that because of radioactive decay, the amount that remains after 140 days is one-half of the original amount. He would like to know how much polonium will remain after running the experiment for 180 days if 10 milligrams are present initially.

3.1.2 Object-Centered Design

3.1.2.1 Behavior

The program should output to the screen a prompt for the name of the radioactive element, its half-life, the initial amount, and a time period. It will then read these values from the keyboard. The program should then compute the amount remaining after the specified time period and output the input values along with this value to the screen.

3.1.2.2 Objects

From our behavioral description, we can identify the following objects:

	Software Objects		
Problem Objects	**Type**	**Kind**	**Name**
screen	output device	variable	cout
prompt	text string	constant	none
name of element	text string	variable	*element*
half-life	real number	variable	*halfLife*
initial amount	real number	variable	*initialAmount*
time period	real number	variable	*time*
keyboard	input device	variable	cin
amount remaining	real number	variable	*amountRemaining*

3.1.2.3 Operations

Again, from our behavioral description, we have the following operations:

i. Output a string (the prompts, descriptive labels, element's name).

ii. Read a string (*element*).

iii. Read numeric values (*halfLife, initialAmount, time*).

iv. Compute *amountRemaining* = *initialAmount* × $(0.5)^{time/halfLife}$.

v. Output numeric values.

C++ provides operations for each of these operations.

3.1.2.4 Algorithm

Next, we organize these objects and operations into an algorithm:

1. Output a prompt for the element's name and it's half-life to `cout`.

2. Read a string from `cin` to store in *element* and a real value to store in *halfLife*.

3. Output a prompt for the initial amount and the time period.

4. Read real values from `cin` to store in *initialAmount* and *time*.

5. Compute *amountRemaining* = *initialAmount* × $(0.5)^{time/halfLife}$.

6. Output *element*, *halfLife*, *initialAmount*, *time*, and *amountRemaining* with appropriate labels to `cout`.

3.1.2.5 Coding, Execution, and Testing

In the preceding chapter we outlined a procedure for translating the algorithm into code:

1. Create a program stub that contains opening documentation, compiler directives that add items in libraries needed for some of the objects and operations, and a main function that contains only the steps of the algorithm as comments.

2. Convert each step of the algorithm into code, adding a *declaration statement* to specify an object's type and name for any software object not already declared.

Space restrictions prevent doing this for each example in this text, but you may find that this procedure helps with developing programs like that in Example 3.1, which shows the finished product and a sample run to find how much of 10 mg of polonium with a half-life of 140 days will remain after 180 days. You are encouraged to enter this program in the version of C++ that you are using, compile and execute it, and experiment with it, making various changes and seeing what happens.

Example 3.1 Radioactive Decay

```
/* This program computes the amount of a radioactive substance
   remaining after an initial amount decays for some time period.

   Written by John Doe for CS 104 -- Project #1 -- Feb. 5, 2012

   Input:   element's name, its half-life, the initial amount,
            and a time period
   Output:  the input items and the amount of the substance
            remaining at the end of the time period
-----------------------------------------------------------------*/
#include <iostream>      // cin, cout, <<, >>
#include <string>        // string
#include <cmath>         // pow()
using namespace std;
```

```
int main()
{
  cout << "Enter the name of your radioactive substance: ";
  string element;
  cin >> element;
  cout << "and its half-life (days): ";
  double halfLife;
  cin >> halfLife;
  cout << "Enter the initial amount (mg) and a time period (days): ";
  double initialAmount, time;
  cin >> initialAmount >> time;

  double amountRemaining = initialAmount * pow(0.5, time / halfLife);

  cout << "\nFor " << element
       << " with half-life " << halfLife << " days\n"
       << initialAmount << " mg" << " will be reduced to "
       << amountRemaining << " mg after " << time << " days\n";
}
```

SAMPLE RUN:
```
Enter the name of your radioactive substance: polonium
and its half-life (days): 140
Enter the initial amount (mg) and a time period (days): 10 180


For polonium with half-life 140 days
10 mg will be reduced to 4.10168 mg after 180 days
```

This program uses two different C++ types, string for the name of the radioactive substance and double for its half-life, initial amount, a time period, and amount remaining after that time. In the remainder of this chapter we will study these and other types provided in C++ and how they can be used to represent a problem's objects.

3.2 TYPES, LITERALS, VARIABLES, AND CONSTANTS

Each data item (i.e., object) in a problem has an associated type; for example, it may be a number, a string of characters, an individual character, a logical value, or perhaps something more complex. Software objects used to represent such problem objects must also have specified types so the compiler knows how much memory needs to be allocated for them and what kinds of operations are permitted on them. For this reason, most programming languages, including C++, require that these types be specified before those objects are used. In this section, we will focus on C++'s **fundamental types**—also known as **primitive** or **intrinsic** types—and in Section 3.3 we will see how data values of these types are stored in memory.

3.2.1 Fundamental Types

The most important fundamental data types provided in C++ are the following:[1]

- **integers:** whole numbers and their negatives: of type `int`

- **integer variations:** types `short`, `long`, and `unsigned`

- **reals**: fractional numbers: of type `float`, `double`, or `long double`

- **characters:** letters, digits, symbols, and punctuation: of type `char`

- **booleans:** logical values `true` and `false`: of type `bool`

A value of one of these types is called a **literal**.[2] For example, `123`, `0`, and `-15` are integer literals; `-45.678` and `3.14159` are real literals; `'A'`, `'a'`, `'0'`, and `'$'` are character literals; and `true` and `false` are boolean literals. We will now examine these types in more detail.

3.2.1.1 Integers

Integer literals are strings of digits that may be preceded by a – sign or a + sign. They are interpreted as

- **octal (base-eight)** integers if they begin with 0 and all digits are octal digits $0, 1, \ldots, 7$;

- **hexadecimal (base-sixteen)** integers if they begin with `0x` — the hexadecimal digits for ten, eleven, $\ldots$, fifteen are A, B, $\ldots$, F or their lowercase equivalents a, b, $\ldots$, f;

- **decimal (base-ten)** integers otherwise.

For example, the literal `345` has the decimal value $345_{10} = 3 \times 10^2 + 4 \times 10^1 + 5 \times 10^0$. However, the literal `0345` has the octal value $345_8 = 3 \times 8^2 + 4 \times 8^1 + 5 \times 8^0 = 229_{10}$, and `0x345` has the hexadecimal value $345_{16} = 3 \times 16^2 + 4 \times 16^1 + 5 \times 16^0 = 837_{10}$. (See the text's website described in the preface for additional details about binary, octal, and hexadecimal number systems.)

Typically, `int` values are stored in 32 bits (= 4 bytes) and can range from -2^{31} (= –2147483648) through $2^{31} - 1$ (= 2147483647). C++ also allows `int` declarations to be modified with one of the key words `short` or `long`:

- `short int` (or just `short`) values usually are usually stored in 16 bits (2 bytes) and can range from -2^{15} (= –32768) through $2^{15} - 1$ (= 32767).

- `long int` (or just `long`) values are the same as `int` values in some versions of C++, while in others they are 64-bit values, ranging from -2^{63} to $2^{63} - 1$.

---

[1] Other types are the `signed char` and `unsigned char` integer types, which are stored in one byte and thus can range from -2^7 (= –128) through $2^7 - 1$ (= 127); the wide character type `wchar_t` for storing Unicode characters; `complex` for complex values; and the `void` type for an empty set of values.

[2] The word *literal* in computing refers to any value entered by a programmer that does not change during program execution—the string of characters you type is (literally) the value you get.

In this text, int will be used in most of the examples.

The internal representation of an integer typically uses one bit as a sign bit, so that the largest positive value of a 32-bit integer is $2^{31} - 1$ and not $2^{32} - 1$. (See the next section for a more detailed explanation.) However, some integer-valued objects never have negative values, and to avoid wasting the sign bit for such integer values, C++ provides the modifier unsigned:

- An unsigned int (or just unsigned) is a nonnegative integer whose size usually is the word size of the particular machine being used, typically 0 through $2^{32} - 1$ (= 4294967295).

- An unsigned short is usually a 16-bit value, ranging from 0 through $2^{16} - 1$ (= 65535).

- An unsigned long is usually a 32-bit or 64-bit value, ranging from 0 through $2^{32} - 1$ or 0 through $2^{64} - 1$.[3]

3.2.1.2 Reals

A value of type float is usually a 32-bit real value; a double is usually a 64-bit real value; and a long double is typically a 96-bit or a 128-bit real value. However, the range of values and the precision of each of these types is implementation dependent and is defined in one of the standard libraries cfloat or climits (described later in this section) that C++ implementations provide.

Like most programming languages, C++ provides two ways to represent real values, fixed-point notation and floating-point notation. A **fixed-point** real literal has the form $m.n$, where either the integer part m or the decimal part n (but not both) can be omitted. For example, 5.0, 0.5, 5., and .5 are all valid fixed-point real literals.

Real numbers are also sometimes expressed in *floating point* notation (also known as *exponential* or *scientific* notation). For example, 23 trillion (= 23,000,000,000,000) might be written more compactly as 2.3×10^{13}. In C++, a **floating-point** real literal has one of the forms xEn or xen, where x is an integer or fixed-point real literal and n is an integer exponent (positive or negative). For example, 23 trillion = 2.3×10^{13} can be written in a variety of forms such as 2.3E13, 0.23E14, 23.e12, and 23E12.

By default, C++ compilers treat all real literals as being of type double.[4] This means that if we compute a real value and assign it to a float variable, then the value stored in the variable will not have the precision of the computed value. For this reason, we will follow the practice of many programmers and *always use the type* double *for real values.*

---

[3] By default, whole numbers (e.g., -20, 0, 1, 13, 345) are treated as int values by the C++ compiler. To instruct it to store a literal value as an unsigned instead of int, append the letter U or u (e.g., 0U, 1U, 13U, 345U). Appending the letter L or l (e.g., -20L, 0L, 1L, 13L, 345L) will cause the value to be stored as a long instead of as an int. Appending both L (or l) and U (or u) in either order causes it to be treated as unsigned long.

[4] To instruct the compiler to process a real literal as a float, an F or f can be appended to it (e.g., 1.0F, 3.1416F, 2.998e8F). Similarly, appending an L or l to a real literal instructs the compiler to treat it as a long double (e.g., 1.0L, 0.1E1L).

3.2.1.3 Characters

We can use the char type for *individual characters* in the C++ character set, which is commonly the ASCII (American Standard Code for Information Interchange) character set (see Appendix A). It includes the uppercase and lowercase letters, common punctuation symbols such as the semicolon (;), comma (,), and period (.), and other special symbols such as +, =, and >.

Characters are represented in memory by numeric codes, and in C++, values of type char are stored using these integer codes (as described in more detail in the next section). **Character literals** are single characters enclosed in single quotes (or apostrophes); for example, 'A', 'z', '#', '8', and '/'.[5] The C++ compiler stores these values using their numeric codes, which in ASCII are 65, 122, 35, 56, and 47, respectively.

Using single quotes to enclose character literals, however, raises the question of how to represent a character literal that is a single quote (apostrophe). Similarly, how do we represent characters such as tabs and an advance to a new line? For such special characters, C++ provides character literals that consist of a backslash character (\) followed by a symbol; for example, '\'' for an apostrophe, '\t' for a tab, and '\n' for an advance to a new line. These "double characters" are called **escape sequences** because the backslash indicates that the character following it is to "escape" from its usual meaning. Table 3.1 lists the escape sequences provided in C++.

3.2.1.4 Other Types

The program in Example 3.1 contains *compiler directives* of the form #include <*something*> before the main part of the program. One of these is

TABLE 3.1   C++ Character Escape Sequences

Character	C++ Escape Sequence
Newline (NL or LF)	\n
Horizontal tab (HT)	\t
Vertical tab (VT)	\v
Backspace (BS)	\b
Carriage return (CR)	\r
Form feed (FF)	\f
Alert (BEL)	\a
Backslash (\)	\\
Question mark (?)	\?
Apostrophe (single quote, ')	\'
Double quote (")	\"
With numeric octal code ooo	\ooo
With numeric hexadecimal code hhh	\xhhh

---

[5] Character literals of the form L'x' where x consists of one or more characters are *wide-character literals* and are used for alternate character sets such as Unicode. They are of type wchar_t.

```
#include <string>
```

Why is this? It is because the type string, like several others that we will study later in Chapter 7, is not one of the fundamental types built into C++ but is a *class* that is defined in the string library. The #include directive makes this type available to the program.

Somewhat similar to character literals, a **string literal** consists of a sequence of characters enclosed in double quotes.[6] For example,

```
"Ohm's Law"
"Enter project id on one line\n\tand its name on the next.\n"
"\nThe project cost = $"
"Einstein said, \"God is subtle but he is not malicious\""
```

are all string literals. As the last three examples illustrate, escape sequences can be used within string literals. Note also that *string literals containing a single character are not char literals;* for example, "X" is a string literal, but 'X' is a char literal. They may not be used interchangeably.

If two string literals are consecutive or are separated only by white space (spaces, tabs, and end-of-lines) they will be concatenated to form a single literal. For example, for

```
"square " "feet"
```

or

```
"square "
"feet"
```

the two string literals will be combined to form

```
"square feet"
```

Readability of output is often improved by double spacing the lines. Using two newline characters can be used to accomplish this. For example, if we output the string literal

```
"First row\n\nSecond row"
```

the first newline (shown as ↵ in the following output) will end the line on which First row appears and the second newline makes the next line a blank line, after which Second row appears:

```
First row↵
↵
Second row
```

---

[6] String literals of the form L"..." are wide string literals; they may contain wide characters.

3.2.2 Identifiers

The program in Example 3.1 uses names for some of the software objects: `element` for the radioactive element, `halfLife` for its half-life, `time` for a time period, and `initialAmount` and `amountRemaining` for amounts of that element. Such names are known as **identifiers**. They should begin with a letter, which may be followed by any number of letters, digits, or underscores.[7] They may not be C++ **keywords** (e.g., `int`, `const`, `double`, etc.), which are reserved for special objects. A complete list of the C++ keywords is given in Appendix B.

It is good programming practice to use meaningful identifiers that suggest what they represent, because such names make programs easier to read and understand. They serve, therefore, as part of the program's documentation and facilitate program maintenance. For example, the identifier `feetPerSecond` is more meaningful than any of the shortened identifiers, `f`, `ft`, or `fPS`, which could represent anything else that contains these letters. You should resist the temptation to use a short identifier just to save a few keystrokes. *Complete words are preferable to abbreviations.*

It is important to remember that *C++ is case sensitive—that is, it distinguishes between uppercase and lowercase.* For example, `length` and `Length` are different identifiers. Similarly, the name of the main function must be `main`, not `Main`.

Naming conventions vary among programmers, but one of the most common is the following:

Variables: Use lowercase names, but with the first letter capitalized for each individual component word after the first.[8] Some examples are `length`, `halfLife`, and `feetPerSecond`.

Constants: Use names with all uppercase characters and with individual component words separated by underscore (_) characters. Some examples are `PI`, `TWO_PI`, and `SPEED_OF_LIGHT`.

Before an identifier can be used in a program, it must be *declared* to inform the compiler of its meaning. We now look first at declarations of variables and then declarations of constants.

3.2.3 Variables

When analyzing a problem, we often discover that there are relationships between the objects in the problem that can be expressed by *formulas*. For example, the formula for the law of radioactive decay used in the program in Example 3.1 might be expressed as follows:

$$a = a_0 \times 0.5^{-t/h}$$

---

[7] Identifiers that begin with an underscore (_) followed by an uppercase letter or that contain two *consecutive* underscores (_ _) are reserved for special use and should be avoided.

[8] This is sometimes referred to as *camel-case* notation because of the uppercase "bumps" that are in the middle of the name.

In this formula, a, a_0, t, and h are variables for which values can be substituted to calculate the value of a.

In computing, the word **variable** refers to a memory location in which values can be stored and later retrieved. One of the tasks of a compiler is to associate such a memory location with each identifier in a program that names a variable. To do this, it must know the name of the variable and the type of values it may have so that it can allocate a memory location of the appropriate size and form. This is the purpose of **variable declarations**, which have the following form:

VARIABLE DECLARATION
FORMS:

```
type variable_name;
type variable_name = initializer_expression;
```

where `type` is one of the fundamental data types (or one of the data types discussed later), and `variable_name` is a valid C++ identifier.

PURPOSE:

Instructs the C++ compiler to reserve memory for a value of the specified `type` and associates it with the name `variable_name`. In the second form, the value of `initializer_expression` will be stored in this memory location; for the first form, the value of `variable_name` is *undefined* (or *indeterminate*).

The declarations

```
string element;
double halfLife;
double initialAmount, time;
```

from the program in Example 3.1 illustrate the first form of variable declaration. The values of these variables are said to be *undefined* because the contents of the memory locations associated with them are not known. If we wanted them to have specific initial values, we could use the second form of a declaration, for example,

```
string element = " ";
double halfLife = 0;
double initialAmount = 0, time = 0;
```

specify that the initial value of `element` is a string containing a blank, and the initial values of `halfLife`, `initialAmount`, and `time` are all 0. For variables such as `element` that are of type `string`, remember that we must `#include` the `string` library as in Example 3.1.

A memory location of appropriate size will be associated with each of these variables. For example, for the `double` variable `time`, a memory location of appropriate size for `doubles` (typically 8 bytes) will be allocated. We might picture this as follows:

time | ? |

Here, the question mark indicates that this variable is **undefined**. Although some versions of C++ will initialize numeric variables to 0, others will simply let their initial values be a **garbage value**—whatever value corresponds to the string of bits in that memory location allocated for it. *You should not assume that uninitialized variables will have a specific value.*

If the declaration of `time` were changed to

```
double time = 0;
```

0 would be used as the initial value:

time | 0 |

For either of the preceding declarations of `time`, if the value 180 were later input as the value for `time`, the value stored in this memory location would change to 180.0:

time | 180.0 |

Similarly, the declaration

```
string element = "polonium";
```

would use "polonium" as the initial value for `element`:

element | polonium |

Where variable declarations are placed is largely a matter of programming style, because C++ allows them to be placed (almost) anywhere before their first use. In this text, variables will usually be declared just prior to their first use because this makes it easier to ensure that the variable is used in a manner consistent with its type.

3.2.4 Constants

In addition to variables, we may also use names for constants. This is especially useful when a program uses universal constants, such as the geometric constant π or the speed of light,

```
const double PI = 3.14159265359,
            SPEED_OF_LIGHT = 2.997925e8; // meters/sec
```

but it can also be used to associate names with other constants to be used in a program:

```
const int UPPER_LIMIT = 1000;
const char
      PERCENT_SIGN = '%',      // using a normal character
      SPACE = ' ',             // using a white space char
      TAB = '\t',              // using an escape sequence
      BELL = '\007';           // using an octal (ASCII) code
```

In general, a **constant declaration** has the following form:

CONSTANT DECLARATION
FORM:


```
const type CONSTANT_NAME = constant-expression;
```

where const is a C++ keyword; type is one of the fundamental data types (or one of the data types discussed later); CONSTANT_NAME is a valid C++ identifier; and constant-expression is any valid expression (as described in later sections) whose value is a constant of type type.

PURPOSE:

Associates a name with a constant. Any attempt to change this value within a program is an error.

There are two main reasons for using names for constants. One is to make it *easier to understand*. To illustrate, consider the following statement:

```
populationChange = (0.1853 - 0.1175) * population;
```

0.1853 and 0.1175 are "magic numbers" because they seem to magically appear without any explanation. However, if we define the constants BIRTH_RATE and DEATH_RATE by

```
const double
      BIRTH_RATE = 0.1853, // rate at which people are born
      DEATH_RATE = 0.1175; // rate at which people die
```

we can rewrite the statement in the more understandable form

```
populationChange = (BIRTH_RATE - DEATH_RATE) * population;
```

A second benefit of using constants is to make a program *easier to modify*. To illustrate, suppose that we used the value 3.1416 for π at several places in a program, and we find that this value doesn't yield the precision we need. To correct the program, we would have to find all occurrences of the old value and replace it. But if we had instead declared the constant PI and used it throughout the program, we would need to make only one change:

```
const double PI = 3.14159265359;
```

Changing the value of PI in this declaration will change its value throughout the program without any further effort on our part.

It is considered good programming practice to *place all declarations of constants at the beginning of* main() (or other function in which they are used). This makes it easy to locate these declarations when it is necessary to modify the values.

C++ provides several predefined constants in its various libraries. For example, the library climits contains

SHRT_MIN, SHRT_MAX	short int minimum, maximum values
INT_MIN, INT_MAX	int minimum, maximum values
UINT_MIN, UINT_MAX	unsigned int minimum, maximum values
LONG_MIN, LONG_MAX	long int minimum, maximum values

and the library cfloat contains

FLT_MIN, FLT_MAX	float minimum, maximum values
DBL_MIN, DBL_MAX	double minimum, maximum values
LDBL_MIN, LDBL_MAX	long double minimum, maximum values

In addition to the minimum and maximum of each of the real types, cfloat contains constants for the precision of each real type, the minimum and maximum exponent permitted in scientific notation, and so on. See Appendix D for descriptions of these and other libraries.

3.3 DATA REPRESENTATION

In Chapter 1 we noted that information is represented in a computer by a binary scheme having only the two binary digits 0 and 1. We also showed how instructions can be represented in base-two and stored in memory. We now look at how literals of the various data types can be represented and stored in binary.

3.3.1 Integers

For integer values, the binary representation is typically stored in one word of memory. To illustrate, suppose that a computer's word size is 32 and that the integer value 180 is to be stored. The base-two representation of 180 is 10110100_2. If 180 is being used as an unsigned literal, its binary digits are stored in the rightmost bits of the memory word and the remaining bits are filled with zeros:

| 0 | 0 | 0 | 0 | 0 | 0 | 0 | 0 | 0 | 0 | 0 | 0 | 0 | 0 | 0 | 0 | 0 | 0 | 0 | 0 | 0 | 0 | 0 | 0 | 1 | 0 | 1 | 1 | 0 | 1 | 0 | 0 |

Unlike `unsigned` values, `int` values may be negative and so must be stored in a binary form in which the sign of the integer is part of the representation. The most common method is to use **two's complement** representation. In this scheme, positive integers are represented in binary form, as just described, with the leftmost bit set to 0 to indicate that the value is positive. Thus, if 180 is being used as an `int` literal, 31 bits are used for the binary digits of the value and one bit for the sign:

For a negative integer, first find the binary representation of its absolute value, complement it (i.e., change each 0 to 1 and each 1 to 0), and then add 1 to the result. For example, to find the two's complement representation of –180 using a string of 32 bits,

1. Represent 180 by a 32-bit binary numeral:

$$00000000000000000000000010110100$$

2. Complement this bit string:

$$11111111111111111111111101001011$$

3. Add 1:

$$11111111111111111111111101001100$$

This string of bits is then stored in memory.

Note that the sign bit in this two's complement representation of a negative integer is 1, indicating that the number is negative.

The range of integers that can be stored in memory is determined by the number of bits used to represent them. For example, with 32-bit representations, the largest `unsigned` value that can be stored is

$$11111111111111111111111111111111_2 = 2^{32} = 4294967296$$

The range of `int` value that can be represented using 32 bits is

$$10000000000000000000000000000000_2 = -2^{31} = -2147483648$$

through

$$01111111111111111111111111111111_2 = 2^{31} - 1 = 2147483647$$

Attempting to use an integer outside this range results in a phenomenon known as **overflow**. Although using more bits to store integers would enlarge this range, it would not solve the problem of overflow; the range of integers is still finite.

3.3.2 Reals

In the binary representation of a real number, digits to the left of the decimal point are coefficients of non-negative powers of two, while those to the right are coefficients of negative powers of two. For example, the expanded form of 11010.011_2 is

$$(1 \times 2^4) + (1 \times 2^3) + (0 \times 2^2) + (1 \times 2^1) + (0 \times 2^0)$$

$$+ (0 \times 2^{-1}) + (1 \times 2^{-2}) + (1 \times 2^{-3})$$

which is equal to

$$16 + 8 + 0 + 2 + 0 + 0 + \frac{1}{4} + \frac{1}{8} = 26.375$$

The **IEEE Floating Point Format** scheme for representing real numbers in computer memory standardized in 1985 by the Institute for Electrical and Electronic Engineers (IEEE) has become almost universal. It specifies how reals can be represented in two formats: *single precision*, using 32 bits, and *double precision*, using 64 bits. Here, only single precision is described; double precision is simply a wider version of it. We begin by representing the number in binary **floating-point form**, which is much like the familiar scientific notation except that the base is two rather than ten:

$$b_1.b_2b_3 \cdots \times 2^k$$

where each b_i is a binary digit, but b_1 must be 1 (unless the number is 0). $b_1.b_2b_3 \cdots$ is called the **mantissa** (or **fractional part** or **significand**) and the exponent k is sometimes called the **characteristic**. For example, we have seen that 26.375 can be written in binary as 11010.011_2, which can be easily rewritten in floating-point form as $1.1010011_2 \times 2^4$, because multiplying (dividing) by 2 is the same as moving the binary point to the right (left). In this floating-point representation, 1.1010011 is the mantissa and 4 is the exponent.

In the IEEE format for single precision values,

- The leftmost bit stores the sign of the mantissa, 0 for positive, 1 for negative.

- The next 8 bits store the binary representation of the exponent + 127; 127 is called a **bias**.

- The rightmost 23 bits store the bits to the right of the binary point in the mantissa (the bit to the left need not be stored because it is always 1).

For 26.375, the stored exponent would be $4 + 127 = 10000011_2$ and the stored mantissa would be $01010110000000000000000_2$:

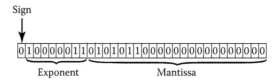

Sign

Exponent   Mantissa

For double precision, an 11-bit exponent with a bias of 1023 and 53 bits for the signed mantissa are used.

Because the binary representation of the exponent may require more than the available number of bits, we see that the **overflow** problem discussed in connection with integers also occurs in storing a real number whose exponent is too large. An 8-bit exponent restricts the range of real values to approximately -10^{38} to 10^{38}, and overflow occurs for values outside this range. A negative exponent that is too small to be stored causes an **underflow**. Real values represented using an 8-bit exponent must be greater than approximately 10^{-38} or less than -10^{-38}, and underflow occurs between these values:

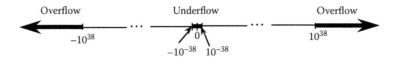

Also, obviously some real numbers have mantissas with more than the allotted number of bits, which means that some of these bits will be lost when these numbers are stored. In fact, *most real numbers do not have finite binary representations and thus cannot be stored exactly in any computer.* For example, the binary representation of the real number 0.7 is

$$(0.1011001100110011001100110011001100110\ldots)_2$$

where the block 0110 is repeated indefinitely. This means that the stored representation will not be exact (e.g., 0.6999999284744263). The error in this representation, known as **round-off error**, means that the precision of real values is limited—approximately 7 significant decimal digits for single precision and 14 for double precision.

3.3.3 Characters and Strings

The schemes used to represent character data assign a numeric code to each of the characters in the character set. One common coding scheme is **ASCII** (American Standard Code for Information Interchange).[9] Characters are represented internally using these numeric codes.

---

[9] See Appendix A for a table of ASCII codes for all characters.

For example, the ASCII code of c is 99 = 01100011$_2$, which can be stored in one 8-bit byte. For a string of characters such as code, individual characters can be stored in consecutive bytes:

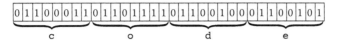

For languages of the world that have many additional characters, the coding scheme **Unicode** was developed and provides codes for more than 65,000 characters. To accomplish this it uses 16-bit codes. For example, the code for c (99—same as in ASCII) would be stored in two bytes

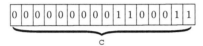

as would the code 960 for the non-ASCII character π (Greek pi):

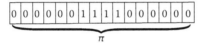

3.3.4 Booleans

There are only two boolean values: false and true. If false is encoded as 0 and true as 1, then a single bit is all that is needed to store a boolean value. However, it is common to use an entire word or a byte with all bits set to 0 for false and any other bit string representing true.

CHAPTER SUMMARY

Key Terms

ASCII	floating-point
bias	fractional part
bool	fundamental type
char	garbage value
character literal	hexadecimal
characteristic	identifier
const modifier	IEEE Floating Point Format
constant declaration	int
decimal	keyword
declaration statement	literal

declare	long double
double	long int
escape sequence	mantissa
false	OCD (object-centered design)
fixed-point	octal
float	overflow
precedence, priority	two's complement
roundoff error	type
short int	underflow
sign bit	Unicode
significand	unsigned
string	variable
true	variable declaration

NOTES

- In a C++ program, types of objects must be declared before those objects are used.

- The fundamental types in C++ include `short`, `int`, and `long` for integers; `unsigned` for non-negative integers; `float` and `double` for real values; `boolean` for logical values; and `char` for individual characters.

- C++ compilers treat all real literals as being of type `double`. Thus, most programmers use `double` for real objects and seldom use `float`.

- The `char` type is used to represent single characters; `char` literals must be enclosed in single quotes.

- Escape sequences such as \n (newline) and \t (tab) are used for characters that have a special purpose.

- The `string` type is provided for processing strings of characters. The `<string>` library must be included before `string` objects can be declared.

- An identifier may begin with a letter or _ (underscore), which may be followed by any number of these characters or digits; it may not be a C++ keyword.

- Using meaningful identifiers that suggest what they represent makes programs easier to read and understand.

- C++ is case sensitive.

- Any name in a program that is not a C++ keyword is an identifier and must be declared before it can be used.

- The `const` modifier is used to declare constants, which are values that cannot be changed during program execution.

- Using named constants instead of the literals they represent improves code readability and facilitates program maintenance.

- Placing constant declarations at the beginning of `main()` in a program makes them easy to locate when modifications are necessary.

- If no initial value is specified in a variable declaration, its value is undefined (or indeterminate).

Style and Design Tips

1. In the examples in this text, certain stylistic guidelines for C++ programs are adopted, and you should write your program in a similar style. In this text the following guidelines are used; others are described in later chapters.

   - Put each statement of the program on a separate line.

   - Use uppercase and lowercase letters in a way that contributes to program readability.

   - Indent and align statements within a block enclosed with curly braces { and }.

   - When a statement is continued from one line to another, indent the continued line(s).

   - Insert blank lines between declarations and statements and between blocks of statements to make clear the structure of the program.

   - Separate the operators and operands in an expression with spaces to make the expression easy to read.

2. Declare constants at the beginning of a program and declare variables near their first use. This makes it easy to find constants when they must be modified.

3. Programs cannot be considered to be correct if they have not been tested.

4. Programs should be readable and understandable.

   - Use meaningful identifiers.

   - Use comments to describe the purpose of a program or other key program segments.

5. Programs should be general and flexible—able to solve a class of problems.

6. Identify any preconditions a program has.

Warnings

1. *Character constants must be enclosed in single quotes. In particular,* string *literals cannot be assigned to variables of type* char.

2. *Values of type* char *are stored as their (integer) numeric codes.* Assigning an integer to a char variable may produce an unexpected symbol.

3. *String constants must be enclosed within double quotes.*

4. *Comments are enclosed within /* and */ or between // and the end of the line.* Be sure that:

   - *Each beginning delimiter /* has a matching end delimiter */. Failure to use these in pairs can produce strange results.*

   - *There is no space between the / and the * or between the two slashes.*

5. *Every { must be matched by a }.*

6. *All identifiers must be declared.* Attempting to use an identifier that has not been declared will produce a compilation error. Also, remember that *C++ distinguishes between uppercase and lowercase letters.*

7. *Using a variable that has not been assigned a value may produce "garbage."*

8. *Keywords, identifiers, and constants may not be broken at the end of a line, nor may they contain blanks (except, of course, a string constant may contain blanks).*

TEST YOURSELF

1. List the fundamental data types provided in C++.

2. List the three integer type variations.

3. List the three real types.

4. A constant of a particular type is called a(n) _____.

5. (True or false) 0123 and 123 represent the same integer value.

6. (True or false) 0xA and 10 represent the same integer value.

7. (True or false) All real literals are treated as being of type double.

8. Character literals must be enclosed in _____.

9. (True or false) '1/n' is a valid character literal.

10. (True or false) '\n' is a valid character literal.

11. '\n' is an example of a(n) _____ sequence.

12. String constants must be enclosed in _____.

For Questions 13–20, tell whether each is a legal identifier. If is it not legal, indicate the reason.

13. calorie

14. W_D_4_0

15. 70mph

16. milesPerHour

17. miles per hour

18. miles_per_hour

19. ps.175

20. type-A

For Questions 21–40, tell whether each is an integer literal, a real literal, a character literal, or a string literal. If it is none of these, indicate the reason.

21. 1234

22. 1,234

23. 1.234

24. -1.

25. 123e4

26. 123-4

27. 0.123E-4

28. +1235

29. 12+34

30. $12.34

31. '12+34'

32. 'one'

33. one

34. "Ohm's law"

35. "1234"

36. 01234

37. 0x1234

38. E4

39. '#1'

40. '\32'

41. Write a declaration for a variable count of type int.

42. Write declarations for variables time of type unsigned, temperature of type float, and scale of type char.

43. Repeat Question 41, but initialize count to zero.

44. Repeat Question 42, but initialize time to 9999, temperature to zero, and scale to a blank.

45. Write constant declarations to associate CELSIUS_FREEZE with the integer 0 and CELSIUS_BOIL with the integer 100.

46. Write constant declarations to associate EARTH with 1.5E10 and MARS with 1.2E12.

EXERCISES

For Exercises 1–20, determine if each is a valid C++ identifier. If it is not, give a reason.

1. x Axis

2. x-Axis

3. x_Axis

4. xAxis

5. carbon14

6. 3M

7. PDQ123

8. angle

9. angel

10. anlge

11. PS.175

12. x

13. 4

14. n/4

15. $M

16. zzzzzz

17. z_z_z_z_z_z

18. A+

19. R2-D2

20. R2_D2

For Exercises 21–40, classify each as an integer literal, a real literal, or neither. If it is neither, give a reason.

21. 5,280

22. 5280

23. "5280"

24. 528.0

25. 5280e0

26. -5280

27. --5280

28. +5280

29. 52+80

30. $52.80

31. 52E80

32. E5280

33. eighty

34. 0.528E0

35. .00005280

36. 5.2e-80

37. +52E+80

38. -(-1)

39. -0

40. -1/2

For Exercises 41–52, determine if each is a valid string literal. If it is not, give a reason.

41. "X"

42. "123"

43. INCH"

44. "square feet"

45. "Print \"MPH\""

46. "isn't"

47. "constant"

48. "$1.98"

49. "DON\'T"

50. "12 + 34

51. "\'twas"

52. "\"A\"\"B\"\"C\""

For Exercises 53–56, write constant declarations to associate each name with the specified constant.

53. 40.0 with the name REGULAR_HOURS and 1.5 with the name OVERTIME_FACTOR

54. Planck's constant 6.6256×10^{-34} with the name PLANCK

55. FAHRENHEIT_FREEZE with the integer 32 and FAHRENHEIT_BOIL with the integer 212

56. 0 with ZERO, * with ASTERISK, and an apostrophe with APOSTROPHE

For Exercises 57–60, write declarations for each variable.

57. length and momentOfIntertia of type double

58. population and year of type unsigned

59. mileage of type double, cost and distance of type unsigned

60. alpha and beta of type long, code of type char, and root of type double

For Exercises 61–62, write declarations to declare each variable to have the specified type and initial value.

61. numberOfDeposits and numberOfChecks to be of type int, each with an initial value of 0; totalDeposits and totalChecks to be of type double, each with an initial value of 0.0; and serviceCharge to be of type double with an initial value of 0.25

62. symbol_1 and symbol_2 to be of type char and with a blank character and a semicolon for initial value, respectively; and debug to be of type char with an initial value of T.

63. Write constant declarations that associate the current year with the name YEAR and 99999.99 with MAXIMUM_SALARY and variable declarations that declare number and prime to be of type int and scale to be of type char.

PROGRAMMING PROBLEMS

1. Write a program to convert a measurement given in feet to the equivalent number of (a) yards, (b) inches, (c) centimeters, and (d) meters (1 ft = 12 in, 1 yd = 3 ft, 1 in = 2.54 cm, 1 m = 100 cm).

2. Write a program to convert a weight given in ounces to the equivalent number of (a) pounds, (b) tons, (c) grams, and (d) kilograms (1 lb = 16 oz, 1 ton = 2000 lb, 1 oz = 28.349523 g, 1 kg = 1000 g).

3. For three resistors connected in series, the total resistance is the sum of the individual resistances and by Ohm's law, the current in the circuit is given by *amps = voltage / total_resistance*. Write a program that inputs the voltage and three resistances and then calculates and displays the current.

4. Proceed as in Problem 3, but for resistors connected in parallel for which the total resistance is computed using the formula

$$total\_resistance = \cfrac{1}{\cfrac{1}{resistor1} + \cfrac{1}{resistor2} + \cfrac{1}{resistor3}}$$

Getting Started with Expressions

CONTENTS

```
<>!*''#
^"`$$-
!*=@$_
%*<>~#4
&[]../
|{,,SYSTEM HALTED
```

"THE WAKAWAKA POEM" BY FRED BREMMER AND STEVE KROEZE
(WHILE STUDENTS AT CALVIN COLLEGE)[1]

---

[1] Several years ago, a magazine poll established "waka" as the proper pronunciation for the angle-bracket characters < and >. Here is a phonetic version of this poem:

Waka waka bang splat tick tick hash,
Caret quote back-tick dollar dollar dash,
Bang splat equal at dollar underscore,
Percent splat waka waka tilde number four,
Ampersand bracket bracket dot dot slash,
Vertical-bar curly-bracket comma comma CRASH.

Arithmetic is being able to count up to twenty without taking off your shoes.

MICKEY MOUSE

A little inaccuracy sometimes saves tons of explanation.

SAKI (H. H. MUNROE)

I N THE PRECEDING CHAPTER we looked at how to represent a problem's objects as *software objects* (also called *program entities*) using C++ type statements. This chapter focuses on the operations provided in C++ that may be used to solve a problem and how they are used to form expressions of various kinds.

4.1 INTRODUCTORY EXAMPLE: EINSTEIN'S EQUATION

4.1.1 Problem

Suppose that the professor for your physics course has assigned a large problem set that is due by the next class meeting and that many of the problems require using Einstein's equation to calculate the amount of energy released by a quantity of matter for a given mass. Because of the time pressure, mistakes will likely be made if all of the calculations are done by hand. It would be nice to have a program that could be used to check answers.

4.1.2 Object-Centered Design

4.1.2.1 Behavior

The program should display on the screen a prompt for the quantity of matter (i.e., its mass). It will then read this value from the keyboard. The program should then use Einstein's equation to compute the energy that can be produced by that quantity of matter and display this value to the screen along with a descriptive label.

4.1.2.2 Objects

From our behavioral description, we can identify the following objects:

	Software Objects		
Problem Objects	**Type**	**Kind**	**Name**
screen	output device	variable	cout
prompt	text string	constant	none
quantity of matter	double	variable	*mass*
keyboard	input device	variable	cin
quantity of energy	double	variable	*energy*
descriptive label	string	constant	none

4.1.2.3 Operations

Again, from our behavioral description, we have the following operations:

  i. Output a string (prompt, descriptive label).

 ii. Read a nonnegative numeric value (*mass*) from the keyboard.

iii. Compute *energy* from *mass*.

iv. Output a numeric value (*energy*) and a string (descriptive label) on the screen.

C++ provides operations for each of these operations except for the third one. For this we must use Einstein's familiar equation

$$e = m \times c^2.$$

where *m* is the mass, *c* is the speed-of-light constant, and *e* is the energy produced. Performing this operation thus requires the following operations:

Exponentiation (c^2)

Multiplication of reals ($m \times c^2$)

Storage of a real ($e = m \times c^2$)

This refinement adds two additional objects to our object list:

		Software Objects	
Problem Objects	**Type**	**Kind**	**Type**
screen	output device	variable	cout
prompt	text string	constant	none
quantity of matter	double	variable	*mass*
keyboard	input device	variable	cin
quantity of energy	double	variable	*energy*
descriptive label	string	constant	none
speed of light	double	constant	*SPEED_OF_LIGHT*
2	int	constant	none

4.1.2.4 Algorithm

Next, we organize these objects and operations into an algorithm:

1. Define the constant *SPEED_OF_LIGHT*.

2. Display to cout a prompt for the mass.

3. Read a non-negative real number from cin into *mass*.

4. Compute *energy = mass × SPEED_OF_LIGHT²*.

5. Output to cout a descriptive label and *energy*.

4.1.2.5 Coding, Execution, and Testing

The program in Example 4.1 implements the preceding algorithm. Two of the sample runs use test data for which the output can be easily checked and the third uses "real" data.

Example 4.1 Mass-to-Energy Conversion

```
/* This program computes energy from a given mass using Einstein's
   mass-to-energy conversion equation.

   Input:  the mass (in kilograms) being converted to energy
   Output: the amount of energy (in kilojoules) corresponding to mass
-------------------------------------------------------------------------*/


#include <iostream>        // cin, cout, <<, >>
#include <cmath>           // pow()
using namespace std;


int main()
{
  const double SPEED_OF_LIGHT = 2.997925e8; // meters/sec

  double mass;
  cout << "To find the amount of energy obtained from a given mass,\n"
          "enter a mass (in kilograms): ";
  cin >> mass;             // get mass

                           // compute energy
  double energy = mass * pow(SPEED_OF_LIGHT, 2);

                           // display energy
  cout << mass << " kilograms of matter will release\n"
       << energy << " kilojoules of energy.\n";
}
```

SAMPLE RUNS:

```
To find the amount of energy obtained from a given mass,
enter a mass (in kilograms): 1
1 kilograms of matter will release
8.98755e+16 kilojoules of energy.


To find the amount of energy obtained from a given mass,
enter a mass (in kilograms): .5
0.5 kilograms of matter will release
4.49378e+16 kilojoules of energy.


To find the amount of energy obtained from a given mass,
enter a mass (in kilograms): 155.5
155.5 kilograms of matter will release
1.39756e+19 kilojoules of energy.
```

This program uses several different C++ expressions. In the rest of this chapter, we will explore some of the rich variety of expressions available in C++.

4.2 NUMERIC EXPRESSIONS

A C++ **expression** consists of one or more data values called *operands* and zero or more *operators* that combine these data values to produce a result. For example,

```
100
```

is an expression that consists of one integer value (100) and no operators; it produces the integer value 100. Similarly,

```
1.2 + 3.4
```

is an expression that consists of two operands, the real values 1.2 and 3.4, one operator (+), and produces the real value 4.6. Expressions that produce an int value are called int expressions; those that produce a double value are called double expressions; and so on. In this section, we will examine the arithmetic operators and functions provided in C++ and how they are used to form numeric expressions.

4.2.1 Operators

In C++, the usual plus (+) and minus (−) signs are used for addition and subtraction. Multiplication is denoted by an asterisk (*), which must be used for every multiplication. For example, to multiply x by 10, we can write 10*x or x*10 but not 10x. A slash (/) is used for both real and integer division. Another operation related to integer division is the **modulus** or **remainder** operation, denoted by percent (%), which gives the remainder in an integer division. The following table summarizes these operators.

Operator	Operation
+	addition, unary plus
−	subtraction, unary minus
*	multiplication
/	real and integer division
%	modulus (remainder in integer division)

For the operators +, −, *, and /, the operands may be of either integer or real type. If both are integer, the result is integer, but if either is real, the result is real. For example,

```
3 + 4 → 7           3 + 4.0 → 7.0
3.0 + 4 → 7.0       3.0 + 4.0 → 7.0
9 / 4 → 2           9 / 4.0 → 2.25
9.0 / 4 → 2.25      9.0 / 4.0 → 2.25
```

Note the difference between integer and real division. In the expression 9 / 4 where both operands are integers, integer division is performed, producing the integer quotient 2.

Any fractional parts are dropped and the resulting integer is the result; no rounding occurs. In the other expressions involving division, at least one of the operands is real, so real division is performed, producing the real result 2.25. A common difficulty for beginning programmers is to remember that *the value of* m/n *is 0 if* m *and* n *are integers with* |m| < |n|; for example, each of 1/2, 1/3, 9/10, -99/100 produces the value 0.

In the case of integer division, C++ provides one operator (/) that gives the integer quotient and another operator (%) that gives the remainder from an integer division.[2] The following are some examples:

```
6 / 4 → 1          6 % 4 → 2
83 / 10 → 8        83 % 10 → 3
148 / 10 → 14      148 % 10 → 8
```

For real values, / performs real division and produces a real quotient whose precision is machine dependent. The % operator cannot be used with real values.

4.2.1.1 Operator Priority

The order in which operators in an expression are applied is determined by a characteristic known as **operator priority** (or **precedence**). For the arithmetic operators, *, /, and % have higher priority than + and −.

Operators with higher priority are applied before those with lower priority. For example, in the expression 3 + 4 * 5, the multiplication is performed before the addition because * has higher priority than +, so the value of this expression is 23. Appendix C gives the precedence levels for all of the C++ operators.

4.2.1.2 Operator Associativity

The operators +, −, *, /, and % are all said to be **left-associative**, which means that in an expression having two operators with the same priority, the left operator is applied first. Thus,

```
9 - 4 - 3
```

is evaluated as

```
(9 - 4) - 3 → 5 - 3 → 2
```

In a later section, we will see that some C++ operators are **right-associative**.

Associativity is also used in more complex expressions containing different operators of the same priority. For example, consider

```
9 * 10 - 8 % 3 * 4 + 5
```

---

[2] Neither i / j nor i % j is defined if j is zero. A run-time error will occur if such an expression is encountered.

There are three high-priority operations, `*`, `%`, and `*`, and so the leftmost multiplication is performed first, giving the intermediate result

```
90 - 8 % 3 * 4 + 5
```

Because of left-associativity, `%` is performed next, giving

```
90 - 2 * 4 + 5
```

followed by the second multiplication:

```
90 - 8 + 5
```

Because the two remaining operations – and +, have equal priority, left associativity causes the subtraction to be performed first, giving

```
82 + 5
```

and then the addition is carried out, giving the final result

```
87
```

4.2.1.3 Using Parentheses

The order in which operations in an expression are performed can be changed by using parentheses, which have highest priority; that is, parenthesized subexpressions are evaluated first in the standard manner. If they are "nested"—that is, if one set of parentheses is contained within another—computations in the innermost parentheses will be performed first.

To illustrate, consider the expression

```
(6 * (12 - 4) % 7) / 4 + 9
```

The subexpression `(12 - 4)` is evaluated first, producing

```
(6 * 8 % 7) / 4 + 9
```

Next, the subexpression `(6 * 8 % 7)` is evaluated left to right, with `*` giving

```
(48 % 7) / 4 + 9
```

followed by `(48 % 7)`, which yields

```
6 / 4 + 9
```

Now the division is performed, giving

```
1 + 9
```

and the addition produces the final result

```
10
```

Even though parentheses may not be required, they should be used freely in expressions containing several operations to make the intended order of evaluation clear and to write complicated expressions in terms of simpler expressions. Be sure, however, that the parentheses balance—*each left parenthesis has a matching right parenthesis later in the expression*—or a compilation error will result.

4.2.1.4 Unary Operators

When + and − are used as **unary operators** (i.e., applied to a single operand) as in −x + 5 and −2 * x, they have higher priority than the corresponding binary operations. Thus, if the value of x is −1, the values of these expressions will be 6 and 2, respectively.

4.2.1.5 Summary

The following rules summarize the evaluation of arithmetic expressions.

OPERATOR PRIORITIES

Higher:	unary +, unary −
	*, −, /
Lower:	binary +, binary −

1. Higher-priority operations are performed before lower-priority ones.
2. Operators with the same priority are applied left to right if they are left-associative and right to left if they are right-associative.
3. If an expression contains parenthesized subexpressions, they are evaluated first, using the order specified in rules 1 and 2. For nested parentheses, the innermost subexpressions are evaluated first.

4.2.1.6 Bitwise Operators

C++ also provides operations that can be used to manipulate the individual bits in the stored binary representations of integers: ~ (negation), & (bitwise and), | (bitwise or), ^ (bitwise exclusive or), << (bitshift left), and >> (bitshift right). They are sometimes used in low-level graphics or operating system programs that must check memory or interact directly with a computer's hardware. More information about these operations can be found on the text's website described in the Preface.

4.2.2 Type Coercion

As noted earlier, in the *mixed-type expression* 6.0 / 4 that involves both integer and real operands, C++ performs real division and produces the real value 1.5. This is because the compiler automatically converts the integer value 4 to a real value 4.0, and then performs the real division operation 6.0 / 4.0. This automatic type conversion by the compiler is known as **type coercion**. It also occurs in mixed-type expressions involving characters and integers or reals; the character values are automatically replaced by their numeric codes (see Section 3.3). For example, if characters are represented in ASCII, the value of the expression 'a' + 3 is 99 because 'a' is replaced by 96, its numeric code.

In general, these type conversions take place automatically in any mixed-type expression with "narrow" values—those stored in a smaller number of bits—being converted to "wider" values that occupy a larger number of bits. This automatic widening of a narrower value to the size of a wider value in an expression is known as **promotion** of the narrower value. It is what permits character, integer, and real values to be freely intermixed in C++ expressions. Two values are said to be **compatible** if (i) they have the same type, (ii) one can be promoted to the type of the other, or (iii) both can be promoted to the same type. For example, int and double are compatible, because int can be promoted to double.

4.2.2.1 Explicit Type Conversions

In addition to automatic type conversions, explicit conversions are also possible. For example, if intVal is an integer value, then either of the expressions double(intVal) or (double)intVal produces the double value equivalent to intVal. Similarly, if doubleVal is of type double, either of int(doubleVal) or (int)doubleVal will truncate the fractional part of doubleVal and produce its integer part.

In general, an expression of the form

```
type(expression)
```

or

```
(type)expression
```

where *type* is a valid C++ type can be used to convert the type of *expression* to *type* (if possible). The first form is sometimes referred to as *functional notation* and the last form as (C-style) *cast notation*.

4.3 ASSIGNMENT EXPRESSIONS

An **assignment expression** uses the assignment operator (=) to assign a value to a variable.

ASSIGNMENT EXPRESSION

FORM:

```
variable = expression
```

where:
    variable is the name of a variable;
    expression is a valid expression (a constant, another variable to which a value has pre-
    viously been assigned, or a formula to be evaluated).

BEHAVIOR:

1. *expression* is evaluated and, if necessary and possible, its value is converted to the
   type of *variable*;
2. The value of *variable* is changed to this value; and
3. This value is the value of the entire assignment expression.

For example, suppose that length and width are real variables, low and high are integer variables, code is a character variable, and isValid is a boolean variable, declared
as follows:

```
double width, length;
int low, high;
char code;
boolean isValid;
```

The compiler will associate memory locations with the six variables as described in
Section 3.3. We might picture them as follows, with question marks indicating unspecified
contents:

width	?
length	?
low	?
high	?
code	?
isValid	?

Now consider the following assignment statements:

```
width = 5.0;
length = 2 * width + 3;
low = 20;
code = 'X';
isValid = true;
```

The first statement assigns the real constant 5.0 to the real variable width, and the second assigns the real constant 13.0 to the real variable length. The next assignment
statements assign the integer constant 20 to the integer variable low, the character X
to the character variable code, and the boolean constant true to the boolean variable
isValid. More precisely, when these assignment statements are executed, the values 5.0,
13.0, and 20, the numeric code for X (88 in ASCII), and the integer 1 or some other nonzero integer (representing true) are stored in the memory locations associated with the

variables width, length, low, code, and isValid, respectively. The variable high is still undefined.

width	5.0
length	13.0
low	20
high	?
code	X (88)
isValid	1

These values are substituted for these variables in any expression containing these variables occurring later in the program. Thus, when the assignment statement

```
high = 6 * low + 5;
```

is executed, the value 20 stored in low's memory location is retrieved and used in evaluating the expression

```
6 * low + 5
```

The resulting value 125 is then assigned to the integer variable high; the value of low remains unchanged.

width	5.0
length	13.0
low	20
high	125
code	X (88)
isValid	1

Now consider the assignment statement

```
width = width + 2.1;
```

in which the variable width appears on both sides of the assignment operator (=). In this case, the current value 5.0 for width is used in evaluating the expression width + 2.1 on the right, yielding the value 7.1, and this value is then assigned to width, replacing the old value 5.0:

width	7.1
length	13.0
low	20
high	125
code	88
isValid	1

In assignment statements, the variable to be assigned a value must appear on the left of the assignment operator and a valid expression on the right.

In assigning values to variables, some mixing of numeric types is permitted. For example, the assignment statement

```
width = 5;
```

could have been used to assign a value to `width`; the compiler would have automatically converted the integer 5 to the real value 5.0 and assigned this value to `width`. However, such mixed-type assignments should be used with caution because significant information can be lost. To demonstrate this, consider the assignment statement

```
low = 3 / width;
```

Here, the expression 3 / `width` will be evaluated and because `width` is of type `double`, the result will be the real value 0.6. However, assigning a real value to an integer variable truncates the real value's fractional part. In this example, therefore, the fractional part .6 is truncated and only the integer portion 0 is assigned to `low`.

As we have noted, characters are stored as integers; for example, the assignment statement

```
code = 'X';
```

stored the numeric representation of the character X (e.g., 88 in ASCII) in `code`'s memory location. This allows us to use assignments like

```
code = 88 + 1;
```

or

```
high = 'A' + 'B';
```

If characters are represented in ASCII, the first statement will assign the value 'Y' (89 in ASCII) to `code` and the second will assign the value 131 (65 + 66) to `high`. Such mixed-type assignments can easily lead to obscure errors and should normally be avoided.

It is important to remember that *an assignment statement is a replacement statement.* It is easy for beginning programmers to forget this and write an assignment statement

```
a = b;
```

when

```
b = a;
```

was intended. These statements produce very different results: The first assigns the value of b to a, leaving b unchanged, and the second assigns the value of a to b, leaving a unchanged.

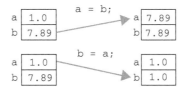

To illustrate further the replacement property of an assignment, suppose that count and number are integer variables whose values are 123 and 99, respectively, and that we wish to interchange these values. To do this, we need an extra integer variable to store the old value of one of the variables before we change them. Suppose we call it temp and use it to store the value of count before we assign count the value of number; then we can assign this stored value to number.

```
temp = count;
count = number;
number = temp;
```

	temp = count;		count = number;		number = temp;
count 123	count 123	→count 99	count 99
number 99	number 99	number 99	number 123
temp ?	temp 123	temp 123	temp 123

4.3.1 Assignment as an Operation

We have seen that for an assignment

    variable = *expression*

three actions result:

1. *expression* is evaluated, producing some value v.

2. The value of *variable* is changed to v.

3. The value v is produced by the = operator.

Up to now, we have focused on the first two actions, but we now consider the third one. Just as the arithmetic expression

```
1 + 2
```

produces the value 3, the assignment

```
number = 99
```

is an expression that produces the value 99; = *is a binary operator whose result is the value assigned to the left operand.* **For example, in the expression**

```
number = number + 1
```

+ is applied first because the = operator has lower priority than almost all other C++ operators,

```
number = (number + 1)
```

producing the result 100. That value is then assigned to number,

```
number = 100
```

which changes the value of number to 100, after which the = operator produces the result 100.

This value-producing property of = means that it can be used inside other arithmetic expressions. For example, the statement

```
temp = 2 * (number = 100) + 1;
```

performs two assignments: first, 100 is assigned to number and the value of the parenthesized expression is 100; then, the resulting expression 2 * 100 + 1 is evaluated and its value 201 assigned to temp. Such statements that use assignment expressions inside other arithmetic expressions, however, are not commonly used and their use is discouraged.

The more common way to use this property of = is to chain several assignment operators together in a single statement such as

```
temp = count = number = 0;
```

Because the assignment operator =, unlike the arithmetic operators we have seen thus far, is right-associative, this means that in this statement, the assignments are made from right to left:

```
temp = (count = (number = 0));
```

First, number is set to 0; and because this = operator produces the value assigned to number (i.e., 0), the statement becomes

```
temp = (count = 0);
```

This assigns 0 to count, producing 0 as its result, and the statement becomes

```
temp = 0;
```

Now, temp is assigned 0, which is returned as the result, and the statement becomes

```
0;
```

which, although a valid C++ statement, does nothing and is discarded. The "side effect" is what is important: All of the variables number, count, and temp have been set to 0.

4.3.2 Assignment Shortcuts

One common form of assignment is one in which the value of a variable is changed by performing some operation on it; for example,

Add *value* to *sum*

Double *number*

These can be written as assignment statements,

```
sum = sum + value;
number = number * 2;
```

but such operations occur so frequently that special shortcut operations are provided for them. Instead of writing the assignment as shown, we can use the shorter forms

```
sum += value;
number *= 2;
```

Each of the arithmetic operators can be used in this way. Any statement of the form

```
variable = variable Δ expr;
```

where Δ is any of the operators +, −, *, /, or %, can be written

```
variable Δ= expr;
```

Each of the following is therefore an acceptable variation of the assignment operator:

```
+=, -=, *=, /=, %=
```

These operations are all right-associative and produce the value assigned as their result. This means that they could be chained together; however, this is not good programming practice because it can be difficult to follow how the resulting expressions are evaluated. For example, if width has the value 3.0 and length the value 5.7, then the statement

```
length *= width += 0.2;
```

is executed as follows:

1. Assign width the value 3.0 + 0.2 = 3.2, and then

2. Assign length the value 5.7 * 3.2 = 18.24.

Chaining operators together in this manner decreases a program's readability and should usually be avoided because such programs are difficult (and perhaps costly) to maintain.

4.3.3 The Increment and Decrement Operations

Algorithms often contain instructions of the form

"Increment count by 1."

that can be encoded as

```
count = count + 1;
```

or, with the preceding shortcut assignment forms, as

```
count += 1;
```

However, this kind of assignment that increments an integer variable by 1 occurs so often that a special unary **increment operator ++** is provided for this operation.[3] It can be used as a postfix operator,

```
variable++
```

or as a prefix operator,

```
++variable
```

where *variable* is an integer variable whose value is to be incremented by 1. Thus, the assignment statement

```
count = count + 1; // or count += 1;
```

can also be written

```
count++;
```

or

```
++count;
```

When ++ is used as a stand-alone operation as in the preceding statements, it makes no difference whether the postfix form or the prefix form is used. There is a subtle difference, however, when they are combined with other operations. To understand this difference,

---

[3] The name C++ stems from this increment operator—C++ is C that has been incrementally improved.

we must remember that these increment expressions are assignment expressions and thus produce values. *They both increment a variable, but the value produced by the postfix form is the original (nonincremented) value of the variable, whereas that produced by the prefix form is the incremented value.*

To illustrate this, consider the following program segments where, as before, count and number are integer variables:

```
//POSTFIX: Use first, then increment          Output
count = 10;
cout << "count =" << count << endl;            count = 10


number = count++;
cout << "number =" << number << endl;          number = 10
cout << "count =" << count << endl;            count = 11
```

and

```
//PREFIX: Increment first, then use           Output
count = 10;
cout << "count =" << count << endl;            count = 10


number = ++count;
cout << "number =" << number << endl;          number = 11
cout << "count =" << count << endl;            count = 11
```

Note that both sets of statements increment the value of count from 10 to 11. However, in the first set of assignments, number is assigned the value 10, but in the second set, number is assigned the value 11.

Just as an integer variable's value can be incremented with the ++ operator, it can be decremented (i.e., 1 is subtracted from it) by using the **decrement operator (--)**. For example, the assignment statement

```
count = count - 1;
```

can be written more compactly as

```
count--;
```

or

```
--count;
```

The differences between the prefix and postfix versions are similar to those for the increment operator.

4.4 INPUT/OUTPUT EXPRESSIONS

In some programming languages, input and output statements contain some special reserved word such as *Read*, *Print*, or *Write* to indicate the operation to be performed. In C++, however, input and output are carried out using special input and output *operators*. Thus, like assignment, we write input expressions to perform input and output expressions to perform output. In this section we take a first look at the basic features of input/output. Other features will follow in later chapters.

4.4.1 I/O Streams

C++ avoids the nitty-gritty details about how I/O is carried out—how characters actually get from the keyboard into a program or how values computed in a program get to the screen—by using **streams**, which connect executing programs with input/output devices. Characters generated by the keyboard enter an *input stream* called an `istream` that transmits them to a program. Similarly, characters that are output by a program enter an *output stream* called an `ostream` that transmits them to the screen.

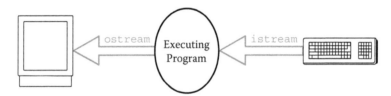

These input and output streams—`istream`s and `ostream`s—are provided by a special C++ **library** named `iostream` that defines three important data objects:

- An `istream` object named `cin` that is associated with the *keyboard*

- An `ostream` object named `cout` that is associated with the *screen*

- An `ostream` object named `cerr` that is associated with the *screen* and is intended for displaying error messages[4]

As you might guess, input and output are the basic operations on these streams.

4.4.2 Input Expressions

The program in Example 4.1 contains the input statement

```
cin >> mass;
```

where `mass` is a real variable. Like arithmetic expressions such as `count + 1` and assignment expressions such as `number = 100` that we have considered,

```
cin >> mass
```

---

[4] A main difference between `cout` and `cerr` is that `cout` is buffered but `cerr` is not. Whereas output to `cerr` goes directly to the screen, output to `cout` goes into a section of memory called an output buffer and appears on the screen only when this buffer is emptied, for example, by `endl`.

is an input expression in which the **input** (or **extraction**) **operator** >> is applied to the two operands cin and mass. That is, >> is a binary operator that acts as follows:

INPUT EXPRESSION

FORM:

```
input_stream >> variable
```

where:
 input_stream is the name of a C++ input stream; and
 variable is a C++ variable for which the input operator >> is defined.

BEHAVIOR:

1. A value whose type is compatible with that of variable must be the next item in input_stream. (If there is none, execution is either suspended or terminated, as described later.)
2. That value is removed from input_stream and assigned to variable.
3. The >> operator produces the modified input_stream as its result.

Because the input operator >> is *left-associative*, input expressions can be chained together. For example, if x and y are variables declared by

```
int x;
double y;
```

the statement

```
cin >> x >> y;
```

will be evaluated as

```
(cin >> x) >> y;
```

in the following manner:

1. If the next value in cin is an integer, it will be removed from cin and assigned to x, and the >> operator produces the value cin, giving the following modified input statement

   ```
   cin >> y;
   ```

2. If the next value in this modified cin is an integer or a real number, it will be removed from cin and assigned to y, and the >> operator produces the value cin, giving the following statement

   ```
   cin;
   ```

3. This last statement contains no operators, only `cin`, which will be discarded, and execution of this input statement is complete.

What happens if either of the necessary inputs in (1) and (2) is not present in `cin` will be described later.

We can describe a typical C++ (interactive) input statement as follows:

C++ INTERACTIVE INPUT STATEMENT

FORM:

```
cin >> variable₁ >> variable₂ >> ... >> variableₙ;
```

where:
    `cin` is the `istream` and `>>` is the input operator provided in the `iostream` library and each `variableᵢ` is a variable for which `>>` is defined.

PURPOSE:

This input statement attempts to read a sequence of n values from the input device with which `cin` is associated, storing them in $variable_1$, $variable_2$,..., $variable_n$, and removing them from the `istream` `cin`.

Note that the user must enter a value (of the appropriate type) for each variable in the statement before execution of the program will resume. Thus, when the following input statement is executed and if `side1`, `side2`, and `side3` are declared as shown,

```
double side1, side2, side3;
cin >> side1 >> side2 >> side3;
```

program execution will be suspended until the user enters values for all three variables. If fewer than three values are entered, for example,

```
2.2  3.3
```

execution will not resume until the user enters a third value, say

```
4
```

The first two values 2.2 and 3.3 will be removed from `cin` and assigned to `side1` and `side2`, respectively; the third value 4 will be removed from `cin`, converted to a real value 4.0, and assigned to `side3`. Program execution then resumes.

The input values can be separated by any **white space** (spaces, tabs, or newlines). For example, if the values are entered on separate lines,

```
2.2
3.3
4
```

they are separated by newlines and the same assignments occur as before.

Because input requires that the correct number and types of values be entered, *it is good practice to provide an informative prompt to the user whenever it is necessary to enter data values.* This is accomplished by preceding input statements with output statements, for example,

```
cout << "Enter the lengths of the triangle's three sides: ";
```

We turn now to a study of output statements.

4.4.3 Output Expressions

C++'s input and output statements are quite similar. To illustrate this, consider the input statement

```
cin >> mass;
```

from Example 4.1 and a shortened version of the output prompt:

```
cout << "Enter a mass (in kilograms): ";
```

cin is the name of an istream object and cout is the name of an ostream object, both of which are defined in the iostream library. The istream input (or extraction) operator is >> and << is the ostream **output** (or **insertion**) **operator**. And very similar to how the expression

```
cin >> mass
```

performs an *input* operation (>>) on two operands (cin and mass), the expression

```
cout << "Enter a mass (in kilograms): "
```

performs an *output* operation (<<) on two operands (cout and "Enter a mass (in kilograms): ").

We can describe a typical C++ output expression as follows:

C++ OUTPUT EXPRESSION
FORM:

```
output_stream << expression
```

where:
output_stream is the name of a C++ output stream and *expression* is any C++ expression for which the output operator << is defined.

BEHAVIOR:
1. *expression* is evaluated.
2. That value is inserted into output_stream.
3. The << operator produces the modified output_stream as its result.

Like >>, the << operator is *left-associative*, which makes it possible to chain output expressions together. To illustrate, consider the following modification of the output statement in Example 4.1:

```
cout << energy << "kilojoules released." << endl;
```

It would be evaluated as follows:

1. The first output expression cout << energy inserts the value of energy into cout and the << operator produces the value cout, giving

   ```
   cout << "kilojoules released." < endl;
   ```

2. The next output expression cout << "kilojoules   released." inserts the string "kilojoules released." into cout and the << operator produces the value cout, giving

   ```
   cout << endl;
   ```

3. The expression endl is evaluated, which inserts a newline character ('\n') into cout and then *flushes* it, causing output to appear on the screen. The << operator produces the value cout, giving

   ```
   cout;
   ```

4. This last statement contains no operators, only cout, which will be discarded and execution of this output statement is complete.

We can thus describe a typical C++ (interactive) output statement as follows:

C++ INTERACTIVE OUTPUT EXPRESSION

FORM:

```
cout << expr₁ << expr₂ << ... << exprₙ;
```

where:
> cout is the ostream and << the output operator provided in the iostream library and each $expr_i$ is a C++ expression for which << is defined.

PURPOSE:

Execution of an output statement inserts the values of $expr_1$, $expr_2$, . . ., $expr_n$ into the ostream cout so they will be displayed on the output device (e.g., the screen) associated with cout.

Note that << only inserts what is being output into cout, character by character with no formatting. In particular, it does not insert any space or newlines between output items unless the items themselves contain them. For example, if the last output statement in Example 4.1 were written as

```
cout << mass << "kilograms of matter will release"
     << energy << "kilojoules of energy.";
```

the output when 1 is entered for mass would be

```
1kilograms of matter will release 8.98755e+16kilojoules of energy.
```

whereas the output statement in Example 4.1,

```
cout << mass << " kilograms of matter will release\n"
     << energy << " kilojoules of energy.\n";
```

adds a space before the words "kilograms" and "kilojoules" and an advance to a new line after the word "release" to make the output more readable:

```
1 kilograms of matter will release
8.98755e+16 kilojoules of energy.
```

Note the \n after the last word "energy." It is there so that any subsequent output will begin on a new line and not be attached at the end of this word. *One or more new line escape sequences* '\n' *or the keyword* endl *must be used to cause output to advance to a newline.*[5] For example,

```
cout << "3\n4\n\nShut the door!" << endl;
```

---

[5] As noted in an earlier footnote, endl, like '\n', produces an advance to a new line but it also flushes the buffer associated with cout so that the output it contains appears immediately on the screen.

will display the output

```
3
4
Shut the door!
```

4.4.4 Output Formatting

There are times when we may want to improve the default output format we have been using up to now. For example, if we are outputting monetary values we might like them rounded to two decimal places and displayed in nicely aligned columns. To accomplish this, C++ provides **format manipulators**, of which the more useful ones are given here; some others are described in later chapters.

FORMAT MANIPULATORS

Manipulators	Description
From iostream:	
showpoint	Display a decimal point and trailing zeros for real values
noshowpoint	For whole real values, hide the decimal point and trailing zeros (default)
fixed	Output real values in fixed-point notation
scientific	Output real values in scientific notation
boolalpha	Output boolean values as strings "true" and "false"
left	Left justify values within output fields
right	Right justify values within output fields (default)
From iomanip:	
setw(*w*)	Output the next value in a field of size *w* (default 1)
setprecision(*p*)	Output *p* fractional digits for all subsequent real values (common default is 6)

PURPOSE:

Format manipulators specify how subsequent items in an output list should appear. *Except for* setw(), *they apply to all subsequent items in the current and following output statements unless modified by other format manipulators;* setw()*applies to only the next output item.*

To illustrate the use of format manipulators, suppose that the truck drivers for a recycling company record the miles traveled, gallons of fuel used, cost of the fuel, and other miscellaneous operating costs for each of their pick-up routes. The program in Example 4.2 inputs these amounts for a route and then calculates miles per gallon, total trip cost, and the cost per mile. It uses two output manipulators from iostream: fixed so that all values are displayed in fixed-point form and showpoint so that decimal points and trailing zeros are displayed; and it uses two from iomanip: setprecision() to round output values to two decimal places and setw()to right-align the values in seven-space zones.

Example 4.2 Recycling Costs

```
/* This program calculates miles per gallon, fuel cost per mile, and
   total cost for a truck on a recycling pick-up route.

   Input:  a route number, total miles traveled, total fuel consumed,
           unit cost of the fuel, and other operating costs
   Output: the miles per gallon, total cost for traveling the route,
           and the cost per mile
-------------------------------------------------------------------*/

#include <iostream>      // cin, cout, <<, >>, fixed, showpoint
#include <iomanip>       // setprecision(), setw()
using namespace std;

int main()
{
  int route;                            // route number
  cout << "Enter your route number: ";
  cin >> route;

  double miles,                         // miles traveled
         gallonsOfFuel,                 // gallons used
         unitFuelCost,                  // fuel cost per gallon
         otherOperatingCost;            // other operating costs
  cout << "Enter:\n\ttotal miles traveled,"
       << "\n\tgallons of fuel used,"
       << "\n\tfuel cost per gallon, and"
       << "\n\tother operating costs (total):"
       << "\n\t---> ";
  cin >> miles >> gallonsOfFuel
      >> unitFuelCost >> otherOperatingCost;

  double milesPerGallon = miles / gallonsOfFuel,
         fuelCost = unitFuelCost * gallonsOfFuel,
         totalTripCost = fuelCost + otherOperatingCost,
         costPerMile = totalTripCost / miles;

  cout << showpoint << fixed << setprecision(2)
       << "For Route #" << route << ":\n"
       << "\n\tMiles per gallon: " << setw(8) << milesPerGallon
       << "\n\tTotal cost:    $" << setw(8) << totalTripCost
       << "\n\tCost per mile: $"  << setw(8) << costPerMile
       << endl;
}
```

SAMPLE RUNS:

```
Enter your route number: 17
Enter:
```

```
        total miles traveled,
        gallons of fuel used,
        fuel cost per gallon, and
        other operating costs (total):
        ---> 100 20 3 1
    For Route #17:

        Miles per gallon: 5.00
        Total cost:     $ 61.00
        Cost per mile: $   0.61


    Enter your route number: 24
    Enter:
        total miles traveled,
        gallons of fuel used,
        fuel cost per gallon, and
        other operating costs (total):
        ---> 111.5 24.5 3.25 33.95
    For Route #24:

        Miles per gallon:  4.55
        Total cost:     $ 113.58
        Cost per mile: $   1.02
```

4.5 BASIC C++ FUNCTIONS AND LIBRARIES

The program in Example 4.1 contains the statement

```
energy = mass * pow(SPEED_OF_LIGHT, 2);
```

to compute the energy released from matter with a given mass, using Einstein's equation $e = mc^2$. Here we see that in addition to simple objects like literals, constants, and variables, an expression's operand may be a value returned by a function. In this example, the standard math library function pow() performs the exponentiation operation.[6]

4.5.1 Libraries

Because of the difficulty of computing exponentials, square roots, logarithms, and other complex operations by hand, many calculators and application software such as Excel provide these as built-in operations. For the same reason, C++ and many other programming languages provide a collection of predefined functions to implement complex operations. For C++, however, this collection is large, and including it in a rather small program like that in Example 4.1 could result in a very large file. For this reason, C++ stores smaller collections of related functions in separate **libraries** so that a program need only include those that are needed.

---

[6] We will use notation of the form functionName() for function names, with the parentheses used to distinguish them from identifiers that do not represent functions.

As the name suggests, a library is a place where functions (and other things such as constant declarations) can be stored, so that a program can "borrow" them. For example, the `iostream` **library** stores declarations of the streams `cin` and `cout` and functions to implement input (`>>`) and output (`<<`) operations. To use the contents of a library, a **#include directive** is added to a program to inform the compiler that the contents of that library must be added to the program. Thus, because the program in Example 4.1 reads numeric values from the keyboard via `cin`,

```
cin >> mass;
```

and outputs values to the screen via `cout` as in the statement

```
cout << mass << " kilograms of matter will release "
     << energy << " kilojoules of energy.\n";
```

the program must contain the directive

```
#include <iostream>                    // cin, cout, <<, >>
```

because the objects `cin` and `cout` and the `>>` and `<<` operations are stored in the `iostream` library. The program also contains the directive

```
#include <cmath>                       // pow()
```

to make the contents of the math library available to the program because the `pow()` function that it uses is in that library. The comments attached to these directives are not mandatory, but it is good programming practice to follow #include directives with comments like those shown that list the items from the libraries that are being used.

When the C++ compiler processes a #include directive, the contents of that file are added to the program. The angle brackets (< and >) around the name of the library together with the line following the #include directives,

```
using namespace std;
```

provide information to the compiler about where to find these libraries.

4.5.2 Numeric Functions

As noted in Chapter 1, C++ was developed by Bjarne Stroustrup, who extended the C language with object-oriented features. However, many powerful features of C have been retained in C++, and this includes the C libraries such as `cmath`.[7] In addition to `pow()`, the **C math library cmath** stores `sqrt()`, `exp()`, and other math-related functions. The **C standard library cstdlib** contains many other commonly used functions such as the absolute value function `abs()` for integers and `exit()`, which can be used to terminate program execution if an error occurs. Becoming familiar with the functions available in

---

[7] C++ libraries whose names begin with "c" such as cmath are C libraries.

these libraries is an important part of learning to program in C++, because they provide many useful operations. We will describe several of these functions throughout the text; others are described in Appendix B.

To use a function to compute a value, referred to as **calling the function**, we simply give the function name followed by parentheses that enclose its **arguments**—constants, variables, or expressions to which the function is to be applied. For example, the program in Example 4.1 uses

```
pow(SPEED_OF_LIGHT, 2)
```

to call the function pow(), sending it SPEED_OF_LIGHT and 2. Computing x^n requires two operands, x and n, so the pow() function requires two arguments. By contrast, if we wanted to find the absolute value of an integer value, we could use the standard library function abs(), which takes a single argument,

```
positiveValue = abs(intValue);
```

but to do so, we must first have included the standard C library cstdlib:[8]

```
#include <cstdlib>
```

Table 4.1 lists most of the mathematical functions provided by cmath; others are described in Appendix B. Each of these functions takes one or more arguments whose type is double (or can be promoted to double) and returns a value of type double. For example, to calculate the square root of 3, we can write sqrt(3.0) or sqrt(3).

TABLE 4.1   cmath Library Functions

Function	Description
fabs(x)	Absolute value of real value x
pow(x, y)	x raised to power y
sqrt(x)	Square root of x
ceil(x)	Least integer greater than or equal to x
floor(x)	Greatest integer less than or equal to x
exp(x)	Exponential function e^x
log(x)	Natural logarithm of x
log10(x)	Base-10 logarithm of x
sin(x)	Sine of x (in radians)
cos(x)	Cosine of x (in radians)
tan(x)	Tangent of x (in radians)
asin(x)	Inverse sine of x
acos(x)	Inverse cosine of x
atan(x)	Inverse tangent of x
sinh(x)	Hyperbolic sine of x
cosh(x)	Hyperbolic cosine of x
tanh(x)	Hyperbolic tangent of x

[8] Some versions of C++ include cstlib automatically.

Some expressions that are more complicated may require more than one function call. For example, if we are solving quadratic equations, we might want to calculate the discriminant $\frac{1}{\sqrt{}}$. We can express this in C++ as

```
sqrt(pow(b, 2) - 4.0 * a * c)
```

Note that if the value of the expression

```
pow(b, 2) - 4.0 * a * c
```

is negative, an error will result because the square root of a negative number is not a real number.

The C library `cstdlib` contains several other general-purpose functions. Table 4.2 lists some of the more useful ones; others are described in Appendix B.

TABLE 4.2 `cstdlib` Library Functions

Function	Description
`abs(i)`	Returns the absolute value of integer i
`rand()`	Returns a pseudorandom integer in the range 0 to RAND_MAX, which is an integer constant (≥ 32767) also defined in `cstdlib`
`srand(seed)`	Uses the integer $seed$ to initialize the sequence of pseudorandom integers returned by `rand()`
`exit(status)`	Terminates program execution and returns control to the operating system; $status = 0$ signals successful termination and any nonzero value signals unsuccessful termination

CHAPTER SUMMARY

Key Terms

`#include` directive

addition operator

assignment operator (=)

C math library (`cmath`)

C standard library (`cstdlib`)

`cerr`

`cin`

compatible

`cout`

decrement operator (--)

division operator (/)

escape sequence

expression

fixed-point

floating-point

format manipulator

increment operator (++)

input operator (>>)

`iostream` library

`istream`

left-associative	output operator (<<)
library	promotion
mixed-type expression	remainder
modulus operator (%)	right-associative
multiplicative operator (*)	short-cut assignment operations
operand	stream
operator	subtraction operator (-)
operator precedence	type conversion
operator priority	unary operator
ostream	white space

NOTES

- It is important to understand the difference between integer and real division. If a and b are both integers with b ≠ 0, a / b gives the integer quotient when a is divided by b, and a % b gives the remainder. If a or b is real, real division is used for a / b and a % b results in an error.

- Promotion is the automatic widening of a narrower value to the size of a wider value. It is used in mixed-type expressions to convert a narrower value to the type of a wider value in the expression.

- Two values are compatible if one of the following is true:

  - They are both of the same type.

  - The type of one value can be promoted to the type of the other value.

  - The types of both can be promoted to the same type.

- In an expression, unary operators + and − have higher priority than *, /, and %, which in turn have higher priority than + (addition) and − (subtraction).

- Associativity determines whether equal-priority operators are applied from left to right or from right to left.

- Parentheses can be used to change the usual order of evaluation in an expression.

- A directive of the form #include <lib> must be used to insert the contents of a C/C++ library before the functions in that library can be used.

- Expressions of the form type(expr) or (type)expr can be used to convert the value of expr to the specified type.

- An assignment statement is a replacement statement: `a = b;` replaces the value of `a` with the value of `b`.

- Assignment is a value-producing operator that returns the value being assigned. This together with right-associativity makes it possible to chain assignments; for example, `a = b = c = d;`.

- An assignment statement of the form `alpha = alpha Δ beta;` can be written more compactly as `alpha Δ= beta;`.

- The increment (`++`) and decrement (`--`) operations are useful for incrementing/decrementing an integer variable by 1.

- I/O is carried out in C++ by operators acting on streams (`istreams` for input and `ostreams` for output), which connect an executing program with an input/output device.

- Input statements of the form

  ```
  an_istream >> variable₁ >> variable₂ >> ... >> variableₙ;
  ```

  are used to input values for variables; `cin` is the `istream` associated with the keyboard. White space (spaces, tabs, or newlines) is used to separate input values. Leading white space will be ignored when reading a value for a variable.

- It is good practice to prompt the user with an informative message when data values are to be entered.

- Output statements of the form

  ```
  an_ostream << expr₁ << expr₂ << ... << exprₙ;
  ```

  are used to output values of expressions; `cout` and `cerr` are `ostreams` associated with the screen.

- The format of output is controlled by inserting format manipulators into output lists.

Style and Design Tips

In the examples in this text, we adopt certain stylistic guidelines for C++ programs, and you should write your program in a similar style. Those at the end of Chapter 3 are important ones and you should review them.

Warnings

1. The type of value stored in a variable should be the same as or promotable to the type of that variable.

2. If an integer value is to be stored in a real variable, the integer will be promoted to a real type. By contrast, if a real value is to be stored in an integer variable, then the real value is truncated, possibly resulting in the loss of information.

3. *Parentheses in expressions must be paired.* That is, for each left parenthesis, there must be exactly one matching right parenthesis that occurs later in the expression.

4. *Both real and integer division are denoted by /; which operation is performed is determined by the type of the operands.*

5. *All multiplications must be indicated by *;* for example, 2*n is valid, but 2n is not.

6. *A semicolon must appear at the end of each expression (assignments, input, output, etc.) that is meant to be a programming statement.*

7. There are many operators in C++—we've barely scratched the surface—and remembering their priorities will be increasingly difficult. For this reason, *it is wise to use parentheses in complex expressions to clearly specify the order in which the operators are to be applied.*

TEST YOURSELF

Section 4.2

Find the value of each of the expressions in Questions 1–8, or explain why it is not a valid expression.

1. 3 − 2 − 1

2. 2.0 + 3.0 / 5.0

3. 2 + 3 / 5

4. 5 / 2 + 3

5. 7 + 6 % 5

6. (7 + 6) % 5

7. (2 + 3 * 4) / (8 − 2 + 1)

8. 12.0 / 1.0 * 3.0

Questions 9–18 assume that two, three, and four are reals with values 2.0, 3.0, and 4.0, respectively; intEight and intFive are integers with values 8 and 5, respectively; and ASCII representation is used for characters (see Appendix A). Find the value of each expression.

9. two + three * three

10. intFive / 3

11. (three + two / four) * 2

12. intEight / intFive * 5.1

13. four * 2 / two * 2

14. intFive * 2 / two * 2

15. 'c' − 1

16. 'b' + 'c' − 'a'

17. int(two * three / four)

18. (int)(two * three / four)

Section 4.3

Questions 1–17 assume that the following declarations have been made:

```
int m, n;
double pi;
char c;
```

Tell whether each is a valid C++ statement. If it is not valid, explain why it is not.

1. pi = 3.0;

2. 0 = n;

3. n = n + n;

4. n + n = n;

5. m = 1;

6. m = "1";

7. m = n = 1;

8. c = '65';

9. c = 65;

10. c = '1';

11. c = "1";

12. m = m;

13. pi = m;

14. m = pi;

15. m++;

16. m + n;

17. ++pi;

For Questions 18–23, assume that the following declarations have been made:

```
int intEight = 8, intFive1 = 5, intFive2 = 5, jobId;
double two = 2.0, three = 3.0, four = 4.0, xValue;
```

Find the value assigned to the given variable or indicate why the statement is not valid.

18. xValue = three + two / four;

19. xValue = intEight / intFive1 + 5.1;

20. jobId = intEight / intFive1 + 5.1;

21. jobId = intFive1++;

22. jobId = ++intFive2;

23. intEight *= 8;

For each of Questions 24–26, write a C++ assignment statement that calculates the given expression and assigns the result to the specified variable.

24. *rate* times *time* to *distance*

25. ◄   to *c*

26. Assuming that *x* is an integer variable, write four different statements that increment *x* by 1.

Section 4.4

1. In C++, input and output are carried out using _____, which connect an executing program with an input/output device.

2. (True or False) C++ has no input or output facilities built into the language.

3. _____ is the stream object associated with the keyboard; its type is _____.

4. _____ and _____ are stream objects associated with the screen; their type is _____.

5. The input operator is _____.

6. The output operator is _____.

7. The value produced by the input expression `cin >> x` is _____.

8. The value produced by the output expression `cout << x` is_____.

9. The input and output operators are (left or right) _____ associative.

10. _____ can be inserted into an output list to format the output of items.

Questions 11–13 assume the declarations

```
int number = 123;
double rate = 23.45678;
```

For each, show precisely the output that the set of statements produces, indicating blanks with ␣, or explain why an error occurs.

11. `cout << number << rate << endl;`

12. `cout << '\n' << setw(5) << number << number + 1`
    `<< setw(5) << number + 2`
    `<< setw(1) << number + 4 << endl;`

13. `cout << showpoint << fixed`
    `<< setw(8) << setprecision(0) << rate << endl`
    `<< setw(8) << setprecision(1) << rate << endl`
    `<< setw(8) << setprecision(2) << rate << endl`
    `<< setw(8) << rate << endl`
    `<< setprecision(1) << rate << endl;`

Questions 14–17 assume the declarations

```
int number1, number2, number3;
double real1, real2, real3;
```

For each, tell what value, if any, will be assigned to each variable, or explain why an error occurs, when the statement is executed with the given input data:

14. `cin >> number1 >> number2 >> number3;`   Input: 11 22
                                                    33 44

15. cin >> real1 >> real2 >> real3;	Input:	1.1 2 3.3 4
16. cin >> number1 >> number2 >> number3;	Input:	1.1 2 3.3 4
17. cin >> number1 >> real1 >> number2; >> real2 >> number3 >> real3;	Input:	1.1 2 3.3 4 5.5 6

Section 4.5

Questions 1–15 assume that r2, r3, and r4 are reals with values 2.0, 3.0, and 4.0, respectively, and i3 and i4 are integers with values 3 and 4, respectively. Find the value of each expression or explain why it is not valid.

1. sqrt(6.0 + 3.0)

2. pow(2.0, 3)

3. floor(2.34)

4. ceil(2.34)

5. sqrt(r2 + r3 + r4)

6. pow(r2, i4)

7. sqrt(pow(4.0, 2))

8. sqrt(pow(-4.0, 2))

9. pow(sqrt(-4.0), 2)

10. ceil(8.0 / 5.0)

11. floor(8.0 / 5.0)

12. pow(r2, i4) / pow(r4, i3)

13. pow(i3, pow(r2, 2))

14. sqrt(i4 * pow(r4, 3))

15. floor(sqrt(i4 - r3))

For Questions 16–19, write C++ expressions that will compute the given expression.

16. The square root of the average of x and y

17. $| a / (b + c) |$ (where $|x|$ denotes the absolute value of x)

18. a^x, computed as $e^{x \ln a}$ (where ln is the natural logarithm function)

19. The real quantity *amount* rounded to the nearest hundredth

EXERCISES

Section 4.2

Find the value of each of the expressions in Exercises 1–19, or explain why it is not a valid expression.

1. 9 − 5 − 3

2. 2 / 3 + 3 / 5

3. 9.0 / 2 / 5

4. 9 / 2 / 5

5. 2.0 / 4

6. (2 + 3) % 2

7. 7 % 5 % 3

8. (7 % 5) % 3

9. 7 % (5 % 3)

10. (7 % 5 % 3)

11. 25 * 1 / 2

12. 25 * 1.0 / 2

13. 25 * (1 / 2)

14. -3.0 * 5.0

15. 5.0 * -3.0

16. 12 / 2 * 3

17. ((12 + 3) / 2) / (8 - (5 + 1))

18. ((12 + 3) / 2) / (8 - 5 + 1)

19. (12 + 3 / 2) / (8 - 5 + 1)

Exercises 20–24 assume that r1 and r2 are reals with values 2.0 and 3.0, respectively; i1, i2, and i3 are integers with values 4, 5, and 8, respectively; and ASCII representation is used for characters (see Appendix A). Find the value of each expression.

20. r1 + r2 * r2

21. i3 / 3

22. i3 / 3.0

23. (r2 + r1) * i1

24. i3 / i2 * 5.1

Write C++ expressions to compute each of the quantities in Exercises 25–28.

25. $10 + 5B - 4AC$

26. The average of m and n (their sum divided by two)

27. Three times the difference $4 - n$ divided by twice the quantity $m^2 + n^2$

28. Using the given values of cost, verify that the statement

```
cost = double(int(cost*100.0 + 0.5)) / 100.0;
```

can be used to convert each of these real values of cost to dollars, rounded to the nearest cent.

a) 12.342

b) 12.348

c) 12.345

d) 12.340

e) 13.0

29. Write an expression similar to that in Exercise 28 that rounds a real amount x to the nearest tenth.

30. Write an expression similar to that in Exercise 28 that rounds a real amount x to the nearest thousand.

Section 4.3

Exercises 1–16 assume that number is an integer variable, xValue and yValue are real variables, and grade is a character variable. Tell whether each is a valid C++ statement. If it is not valid, explain why it is not.

1. xValue = 2.71828;

2. 3 = number;

3. grade = 'B+';

4. number = number + number;

5. xValue = 1;

6. grade = A;

7. number + 1 = number;

8. xValue = '1';

9. xValue = yValue = 3.2;

10. yValue = yValue;

11. xValue = 'A';

12. grade = grade + 10;

13. xValue /= yValue;

14. xValue = number;

15. number = yValue;

16. xValue = yValue++;

For Exercises 17–29, assume that the following declarations have been made:

```
int int16 = 16, int10 = 10, number;
double real4 = 4.0, real6 = 6.0, real8 = 8.0, xCoord;
char numeral = '2', symbol;
```

Find the value assigned to the given variable or indicate why the statement is not valid.

17. xCoord = (real4 + real6) * real6;

18. xCoord = (real6 + real4 / real8) * 2;

19. xCoord = int16 / int10 + 5;

20. number = int16 / int10 + 5;

21. `symbol = 4;`

22. `symbol = numeral;`

23. `symbol = '4';`

24. `symbol = real8;`

25. `real4 = 2;`

26. `real4 = '2';`

27. `real4 = numeral;`

28. `int16 = int16 + 2;`

29. `number = 1 + numeral;`

For each of Exercises 30–34, write an assignment statement that changes the value of the integer variable `number` by the specified amount.

30. Increment `number` by 77.

31. Decrement `number` by 3.

32. Increment `number` by twice its value.

33. Add the rightmost digit of `number` to `number`.

34. Decrement `number` by the integer part of the real value `xCoord`.

For each of Exercises 35–40, write a C++ assignment statement that calculates the given expression and assigns the result to the specified variable. Assume that all variables are of type `double`.

35. *rate* times *time* to *distance*

36. *xCoord* incremented by an amount *deltaX* to *xCoord*

37. $total\_resistance = \dfrac{1}{\frac{1}{resistor1} + \frac{1}{resistor2} + \frac{1}{resistor3}}$ to *resistance*

38. 5/9 of the difference *fahrenheit* – 32 to *celsius*

39. Area of a triangle with a given *base* and *height* (one-half *base* times *height*) to *area*

40. *amount* rounded to the nearest integer to *amount*

For each of Exercises 41–43, give values for the integer variables a, b, and c for which the two given expressions have different values:

41. a * (b / c) and a * b / c

42. a / b and a * (1 / b)

43. (a + b) / c and a / c + b / c

Section 4.4

Exercises 1–8 assume the declarations

```
double alpha = -567.392, beta = 0.0004;
int rho = 436;
```

For each, show precisely the output that each of the statements produces, indicating blanks with ␣, or explain why an error occurs.

1. cout << rho << rho + 1 << rho + 2;

2. cout << "alpha ="
          << setw(9) << setprecision(3) << alpha << endl
          << setw(10) << setprecision(5) << beta << endl
          << setw(7) << setprecision(4) << beta << endl;

3. cout << setprecision(1) << setw(8) << alpha << endl
          << setw(5) << rho << endl
          << "Tolerance:"
          << setw(8) << setprecision(5) << beta << endl;

4. cout << "alpha =" << setw(12) << setprecision(5)
          << alpha << endl
          << "beta =" << setw(6) << setprecision(2) << beta << endl
          << "rho =" << setw(6) << rho << endl << setw(15)
          << setprecision(3)
          << alpha + 4.0 + rho << endl;

5. cout << "Tolerance =" << setw(5)
          << setprecision(3) << beta;
   cout << setw(2) << rho << setw(4) << alpha;

6. cout << setw(8) << setprecision(1) << 10 * alpha
          << setw(8) << ceil(10 * alpha);
   cout << setprecision(3) << setw(5) << pow(rho / 100, 2.0)
          << setw(5) << sqrt(rho / 100);

7. cout << "rho =" << setw(8) << setprecision(2) << rho
   << "*****";

8. cout << setw(10) << alpha << setw(10) << beta;

For Exercises 9 and 10, assume the declarations

```
int i = 15, j = 8;
char c = 'c', d = '-';
double x = 2559.50, y = 8.015;
```

Show precisely the output that each of the statements produces; indicate blanks with ⌴, or explain why an error occurs.

9. cout << setw(j) << setprecision(2) << "new balance ="
        << x << ' ' << setw(i % 10) << c
        << setw(j) << setprecision(j - 6) << y;

10. cout << "i =" << setw(i) << i
        << "j =" << setw(j) << setprecision(j) << j << endl
        << setw(j) << i << ' '
        << setw(i) << j;

For Exercises 11–14, assume the declarations

```
int n1 = 39, n2 = -5117;
char c = 'F';
double r1 = 56.7173, r2 = -0.00247;
```

For each exercise, write output statements that use these variables to produce the given output. (The underlining dashes are shown here only to help you determine the spacing.)

11.   56.7173   F   39

     -5117PDQ-0.00247
    ----

12. 56.717   -0.0025***39   F

      56.72   39-5117
    ----

13. Roots are   56.717 and -0.00247

```
14. Approximate angles:    56.7 and -0.0
    ----------------------------------
    Magnitudes are        39 and  5117
    ----------------------------------
```

For Exercises 15–21, assume that a, b, and c are integer variables and x, y, and z are real variables. Tell what value, if any, will be assigned to each of these variables, or explain why an error occurs, when the input statements are executed with the given input data:

```
15. cin >> a >> b >> c          Input:  1 2    3
        >> x >> y >> z;                 4 5.5 6.6
```

```
16. cin >> a >> b >> c;         Input:  1
    cin >> x >> y >> z;                 2
                                        3
                                        4
                                        5
                                        6
```

```
17. cin >> a >> x;              Input:  1 2.2
    cin >> b >> y;                      3 4.4
    cin >> c >> z;                      5 6.6
```

```
18. cin >> a >> b >> c;         Input:  1 2.2
    cin >> x >> y >> z;                 3 4.4
                                        5 6.6
```

```
19. cin >> a;                   Input:  1 2    3
    cin >> b >> c;                      4 5.5 6.6
    cin >> x >> y;
    cin >> z;
```

```
20. cin >> a                    Input:  1 2    3
        >> b >> c                       4 5.5 6.6
        >> x >> y
        >> z;
```

```
21. cin >> a >> b;              Input:  1   2     3
    cin >> c >> x >> y >> z;            4   5.5   6.6
                                        7   8.8   9.9
                                       10  11.11 12.12
                                       13  14.14 15.15
```

Section 4.5

For Exercises 1–11, assume that r1 and r2 are reals with values 2.0 and 3.0, respectively, and i1, i2, and i3 are integers with values 4, 5, and 8, respectively. Find the value of the expression or explain why it is not valid.

<table>
<tr><td>1. <code>sqrt(pow(4.0,2))</code></td><td>5. <code>floor(8.0 / 5.0)</code></td></tr>
<tr><td>2. <code>sqrt(pow(-4.0,2))</code></td><td>6. <code>pow(i1,2) / pow(r1,2)</code></td></tr>
<tr><td>3. <code>pow(sqrt(-4.0),2)</code></td><td>7. <code>pow(i2,2) / pow(r1,2)</code></td></tr>
<tr><td>4. <code>ceil(8.0 / 5.0)</code></td><td>8. <code>sqrt(r1 + r2 + i1)</code></td></tr>
</table>

For Exercises 9–12, assume that the following declarations have been made:

```
int int1 = 10, int2 = 16, number;
double rVal = 5, xCoord;
```

Find the value assigned to the given variable or indicate why the statement is not valid.

9. `xCoord = pow(int1,2) / sqrt(int2);`

10. `number = pow(int1,3) / pow(int1,2);`

11. `number = ceil(pow(int2 % int1, 2) / rVal);`

12. `number = floor(pow(int2 % int1, 2) / rVal);`

For Exercises 13–17, write an expression to compute the given quantity.

13. The square root of the average of m and n

14. $| A / (m + n) |$ (where $| x |$ denotes the absolute value of x)

15. a^x, computed as $e^{x \ln a}$ (where ln is the natural logarithm function)

16. The real quantity *amount* rounded to the nearest hundredth

17. $\dfrac{2v^2 \sin a \cos a}{g}$

PROGRAMMING PROBLEMS

1. Write a program to read the lengths of the two legs of a right triangle, and to calculate and display the area of the triangle (one-half the product of the legs) and the length of the hypotenuse (square root of the sum of the squares of the legs).

2. Write a program to read values for the coefficients a, b, and c of the quadratic equation $ax^2 + bx + c = 0$, and then find the two roots of this equation by using the quadratic formula

$$\frac{-b \pm \sqrt{b^2 - 4ac}}{2a}$$

Execute the program with several values of a, b, and c for which the quantity $b^2 - 4ac$ is nonnegative, including $a = 4$, $b = 0$, $c = -36$; $a = 1$, $b = 5$, $c = -36$; and $a = 2$, $b = 7.5$, $c = 6.25$.

3. Write a program to convert a measurement given in feet to the equivalent number of (a) yards, (b) inches, (c) centimeters, and (d) meters (1 foot = 12 inches, 1 yard = 3 feet, 1 inch = 2.54 centimeters, 1 meter = 100 centimeters). Read the number of feet and display, with appropriate labels, the number of feet and the corresponding number of yards, inches, centimeters, and meters.

4. The horizontal displacement x and the vertical displacement y (in feet) of a rocket at t seconds after firing are given by

$$x = v_0 t \cos\theta$$

$$y = v_0 t \sin\theta - 16t^2$$

where v_0 is the initial velocity (ft/sec) and θ is the angle (in radians) at which the rocket is fired. Write a program that reads values for v_0, θ, and t; calculates x and y using these formulas; and displays these values.

5. The current in an alternating current circuit that contains resistance, capacitance, and inductance in series is given by

$$I = \frac{E}{\sqrt{R^2 + (2\pi fL - 1/(2\pi fC))^2}}$$

where I = current (amperes), E = voltage (volts), R = resistance (ohms), L = inductance (henrys), C = capacitance (farads), and f = frequency (hertz). Write a program that reads values for the voltage, resistance, capacitance, and frequency and then calculates and displays the current.

6. Angles are often measured in degrees (°), minutes ('), and seconds ("). There are 360 degrees in a circle, 60 minutes in one degree, and 60 seconds in one minute. Write a program that reads two angular measurements given in degrees, minutes, and seconds, and then calculates and displays their sum. Use the program to verify each of the following:

$$74°29'13" + 105°8'16" = 179°37'29"$$

$$7°14'55" + 5°24'55" = 12°39'50"$$

$$20°31'19" + 0°31'30" = 21°2'49"$$

7. In order for a shaft with an allowable shear strength of S lb/in² to transmit a torque of T in-lbs, it must have a diameter of at least $D = \sqrt[3]{\dfrac{16T}{S}}$ inches. If P horsepower is applied to the shaft at a rotational speed of N rpm, the torque is computed by $T = 63000\dfrac{P}{N}$.

Write a program that reads values for P, N, and S and then calculates and displays the torque developed and the required diameter to transmit that torque. Execute the program with the following inputs:

P (HP)	N (rpm)	S (psi)
20	1500	5000
20	50	5000
270	40	6500

8. The period of a pendulum is given by the formula

$$P = 2\pi \sqrt{\frac{L}{g}} \left(1 + \frac{1}{4} \sin^2 \left(\frac{\alpha}{2} \right) \right)$$

where $g = 980$ cm/sec^2, L = pendulum length (cm), and α = angle of displacement. Write a program to read values for L and α for a pendulum and then calculate and display its period. Execute your program with the following inputs:

L (cm)	α (degrees)
120	15
90	20
60	405
74.6	10
83.6	12

9. Write a program that reads the amount of a purchase, the amount received in payment (both amounts in cents), and then computes and displays the change in dollars, half-dollars, quarters, dimes, nickels, and pennies.

Control Structures

CONTENTS

> When you get to the fork in the road, take it.
>
> YOGI BERRA

> But what has been said once can always be repeated.
>
> ZENO OF ELEA

> Progress might be a circle, rather than a straight line.
>
> EBERHARD ZEIDLER

> It's déjà vu all over again.
>
> YOGI BERRA

IN ALL OF THE EXAMPLE programs in the preceding chapters, the flow of execution has been **sequential**—the first statement of main() is executed, then the second statement, and so on until the last statement has been executed and the closing curly brace } reached. The following **flow diagram** shows this straight-line pattern of execution that characterizes sequential execution and indicates clearly that the statements are executed in the order in which they are given with each statement being executed exactly once.

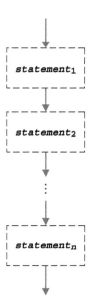

Although sequence is a fundamental **control structure**, it is not powerful enough to solve all problems. Some problems require selecting one of several alternative actions, which means that programs for solving them will need a **selection structure**. In this chapter we will consider one selection structure that C++ provides: the `if` statement. Another, the `switch` statement, is described in Chapter 8.

For some problems, however, another control structure besides sequence and selection is needed, one that makes it possible to execute one or more statements repeatedly—a **repetition structure**. We will consider two such repetition structures provided in C++, the `for` and `while` statements, in the last part of this chapter.

Sequence, selection, and repetition are the only control structures needed. A fundamental result of theoretical computer science states that *any problem solvable by a computer program can be solved by one that is made up of sequence, selection, and repetition structures.*[1]

5.1 INTRODUCTORY EXAMPLE: AIR QUALITY INDEX (AQI)

5.1.2 Problem

Pollution levels are measured at three locations in a city and the Air Quality Index (AQI), also known as the Air-Pollution Index (API), is the integer average of these three readings. An index of less than 50 parts per million indicates a safe condition, whereas indexes of 50 parts per million or greater indicate a hazardous condition.[2] The city's environmental

---

[1] This theorem is usually credited to a 1966 paper by Corrado Böhm and Giuseppe Jacopini.

[2] Here we are considering a simplified AQI that has only two categories, but in practice, there are usually more than two. For example, as we will see in Chapter 8 where we consider selection structures with more than two alternatives, there are 6 categories for the United States AQI.

statistician would like a program that calculates these air quality indexes and determines the appropriate pollution condition, safe or hazardous.

5.1.3 Object-Centered Design

5.1.3.1 Behavior

The program should display on the screen a prompt for three pollution readings (parts per million). It will then read these values from the keyboard and calculate the air quality index, which is the average of the readings rounded to the nearest integer. The program should display the air quality index to the screen, compare it with the cutoff value that separates hazardous and safe conditions, and display the corresponding condition.

5.1.3.2 Objects

From our behavioral description, we can identify the following objects:

	Software Objects		
Problem Objects	**Type**	**Kind**	**Name**
screen	output device	variable	cout
prompt	text string	constant	none
keyboard	input device	variable	cin
pollution levels	double	variables	*level1, level2, level3*
air quality index	int	variable	*index*
cutoff value	int	constant	*CUTOFF*

5.1.3.3 Operations

Again, from our behavioral description, we have the following operations:

i. Output a string (prompt, descriptive label)

ii. Read real values (*level1, level2, level3*) from the keyboard

iii. Compute *index* from *level1, level2, level3*: add the levels, divide by 3, and round by adding 0.5

iv. Output a numeric value (*index*) and a string (pollution condition) on the screen

5.1.3.4 Algorithm

The next step is to organize these objects and operations into an algorithm:

1. Define the constant *CUTOFF*.

2. Display to cout a prompt for the three pollution levels.

3. Read three real values from cin into *level1, level2, level3*.

4. Compute *index* = (*level1* + *level2* + *level3*) / 3 + 0.5.[3]

---

[3] A real number can be rounded to the nearest integer by adding 0.5 and truncating the decimal part. In general, to round to the nearest nth decimal place, multiply by 10^n, add 0.5, truncate the decimal part, and divide by 10^n.

5. Output to cout a descriptive label and index.

6. Output to cout "Safe" if *index* < *CUTOFF* and "Hazardous" otherwise.

5.1.3.5 Coding, Execution, and Testing

The program in Example 5.1 implements the preceding algorithm. Note the rounding of the average pollution reading in the computation of index.

Example 5.1 Air Quality Indexes—Version 1

```
/* This program reads three pollution levels, calculates an air
   quality index (AQI) as their integer average, and then displays
   an appropriate air-quality message.

   Input:     the three pollution levels
   Constant:  the cutoff value (parts per million)
   Output:    the air quality index and a "safe condition" message if
              this index is less than the cutoff value, otherwise a
              "hazardous condition" message
------------------------------------------------------------------*/

#include <iostream>                     // cin, cout, <<, >>
using namespace std;

int main()
{
  const int CUTOFF = 50;                // safe pollution level cutoff

  double level1, level2, level3;
  cout << "Enter 3 pollution readings (parts per million): ";
  cin >> level1 >> level2 >> level3;    // get pollution levels

                                        // compute AQI
  int index = (level1 + level2 + level3) / 3 + 0.5;
                                        // display AQI
  cout << "AQI: " << index << " -- ";
                                        // display condition
  if (index < CUTOFF)
    cout << "Safe condition\n";
  else
    cout << "Hazardous condition\n";
}
```

SAMPLE RUNS:

```
Enter 3 pollution readings (parts per million): 30 40 50
AQI:  40 -- Safe condition


Enter 3 pollution readings (parts per million): 40 50 60
AQI:  50 -- Hazardous condition
```

```
Enter 3 pollution readings (parts per million): 44.5 53.8 61
AQI:  53 -- Hazardous condition

Enter 3 pollution readings (parts per million): 39.2 52 54.5
AQI:  49 -- Safe condition
```

The expression index < CUTOFF used in the if statement of the preceding program to determine which pollution condition to display is known as a *boolean* expression. Because they form such an essential part of selection and repetition structures, we will look next at boolean expressions in detail.

5.2 BOOLEAN EXPRESSIONS

In the mid-1800s, George Boole, a British mathematician and philosopher, developed a system of algebraic logic, which has since come to be known as *boolean logic* and is the basis for modern digital computer logic. The logical expressions in this system, which have either the value *true* or the value *false*, have thus come to be known as **boolean expressions**. In computing, they are also often called **conditions**.

All modern programming languages provide boolean expressions, and in this section we consider them in C++. We will consider *simple* boolean expressions first and then *compound* expressions constructed by applying logical operators to modify or to combine other boolean expressions.

5.2.1 Simple Boolean Expressions

The bool type has two literals, false and true, and boolean expressions have these as their values.[4] The simplest boolean expressions test some *relationship* between two values such as whether one of them is less than the other. For example, the program in Example 5.1 contains the boolean expression

```
index < CUTOFF
```

which compares the operands index and CUTOFF using the *less-than* relationship, and produces the value true if the value of index is less than CUTOFF and the value false otherwise. Similarly, the boolean expression

```
count == 5
```

tests the *equality* relationship between the (variable) operand count and the (literal) operand 5, producing the value true if the value of count is 5 and the value false if it is not. *Note: Be sure to use the == operator for equality comparisons and not = (assignment) because an error will almost surely result otherwise.*[5]

---

[4] For upward compatibility with C, integers may also be used as boolean values: 0 in place of false and any nonzero value for true, with 1 being the most common.

[5] For example, in an if statement of the form if (x = 1) *something*, the "condition" x = 1 is actually an assignment expression and thus has the value 1, which represents true. Consequently, *something* will *always* be executed.

The operators given in the following table are called **relational operators** because they test a relationship between two operands:

Relational Operator	Relation Tested
<	Is less than
>	Is greater than
==	Is equal to
!=	Is not equal to
<=	Is less than or equal to
>=	Is greater than or equal to

They are used in boolean expressions of the form

> $expression_1$ $relational\_operator$ $expression_2$

where *expression₁* and *expression₂* have compatible types. For example, if root, a, b, and c are of type double, count is int, and answer is of type char, then the following are valid boolean expressions formed using these relational operators:

```
root < 1
b * b >= 4.0 * a * c
count == 100
answer != 'T'
```

If root has the value 0.7, then the expression

```
root < 1
```

has the value true. Similarly, if count has the value 99, then the expression

```
count == 100
```

produces the value false.

As illustrated by the expression answer != 'T', char values may also be compared using relational operators. As we saw in Chapter 3, characters are represented in memory by numeric codes, commonly ASCII (see Appendix A), and when two characters are compared, it is their numeric codes that are compared. For example, if char1 and char2 are character variables initialized by

```
char char1 = 'A', char2 = 'B';
```

the boolean expression

```
char1 < char2
```

will be true because the code of A (65) is less than the code of B (66). Similarly, this expression would be true if char1 and char2 were assigned the corresponding lower case letters,

```
char1 = 'a';
char2 = 'b'
```

because the code of a (97) is less than the code of b (98). However, if char2 were assigned the value 'A' instead,

```
char1 = 'a';
char2 = 'A';
```

the boolean expression char1 < char2 would be false, because the code of a (97) is not less than the code of A (65).

5.2.2 Compound Boolean Expressions

Some relationships are too complex to be expressed in C++ using only the relational operators. For example, if a temperature is restricted to the range 0 through 100,

$$0 \leq \text{temperature} \leq 100$$

this *cannot* be correctly represented in C++ by

```
0 <= temperature <= 100
```

even though this is a valid C++ expression. Suppose, for example that temperature has the value 105, which is not in the range 0 through 100, so we might expect this expression to be false. However, because relational operators are left-associative, this expression will be processed as

```
(0 <= 105) <= 100
```

with the subexpression 0 <= 105 evaluated first. This has the value true, which is represented in C++ as the integer 1. Thus, the preceding expression becomes

```
(1 <= 100)
```

which, of course, is true; however, the corresponding mathematical inequality

$$0 \leq 105 \leq 100$$

is false.

To avoid this difficulty, we must rewrite the original inequality $0 \le temperature \le 100$ as

$$(0 \le temperature) \text{ and } (temperature \le 100)$$

which can be coded correctly in C++ using the **logical operators** given in the following table that combine boolean expressions to form **compound boolean expressions**.

Logical Operator	Logical Expression	Name of Operation	Description
!	!p	*Not (Negation)*	!p is false if p is true; !p is true if p is false.
&&	p && q	*And (Conjunction)*	p && q is true if both p and q are true; it is false otherwise.
||	$p \|\| q$	*Or (Disjunction)*	$p \|\| q$ is true if either p or q or both are true; it is false otherwise

The following **truth tables** summarize these definitions by displaying the values of each logical expression for all possible values of p and q:[6]

p	!p
true	false
false	true

p	q	p && q	$p \|\| q$
true	true	true	true
true	false	false	true
false	true	false	true
false	false	false	false

We can thus use the && operator to represent the mathematical expression

$$0 \le temperature \le 100$$

by the compound boolean expression

```
(0 <= temperature) && (temperature <= 100)
```

This expression will correctly test the condition that `temperature` be in the range `0` through `100` for all possible values of `temperature`.

5.2.3 Operator Precedence

Boolean expressions that contain an assortment of arithmetic operators, boolean operators, and relational operators are evaluated using the following priority (or precedence) and associativity rules:

---

[6] *Note: Be sure to use the && and || operators for logical operations, and not & and | (bitwise operators) because an error will almost surely result otherwise.*

Operator	Priority	Associativity
!, ~	highest	Right
/, *, %		Left
+, −		Left
<, >, <=, >=		Left
==, !=		Left
&		Left
^		Left
|		Left
&&		Left
||		Left
=, +=, *=, ...	lowest	Right

Operators with higher priority are applied before those with lower priorities. To illustrate, consider the boolean expression for a condition that $ax^2 + bx + c = 0$ be a quadratic equation with a real solution:

```
a != 0 && b*b >= 4*a*c
```

The multiplication operator * has highest priority so it is applied first, producing some real values r1 for b*b and r2 for 4*a*c and the expression becomes

```
a != 0 && r1 >= r2
```

In this expression, >= has the highest priority so it is applied next, producing some intermediate boolean value b1 and the expression

```
a != 0 && b1
```

Now, != has higher priority than && so it is applied next, producing an intermediate boolean value b2 and the expression

```
b2 && b1
```

Finally, the && operator is applied to the two boolean values b1 and b2 to produce the expression's value.

Because of the difficulty in remembering so many priority levels—and there are quite a lot more (see Appendix C)—it is helpful to remember the following:

- Unary operators (−, +, !) have the highest priority.

- Arithmetic operators are next with *, /, % higher than + and −.

- Relational operators have lower priority than the preceding operators.

- Logical operators are next with && higher than ||.

- Assignments have lowest priority.

- *When in doubt, use parentheses to clarify the order in which the operators are to be applied.*

5.2.4 Short-Circuit Evaluation

Sometimes it is possible to determine the truth or falsity of a logical expression without evaluating it in its entirety. For example, if p is false, then regardless of the value of q, p && q will be false so it is not necessary to evaluate q. Similarly, if p is true, then regardless of the value of q, the condition p || q will be true so it need not be evaluated. This is referred to as **short-circuit evaluation** and has two important consequences:

1. One boolean expression can *guard* another that is potentially unsafe.

2. Time can be saved in evaluating complex expressions.

To illustrate the first benefit, consider the following compound boolean expression:

```
(a != 0) && (a + 1/a < 100)
```

No division-by-zero error can occur in the second part of this expression because if a is 0, then the first expression (a != 0) is false and the second expression (a + 1/a < 100) is not evaluated. Similarly, no division-by-zero error will occur in evaluating the condition

```
(a == 0) || (a + 1/a < 100)
```

because if a is 0, the first expression (a == 0) is true and the second expression is not evaluated.

5.2.5 Preconditions and the assert() Mechanism

In many problems, it may be necessary to impose *preconditions* on values to guard against incorrect or irrelevant results. For example, in the air quality index problem we considered in Section 5.1, we might impose the following precondition:

level1, *level2*, and *level3* must all be nonnegative.

This precondition should be true before program execution is allowed to continue.

We can construct boolean expressions that represent such preconditions. For example, the preceding precondition can be represented by

```
level1 >= 0 && level2 >= 0 && level3 >= 0
```

and C++ provides the assert() **mechanism** (defined in the C library cassert) as a convenient way to check a precondition. For example, Example 5.2 is a modification of the program in Example 5.1 that checks for nonnegative inputs by inserting an assertion immediately after the input statement.

```
assert(level1 >= 0 && level2 >= 0 && level3 >= 0);
```

When this statement is executed, if the precondition is true, execution proceeds normally; but if it is false, execution of the program will be terminated and a diagnostic message such as the following will be displayed:

```
failed assertion: level1 >= 0 && level2 >= 0 && level3 >= 0
```

Example 5.2 Air Quality Indexes—Version 1

```
/* This program reads three pollution levels and checks that they are
   all nonnegative. If not, it terminates execution and displays an
   error message. Otherwise, it calculates an air quality index (AQI)
   as the integer average of the levels and then displays an
   appropriate air-quality message.


   Input:    the three pollution levels
   Constant: the cutoff value (parts per million)
   Output:   an error message if any pollution level is negative; for
             nonnegative levels: the air quality index and a "safe
             condition" message if this index is less than the cutoff
             value, otherwise a "hazardous condition" message
-------------------------------------------------------------------*/

#include <iostream>              // cin, cout, <<, >>
#include <cassert>               // assert()
using namespace std;

int main()
{
  const int CUTOFF = 50;                // safe pollution level cutoff

  double level1, level2, level3;
  cout << "Enter 3 pollution readings (parts per million): ";
  cin >> level1 >> level2 >> level3;    // get pollution levels
  assert(level1 >= 0 && level2 >= 0 && level3 >= 0);
                                        // compute AQI
  int index = (level1 + level2 + level3) / 3 + 0.5;
                                        // display AQI
  cout << "AQI: " << index << " -- ";
                                        // display condition
  if (index < CUTOFF)
    cout << "Safe condition\n";
  else
    cout << "Hazardous condition\n";
}
```

In general, the `assert()` mechanism can be described as follows:

THE `assert( )` MECHANISM

FORM:

```
assert(boolean_expression);
```

where:
   boolean_expression is any valid expression evaluating to true or false.

PURPOSE:

Checks whether *boolean_expression* is true. If it is, execution proceeds normally. If it is false, execution is terminated and a diagnostic message is displayed.

5.2.6 Boolean Character Functions

The C library `cctype` provides a number of boolean-valued functions for performing useful checks on character values. Some of the more useful of these are shown in Table 5.1. Also included at the end of the table are two functions for changing the case of a character.

TABLE 5.1   Character Operations in `cctype`

Operation	Description
`isalnum(`*ch*`)`	true if *ch* is a letter or a digit, false otherwise
`isalpha(`*ch*`)`	true if *ch* is a letter, false otherwise
`iscntrl(`*ch*`)`	true if *ch* is a control character, false otherwise
`isdigit(`*ch*`)`	true if *ch* is a decimal digit, false otherwise
`isgraph(`*ch*`)`	true if *ch* is a printing character except space, false otherwise
`islower(`*ch*`)`	true if *ch* is lower case, false otherwise
`isprint(`*ch*`)`	true if *ch* is a printing character, including space, false otherwise
`ispunct(`*ch*`)`	true if *ch* is a punctuation character (not a space, an alphabetic character, or a digit), false otherwise
`isspace(`*ch*`)`	true if *ch* is a white space character (space, '\f', '\n', '\r', '\t', or '\v'), false otherwise
`isupper(`*ch*`)`	true if *ch* is upper case, false otherwise
`isxdigit(`*ch*`)`	true if *ch* is a hexadecimal digit, false otherwise
`toupper(`*ch*`)`	returns the uppercase equivalent of *ch* (if *ch* is lower case)
`tolower(`*ch*`)`	returns the lowercase equivalent of *ch* (if *ch* is upper case)

5.3 EXAMPLE: DIGITAL CIRCUITS—A BINARY HALF-ADDER

As noted earlier, George Boole laid the foundations of circuit design in the mid-1800s. Boole formalized several axioms of logic and developed an algebra for writing logical expressions.[7] In the late 1930s as mathematicians, engineers, and physicists worked at

---

[7] More information about boolean algebras can be found on the text's website (see the Preface).

building the arithmetic and logic circuitry of the early computers, the axioms and theorems of Boole's algebra became extremely important.

These circuits utilize three basic electronic components: the *AND gate*, the *OR gate*, and the *NOT gate* or *inverter*, whose symbols are as follows:

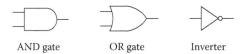

| AND gate | OR gate | Inverter |

Inputs to these gates (pictured as short line segments on their left) are electrical voltages and are interpreted as 1 (true) when they exceed a certain threshold value or 0 (false) if they drop below that threshold. The outputs from these gates (pictured as short line segments on their right) are similarly interpreted in binary (0 or 1). For an AND gate, 1 is produced only when both inputs are 1. An OR gate produces a 1 when at least one of the inputs is a 1. The output of a NOT gate is the opposite of its input. These three basic components thus behave in the same manner as the AND, OR, and NOT operators from logic, which means that we can use boolean expressions to design and analyze circuits.

To illustrate, consider how we would design a circuit that carries out binary addition:

+	0	1
0	0	1
1	1	10

We see that the inputs for our circuit will be the two binary digits being added and there will be two binary outputs: a sum bit, which is the rightmost bit of the entries in the table, and a carry bit that will be 1 for the sum 1 + 1 and 0 otherwise. If we denote the inputs by digit1 and digit2 and the outputs by sum and carry, we obtain the following version of the preceding table known as a *truth table*:

digit1	digit2	carry	sum
0	0	0	0
0	1	0	1
1	0	0	1
1	1	1	0

Note the patterns in the last two columns. The carry output is 1 (true) only when digit1 and digit2 are both 1 (true), so we can use an AND gate for it. The sum output is 1 (true) only when exactly one of digit1 and digit2 is 1 (true); that is, digit1 or digit2 is 1 (true) but not both. We can express these properties using the following boolean expressions,

```
bool carry = digit1 && digit2,
     sum = (digit1 || digit2) && !(digit1 && digit2);
```

and design the following circuit, called a **binary half-adder**, that adds two binary digits:

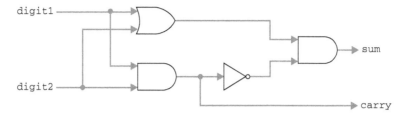

Simulating the operation of such a circuit with a program is straightforward because we need only code the boolean expressions that model the circuit and then execute the program with the various combinations of inputs for the circuit. For example, the program in Example 5.3 simulates the action of a binary half-adder. The sample runs that are shown read a binary digit (0 or 1) for each of digit1 and digit2 and output the resulting value for sum and carry for each of the four different combinations of inputs for digit1 and digit2. Once we have studied the repetition structures later in this chapter, we will be able to do this without having to repeatedly re-execute the program.

Example 5.3 A Binary Half-Adder

```
/* This program calculates the outputs from boolean expressions that
   represent the logical circuit for a binary half-adder.

   Input:  two binary digits
   Output: two binary values representing the sum and carry that
           result when the input values are added
-----------------------------------------------------------------*/

#include <iostream>     // cout, cin, <<, >>
using namespace std;

int main()
{
  cout << "Enter two binary inputs: ";
  short digit1, digit2; // the two binary inputs
  cin >> digit1 >> digit2;

  // the two circuit outputs
  bool sum = (digit1 || digit2) && !(digit1 && digit2),
       carry = (digit1 && digit2);
  cout << "Carry = " << carry << " Sum = " << sum << "\n\n";
}
```

SAMPLE RUNS:

```
Enter two binary inputs: 0 0
Carry = 0 Sum = 0
```

```
Enter two binary inputs: 0 1
Carry = 0 Sum = 1

Enter two binary inputs: 1 0
Carry = 0 Sum = 1

Enter two binary inputs: 1 1
Carry = 1 Sum = 0
```

5.4 SELECTION: THE IF STATEMENT

As noted at the beginning of this chapter, some problems require a structure that selects one of several alternative actions. In the simplest case of two alternatives, the selection is based on a boolean expression with one alternative selected when its value is true and the other when it is false. In a flow diagram this is commonly pictured as a diamond-shaped box that contains the boolean expression with one corner labeled true and one labeled false:

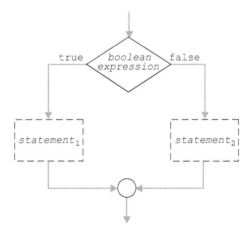

As this diagram clearly indicates, execution will pass through $statement_1$ or through $statement_2$, *but not both*, before proceeding to the next statement. Which route it takes will be determined by the truth or falsity of the boolean expression.

This selection structure can be implemented with a C++ if **statement** of the form

```
if (boolean_expression)
    statement₁
else
    statement₂
```

If the boolean expression is true, then the first statement is executed and the trailing else and its statement are skipped, but if the boolean expression is false, the first statement is skipped and the statement in the else part is executed. This statement does, therefore, implement the two-alternative selection structure pictured earlier.

The `if` statement in our opening example of this chapter (Example 5.1) has this form:

```cpp
if (index < CUTOFF)
   cout << "Safe condition\n";
else
   cout << "Hazardous condition\n";
```

Indenting the two alternatives as shown here to set them off is a good programming practice and one that we will follow in this text.

As another example, consider the problem of finding the larger of two real values. This can be easily done using an `if` statement as the following code segment demonstrates.

```cpp
double maximum;
if (value1 > value2)
    maximum = value1;
else
    maximum = value2;
```

Sometimes the second alternative in a selection structure is empty as in the following diagram where a specified *statement* is either executed or bypassed:

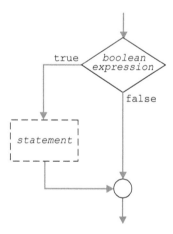

In this case, the `else` clause in the `if` statement may be omitted:

```cpp
if (boolean_expression)
   statement
```

or it may be an empty statement consisting only of a semicolon,

```cpp
if (boolean_expression)
   statement
else;
```

For example, the following is an alternative way of finding the maximum of two values:

```
double maximum = value1;
if (value1 < value2)
   maximum = value2;
```

In some problems, one or both of the statements in an if statement may themselves be other if statements. To illustrate, suppose that in the problem in Section 5.1, there are three different pollution conditions—good, fair, poor—as determined by two different cutoff values; for example,

```
const int CUTOFF1 = 40,  // cutoff for good condition
          CUTOFF2 = 70;  // cutoff for fair condition
```

We could use the following "nested if" statement to determine the condition that corresponds to a pollution level:

```
if (index < CUTOFF1)
  cout << "Good condition\n";
else
  if (index < CUTOFF2)
    cout << "Fair condition\n";
  else
    cout << "Bad condition\n";
```

It is more common, however, to reformat this as follows to more clearly display the three different alternatives:

```
if (index < CUTOFF1)
  cout << "Good condition\n";
else if (index < CUTOFF2)
  cout << "Fair condition\n";
else
  cout << "Bad condition\n";
```

In general, the form of an if statement can be summarized as follows:

if STATEMENT

if FORM (NO else):

```
  if (boolean_expression)
    statement
```

`if-else` **Form:**

```
if (boolean_expression)
   statement₁
else
   statement₂
```

`if-else if` **Form:**

```
if (boolean_expression₁)
   statement₁
else if (boolean_expression₂)
   statement₂

       .
       .
       .

else if (boolean_expressionₙ)
   statementₙ
else
   statementₙ₊₁
```

where

`if` and `else` are keywords;

$statement_1$, $statement_2$, ... are C++ statements (and may be compound, as described next).

Purpose:

In the first form, if the `boolean_expression` is true, then $statement$ will be executed; otherwise it will be bypassed.

In the second form, if the `boolean_expression` is true, $statement_1$ will be executed and $statement_2$ bypassed; otherwise, $statement_1$ will be bypassed and $statement_2$ executed.

In the last form, if $boolean\_expression_i$ is the first of the boolean expressions that is true, $statement_i$ will be executed and all of the others bypassed. If none of the $boolean\_expression_i$ is true, $statement_{n+1}$ (if present) will be executed.

5.4.1 Blocks

When a group of statements must be selected for execution, we enclose them between curly braces { and } to form a single statement. This is called a **block** or a **compound statement**:

BLOCK (COMPOUND STATEMENT)

Forms:

```
{
   statement₁
```

```
    statement₂
        ·
        ·
        ·
    statementₙ
}
```

where
>   each `statement` is a C++ statement (and may itself be compound).

PURPOSE:

The sequence of statements is treated as a single statement in which `statement₁`, `statement₂`, ..., `statementₙ` are executed in order.

Note that the block does not require a semicolon after the final curly brace; it is a complete statement by itself.

When one of the alternatives in an `if` statement consists of more than one statement, a block must be used. The quadratic equation solver in the following example illustrates this.

5.4.2 Example: Quadratic Equation Solver

A quadratic equation has the form $ax^2 + bx + c = 0$ where the coefficients a, b, and c are real numbers with $a \neq 0$ and can be solved using the *quadratic formula*, which states that the roots are given by

$$\frac{-b \pm \sqrt{b^2 - 4ac}}{2a}$$

provided that the discriminant $b^2 - 4ac$ is nonnegative. The program in Example 5.4 solves a quadratic equation using this method.

Example 5.4 Quadratic Equation Solver

```
/* This program solves quadratic equations using the quadratic
   formula.

   Input:  the three coefficients of a quadratic equation
   Output: the roots of the equation or a message that there are
           no real roots
 -----------------------------------------------------------------*/

#include <iostream> // cout, cin, <<, >>
#includ <cassert>   // assert()
#include <cmath>     // sqrt()
using namespace std;
```

```
int main()
{
  double a, b, c;
  cout << "Enter the coefficients of a quadratic equation: ";
  cin >> a >> b >> c;
  assert (a != 0);

  double discriminant = b*b - 4*a*c,
         root1, root2;

  if (discriminant >= 0)
  {
    root1 = (-b + sqrt(discriminant)) / (2*a);
    root2 = (-b - sqrt(discriminant)) / (2*a);
    cout << "Roots are " << root1 << " and " << root2 << endl;
  }
  else
    cout << "There are no real roots, only complex ones" << endl;
}
```

SAMPLE RUNS:

```
Enter the coefficients of a quadratic equation: 1 4 3
Roots are -1 and -3

Enter the coefficients of a quadratic equation: 2 -8 8
Roots are 2 and 2

Enter the coefficients of a quadratic equation: 1 2 3
There are no real roots, only complex ones
```

5.4.3 Style

There is no one universal style used by programmers in writing if statements, but that used in this text is a common one and is intended to promote readability:

1. Align the if and the else.

2. Use white space and indentation to clearly identify the alternatives in the if statement.

When an alternative is a single statement, we will usually indent it on the line below the if or the else; for example,

```
if (index < CUTOFF)
  cout << "Safe condition\n";
else
  cout << "Hazardous condition\n";
```

For an alternative that consists of a block of statements, the curly braces will be placed on separate lines, aligned with the `if` and `else`, and the statements they enclose will be indented; for example,

```
if (discriminant >= 0)
{
   root1 = (-b + sqrt(discriminant)) / (2*a);
   root2 = (-b - sqrt(discriminant)) / (2*a);
   cout << "Roots are " << root1 << " and " << root2 << endl;
}
else
   cout << "There are no real roots, only complex ones" << endl;
```

In another frequently used style, opening curly braces are placed on the same line as the `if` (or `else`) keyword and the closing curly brace is aligned with that keyword instead of with the opening curly brace:[8]

```
if (discriminant >= 0) {
   root1 = (-b + sqrt(discriminant)) / (2*a);
   root2 = (-b - sqrt(discriminant)) / (2*a);
   cout << "Roots are " << root1 << " and " << root2 << endl;
}
else
   cout << "There are no real roots, only complex ones" << endl;
```

For `if-else-if` statements (the third form), each `else` and the following `if` will be placed on the same line, thus forming `else if`, and the statements in each part indented; for example,

```
if (index < CUTOFF1)
   cout << "Good condition\n";
else if (index < CUTOFF2)
   cout << "Fair condition\n";
else
   cout << "Bad condition\n";
```

5.5 REPETITION: THE FOR AND WHILE STATEMENTS

In earlier preceding chapters and at the beginning of this chapter, our programs required only *sequential* processing of instructions, but in the preceding section, we saw examples of

---

[8] This is known as the *One True Brace (OTB)* style. A variation of it puts the keyword `else` on the same line as the closing curly brace:

```
if (condition) {
  statements
} else {
  statements
}
```

problems whose solutions require *selection*. In this section we introduce the third control structure, *repetition*.

The **factorial** of a nonnegative integer n, denoted by $n!$, is defined by

$$n! = \begin{cases} 1 & \text{if } n = 0 \\ 1 \times 2 \times \cdots \times n & \text{if } n > 0 \end{cases}$$

To calculate $n!$ by hand, we would probably begin with 1, multiply it by 2, multiply that product by 3, and so on, until we multiply by n. For example, to find 5!, we might do the following computation in which we keep a running product and increment a count from one step to the next:

$$
\begin{array}{rl}
1 & \longleftarrow \text{initial running product} \\
\times\, 2 & \longleftarrow \text{initial count} \\
\hline
2 & \longleftarrow \text{new running product} \\
\times\, 3 & \longleftarrow \text{new count} \\
\hline
6 & \longleftarrow \text{new running product} \\
\times\, 4 & \longleftarrow \text{new count} \\
\hline
24 & \longleftarrow \text{new running product} \\
\times\, 5 & \longleftarrow \text{new count} \\
\hline
120 & \longleftarrow \text{new running product}
\end{array}
$$

If we analyze this computation of 5!, we see three parts in this repetition mechanism (also called a **loop**):

1. *Initialization:* Assign starting values to the running product and count.

2. *Repeated execution:* Repeatedly multiply the product by the count and increment the count.

3. *Termination:* Stop when the count reaches our final value 5.

We can describe the computation in general as follows:

Algorithm for Factorial Computation

1. Initialize *product* to 1.

2. Repeat the following for each value of *count* in the range 2 to n

    Multiply *product* by *count*.

One of the statements provided in C++ for executing a statement more than once is the for **statement**.[9] Here is an example of how it can be used to compute $n!$:

```
int product = 1;
for (int count = 2; count <= n; count++)
   product *= count;
```

It executes a statement *repeatedly*, once for each number in the range 2 through n. The following flow diagram shows its behavior more precisely:

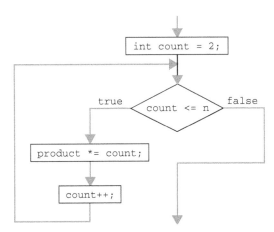

When execution reaches the loop, the variable count is created and initialized to 2. The value of count is then tested against n. If count exceeds n, repetition ceases and execution continues with the statement that follows the for statement. However, if count is less than or equal to n, then the *= statement controlled by the loop is executed, so product is multiplied by count. The value of count is then incremented by the expression count++, after which execution returns to the top of the loop where the boolean expression count <= n is re-evaluated and the cycle starts again. This cyclic behavior continues so long as the boolean expression evaluates to true.

Another way to understand the execution of such a loop is with a **trace table**, which traces the execution of the loop's statements, one at a time. The following is a trace table for n = 5:

Time	Statement Executed	Product	Count	Comment
0	int count = 2;	1	2	loop initialization
1	count <= n	1	2	true, loop executes
2	product *= count;	2	2	product updated
3	count++;	2	3	count incremented
4	count <= n	2	3	true, loop executes
5	product *= count;	6	3	product updated

---

[9] Other forms of the for statement as well as other repetition statements will be studied in Chapter 9.

6	count++;	6	4	count incremented
7	count <= n	6	4	true, loop executes
8	product *= count;	24	4	product updated
9	count++;	24	5	count incremented
10	count <= n	24	5	true, loop executes
11	product *= count;	120	5	product updated
12	count++;	120	6	count incremented
13	count <= n	120	6	false, repetition ceases

The usual form of a for statement is the following:

for STATEMENT

FORM:

```
for (init_declaration; loop_condition; step_expression)
    statement
```

where
for is a C++ keyword;
`init_declaration` declares and initializes a variable that controls the repetitions;
`loop_condition` is a boolean expression that determines whether repetition is to continue;
`step_expression` specifies how the value of the control variable is to change; and
`statement`, called the *loop body*, is a C++ statement (simple or compound).

BEHAVIOR:

1. `init_declaration` initializes the value of the control variable.
2. If `loop_condition` is true, the following occur:
   a. `statement` is executed.
   b. `step_expression` is evaluated.
   c. Control returns to step 2.

   Otherwise execution continues with statements that follow the for statement.

The small number of restrictions on the parts of a for statement gives it considerable flexibility. For example, we can count down from larger numbers to smaller as in the following factorial calculation:

```
int product = 1;
for (int count = n; count > 1; count--)
   product *= count;
```

We can also use real numbers for the initial value, final value, and step size as in

```
for (double x = 0.0; x <= 2.0; x += 0.5)
   cout << x*sqrt(x) - 1 << " " ;
```

which will produce output like the following:

```
-1 -0.646447 0 0.837117 1.82843
```

5.5.2 Processing Several Input Values

The program in Example 5.4 for solving quadratic equations suffers from one major drawback: It solves only one equation. To solve other equations, the program must be re-executed, which may involve retyping its name, re-clicking on its icon, or whatever is necessary for a particular system.

A more user-friendly program would permit the user to process any number of equations before it terminated. One way to do this is to have the user specify in advance how many values are to be processed and then wrap the body of the program in a `for` loop that counts from 1 to that number:

```cpp
// ...
int main()
{
  cout << "How many quadratic equations do you wish to solve? ";
  int numEquations;
  cin >> numEquations;

  for (int i = 1; i <= numEquations; i++)
  {
    cout << "\nEnter the coefficients of a quadratic equation: ";
    double a, b, c;
    cin >> a >> b >> c;
    assert (a != 0);

    double discriminant = b*b - 4*a*c,
           root1, root2;

    if (discriminant >= 0)
    {
      root1 = (-b + sqrt(discriminant)) / (2*a);
      root2 = (-b - sqrt(discriminant)) / (2*a);
      cout << "Roots are " << root1 << " and " << root2 << endl;
    }
    else
      cout << "There are no real roots, only complex ones" << endl;
  }
}
```

A sample run might appear as follows:

```
How many quadratic equations do you wish to solve? 3

Enter the coefficients of a quadratic equation: 1 4 3
Roots are -1 and -3
```

```
Enter the coefficients of a quadratic equation: 2 -8 8
Roots are 2 and 2

Enter the coefficients of a quadratic equation: 1 2 3
There are no real roots, only complex ones
```

By enclosing the critical portion of a driver program within a loop, we can process several input data values without having to re-execute the program.

One problem with this approach is that it requires knowing how many input values there will be. When the set of input values is very large, counting the values may be inconvenient or even impractical. A common alternative that does not require this is to use a **while loop**, which we only introduce briefly here. This repetition structure, along with other kinds of loops, is studied in more detail in Chapter 9.

5.5.3 Repeated Execution: The while Statement

While loops are implemented in C++ using a **while statement** of the following form:

while STATEMENT
FORM:

```
while (loop_condition)
   statement
```

where
 while is a C++ keyword;
 loop_condition is a boolean expression;
 statement, called the *loop body*, is usually a compound statement

BEHAVIOR:

When execution reaches a while statement:

1. *loop_condition* is evaluated.
2. If *loop_condition* is true:
    a. The loop's body *statement* is executed.
    b. Control returns to step 1.

 Otherwise execution continues with statements that follow the while statement.

Example 5.5 illustrates how a while loop can be used to repeatedly solve quadratic equations, stopping when the user enters the "stop indicator" 0 for the coefficient of x^2.

Example 5.5 Quadratic Equation Solver—Revised

```
/* This program solves quadratic equations using the quadratic formula.

   Input:   the three coefficients of a quadratic equation
   Output:  the roots of the equation or a message that there are no
            real roots
-------------------------------------------------------------------*/

#include <iostream>        // cout, cin, <<, >>
#include <cmath>           // sqrt()
using namespace std;

int main()
{
  double a, b, c;
  cout << "Enter coefficients of quadratic equation (0's to stop): ";
  cin >> a >> b >> c;

  while (a != 0)
  {
    double discriminant = b*b - 4*a*c,
           root1, root2;

    if (discriminant >= 0)
    {
      root1 = (-b + sqrt(discriminant)) / (2*a);
      root2 = (-b - sqrt(discriminant)) / (2*a);
      cout << "Roots are " << root1 << " and " << root2 << endl;
    }
    else
      cout << "There are no real roots, only complex ones" << endl;

    cout << "Enter coefficients of quadratic equation (0's to stop): ";
    cin >> a >> b >> c;
  }
}
```

SAMPLE RUNS:

```
Enter coefficients of quadratic equation (0's to stop): 1 0 -2
Roots are 1.41421 and -1.41421

Enter coefficients of quadratic equation (0's to stop): 2 10 8
Roots are -1 and -4

Enter coefficients of quadratic equation (0's to stop): 1 1 1
There are no real roots, only complex ones

Enter coefficients of quadratic equation (0's to stop): 0 0 0
```

CHAPTER SUMMARY

Key Terms

assert()	loop body
binary half-adder	nested if statements
block	operator precedence
boolean expression	precondition
cctype library	relational operator
compound boolean expression	repetition structure
compound statement	selection structure
condition	separate compilation
control structure	sequential execution
flow diagram	short-circuit evaluation
for statement	termination (or exit) condition
if statement	trace table
logical operator	while loop
loop	while statement

NOTES

- Be sure to use the == operator for equality comparisons, not = (assignment).

- Be sure to use the && and || operators for logical operations, not & and | (bitwise operators).

- Some problems require control mechanisms that are more powerful than sequential execution, namely, selection and repetition. These three control structures are sufficient for any program.

- *Sequential execution* refers to execution of a sequence of statements in the order in which they appear, so that each statement is executed exactly once.

- A *block* (or *compound statement*) groups a sequence of statements into a single statement by enclosing them in curly braces ({ and }).

- *Selective execution* refers to selecting and executing exactly one of a collection of alternative actions.

- The assert() mechanism evaluates a boolean expression and terminates the program if it is false. It is useful for checking a condition that must be true at a given point in a program.

- The `if` statement is the most common selection structure for selecting between two or more alternatives.

- When one of the alternatives in an `if` statement contains another `if` statement, the second `if` statement is said to be *nested* in the first. In this case, an `else` clause is matched with the nearest preceding unmatched `if`.

- There are three parts to a repetition mechanism (also called a loop): *initialization*, *repeated execution*, and *termination*.

- A for loop is one of the structures provided in C++ for loops in which the number of repetitions can be determined in advance.

- A while loop is commonly used to implement input loops.

Style and Design Tips

- *Indent and align the statements within a block (compound statement).*

- *Identify any preconditions a program has, and check them using the* `assert()` *mechanism (or an* `if` *statement).* Preconditions are assumptions made in a program, often one or more restrictions on what comprises a valid input value.

- *If a problem requires the selection of one or more operations, use a selection statement like the* `if` *statement.*

- *Align the* `if`s *and* `else`s *in an* `if` *statement. Also, use white space and indentation to set off the various alternatives.*

- *If a problem requires the repetition of one or more operations, and the number of repetitions can be computed in advance, a* `for` *statement is appropriate.*

- *If a problem involves a set of data values whose size is not known in advance, a* `while` *statement can be used to read and process the values.*

Warnings

1. *When real quantities that are algebraically equal are compared with* `==`, *the result may be false because most real numbers are not stored exactly.* For example, x * (1/x) and 1.0 are algebraically equal, but the boolean expression x * (1/x) == 1.0 may be false for some real numbers x.

2. *It is easy to forget that in C++,* `=` *is the assignment operator, and to incorrectly encode an instruction of the form*

    If variable is equal to value, then
       statement

as

```
if (variable = value)
    statement
```

Instead of testing whether *variable* is equal to *value*, the condition in this if statement assigns *value* to *variable*. If *value* happens to be zero, which C++ interprets as false, the *statement* will not be executed, regardless of the value of *variable*. If *value* is nonzero, which C++ interprets as true, *statement* will be executed, regardless of the value of *variable*.

3. *Be careful in writing compound boolean expressions to use the logical operators && and == and not the bitwise operators & and |.*

4. *Each { must have a matching }.* To make it easier to find matching braces, we align each { with its corresponding }.

5. *In a nested* if *statement, each* else *clause is matched with the nearest preceding unmatched* if*. Use indentation and alignment to show such associations.*

6. *When using repetition, care must be taken to avoid infinite looping; be sure that the boolean expression controlling repetition eventually becomes false for* for *and* while *statements.*

7. *In a* for *loop, neither the control variable nor any variable involved in the loop condition should be modified within the body of the* for *loop, since it is intended to run through a specified range of consecutive values. Strange or undesirable results may be produced otherwise.*

TEST YOURSELF

Section 5.2

1. What are the three basic control structures?

2. The two bool literals are _____ and _____.

3. List the six relational operators.

4. List the three logical operators.

For Questions 5–9 assume that p, q, and r are boolean expressions with the values true, true, and false, respectively. Find the value of each boolean expression.

5. p && !q   7. p && !(q || r)   9. p || q && r

6. p && q || !r   8. !p && q

For Questions 10–14, assume that number, count, and sum are integer variables with values 3, 4, and 5, respectively. Find the value of each boolean expression, or indicate why it is not valid.

10. sum - number <= 4

11. number*number + count*count == sum*sum

12. `number < count || count < sum`

13. `0 <= count <= 2`

14. `(number + 1 < sum) && !(count + 1 < sum)`

15. Write a boolean expression to express that x is nonzero.

16. Write a boolean expression to express that x is strictly between –10 and 10.

17. Write a boolean expression to express that both x and y are positive or both x and y are negative.

Section 5.4

Questions 1–3 refer to the following `if` statement. For each, describe the output produced for the given values of x and y.

```
if (x >= y)
    cout << x;
else
    cout << y;
```

1. x is 6 and y is 5.

2. x is 5 and y is 5.

3. x is 5 and y is 6.

Questions 4–6 refer to the following `if` statement. For each, describe the output produced for the given values of x and y.

```
if (x >= 0)
    if (y >= 0)
        cout << x + y;
    else
        cout << x - y;
else
    cout << y - x;
```

4. x is 5 and y is 5.

5. x is 5 and y is –5.

6. x is –5 and y is 5.

Questions 7–11 refer to the following `if` statement. For each, describe the output produced for the given values of n.

```
if (n >= 90)
   cout << "excellent\n";
else if (n >= 80)
   cout << "good\n";
else if (n >= 70)
   cout << "fair\n";
else
   cout << "bad\n";
```

7. n is 100.

8. n is 90.

9. n is 89.

10. n is 70.

11. n is 0.

12. Write a statement that displays "Out of range" if number is negative or is greater than 100.

13. Write an efficient if statement to assign n the value 1 if $x \leq 1.5$, 2 if $1.5 < x < 2.5$, and 3 otherwise.

Section 5.5

For Questions 1–10, describe the output produced.

1.
```
for (int i = 0; i <= 5; i++)
   cout << "Hello\n";
```

2.
```
for (int i = 1; i < 4; i++)
   cout << "Hello";
```

3.
```
for (int i = 1; i <= 5; i += 2)
   cout << "Hello\n";
```

4.
```
for (int i = 6; i > 0; i--)
   cout << i*i << endl;
```

5.
```
for (int i = 6; i <= 6; i++)
   cout << "Hello\n";
```

6.
```
for (int i = 6; i <= 5; i++)
   cout << "Hello\n";
```

7.
```
for (int i = 1; i <= 10; i++)
{
   cout << i << endl;
   i++;
}
```

8.
```cpp
int i = 4;
while (i >= 0)
{
  i--;
  cout << i << endl;
}
cout << "\n*****\n";
```

9.
```cpp
int i = 1;
while (i < 100)
{
  cout << i << endl;
  i = 2*i;
}
```

10.
```cpp
int i = 0;
while (i < 10)
{
  cout << i;
  if (i % 2 == 0)
    cout << " E";
  else
    cout << " O";
  ++i;
}
cout << endl;
```

11. How many lines of output are produced by the following?

```cpp
for (int i = 1; i <= 50; i += 2)
{
  cout << i << " ";
  if (i % 5 == 0)
    cout << endl;
}
```

EXERCISES

Section 5.2

For Exercises 1–10, assume that m and n are integer variables with the values –5 and 8, respectively, and that x, y, and z are real variables with the values –3.56, 0.0, and 44.7, respectively. Find the value of the boolean expression.

1. `m <= n`

2. `2 * abs(m) <= 8`

3. `x * x < sqrt(z)`

4. `int(z) == (6 * n - 4)`

5. `(x <= y) && (y <= z)`

6. `!(x < y)`

7. `!((m <= n) && (x + z > y))`

8. `!(m <= n) || !(x + z > y)`

9. `!((m <= n) || (x + z > y))`

10. `!((m > n) && !(x < z))`

For Exercises 11–16, use truth tables to display the values of the boolean expression for all possible (boolean) values of a, b, and c:

11. `a || !b`

12. `!(a && b)`

13. `!a || !b`

14. `(a && b) || c`

15. `a && (b || c)`

16. `(a && b) || (a && c)`

For Exercises 17–25, write C++ boolean expressions to express the condition:

17. x is greater than 3.

18. y is strictly between 2 and 5.

19. r is negative and z is positive.

20. Both `alpha` and `beta` are positive.

21. `alpha` and `beta` have the same sign (both are negative or both are positive).

22. $-5 < x < 5$.

23. a is less than 6 or is greater than 10.

24. p is equal to q, which is equal to r.

25. x is less than 3, or y is less than 3, but not both.

Exercises 26–28 assume that a, b, and c are boolean values.

26. Write a C++ boolean expression that is true if and only if a and b are true and c is false.

27. Write a C++ boolean expression that is true if and only if a is true and at least one of b or c is true.

28. Write a C++ boolean expression that is true if and only if exactly one of a and b is true.

Exercises 1–4 refer to the following `if` statement:

```
if (x * y >= 0)
   cout << "yes\n";
else
   cout << "no\n";
```

1. Describe the output produced if x is 5 and y is 6.

2. Describe the output produced if x is 5 and y is –6.

3. Describe the output produced if x is –5 and y is 6.

4. Describe the output produced if x is –5 and y is –6.

Exercises 5–7 refer to the following `if` statement. Describe the output produced for the given value of n.

```
if (abs(n) <= 4)
   if (n > 0)
      cout << 2*n + 1;
   else
      cout << 2*n;
else
   cout << n << " out of range";
```

5. n is 2.

6. n is –7.

7. n is 0.

For Exercises 8–12, write `if` statements that will do what is required.

8. If `taxCode` is 'T', increase `price` by adding `taxRate` percentage of `price` to it.

9. If code is 1, input values for x and y and calculate and display the sum of x and y.

10. If a is strictly between 0 and 5, set b equal to $1/a^2$; otherwise set b equal to a^2.

11. Given a distance, compute a cost, according to the following table:

Distance	Cost
0 through 100	$5.00
More than 100 but not more than 500	$8.00
More than 500 but less than 1000	$10.00
1000 or more	$12.00

12. Given a wind chill, display its classification according to the following table:

Wind Chill	Classification
10°F or above	Not dangerous or unpleasant
–10°F or higher but less than 10°F	Unpleasant
–30°F or above but less than –10°F	Possible frostbite
–70°F or higher but below –30°F	Dangerous, frostbite likely
Less than –70°F	Flesh may freeze in half a minute

Section 5.5

For Exercises 1–6, describe the output produced.

```cpp
1. for (int i = 0; i <= 3; i++)
     cout << i << " squared = " << i*i << endl;

2. for (int i = 5; i > 0; i--)
     cout << i << " squared = " << i*i << endl;

3. int k = 5;
   for (int i = k; i <= 5; i++)
   {
     cout << i + k << endl;
     k = 1;
   }

4. int s = 5;
   for (int i = s; i < 5; i++)
     cout << i + s << endl;
   cout << "***\n";

5. int s = 1;
   while (s <= 10)
   {
     cout << 2*s << endl;
     s *= 2;
   }

6. int s = 20;
   while (s/2 < 10)
   {
     cout << 2*s << endl;
     s /= 2;
   }
```

For Exercises 7–14, write C++ statements to do what is asked for.

7. For a positive integer n, find the sum $1 + 2 \ldots + n$ and output this sum.

8. For two integers m and n with $m \leq n$, find the sum $m + m + 1 + \ldots + n$ and output this sum.

9. Display the squares of the first 100 positive integers in increasing order.

10. Display the cubes of the first 50 positive integers in decreasing order.

11. Display the square roots of the first 25 odd positive integers.

12. Display a list of points (x, y) on the graph of $y = x^3 - 3x + 1$ for x ranging from -2 to 2 in steps of 0.1.

13. Display the value of x, starting at 10.0, and decrease x by 0.5 as long as x is positive.

14. Calculate and display the squares of consecutive positive integers until the difference between a square and the preceding one is greater than 50.

PROGRAMMING PROBLEMS

Sections 5.2–5.3

1. In a certain region, pesticides can be sprayed from an airplane only if the temperature is at least 70°, the relative humidity is between 15% and 35%, and the wind speed is at most 10 miles per hour. Write a program that accepts three numbers representing temperature, relative humidity, and wind speed; assigns the value true or false to the boolean variable okToSpray according to these criteria; and displays this value.

2. Write a program that reads three real numbers, assigns the appropriate boolean value to the following boolean variables, and displays these values.

   triangle: true if the real numbers can represent lengths of the sides of a triangle (the sum of any two of the numbers must be greater than the third); false otherwise.

   equilateral: true if triangle is true and the triangle is equilateral (the three sides are equal); false otherwise.

   isosceles: true if triangle is true and the triangle is isosceles (at least two sides are equal); false otherwise.

   scalene: true if triangle is true and the triangle is scalene (no two sides are equal); false otherwise.

3. A *binary full-adder* has three inputs: the two bits a and b being added, and a "carry-in" bit cIn (representing the carry bit that results from adding the bits to the right of a and b in two binary numbers). It can be constructed from two binary half-adders and an OR gate:

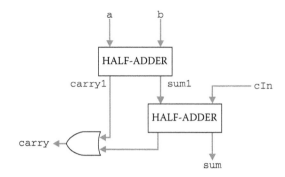

a) Write boolean expressions for

   i) sum1 and carry1 in terms of a and b

   ii) sum and carry in terms of cIn, sum1, and carry1

b) Write a program to implement this binary full-adder, and use it to verify the results shown in the following table:

a	b	cIn	sum	carry
0	0	0	0	0
0	0	1	1	0
0	1	0	1	0
0	1	1	0	1
1	0	0	1	0
1	0	1	0	1
1	1	0	0	1
1	1	1	1	1

4. An adder to calculate binary sums of two-bit numbers

```
      a2 a1
   +  b2 b1
   cOut s2 s1
```

where s1 and s2 are the sum bits and cOut is the carry-out bit, can be constructed from a binary half-adder and a binary full-adder:

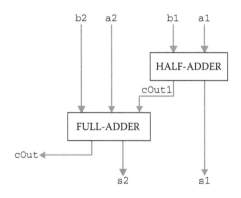

a) Write boolean expressions for

    i) `s1` and `cOut1` in terms of `a1` and `b1`

    ii) `s2` and `cOut` in terms of `a2`, `b2`, and `cOut1`

b) Write a program to implement this adder and use it to demonstrate that $00 + 00 =$ 000, $01 + 00 = 001$, $01 + 01 = 010$, $10 + 01 = 011$, $10 + 10 = 100$, $11 + 10 = 101$, and $11 + 11 = 110$.

Section 5.4

1. Write a driver program to test your distance-cost calculator from Exercise 11.

2. Write a driver program to test your wind-chill classifier from Exercise 12.

3. Modify the program in Example 5.4 for solving quadratic equations so that if the discriminant is negative, the complex roots of the equation are displayed. If the discriminant D is negative, these roots are given by

$$\frac{-b \pm \sqrt{-D}\ i}{2a}$$

where $i^2 = -1$. (This does not require using C++'s `complex` type provided in its complex library, but you may use it if you prefer.)

4. Write a program that reads values for the coefficients $A, B, C, D, E,$ and F of the equations

$$Ax + By = C$$

$$Dx + Ey = F$$

of two straight lines. Then determine whether the lines are parallel (their slopes are equal) or they intersect. If they intersect, determine whether the lines are perpendicular (the product of their slopes is equal to -1).

5. Suppose the following formulas give the safe loading L in pounds per square inch for a column with slimness ratio S:

$$L = \begin{cases} 16500 - .475S^2 & \text{if } S < 100 \\ \dfrac{17900}{1 + (S^2 / 17900)} & \text{if } S \geq 100 \end{cases}$$

Write a program that reads a slimness ratio and then calculates and displays the safe loading.

6. Suppose that a gas company bases its charges on consumption according to the following table:

Gas Used	Rate
First 70 cubic meters	$5.00 minimum cost
Next 100 cubic meters	5¢ per cubic meter
Next 230 cubic meters	2.5¢ per cubic meter
Over 400 cubic meters	1.5¢ per cubic meter

Meter readings are four-digit numbers that represent cubic meters. Write a program in which the meter reading for the previous month and the current meter reading are entered and then the amount of the bill is calculated. *Note:* The current reading may be less than the previous one; for example, one month's reading might be 9897 and the next month's reading 0103.

Section 5.5

1. Write a driver program to test your summation code from Exercise 7.

2. Write a driver program to test your summation code from Exercise 8.

3. The sequence of *Fibonacci numbers* begins with the integers 1, 1, 2, 3, 5, 8, 13, 21, . . . where each number after the first two is the sum of the two preceding numbers. Write a program that reads a positive integer n and uses a `for` loop to generate and display the first n Fibonacci numbers.

4. Ratios of consecutive Fibonacci numbers 1/1, 1/2, 2/3, 3/5, . . . approach the *golden ratio.* Modify the program in Problem 3 so that it also displays the decimal values of the ratios of consecutive Fibonacci numbers.

5. A certain product is to sell for `unitPrice` dollars. Write a program that reads values for `unitPrice` and `totalNumber` and then produces a table showing the total price of from 1 through `totalNumber` units. The table should have a format like the following:

```
Number of Units        Total Price
===============        ===========
       1                  $1.50
       2                  $3.00
       3                  $4.50
       4                  $6.00
       5                  $7.50
```

6. Write a program to read a set of numbers and calculate and display the mean, variance, and standard deviation of the set of numbers. The mean and variance of numbers $x_1, x_2, \ldots, x_n$ can be calculated using the formulas

$$\text{mean} = \frac{1}{n}\sum_{i=1}^{n} x_i , \quad \text{variance} = \frac{1}{n}\sum_{i=1}^{n} x_i^2 - \frac{1}{n^2}\left(\sum_{i=1}^{n} x_i\right)^2$$

The *standard deviation* is the square root of the variance.

7. Two measures of central tendency other than the (arithmetic) mean (defined in Problem 6) are the *geometric mean* and the *harmonic mean* defined for a list of positive numbers $x_1, x_2, \ldots, x_n$ as

$$\text{geometric mean} = \sqrt[n]{x_1 \cdot x_2 \cdots x_n}$$

$$= \text{the } n\text{th root of the product of the numbers}$$

$$\text{harmonic mean} = \frac{n}{\dfrac{1}{x_1} + \dfrac{1}{x_2} + \cdots \dfrac{1}{x_n}}$$

Write a program that reads a list of numbers and calculates their arithmetic mean, geometric mean, and harmonic mean. These values should be displayed with appropriate labels.

8. Suppose that at a given time, genotypes AA, AB, and BB appear in the proportions x, y, and z, respectively, where $x = 0.25$, $y = 0.5$, and $z = 0.25$. If individuals of type AA cannot reproduce, the probability that one parent will donate gene A to an offspring is

$$p = \frac{1}{2}\left(\frac{y}{y+z}\right)$$

because $y/(y + z)$ is the probability that the parent is of type AB and $1/2$ is the probability that such a parent will donate gene A. Then the proportions x', y', and z' of AA, AB, and BB, respectively, in each succeeding generation are given by

$$x' = p^2, \; y' = 2p(1-p), \; z' = (1-p)^2$$

and the new probability is given by

$$p' = \frac{1}{2}\left(\frac{y'}{y'+z'}\right)$$

Write a program to calculate and display the generation number and the proportions of AA, AB, and BB under appropriate headings for 30 generations. (Note that the proportions of AA and AB should approach 0, because gene A will gradually disappear.)

9. Write a program that uses an input loop to read data values as shown in the following table, calculates the miles per gallon in each case, and displays the values with appropriate labels:

Miles Traveled	Gallons of Gasoline Used
231	14.8
248	15.1
302	12.8
147	9.25
88	7
265	13.3

10. Write a program that uses an input loop to read several values representing miles, converts miles to kilometers (1 mile = 1.60935 kilometers), and displays all values with appropriate labels.

11. Suppose that two hallways, one 8 feet wide and the other 10 feet wide, meet at a right angle and that a ladder is to be carried around the corner from one hallway into the other. Using the similar triangles in the following diagram, we see that

$$L = x + \frac{10x}{\sqrt{x^2 - 64}}$$

Write a program that initializes x to 8.1 and then increments it by 0.1 to find to the nearest 0.1 foot the length of the longest ladder that can be carried around the corner. (*Note*: This length is the same as the minimum value of the distance L.)

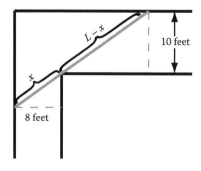

Functions and Libraries

Home computers are being called upon to perform many new functions, including the consumption of homework formerly eaten by the dog.

DOUG LARSON

The function of education is to teach one to think intensively and to think critically . . . . Intelligence plus character—that is the goal of true education.

MARTIN LUTHER KING, JR.

Doing research on the Web is like using a library assembled piecemeal by pack rats and vandalized nightly.

ROGER EBERT

I never met a library I didn't like.

UNKNOWN

A MAJOR THEME OF THIS CHAPTER is writing code that is *reusable*. As we show first, reusability may be on a "local" level in which code segments are *encapsulated in a function* that is a separate part of a program but can be used anywhere else in the program. Also, because it is a separate piece of code, it can be used in other programs by copying and pasting this code into them. For reusability on a larger scale, functions (and other program components such as constant declarations) can be stored in a *library* from which any program can retrieve and use them, just as the programs in Examples 5.4 and 5.5 used the function `sqrt()` from the `cmath` library.

A second theme of this chapter is *information hiding*, also known as *abstraction*, which refers to separating unnecessary details about a computation from what others need to know in order to use it. Here again we will see that this separation can be done on a local scale for functions that are part of a program, and that libraries make true information hiding possible.

Also, functions make it possible to *modularize* programs, which facilitates solving complex problems. We divide a problem into a number of simpler problems, develop functions to solve these simpler problems, and then combine them with a main "driver" program to produce a complete program that solves the original problem. In this chapter, however, we will look only at fairly simple functions in detail and give several examples, because understanding how to construct and use functions is fundamental to writing reusable code. We will also see how libraries are constructed. It must be realized, however, that this first look at functions and libraries is only an introduction. Many more features and uses will be given in later chapters.

6.1 INTRODUCTORY EXAMPLE: TEMPERATURE CONVERSION WITH EXPRESSIONS

Many problems involve the use of formulas in their solutions. For example, in Section 4.1, we used Einstein's formula $e = m \times c^2$ to compute the energy released by a quantity of matter and found that converting this formula into a C++ expression was quite straightforward. In this chapter we begin with a temperature-conversion problem that is also easy to solve in C++, but our first version of a program to perform these conversions will undergo a series of modifications with a view to increasing reusability.

6.1.1 Problem: Temperature Conversion

Fahrenheit and Celsius are two scales used to measure temperature. We will develop a program to convert temperatures in Celsius to the equivalent Fahrenheit temperatures.

6.1.2 Object-Centered Design

6.1.2.1 Behavior

Our program will repeatedly display on the screen a prompt for a Celsius temperature. Each time, the user will enter a numeric value (the Celsius temperature) from the keyboard that the program will read. It will then compute the Fahrenheit temperature equivalent of that Celsius temperature and display this Fahrenheit temperature on the screen along with appropriate descriptive text.

6.1.2.2 Objects

From the statement of the desired behavior, we can identify the following objects:

	Software Objects		
Problem Objects	**Type**	**Kind**	**Name**
screen	output device	variable	`cout`
prompt	text string	constant	none
Celsius temperature	`double`	variable	*tempCelsius*
keyboard	input device	variable	`cin`
Fahrenheit temperature	`double`	variable	*tempFahrenheit*
descriptive text	text string	constant	none

6.1.2.3 Operations

The formula for converting temperature measured in Fahrenheit to Celsius is

$$F = 1.8 \times C + 32$$

where C is the Celsius temperature and F is the corresponding Fahrenheit temperature. We now have two additional data objects—the numeric values 1.8 and 32. However, we won't usually include constant values such as these in an object list because they are always provided in C++.

From our statement of how our program is to behave and this formula, we can identify the following operations:

 i. Display a string on the screen.

 ii. Read a number from the keyboard.

 iii. Compute the Fahrenheit equivalent of a Celsius temperature using multiplication and addition of reals.

 iv. Display a string and a number on the screen.

6.1.2.4 Algorithm

Because each of these operations is provided in C++, we can proceed to organize them and the objects into an algorithm:

Repeat the following:

 1. Display a prompt for a Celsius temperature to `cout`.

 2. Input *tempCelsius* from `cin`.

 3. Calculate *tempFahrenheit* = 1.8 * *tempCelsius* + 32.

 4. Output *tempFahrenheit* with appropriate descriptive text to `cout`.

6.1.2.5 Coding and Testing

The program in Example 6.1 implements this algorithm. It uses a while loop to control the repetition with the user entering any value less than the smallest possible temperature −273.15 degrees Celsius to stop.

Example 6.1 Converting Temperatures—Version 1

```
/* This program converts temperatures from Celsius to Fahrenheit,
   using the standard Celsius-to-Fahrenheit conversion formula.
   Entering any value less than -273.15 Celsius (approx. absolute
   zero) terminates the program.

   Input:  tempCelsius
   Output: tempFahrenheit
------------------------------------------------------------------*/

#include <iostream>
using namespace std;

int main()
{
  cout << "Program to convert Celsius temperatures to Fahrenheit.\n";

  double tempCelsius,        // Celsius temperature
         tempFahrenheit;     // Fahrenheit temperature

  cout <<"\nEnter a Celsius temperature (< -273.15 to stop): ";
  cin >> tempCelsius;
  while (tempCelsius >= -273.15)
  {
    tempFahrenheit = 1.8 * tempCelsius + 32;

    cout << tempCelsius << " degrees Celsius is equivalent to "
         << tempFahrenheit << " degrees Fahrenheit\n";
    cout <<"\nEnter a Celsius temperature (< -273.15 to stop): ";
    cin >> tempCelsius;
  }
}
```

SAMPLE RUN:

```
Program to convert Celsius temperatures to Fahrenheit.

Enter a Celsius temperature (-273.15 or less to stop): 100
100 degrees Celsius is equivalent to 212 degrees Fahrenheit

Enter a Celsius temperature (-273.15 or less to stop): 0
0 degrees Celsius is equivalent to 32 degrees Fahrenheit
```

```
Enter a Celsius temperature (-273.15 or less to stop): -72.3
-72.3 degrees Celsius is equivalent to -98.14 degrees Fahrenheit

Enter a Celsius temperature (-273.15 or less to stop): -333
```

6.2 INTRODUCTORY EXAMPLE: TEMPERATURE CONVERSION WITH A FUNCTION

To incorporate code from a program like that in Example 6.1 for converting Celsius temperatures to Fahrenheit into another program, one would have to first search through the program to find it and then copy and paste it into one's program. In some cases, this code may be somewhat scattered and it would require some effort to find all of it. It would certainly be more convenient if we could construct a function that *encapsulates* that code, and that could be attached to another program where it could be called where needed, as illustrated in Example 6.2.

Example 6.2 Converting Temperatures—Version 2

```cpp
/* This program converts temperatures from Celsius to Fahrenheit,
   using the standard Celsius-to-Fahrenheit conversion formula.
   Entering any value less than -273.15 Celsius (approx. absolute
   zero) terminates the program.

   Input:  tempCelsius
   Output: tempFahrenheit
-------------------------------------------------------------------*/

#include <iostream>
using namespace std;

double celsiusToFahrenheit(double tempCels); // function prototype

int main()
{
  cout << "Program to convert Celsius temperatures to Fahrenheit.\n";

  double tempCelsius,      // Celsius temperature
         tempFahrenheit;   // Fahrenheit temperature

  cout <<"\nEnter a Celsius temperature (< -273.15 to stop): ";
  cin >> tempCelsius;
  while (tempCelsius >= -273.15)
  {
    tempFahrenheit = celsiusToFahrenheit(tempCelsius);
```

```
    cout << tempCelsius << " degrees Celsius is equivalent to "
        << tempFahrenheit << " degrees Fahrenheit\n";
    cout <<"\nEnter a Celsius temperature (< -273.15 to stop): ";
    cin >> tempCelsius;
  }
}


double celsiusToFahrenheit(double tempCels) // function definition
{
  return 1.8 * tempCels + 32;
}
```

The program in Example 6.2 produces exactly the same output as that in Example 6.1, but the flow of execution is very different. Here the Celsius-to-Fahrenheit conversion is performed using a function that is defined following the main function. The following diagram illustrates the flow of control in this new program.

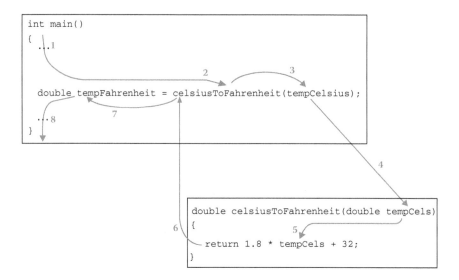

Execution of the program starts at the beginning of main() and proceeds in the usual manner (1), until it reaches the **call** to function celsiusToFahrenheit() (2).[1] At that time, the function's **argument** tempCelsius is evaluated (3) and copied into the **parameter** tempCels in the function celsiusToFahrenheit() (4). Control is transferred from the main function to the function celsiusToFahrenheit(), which begins executing (5). The expression

```
1.8 * tempCels + 32.0
```

[1] See the text's website (given in the Preface) about why we speak of function "calls" and "libraries."

is evaluated, and because tempCels contains a copy of the value of tempCelsius, the resulting value is the Fahrenheit equivalent of the value of tempCelsius. The **return statement** makes this value the **return value** of celsiusToFahrenheit() and transfers execution back to the main function (6). There, the return value of celsiusToFahrenheit() is assigned to tempFahrenheit (7), and execution proceeds normally through the rest of the main function (8).

In the next section we will describe how such C++ functions are constructed, using the program in Example 6.2 to illustrate the discussion.

6.3 PROGRAMMER-DEFINED FUNCTIONS

In earlier chapters we saw examples of functions provided in C++; for example, pow() and sqrt() from C++'s cmath library and assert() from its cassert library. In the preceding section we saw an example of how programmers can add their own functions such as celsiusToFahrenheit() to a program by including its prototype before the main function and its definition after it. In this section we will look at such programmer-defined functions in more detail.

The definition of a function can be described in general as follows:

C++ FUNCTION DEFINITION

FORM:

```
return_type function_name(parameter-declaration_list)
{
    statement_list
}
```

where:

    return_type is the type of value the function returns or the keyword void if the function does not return a value;

    function_name is an identifier that names the function;

    parameter-declaration_list is a list, possibly empty, of comma-separated declarations of the function's parameters;

    statement_list describes the behavior of the function.

PURPOSE:

A call to the function causes the statements in it to be executed. After this, execution returns to the calling program unit when a return statement is executed or the end of the function's statement list is reached.

Thus, in the program in Example 6.2, the lines

```
double celsiusToFahrenheit(double tempCels)
{
    return 1.8 * tempCels + 32;
}
```

constitute the definition of the function `celsiusToFahrenheit()`. The first occurrence of the keyword `double` specifies the return type of the function; the identifier `celsius-ToFahrenheit` names the function; and inside the parentheses that follow, the parameter declaration list consists of one parameter of type `double` whose name is `tempCels`. The curly braces enclose the **body** of the function, which in this example consists of only a `return` statement. Its general form is

```
return expression;
```

where the type of *expression* matches the function's return type.

6.3.1 Function Design

The steps we've used to design programs can also be used for functions—one of the reasons they are often called **subprograms:**

1. Describe its behavior.

2. Identify its objects.

3. Identify its operations.

4. Formulate an algorithm.

5. Code the algorithm.

6. Execute, test, and debug it.

6.3.1.1 Behavior

The key behavior of a function consists of *receiving values* from some other function such as `main()` and *returning a value* to the function that called it (or for void functions, simply returning control to that function but no value).[2] For example, `celsiusToFahrenheit()` receives a Celsius temperature and returns to its caller the equivalent Fahrenheit temperature. But before we can write code for a function, we must describe this receive and return information more precisely, just as we have done with a program's objects up to now. This description is often referred to as the function's **specification**.

6.3.1.2 A Function's Specification

The objects received and returned by a function have the same properties as other programming objects we have been considering. But they also have an additional characteristic: received values flow from the caller *into* the function, and returned values flow from the function *out* to the caller. We will call this characteristic the **movement** of the objects, and add it as a property in our object lists for functions as illustrated in the following specification for our `celsiusToFahrenheit()` function:

---

[2] As described in Chapter 10, a function may also call itself, a property known as *recursion*.

Problem Objects	Software Objects			
	Type	Kind	Movement	Name
a Celsius temperature	double	varying	received (in)	tempCels
the equivalent Fahrenheit temperature	double	varying	returned (out)	none

Thus, in addition to a name, most functions also have the following:

- **parameters**: variables declared within the function heading's parentheses to hold the values received by the function; if no values are *received*, the parentheses are left empty;[3]

- a **return type**: the type of the value returned by the function or the **keyword void**, indicating the absence of type, if the function does not return a value.

A function's specification provides this information. For our `celsiusToFahrenheit()` example, it tells us that

- it has one parameter of type `double`, and

- its return type is `double`.

Once we have this information, we can begin coding the function. We select a meaningful name such as `tempCels` for its parameter and write a **function stub**, consisting of a heading with an empty function body:

```
double celsiusToFahrenheit(double tempCels)
{
}
```

Now, all that remains is to design an algorithm that solves the problem, and then encode that algorithm within the body of this function stub.

6.3.1.3 A Function's Algorithm
Designing the algorithm for a function is almost the same as designing the algorithm for a program. The differences are:

- In a program, *inputting* data values needed in solving a problem is an early step in the algorithm; but in a function, the *parameters already contain these values.*

- In a program, data values computed as part of the solution are usually *output*; but in a function, they are *returned* to the calling function.

---

[3] For example, the random-number generator function `rand()` from the `cstlib` library has no parameters as do most functions that simply display some instructions to the user.

Thus, when the function `celsiusToFahrenheit()` begins execution, we can assume that parameter `tempCels` already contains the Celsius temperature to be converted. This leaves the following operations to be performed:

Real multiplication and addition (`1.8 * tempCels + 32`)

Return a real value

They can be combined as in the following very simple algorithm for function `fahrToCelsius()`:

Algorithm for Celsius-to-Fahrenheit Conversion Function

Return $1.8 \times tempCels + 32$

6.3.1.4 Coding a Function

Once we have a function stub and an algorithm, we obtain a function definition by inserting C++ statements into the stub that implement the algorithm. In our example, it is straightforward to convert our one-step algorithm into a `return` statement:

```
double celsiusToFahrenheit(double tempCels)
{
   return 1.8 * tempCels + 32;
}
```

As this example demonstrates,

- the specification of a function determines the form of a function's heading;
- its algorithm determines the content of the function's body.

For a simple function like `celsiusToFahrenheit()`, formulating the specification and designing the algorithm may seem like unnecessary busy work, but they are essential for more complicated problems. Constructing a function heading from a well-defined specification is almost mechanical, as is coding the body of the function from a well-defined algorithm.

6.3.1.5 Testing, Execution, and Debugging

Like programs, a function should be rigorously tested to ensure its correctness. This is especially true for real-world software projects that are often large and complex, and must be thoroughly checked for logical errors before they are incorporated into the project.

A function is tested by writing a program that calls the function with a collection of test values and displays the values returned by the function. Such a program is called a **driver program** because it "test drives" the function. Example 6.3 shows a simple driver program for `celsiusToFahrenheit()`. The sample run shows that `celsiusToFahrenheit()` is performing correctly for the test values we chose.

Example 6.3 A Sample Driver Program

```cpp
#include <iostream>        // cin, cout, <<, >>
using namespace std;

double celsiusToFahrenheit(double tempCels);   // function prototype

int main()
{
   cout << "100C => " << celsiusToFahrenheit(100) << "F\n"
        << "0C => " << celsiusToFahrenheit(0) << "F\n"
        << "10C => " << celsiusToFahrenheit(10) << "F\n";
}

double celsiusToFahrenheit(double tempCels)
{
   return 1.8 * tempCels + 32;
}
```

SAMPLE RUN:

```
100C => 212F
0C => 32F
10C => 50F
```

6.3.2 Function Prototypes

A general rule in C++ is that things must be declared before they are used so that the compiler can check whether each is being used in a manner consistent with its type. This principle applies to functions also; the compiler must be able to check whether they are being called correctly. This is the purpose of the line

```cpp
double celsiusToFahrenheit(double tempCels);
```

before the main function. It is called a **function prototype** and plays the same role as a declaration of a variable or constant in that it provides information that the compiler needs so that it can check that the function is being used correctly:

- its *return type*

- its *name*

- the *number of parameters* it has

- the *type* of each parameter

The (simplified) syntax of a function prototype is as follows:

C++ FUNCTION PROTOTYPE

FORM:

```
return_type function_name(parameter-declaration_list);
```

where
> *return_type*, *function_name*, and *parameter-declaration_list* play the same roles as in a function definition. One difference is that parameter names are optional—only their types are required—but it is good practice to include them to indicate what the parameters represent.

PURPOSE:

Declares a function, providing the information that the compiler needs to check the correctness of function calls—the number and types of the parameters and the return type. Note that like other declarations, a semicolon is placed at the end of a function prototype to make it a statement.

We see, therefore, that for a function's prototype, we can simply use the heading of the function's definition and append a semicolon. This is what we did in our example program, using

```
double celsiusToFahrenheit(double tempCels);
```

to prototype `celsiusToFahrenheit()`, informing the compiler that its return type is `double`, its name is `celsiusToFahrenheit`, and it has one parameter and its type is `double`. Thus, if we were to call the function incorrectly as in

```
tempFahrenheit = celsiusToFahrenheit(tempCelsius, 0);
```

the compiler can generate an error message that the function may not be called with two arguments. In this text we will place function prototypes before `main()` in the program because this is consistent with the use of libraries (see Section 6.4) and makes clear that such functions have been declared before they are called—in `main()` or in some other function.[4]

6.3.3 Calling a Function

A function call like those we have been considering returns a value, which means that it is a kind of expression and thus can be used anywhere that an expression whose type matches that function's return type can be used. For example, the return type of `celsiusToFahrenheit()` is `double`, which means that a call to `celsiusToFahrenheit()`

---

[4] C++ does permit function prototypes to be placed inside a calling function, but this tends to clutter and obscure the structure of these functions. Placing them before `main()` and other function definitions avoids this.

can be used to give a value to a variable of type `double` such as `tempFahrenheit` in the program in Example 6.2:

```
double tempFahrenheit = celsiusToFahrenheit(tempCelsius);
```

Similarly, the output statement

```
cout << "100C => " << celsiusToFahrenheit(100) << "C\n"
     << "0C => " << celsiusToFahrenheit(0) << "C\n"
     << "10C => " << celsiusToFahrenheit(10) << "C\n";
```

in the driver program in Example 6.3 has three calls to `celsiusToFahrenheit()` and is a valid statement because the `<<` operator can be used to output `double` values.

6.3.4 Local Variables

The function `celsiusToFahrenheit()` is a very simple function; its body consists of a single statement, and uses only one variable—the parameter `tempCelsius`—and two constants. Many functions, however, are more complex, having more statements and using several variables and/or constants to compute their return values.

For example, suppose we wish to construct a function to compute wind chill indexes. The formula for the wind chill index used by the National Weather Service is

$$\text{wind chill} = 35.74 + 0.6215 \times t - 35.75 \times v^{0.16} + 0.4275 \times t \times v^{0.16}$$

which contains five different constants and two variables: t, the temperature in degrees Fahrenheit, and v, the wind speed in miles per hour. The function `windChill()` in Example 6.4, accordingly, has two parameters: `tempFahr`, representing t in the formula, and `windSpeed`, representing v. It uses the preceding formula to compute the wind chill index, but for efficiency, breaks it up into two steps so that $v^{0.16}$ is computed only once:

- Compute *multiplier* = $-35.75 \times 0.4275 \times t$

- Compute wind chill = $35.74 + 0.6215 \times t + multiplier \times v^{0.16}$

The variable `multiplier` is used to store the result computed in the first step. Example 6.4 also shows a simple driver program to test `windChill()`.

Example 6.4 Computing Wind Chill

```
/* This is a driver program to test the windChill() function.
   ------------------------------------------------------------------*/

#include <iostream>          // cin, cout, <<, >>
using namespace std;
```

```
double windChill(double tempFahr, double windSpeed); // prototype

int main()
{
  double temp,            // Fahrenheit temperature
  wind;                   // wind speed (mph)

  cout << "Enter Fahrenheit temperature and wind speed (mph): ";
  cin >> temp >> wind;
  cout << "Wind chill index is " << windChill(temp, wind) << endl;
}

/* windChill computes wind chill.

    Receive: tempFahr, a Fahrenheit temperature, and windSpeed,
             in miles per hour
    Return:  the wind chill
-----------------------------------------------------------------*/

#include <cmath>                 // pow()
using namespace std;

double windChill(double tempFahr, double windSpeed) // definition
{
  double multiplier = -35.75 + 0.4275 * tempFahr;
  return 35.74 + 0.6215 * tempFahr + multiplier * pow(windSpeed, 0.16);
}
```

SAMPLE RUNS:

```
Enter Fahrenheit temperature and wind speed (mph): 10 5
Wind chill index is 1.23564

Enter Fahrenheit temperature and wind speed (mph): 0 20
Wind chill index is -21.9952

Enter Fahrenheit temperature and wind speed (mph): -10 45
Wind chill index is -44.0695
```

Variables such as `multiplier` along with parameters such as `tempFahr` and `windSpeed` and constants declared within a function are said to be **local** to that function because they are *defined only while the function is executing* and are undefined both before and after its execution.[5] This means that they can be accessed only within the function; an

---

[5] These are also called **automatic** objects—their existence automatically begins when the function begins executing and automatically ends when it is finished. If we prepend the keyword `static` to the declaration of a local object, it becomes a **static** object that will retain its value from one function call to the next.

error results if an attempt is made to use them outside the function. One consequence of this is that *an identifier outside the function may have the same name as a parameter or a local variable or constant inside the function.*

6.3.5 Control Structures in Functions

Thus far, sequence has been the only control structure used in our examples of functions. However, selection and repetition structures may also be used. The program in Example 6.5 illustrates this by encapsulating in a function the code in Section 5.5 that used a for loop to calculate factorials.

Example 6.5 Computing Factorials

```
/* This is a driver program to test the factorial() function.
   ----------------------------------------------------------------*/

#include <iostream>      // cin, cout, <<, >>
using namespace std;

unsigned factorial(unsigned n);      // prototype

int main()
{
  unsigned number;         // integer whose factorial is computed

  cout << "Enter a nonnegative integer: ";
  cin >> number;
  cout << number <<"! = " << factorial(number) << endl;
}

/* factorial computes the factorial of a nonnegative integer

    Receive: n, a nonnegative integer
    Return:  n!
   ----------------------------------------------------------------*/

unsigned factorial(unsigned n)      // definition
{
  unsigned nfact = 1;
  for (unsigned i = 1; i <= n; i++)
    nfact *= i;

  return nfact;
}
```

SAMPLE RUNS:

```
Enter a nonnegative integer: 0
0! = 1
```

```
Enter a nonnegative integer: 4
4! = 24

Enter a nonnegative integer: 10
10! = 3628800
```

6.3.6 Functions That Return Nothing

Suppose that a problem involves monetary calculations and we need to display an amount in the form $dd.cc. We could just use an output statement containing several format manipulators:

```
int main()
{
  // ...
  cout << fixed << showpoint
       << right << setprecision(2)
       << '$' << dollarAmount;
}
```

but putting such details in a main function can make a program messy and cluttered, especially when they are needed at several different places. It would be preferable to use a simpler statement such as

```
int main()
{
  // ...
  printAsMoney(dollarAmount);
}
```

where printAsMoney() would act like a function with a parameter that receives the value of dollarAmount and outputs it in the required format, but doesn't return a value.

In some programming languages we could do this with subprograms called *procedures* or *subroutines*, which are different from functions. However, in C++, the *only subprograms are functions*. This means that we will have to write a function named printAsMoney(). In place of the usual "Receive–Return" specification, we might use

Receive: a real value dollars

Output: the value of dollars, appropriately formatted as a monetary value

From this specification, we see that printAsMoney() will have a double (or float) parameter named dollars, but what do we use as the return type for a function that returns nothing?

For these special kinds of functions that return nothing to the caller, C++ provides the keyword void to denote the absence of any type. So we simply use it as the return type for printAsMoney():

```
void printAsMoney(double dollars);
```

Void functions in C++ are therefore the counterparts of procedures and subroutines in other programming languages.

Because void functions do not return values, they cannot be used in expressions like functions we have considered up to now. For example, if we wrote

```
cout << "Amount due: "
     << printAsMoney(dollarAmount)      // ERROR!
     << endl;
```

an error would result because printAsMoney(dollarAmount) returns no value. Instead, void functions are called with statements that have the form

```
function_Name(argument_list);
```

Thus, in place of the preceding output statement, we could write

```
cout << "Amount due: ";
printAsMoney(dollarAmount);
cout << endl;
```

Example 6.6 gives the complete definition of printAsMoney(). It also shows a program that inputs the amount of a purchase and the amount received, and uses printAs-Money() to display the amount returned to the customer.

Example 6.6 Monetary Transactions

```
/* Program to compute the amount to be returned for a purchase.

 Input:  purchase, payment
 Output: amount returned to customer (via printAsMoney())
 ------------------------------------------------------------------*/

#include <iostream>    // cin, cout, <<, >>, ...
#include <iomanip>     // setprecision, ...
using namespace std;

void printAsMoney(double dollars);      // prototype

int main()
```

```cpp
{
  double purchase,                    // amount of purchase
         payment;                     // amount paid

  cout << "Enter amount of purchase: ";
  cin >> purchase;
  cout << "Enter amount paid (>= purchase): ";
  cin >> payment;
  if (payment >= purchase)
  {
    cout << "Amount to return is: ";
    printAsMoney(payment - purchase);
    cout << endl;
  }
  else
    cout << "Inadequate payment!";
  cout << endl;
}

/* printAsMoney displays an amount in monetary format.

   Receive: dollars, the double value to be displayed
   Output:  dollars in monetary format
-----------------------------------------------------------*/

void printAsMoney(double dollars)
{
  cout << fixed << showpoint
       << setprecision(2)
       << '$' << dollars;
}
```

SAMPLE RUNS:

```
Enter amount of purchase: 4.01
Enter amount paid (>= purchase): 5.00
Amount to return is: $0.99

Enter amount of purchase: 9.00
Enter amount paid (>= purchase): 20.00
Amount to return is: $11.00
```

Note how hiding the money-format details in a `void` function `printAsMoney()` makes the main function easier to read. As noted in the introduction to this chapter, such a view of something that is simplified by hiding some of its details is referred to as *abstraction* or *information hiding.* Our `printAsMoney()` function "hides" the details of how the display-a-monetary-value operation is accomplished (at least as far as readers of the main function are concerned).

Because functions are so fundamental in C++ programming, we will summarize some of the important ideas about them that we have seen:

- For each value to be received from a caller, a function must declare a variable, called a *parameter*, within the parentheses of the function heading to hold that value. For example, in the definition of the function celsiusToFahrenheit(),

  ```
  double celsiusToFahrenheit(double tempCels)
  {
     return (tempFahr - 32) / 1.8;
  }
  ```

  the variable tempCels is a parameter of celsiusToFahrenheit().

- An *argument* is a value supplied to a function when it is called. For example, in the statement

  ```
  tempFahrenheit = celsiusToFahrenheit(tempCelsius);
  ```

  tempCelsius is an argument to the function celsiusToFahrenheit(). During execution, the value of the argument tempCelsius will be *passed* (i.e., copied) to the parameter tempCels.[6]

- If one function f() calls another function g(), execution is transferred from f() to g() and then back to f(). To illustrate, consider again the main function's call of function celsiusToFahrenheit() in Example 6.2:

  ```
  //...
  int main()
  {
     //...
     tempFahrenheit = celsiusToFahrenheit(tempCelsius);
     //...
  }
  ```

  When the call to celsiusToFahrenheit() is encountered, execution proceeds as follows:

  1. The value of the argument tempCelsius is passed from the main function to celsiusToFahrenheit() and copied into the parameter tempCels.

---

[6] Chapter 10 will describe a different kind of parameter called a *reference parameter* that shares the same memory location as its corresponding argument. This means that the value of the argument is not copied to the parameter because they share the same memory location and thus have the same value. It also means that modifying the parameter in the function will also modify the corresponding argument.

2. Control then transfers from the line containing the function call (in `main()`) to the first statement of `celsiusToFahrenheit()`, which begins execution using the value of its parameter.

3. When a `return` statement (or the final statement of the function) is executed, control transfers back to the caller (i.e., the main function), where execution resumes.

- *Local variables, constants, and parameters of a function are defined only while that function is executing. They can be accessed only within the function—not from outside.*

- The keyword `void` is used to specify the return type of a function that does not return any values. Such functions are called with statements of the form
  `function_ Name(argument_list);`

The ability to define functions is a powerful tool in programming. If some problem requires that an operation not provided in C++ be applied to some item in that problem, we can simply

1. define a function to perform that operation, and

2. apply that function to the item

as we did with the temperature-conversion operation `celsiusToFahrenheit()`. Also, as we will see in the next section, such functions can then be stored in a library, from which they can be retrieved when needed.

6.4 AN INTRODUCTION TO LIBRARIES

We have seen that if a problem requires some new operation, we can define a function to perform it and then call it just as if were provided in C++. *In essence, we are extending C++'s collection of operations.* However, although such programmer-defined functions provide a significant improvement over using only expressions, they do not in themselves make it easy to reuse our work. For example, to use the `celsiusToFahrenheit()` function from Example 6.2 in another program, we would have to copy its prototype and its definition from that program and insert them at the appropriate places in the new program. We might use the *copy-and-paste* capabilities of a text editor or word processor to do this or we might simply make a copy of the program in Example 6.2, delete everything between the prototype and definition of `celsiusToFahrenheit()`, and then write the new program between this prototype and definition.

To facilitate reusability, C++'s parent language C provides several standard **libraries**, which are simply files containing items that can be shared by different programs and by other libraries. For example, we have used the function `pow()` from C's standard library `cmath` to compute powers, and in the preceding chapter we used the `assert()` function from C's `cassert` library to check whether certain conditions are true. To use one of these functions, for example, `pow()`, we need only include the library in the program:

```
#include <cmath>
```

C's libraries along with those added in C++, several of which are described in Appendix D, are one feature that have made these programming languages so powerful, because they make it possible to share commonly used functions between different programs. This section gives a brief description of how such libraries are constructed and how we can build our own libraries.

6.4.1 Constructing a Library

If we examine the standard libraries, we see that the contents of each are *related* in some way. Having an *organizing principle*, therefore, is a first step in constructing a library so that we know what kind of things it should have. For example, the `cmath` library contains useful mathematical functions, and the `iostream` library contains items that are related to input and output streams. Thus, if we intend to store the temperature-conversion function `celsiusToFahrenheit()` in a library, we might decide that it should contain items that are in some way related to heat and temperature, and call the library `Heat`.

Once we have an organizing principle, we must decide what items to put in our library. For example, if our `Heat` library is to contain our `celsiusToFahrenheit()` function, then it should also include the inverse operation for Fahrenheit-to-Celsius conversions. We might also include functions to convert temperatures from each scale to Kelvin and back. With careful planning and anticipating what items might be most useful, we hope to develop a library that is useful in the long term rather than short.

Although our examples have only been functions, libraries may also include other items. For example, we might also want to include important heat-related constants in our library such as absolute zero (0K on the Kelvin scale), the heat of fusion of water (79.71 calories/gram), and the heat of vaporization of water (539.55 calories/gram). The important things to remember are that items should be consistent with the library's organizing principle and that they should be useful to users (and thus avoid having to "reinvent the wheel").

Once the contents of the library have been determined, we are ready to begin building it. A library is a collection of files:

- A **header file** that contains the *declarations* and *prototypes* of the items in the library and is inserted into a program (or another library) using a `#include` directive. It serves as an *interface* between the library and a program that uses it, and is thus sometimes called the library's **interface file**.

- An **implementation file** that contains the *definitions* of items not defined in the header file. It must be separately compiled and then linked to a program needing to access its contents (as described later in this section).

- A **documentation file** that contains *documentation* for the items in the library.[7]

---

[7] An alternative is to document each item in the header file. However, this does tend to clutter the header file and reduce its overall readability, so we will put this documentation in a separate file.

6.4.1.1 Building the Header File

A header file contains declarations of the items in the library. It is so named because a function is usually declared by giving its *heading*, i.e., its *prototype*. Thus, just as iostream contains declarations of items for input and output, our header file will contain declarations of items needed to process heat-related values. A first version is shown in Example 6.7. Note that its name, Heat.h, contains the extension .h to indicate that it is a header file.

Example 6.7 Header File for Library Heat

```
/* Heat.h provides an interface for a library of heat-related
   constants and functions.

   Created by: Jane Doe, January, 2012, at Dooflingy Engineering.
   Modification History: Kelvin items added February, 2012 -- JD.
   ----------------------------------------------------------------*/

const double KELVIN_ABSOLUTE_ZERO = 0;        // absolute 0 Kelvin

const double HEAT_OF_FUSION = 79.71;          // calories per gram

const double HEAT_OF_VAPORIZATION = 539.55;   // calories per gram

double celsiusToFahrenheit(double tempCels); // Celsius -> Fahrenheit

double fahrenheitToCelsius(double tempFahr); // Fahrenheit -> Celsius

double fahrenheitToKelvin(double tempFahr);  // Fahrenheit -> Kelvin

double kelvinToFahrenheit(double tempKelv);  // Kelvin -> Fahrenheit

double celsiusToKelvin(double tempCels);     // Celsius -> Kelvin

double kelvinToCelsius(double tempKelv);     // Kelvin -> Celsius
```

Including this header file using the directive #include "Heat.h" before a program's main function inserts the declarations of the library's constants and function prototypes, so these items can be used subsequently in the program. When the program is compiled, these declarations will be inserted into the program at that point and will be compiled along with the rest of the program.

6.4.1.2 Building the Implementation File

A library's implementation file stores the *definitions* of the functions declared in the header file.[8] This file is so named because these definitions implement those functions. Part of the implementation file for our library Heat is given in Example 6.8.

---

[8] To improve on the execution time of programs that use a library, functions that are sufficiently simple (e.g., whose bodies consist of 3–5 operations) are sometimes defined in the header file, provided that the inline modifier precedes the heading of the function. These are explained in more detail in Chapter 10.

Example 6.8 Implementation File for Library `Heat`

```
/* Heat.cpp provides the function implementations for Heat, a
   library of heat-related constants and functions.

   Created by: Jane Doe, January, 2012, at Dooflingy Engineering.
   Modification History: Kelvin items added February, 2012 -- JD.
   ----------------------------------------------------------------*/

#include "Heat.h"

double celsiusToFahrenheit(double tempCels)
{
  return 1.8 * tempCels + 32;
}

double fahrenheitToCelsius(double tempFahr)

{
  return (tempFahr - 32) / 1.8;
}

// . . . Definitions of other functions omitted to save space . . .
```

It is important to note that, unlike the header file, a library's implementation file must be compiled and thus its name, `Heat.cpp`, has the same extension `.cpp` as the source files we've been considering in earlier chapters. Notice also that this implementation file contains the line

```
#include "Heat.h"
```

in addition to the definitions of the various temperature-related functions. Before it compiles the code in the implementation file, the compiler will insert the function prototypes from the header file at the beginning of the implementation file. This enables it to check that the function prototypes and definitions are consistent. If inconsistencies are detected, error messages will be displayed to alert us to the problem.

It is important to note that:

- Items in the implementation file that define items declared in the header file can be accessed in any program that (i) uses the `#include` directive to insert the header file, and (ii) is linked to the implementation file.

- In addition to definitions of the functions declared in the header file, an implementation file may use other items—constants, variables, functions—that are not declared in the header. Such items cannot be accessed outside of the implementation file, even by a program that includes the header file.

Stated simply, items declared in the header file can be thought of as **public** information, whereas those declared in the implementation file are **private** within the library.

6.4.1.3 Building the Documentation File

As the name suggests, the documentation file provides information to the user of the library about the items in it. In this text we will construct it as a copy of the header file, but annotate it with documentation that describes each object and provides the specification for each function prototype. Example 6.9 shows part of this documentation file for library Heat. We have named it Heat.txt because it is a text document, but other extensions such as *.doc* would also be appropriate.

Example 6.9 Documentation File for Library Heat

```
/* Heat.txt provides the documentation for Heat, a library of
   heat-related constants and functions.

   Created by: Jane Doe, January, 2012, at Dooflingy Engineering.
   Modification History: Kelvin items added February, 2012 -- JD.
   --------------------------------------------------------------------*/

// Absolute zero on the Kelvin scale
   const double KELVIN_ABSOLUTE_ZERO = 0;        // absolute 0 Kelvin

// The amount of heat needed to change water from liquid to solid
   const double HEAT_OF_FUSION = 79.71;          // calories per gram

// The amount of heat needed to change water from liquid to gas
   const double HEAT_OF_VAPORIZATION = 539.55;   // calories per gram

/* celsiusToFahrenheit converts a temperature from Celsius
   to Fahrenheit.

   Receive: A Celsius temperature
   Return:  The equivalent Fahrenheit temperature
   --------------------------------------------------------------------*/
double celsiusToFahrenheit(double tempCels);

/* fahrenheitToCelsiusconverts a temperature from Fahrenheit
   to Celsius.

   Receive: A Fahrenheit temperature
   Return:  The equivalent Celsius temperature
   --------------------------------------------------------------------*/
double fahrenheitToCelsius(double tempFahr)

// . . . Other functions omitted to save space . . .
```

Such a documentation file serves a secondary purpose as an annotated copy of the header file in that it serves as a *backup* for the header file.

6.4.2 Using a Library in a Program

Once our library `Heat` has been constructed, we can use it in a program like that in Example 6.10 for solving a temperature-conversion problem.

Example 6.10 Converting a Temperature Using a Library

```
/* This program converts temperatures from Celsius to Fahrenheit, using
   function fahrenheitToCelsius() that is stored in library Heat.

   Input:   tempFahrenheit
   Output:  tempCelsius
-------------------------------------------------------------------*/

#include <iostream>              //cin, cout, <<, >>
using namespace std;
#include "Heat.h"                // our library's header file

int main()
{
  cout << "Program to convert Celsius temperatures to Fahrenheit.\n";

  double tempCelsius,       // Celsius temperature
         tempFahrenheit;    // Fahrenheit temperature

  cout << "\nEnter a Celsius temperature (< -273.15 to stop): ";
  cin >> tempCelsius;
  while (tempCelsius >= -273.15)
  {
    tempFahrenheit = 1.8 * tempCelsius + 32;

    cout << tempCelsius << " degrees Celsius is equivalent to "
         << tempFahrenheit << " degrees Fahrenheit\n";
    cout << "\nEnter a Celsius temperature (< -273.15 to stop): ";
    cin >> tempCelsius;
  }
}
```

Execution of this program is identical to that in Example 6.2. In this program, however, the prototype of `celsiusToFahrenheit()` before the main function has been replaced by the line

```
#include "Heat.h"
```

and the definition of `celsiusToFahrenheit()` is no longer present. The prototype and the definition of `celsiusToFahrenheit()` are not given in this file, because the prototype is in the header file of library `Heat` and the definition is in the library's implementation file.

It is important to note and understand the difference between the `#include` directive for one of the C++ standard libraries such as

```
#include <iostream>
```

and that used to include a programmer-defined library such as

```
#include "Heat.h"
```

In the first case where the name of the library is surrounded by *angle brackets* (< and >), the C++ compiler will search for it in the special system `include` directories. By contrast, if the name of a library's header file is enclosed in *double quotes*, the C++ compiler will search for it in the directory that contains the source file being compiled.[9]

6.4.3 Translating a Library

Translation of a program consists of two separate steps:

1. **Compilation:** translating a source program into an equivalent machine-language program called an *object program* that is stored in an *object file*. If the program uses a programmer-defined library, it too must be compiled and stored in a separate object file.

2. **Linking:** calls to library functions from a program are linked to the function definitions in the library, creating an *executable program*.

The following diagram illustrates this process:

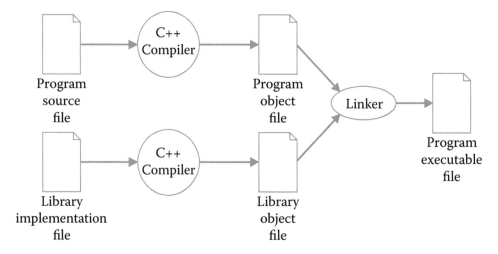

---

[9] C++ also permits the programmer to store a library in a different directory (e.g., a library directory) and then instruct the compiler to search that directory when looking for files named by `#include` directives. The details of how to do this vary from one system to another.

Compiling libraries separately significantly reduces the time to translate a source program that includes a library because the library needs no recompilation, only linking to the program's object file. Also, any errors detected in the translation are most likely in the source program (because the library has already been compiled).

How this translation is carried out varies from one system to another, but there are two basic approaches. In integrated development environments such as *Visual C++*, translation is coordinated by creating a *project* and then adding source files and library files to it. Menu selections can then be used to carry out the compiling and linking to produce a binary executable and then execute it. In *command-line* environments, a user interacts with the computer by typing commands. For example, with GNU C++ on a Unix/Linux system, compiling and linking the source code in Example 6.10 and `Heat.cpp` requires several commands. We begin by entering commands to separately compile the source file and the library's implementation file:

```
g++ -c example6-10.cpp
g++ -c Heat.cpp
```

creating object files `example6-10.o` and `Heat.o`. Then we link these two object files together to produce a binary executable file named `example6-10.exe` with the command

```
g++ example6-10.o Heat.o -o example6-10.exe
```

To execute the file that results, we simply type its name as a command:

```
example6-10.exe
```

6.4.4 Object-Centered Design: Incorporating Functions and Libraries

Until now, we have designed software solutions to problems using object-centered design (OCD), which consists of the following: (1) describe the program's behavior; (2) identify the problem's objects; (3) identify the operations needed; and (4) arrange them in a way that solves the problem. Now we consider how to incorporate functions and libraries into OCD.

Because functions can be viewed as operations added to those provided in C++, they fit naturally into that part of OCD:

Identify the operations needed to solve the problem.

If an operation is not provided in C++, write a function to implement it; and if it seems it might be reusable in other problems, store it in a library.

6.4.5 Benefits of Using Libraries

We end our discussion of libraries by describing some of their main properties and benefits that make them so important in the object-oriented approach to programming.

6.4.5.1 Libraries Hide Implementation Details

In our programs, we can use the items in a library without being concerned at all with how they are defined in its implementation file. All that is needed is the information in the header and documentation files, and access to an object file generated by compiling the implementation file so that its contents can be linked to our programs. The details in the implementation file can remain "hidden," allowing us to use a library without being concerned about these details. For example, we have used input/output operations provided in the `iostream` library in every one of our programs without being concerned with the details of how they are implemented.

6.4.5.2 Libraries Facilitate Program Maintenance

Separate compilation of programs and libraries makes it possible to change the implementation file of a library and recompile it only without having to change or recompile programs or other libraries that use the library. For example, if function `celsiusTo-Fahrenheit()` in the implementation file of the library `Heat` is changed, then only the implementation file needs to be recompiled. A program that uses this function does not need recompiling, because the prototype of `celsiusToFahrenheit()` has not changed, only its definition. All that is needed is to relink the program to the new object file of library `Heat`.

By contrast, if the interface in the header file is altered, then both the library's implementation file and all programs and other libraries that name that header file in a `#include` directive must be recompiled and relinked.

6.4.5.3 Libraries Support Independent Coding

Libraries enable *modularity* in software design by allowing related items to be grouped together into independent units, which is especially useful in large software projects. Once the objects and operations required in the project are identified and categorized, libraries can be designed to house the items for the various kinds of objects and header files for these libraries constructed.

After this, work on the project can be divided up among different groups of programmers assigned to construct the implementation files of the various libraries and other groups working on programs that will use the libraries in the software project. Such partitioning of the work into manageable "chunks" that can be coded in parallel can speed up completion of the entire project. Also, because the work in one part of the project proceeds independently of the work in another, there is less likelihood of errors.

6.4.5.4 Libraries Simplify Testing

Separate libraries can be developed and tested independently (using driver programs) by different teams of programmers. This means that each library can be tested more thoroughly (and rapidly) than can a single, very large program.

6.4.5.5 Libraries Extend a Programming Language

Libraries provide items—functions, constants, and so on—available to any program (or other library). A programmer does not have to "reinvent the wheel" each time these items are needed.

6.5 INTRODUCTION TO NUMERICAL METHODS

Problems and mathematical models in science and engineering, plus a wide variety of other areas, involve solving ordinary algebraic equations, differential equations, systems of equations, and so on. In many cases, **numerical methods** that can be implemented by computer programs are used to find solutions. Here are some of the major types of problems in which numerical methods are routinely used:

1. For a few algebraic equations, a formula (e.g., the quadratic formula for quadratic equations) gives the exact roots, but for most equations, a numerical method is used to *find approximate roots.*

2. The solution of many problems such as finding the area under a curve, calculating probabilities of certain events, and calculating work done by a force, require the *evaluation of an integral.* For some, the methods of integral calculus can be used, but for others a numerical technique is used to find approximate values.

3. *Differential equations* play an important role in many applications, and a variety of effective and efficient numerical methods have been developed to obtain solutions.

4. *Curve fitting* refers to analyzing pairs of data values to determine whether the items in these pairs are related by some algebraic equation. For example, the least-squares method described in Programming Problem 10 of Section 9.4 is a method for fitting a line to a set of points.

5. Numerical methods are commonly used to find a collection of values that satisfies a collection of linear equations with several unknowns simultaneously. Such *linear systems* will be described later when we consider matrices.

In this section, a few of the numerical methods in the first three categories are described.

6.5.1 Solving Equations—The Bisection Method

In many applications it is necessary to find a zero or root of a function, that is, to solve an equation of the form $f(x) = 0$. For some functions it may be very difficult or even impossible to find this solution exactly. Examples include the function

$$f(v) = 50 \cdot 10^{-9}(e^{40v} - 1) + v - 20$$

which may arise in a problem of determining the d-c operating point in an electrical circuit, or a function of the form

$$f(x) = x \tan x - a$$

for which a zero must be found to solve some heat conduction problem.

One simple numerical method for finding an approximate solution of an equation $f(x) = 0$ is the **bisection method**. (The programming problems describe another commonly used method known as the *Newton-Raphson* method.) In the bisection method, we begin with an interval $[a, b]$ where the function values $f(a)$ and $f(b)$ at the endpoints have opposite signs. If f is continuous in this interval—that is, there is no break in the graph of $y = f(x)$ in this interval—then the graph of f must cross the x-axis at least once between $x = a$ and $x = b$, and thus there must be at least one solution of the equation $f(x) = 0$ between a and b. To locate one of these solutions, we first bisect the interval $[a, b]$ and determine in which half f changes sign, thereby locating a smaller subinterval containing a solution of the equation. We bisect this subinterval and determine in which half of it f changes sign; this gives a still smaller subinterval containing a solution.

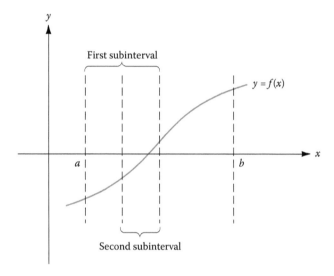

Repeating this process gives a sequence of subintervals, each of which contains a solution of the equation and has a length one-half that of the preceding interval. Note that at each step, the midpoint of a subinterval of length L is within $L / 2$ of the exact solution:

The program in Example 6.11 uses the bisection method to find an approximate solution of

$$f(x) = x^3 + x - 5 = 0$$

It generates successive approximations to a solution, terminating when an interval is obtained whose length guarantees a specified accuracy entered by the user.

Example 6.11 Bisection Method

```
/* Program finds an approximate solution of the equation f(x) = 0
   in a given interval, using the bisection method.

   Input:  desired accuracy of approximation, endpoints of an interval
           containing a solution
   Output: prompts to the user and the approximate solution
------------------------------------------------------------------*/

#include <iostream>
using namespace std;

double f(double x);

int main()
{
  cout << "\nThis program uses the bisection method to find an\n"
       << "approximate solution to the equation f(x) = 0.\n";

  double desiredAccuracy;        // the accuracy desired

  cout << "\nEnter the accuracy desired (e.g. .001): ";
  cin >> desiredAccuracy;

  double left, right;            // get interval containing a solution

  do
  {
    cout << "Enter endpoints of an interval containing a solution: ";
    cin >> left >> right;
  }
  while (f(left) * f(right) >= 0.0);

  double width = right - left,  // the interval width
         midPt,                 // the midpoint of the interval
         fMid;                  // value of f() at midpoint

  while (width/2.0 > desiredAccuracy)
  {
    midPt = (left + right) / 2.0;
    fMid = f(midPt);

    if (f(left) * fMid < 0.0 )  // solution is in left half
      right = midPt;
    else                        // solution is in right half
        left = midPt;

    width /= 2.0;               // split the interval
  }
```

```
            cout << "\n-->" << midPt << " is an approximate solution of "
                 << " f(x) = 0, to within " << desiredAccuracy << endl;
        }


        // Function for which a root is being found
        double f(double x)
        {
            return x*x*x + x - 5;
        }
```

SAMPLE RUN:

```
This program uses the bisection method to find an
approximate solution to the equation f(x) = 0.

Enter the accuracy desired (e.g. .001): 1e-4
Enter endpoints of an interval containing a solution: 0 1
Enter endpoints of an interval containing a solution: 0 2

-->1.51599 is an approximate solution of f(x) = 0, to within 0.0001
```

6.5.2 Numerical Integration—Approximating Areas of Regions

One important problem in calculus is finding the area of a region bounded below by the x-axis, above by the graph of a function $y = f(x)$, on the left by a vertical line $x = a$, and on the right by a vertical line $x = b$:

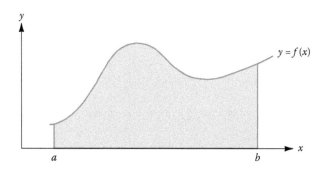

More generally, the problem is to approximate the integral

$$\int_a^b f(x)\, dx$$

One common method is to divide the interval $[a, b]$ into n subintervals, each of width $\Delta x = (b - a)/n$ using $n - 1$ equally spaced points $x_1, x_2, ..., x_{n-1}$. Locating the corresponding points on the curve and connecting consecutive points using line segments forms n trapezoids:

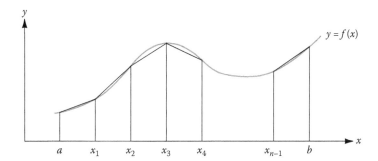

The sum of the areas of these trapezoids is approximately the area under the graph of f; thus, this method is known as the *trapezoidal method* for approximating integrals. (The programming problems describe *Simpson's method*, which uses parabolas instead of trapezoids.) The formula for the area of a trapezoid gives

$$\frac{1}{2}\ x(y_{i-1} + y_i)$$

for the area of the ith trapezoid, where y_{i-1} and y_i are the values of the function f at x_{i-1} and x_i, respectively. Summing these and combining terms gives the formula

$$x\left(\frac{y_0 + y_n}{2} + y_1 + y_2 + \ \cdots \ y_{n-1}\right)$$

or, written more concisely using summation (Σ) notation,

$$\Delta x\left(\frac{y_0 + y_n}{2} + \sum_{i=1}^{n-1} y_i\right)$$

The program in Example 6.12 implements the trapezoidal method.

Example 6.12 Trapezoidal Approximation of an Integral

```
/* This program approximates the area under the graph of a function
   f() over an interval [a, b] using the trapezoidal rule -- and
   thus approximates the definite integral of f() from a to b.

   Input:   the endpoints a and b of the interval, and the number of
            subintervals to use
   Output:  approximation to the integral of f() on [a, b]
   ----------------------------------------------------------------*/
```

```
#include <iostream>        // cin, cout, <<, >>
using namespace std;

double f(double x);        // function to be integrated

int main()
{
  cout << "Trapezoidal approximation of an integral:\n";

  int n;                   // number of subintervals
  double a, b,             // endpoints of interval
         deltaX,           // length of subintervals
         x, y,             // point on graph of f()
         sum;              // sum of areas of trapezoids

  cout << "Enter interval endpoints and the # of subintervals: ";
  cin >> a >> b >> n;

  deltaX = (b - a) / n;
  sum = 0;
  x = a;

  for (int i = 1; i <= n - 1; i++)
  {
    x += deltaX;
    y = f(x);
    sum += y;
  }
  sum = deltaX * ((f(a) + f(b))/2 + sum);

  cout << "Approximate value using " << n << " subintervals: "
       << sum << endl;
}

// Function being integrated
double f(double x)
{ return x*x + .1; }
```

SAMPLE RUNS:

```
Trapezoidal approximation of an integral:
Enter interval endpoints and the # of subintervals: 0 1 10
Approximate value using 10 subintervals: 1.335

Trapezoidal approximation of an integral:
Enter interval endpoints and the # of subintervals: 0 1 50
Approximate value using 50 subintervals: 1.3334

Trapezoidal approximation of an integral:
```

```
Enter interval endpoints and the # of subintervals: 0 1 100
Approximate value using 100 subintervals: 1.33335
```

6.5.2.1 Application: Road Construction

There are many problems where it is necessary to compute the area under a curve (or the more general problem of calculating an integral). One such problem is the following road construction problem. A construction company has contracted to build a highway for the state highway commission. Several sections of this highway must pass through hills from which large amounts of dirt must be excavated to provide a flat and level roadbed.

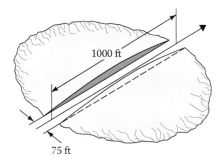

To estimate the construction costs, the company needs to know the volume of dirt that must be excavated from the hill.

To estimate the volume of dirt to be removed, we can assume that the height of the hill does not vary from one side of the road to the other. The volume can then be calculated as

(volume = cross-sectional area of the hill) × (width of the road)

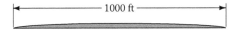

The cross-sectional area of the hill can be computed using the trapezoidal method. (A program that solves this problem is included on the website for this text.)

6.5.3 Numerical Solutions of Differential Equations

Equations that involve derivatives or differentials are called **differential equations**. They arise in a large number of problems in science and engineering. It is very difficult or even impossible to solve many differential equations exactly, but it may be possible to find an approximate solution using a numerical method. There are many such methods, and we describe one of the simpler ones, **Euler's method**, here. (Another popular method and one of the most accurate, the **Runge-Kutta method**, is described in the programming problems.)

We are given a *first-order differential equation*:

$$y' = f(x, y)$$

that satisfies a given *initial condition*:

$$y(x_0) = y$$

Euler's method for obtaining an approximate solution over some interval $[a, b]$, where $a = x_0$, is as follows:

1. Select an x-increment Δx.

2. For $n = 0, 1, 2, \ldots$, do the following:

    a. Set $x_{n+1} = x_n + \Delta x$.

    b. Find the point $P_{n+1}(x_{n+1}, y_{n+1})$ on the line through $P_n(x_n, y_n)$ with slope $f(x_n, y_n)$.

    c. Output y_{n+1}, which is the approximate value of y at x_{n+1}.

The following diagram illustrates Euler's method:

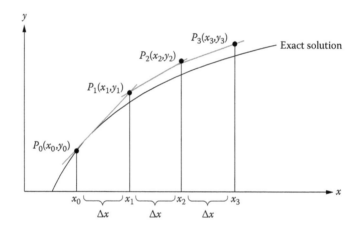

The program in Example 6.13 uses Euler's method to find an approximate solution for

$$y' = 2xy$$

$$y(0) = 1$$

Sample runs with $a = 0$, $b = 1$, $\Delta x = 0.2$ and $a = 0$, $b = 0.5$, $\Delta x = 0.05$ are shown.

Example 6.13 Euler's Method for Solving a Differential Equation

```
/* This program uses Euler's method to obtain an approximate solution
   to a first-order differential equation of the form y' = f(x, y)

   Input:   initial values for x and y, an increment deltaX, and
            number of iterations
```

```
   Output: a sequence of points (x, y) that approximate the solution
           curve
--------------------------------------------------------------------*/


#include <iostream>        // cin, cout, <<, >>
#include <iomanip>         // output formatters
using namespace std;


double f(double x, double y);


int main()
{
  cout << "Euler's method for solving a differential equation:\n";

  double x, y,              // current x value, y value
         xNext, yNext,      // next x value, y value
         deltaX;            // x-increment used
  int numIterations;        // number of iterations.

  cout << "Enter x0 and y0, x-increment to use, and "
       << "number of values to calculate:\n";
  cin >> x >> y >> deltaX >> numIterations;

  cout << " X             Y\n"
       << "====================\n"
       << fixed << setprecision(5) << showpoint
       << setw(10) << x << setw(10) << y << endl;

  // Iterate with Euler's method

  for (int n = 1; n <= numIterations; n++)
  {
    xNext = x + deltaX;
    y = y + f(x, y) * deltaX;
    x = xNext;
    cout << setw(10) << x << setw(10) << y << endl;
  }
}


double f(double x, double y)
{ return 2.0 * x * y; }
```

SAMPLE RUNS:

```
Euler's method for solving a differential equation:
Enter x0 and y0, x-increment to use, and number of values to calculate:
0 1 0.2 5
```

```
        X            Y
====================
  0.00000     1.00000
  0.20000     1.00000
  0.40000     1.08000
  0.60000     1.25280
  0.80000     1.55347
  1.00000     2.05058

Euler's method for solving a differential equation:
Enter x0 and y0, x-increment to use, and number of values to calculate:
0 1 .05 10
        X            Y
====================
  0.00000     1.00000
  0.05000     1.00000
  0.10000     1.00500
  0.15000     1.01505
  0.20000     1.03028
  0.25000     1.05088
  0.30000     1.07715
  0.35000     1.10947
  0.40000     1.14830
  0.45000     1.19423
  0.50000     1.24797
```

CHAPTER SUMMARY

Key Terms

#include directive	function
abstraction	function declaration
argument	function definition
body of a function	function heading
call to a function	function prototype
compilation	function stub
documentation file	header file
driver program	implementation file
encapsulation	information hiding
executable program	interface file
executable file	library

linking	received value
local variable	`return` statement
object file	return value
object program	separate compilation
parameter	specification
private	subprogram
public	`void` keyword

NOTES

- A function provides a way to encapsulate code so that it can be reused.

- Like variables and constants, a function must be declared before it can be used. The declaration of a function is also called a *prototype* and has the form

    ```
    returnType functionName(parameter_declaration_list);
    ```

- The general form of a function definition is

    ```
    function heading
    function body
    ```

    where the heading has the general form

    ```
    returnType functionName(parameter_declaration_list)
    ```

    and the body is a sequence of statements enclosed in curly braces ({ and }):

    ```
    {
      statements
    }
    ```

- When a function is called, the arguments are associated with the parameters from left to right—the first argument with the first parameter, the second argument with the next parameter, and so on, until the matching is complete. There should be the same number of arguments as parameters, and each argument's type must be compatible with the type of the corresponding parameter.

- Execution transfers from a function back to the caller when a `return` statement is encountered or the end of the function is reached.

- No `return` statement is required for functions whose return type is `void`, but it may be used in the form `return;` to return before the end of the function is reached.

- Like programs, functions should be tested to ensure correctness.

- A function's documentation should include its *specification*—descriptions of its parameters and their types, values input to the function, what it returns, values it outputs, preconditions, and postconditions.

- Local variables, constants, and parameters exist only while a function is executing and thus can be accessed only within the function. This means that other functions may reuse the name of a local item for some other purpose without causing a conflict.

- A library is a collection of files:

  - A *header file* that contains declarations and prototypes of a library's items that is also called the *interface file* because it acts as an interface between the library and client programs. It is inserted into a program (or another library) using a directive of the form #include "Library.h".

  - An *implementation file* that contains definitions of items not defined in the header file. It must be separately compiled and then linked to a client program. Items defined in it that are declared in the header file can be accessed in any program that #includes the header file and links to this implementation file. Items defined here but not declared in the header file cannot be accessed outside of this implementation file, even by a program that #includes the header file.

  - A *documentation file* that contains documentation for the library's items.

Style and Design Tips

- *Functions should be documented in the same way as programs.* The documentation should include the following:

  - A statement of what it does

  - Its specification, which consists of those of the following that apply:
    - what it receives (i.e., its parameters);
    - what is input to the function;
    - preconditions—restrictions or limitations on the parameters' values in order for the function to work properly;
    - what it returns;
    - what it outputs;
    - postconditions—effects produced by the function.

In this text, to avoid cluttering the header and implementation files, we place the documentation for functions stored in a library in a separate documentation file.

- *Functions are separate program components, and the white space in a program should reflect this.* In this text, we

- Insert appropriate documentation before each function defined in the main function's file, to separate it from other program components;

- Indent the declarations and statements within each function.

- *All guidelines for programming style apply to functions.*

- *Once a problem has been analyzed to identify the problem's objects and the operations needed to solve it, an algorithm should be constructed that specifies the order in which the operations are applied to the objects.*

- *Operations that are not predefined (or are nontrivial) in C++ should be encoded as functions, separate from the main function.*

- *A function that encodes an operation should be designed in the same manner as the main function.*

- *A function that returns no values should have its return type declared as* `void`.

- *A function that receives no values should have no parameters within the parentheses of the function heading.*

- *If a function is sufficiently general that it might someday prove useful in solving a different problem, a library should be constructed to store that function, rather than declaring and defining it in the program's source file.*

- *A library's files should be documented in much the same way as programs. For each library, provide a special documentation file that describes clearly, precisely, and completely the contents of the library and how to use items in it, any special algorithms it implements, and other useful information such as the author, a modification history, and so on.* This documentation file should be kept in the same place as the other files of a library, so that users can refer to that file in order to understand and use the objects and functions stored in the library.

- *When the header file of a less commonly used library is inserted in a program (using* `#include`*), a comment should be used to explain its purpose,* for example, what is being used from it in the program.

- *Libraries provide the following benefits:*

  - *A library extends the language,* because its objects can be made available to any program or to another library.

  - *The items in a library's interface can be used without being concerned about the details of their implementation.*

  - *Programs and libraries can be compiled separately.* Changing the implementation file of a library requires recompilation of only that implementation file.

- *Libraries provide another level of modularity in software design;* related functions and other objects can be grouped together in independent libraries.

Warnings

1. *When a function is called, the list of arguments is matched against the list of parameters from left to right, with the leftmost argument associated with the leftmost parameter, the next argument associated with the next parameter, and so on. The number of arguments* must be the same as the number of parameters (for exceptions see the description of default arguments on the text's website) *and the type of each argument must be compatible with the type of the corresponding parameter.*

2. *Identifiers defined within a function (e.g., parameters, local variables, and local constants) are defined only during the execution of that function; they are undefined both before and after its execution.* Any attempt to use such identifiers outside the function (without redeclaring them) is an error.

3. *If a function changes the value of a parameter, the value of the corresponding argument is not altered.* A parameter is a completely separate variable into which the argument value is copied. Any change to the parameter changes the copy, not the corresponding argument.[10]

4. *A function must be declared before it is called.*

5. *If a function needs things from a library, the header file of that library must be inserted (using #include) before the definition of that function.*

6. *The implementation file of a library should always insert the header file of that library (using #include) so that the compiler can verify that each function's prototype is consistent with its definition.* Failure to follow this rule is a common source of linking errors.

7. *A function that is defined in the implementation file of a library but not declared in that library's header file cannot be called outside the library* (but it can be called inside the library).

TEST YOURSELF

Section 6.3

1. In addition to input and output objects, what two other kinds of objects are usually included in the description of a function's behavior?

2. In the function heading double sum(int a, char b), a and b are called _____.

3. For a function whose heading is double sum(int a, char b), the type of the value returned by the function is _____.

4. The keyword _____ is used to indicate the return type of a function that returns no value.

---

[10] Except as noted in Footnote 5.

5. A function stub is a function definition in which the function's body contains _____.

Questions 6–10 deal with the following function definition:

```
int what(int n)
{
   return (n * (n + 1)) / 2;
}
```

6. If the statement number1 = what(number2); appears in the main function, number2 is called a(n) _____ in this function call.

7. If the statement int number = what(3); appears in the main function, the value assigned to number will be _____.

8. (True or false) The value assigned to number by the statement int number = what(2+3); in the main function will be 15.

9. (True or false) The value assigned to number by the statement int number = what(1, 5); in the main function will be 3.

10. Write a prototype for function what().

11. Write a function definition that calculates values of $x^2 + \sqrt{x}$.

12. Write a function definition that calculates the integer average of two integers.

13. Write a function definition that displays three integers on three lines separated by two blank lines.

Section 6.4

1. (True or false) Libraries were first provided in the language C++.

2. What are the three types of files included in a library?

3. What are the main benefits of using libraries?

4. A library's _____ file contains the declarations and prototypes of the items in the library.

5. A library's _____ file contains the definitions of the functions in the library.

6. A library's _____ file is sometimes called its interface file.

7. The items in a library's header file contain _____ information, whereas the items in its implementation file contain _____ information (public or private).

8. If a program contains #include _____, the compiler will search for the file lib in a special system include directory.

9. If a program contains #include _____, the compiler will search for the file lib in the directory that contains the program being compiled.

10. What two steps are required to translate a program that uses libraries?

11. (True or false) A library's header file is usually compiled separately from a program that uses the library.

12. (True or false) If a library's header file is modified then all programs that use the library must be recompiled.

13. _____ makes it possible to use a library without knowing all the details of how it is implemented.

EXERCISES

Section 6.3

For Exercises 1–13, you are to write functions to compute and return various quantities. To test these functions, you should write driver programs as instructed in the Programming Problems.

1. Find the circumference of a circle with a given radius. ($C = 2\pi r$).

2. Find the area of a circle with a given radius. ($A = \pi r^2$).

3. Find the perimeter of a rectangle, given the lengths of the sides. ($P = 2l + 2w$).

4. Find the area of a square, given the lengths of the sides. ($A = s^2$).

5. Find the perimeter of a triangle, given the lengths of the three sides. ($P = s_1 + s_2 + s_3$).

6. Find the area of a triangle, given the lengths of the three sides. (The area of a triangle can be found by using *Hero's formula*, $area = \sqrt{s(s-a)(s-b)(s-c)}$ where a, b, and c are the lengths of the sides and s is one half of the perimeter.)

7. Find the sum $1 + 2 \ldots + n$ for a given positive integer n.

8. Find the sum $m + m + 1 + \ldots + n$ for two given integers m and n with $m \le n$.

9. U.S. dollars are typically converted to another country's currency by multiplying the U.S. dollars by an *exchange rate*, which varies over time. For example, if on a given day, the U.S.-to-Canada exchange rate is 1.22, then $10.00 in U.S. currency can be exchanged for $12.20 in Canadian currency. Write a function US_to_Canadian() that, given a dollar amount in U.S. currency and the exchange rate, returns the equivalent number of dollars in Canadian currency.

10. Proceed as in Exercise 9, but write a function Canadian_to_US() that, given a dollar amount in Canadian currency and the exchange rate, returns the equivalent number of dollars in U.S. currency.

11. The number of bacteria in a culture can be estimated by $N \cdot e^{kt}$, where N is the initial population, k is a rate constant, and t is time. Write a function to calculate the number of bacteria present for given initial population, rate, and time.

12. The wind chill index, described in the text, was developed in 1941, with the latest revision of the formula for it published in 2001. It is a measure of discomfort due to the combined cold and wind, and is based on the rate of heat loss due to various combinations of temperature and wind. The *heat index*, developed in 1979, is a measure of discomfort due to the combination of heat and high humidity, and is based on studies of evaporative skin cooling for combinations of temperature and humidity. It is computed using the following formula:

$$\text{heat index} = -42.379 + 2.04901523 \times t + 10.14333127 \times r$$

$$- 0.22475541 \times t \times r - (6.83783\text{E} - 3) \times t^2$$

$$- (5.48171\text{E} - 2) \times r^2 + (1.22874\text{E} - 3) \times t^2 \times r$$

$$+ (8.5282\text{E} - 4) \times t \times r^2 - (1.99\text{E} - 6) \times t^2 \times r^2$$

where t is the temperature in degrees Fahrenheit and r is the relative humidity. Write a function to compute the heat index for given temperature and relative humidity.

13. Write a function that for a given distance, returns a cost, according to the following table:

Distance	Cost
0 through 100	$5.00
More than 100 but not more than 500	$8.00
More than 500 but less than 1000	$10.00
1000 or more	$12.00

14. A quadratic equation of the form $ax^2 + bx + c = 0$ has real roots if the discriminant $b^2 - 4ac$ is non-negative. Write a function that receives the coefficients a, b, and c of a quadratic equation, and returns true if the equation has real roots and false otherwise.

15. A *prime number* is an integer $n > 1$ whose only positive divisors are 1 and n itself. Write a boolean-valued function that determines whether an integer is a prime number.

16. The *greatest common divisor* (GCD) of two integers a and b, not both of which are zero, is the largest positive integer d that divides both a and b. The *Euclidean Algorithm* for finding this greatest common divisor of a and b, GCD(a, b), is as follows: If $b = 0$, GCD(a, b) is a. Otherwise divide a by b to obtain quotient q and remainder r, so that $a = bq + r$. Then GCD(a, b) = GCD(b, r). Replace a by b and b by r and repeat this procedure. Because the remainders are decreasing, a remainder of 0 will eventually result. The last nonzero remainder is then GCD(a, b). For example,

$$\begin{aligned}
1260 &= 198 \cdot 6 + 72 \\
198 &= 72 \cdot 1 + 54 \\
72 &= 54 \cdot 1 + 18 \\
54 &= 18 \cdot 3 + 0
\end{aligned}
\qquad
\begin{aligned}
\text{GCD}(1260, 198) &= \text{GCD}(198, 72) \\
&= \text{GCD}(72, 54) \\
&= \text{GCD}(54, 18) \\
&= 18
\end{aligned}$$

Note: If either *a* or *b* is negative, replace it with its absolute value.

Write a function to calculate the GCD of two integers.

17. Write a function that, given a positive integer *n*, displays the squares, cubes, and square roots of the first *n* positive integers.

18. Write a function that, given real numbers *a*, *b*, and *deltaX*, displays a list of points (*x*, *y*) on the graph of $y = x^3 - 3x + 1$ for *x* ranging from *a* to *b* in increments of *deltaX*.

19. A certain city classifies a pollution index less than 35 as "pleasant," 35 through 60 as "unpleasant," and above 60 as "hazardous." Write a function that displays the appropriate classification for a pollution index.

20. A wind chill of 10°F or above is not considered dangerous or unpleasant; a wind chill of –10°F or higher but less than 10°F is considered unpleasant; if it is –30°F or above but less than –10°F, frostbite is possible; if it is –70°F or higher but below –30°F, frostbite is likely and outdoor activity becomes dangerous; if the wind chill is less than –70°F, exposed flesh will usually freeze within half a minute. Write a function that displays the appropriate weather condition for a wind chill index.

PROGRAMMING PROBLEMS

Section 6.3

For Problems 1–20, write a driver program to test the function.

1. Exercise 1: circumference of a circle

2. Exercise 2: area of a circle

3. Exercise 3: perimeter of a rectangle

4. Exercise 4: area of a rectangle

5. Exercise 5: perimeter of a triangle

6. Exercise 6: area of a triangle

7. Exercise 7: sum of first *n* positive integers

8. Exercise 8: sum of consecutive integers

9. Exercise 9: convert U.S. to Canadian currency

10. Exercise 10: convert Canadian to U.S. currency

11. Exercise 11: number of bacteria in a culture

12. Exercise 12: heat index

13. Exercise 13: cost for a given distance

14. Exercise 14: quadratic equation checker. Test it with the following values for a, b, and c:

    1, 25, 6;   1, 22, 1;   1, 0, 4;   1, 1, 1;   2, 1, 2

15. Exercise 15: check if a number is a prime

16. Exercise 16: find the greatest common divisor of two integers

17. Exercise 17: display squares, cubes, square roots of consecutive positive integers

18. Exercise 18: display list of points on a graph

19. Exercise 19: display classification of a pollution index. Test it with the following data: 20, 45, 75, 35, 60.

20. Exercise 20: display weather condition for a wind chill index. Test it with the following data: –80, 10, 0, –70, –10, –5, 10, –20, –40.

21. Write a modification of the quadratic-equation function of Exercise 14 that returns 0 if the quadratic equation has no real roots (discriminant is negative), 1 if it has a repeated real root (discriminant is 0), and 2 if it has two distinct real roots (discriminant is positive). Add this quadratic-checker function to a driver program and test it using the values in Exercise 14 for a, b, and c.

Section 6.4

1. Construct a library `Exchange` that contains the monetary-conversion functions from Exercises 9 and 10 in Section 6.3. Write a driver program to test your library.

2. Construct a library `Time` that contains time-conversion functions that convert seconds to minutes, minutes to hours, hours to days. Write a driver program to test your library.

3. Construct a library `Geometry` that contains the functions from Exercises 1–6 in Section 6.3. Write a driver program to test your library.

4. Write a program to read one of the codes C for circle, R for rectangle, or T for triangle, followed by the radius of the circle, the sides of the rectangle, or the sides of the triangle, respectively. Using the functions in the library `Geometry` of Problem 3, the program should then calculate and display with appropriate labels the perimeter and the area of that geometric figure.

5. Construct a library `Cylinder` containing functions to compute the total surface area, lateral surface area, and volume of a right-circular cylinder. For a cylinder of radius r and height h, these can be calculated using:

$$\text{Total Surface Area} = 2\pi r(r + h)$$
$$\text{Lateral Surface Area} = 2\pi rh$$
$$\text{Volume} = \pi r^2 h$$

Write a driver program to test your library.

6. Construct a library `Metric` that contains functions to convert English-system measurements into their metric-system counterparts. Your library should provide the following conversions:

   - Inches to centimeters (1 in. = 2.54 cm)

   - Feet to centimeters (1 ft. = 12 in.)

   - Feet to meters (1 meter = 100 cm)

   - Yards to meters

   Write a driver program to test your library.

7. Extend the metric library in Problem 6 to include (at least) the following conversions and extend your driver program to test them:

   - Feet to decimeters (1 meter = 10 decimeters)

   - Inches to millimeters (1 centimeter = 10 millimeters)

   - Miles to kilometers (1 mile = 1,760 yards, 1 kilometer = 100 meters)

Section 6.5

Root Finding

1. The steady state of a certain circuit with a coil wound around an iron core is obtained by solving the equation $f(\Phi) = 0$ for the flux Φ, where $f(\Phi) = 20 - 2.5\Phi - 0.015\Phi^3$. It is easy to check that the function f changes sign in the interval [6, 7]. Use the bisection method to find a solution to the equation in this interval.

2. The state of an imperfect gas is given by van der Waal's equation

$$\left(p + \frac{\alpha}{v^2} \right)(v - \beta) = RT$$

where p = pressure (atm), v = molar volume (l/mole), T = absolute temperature (°K), and R = gas constant (0.0820541 atm/mole °K). For carbon dioxide, α = 3.592 and β = 0.04627. Assume that p = 0.9 atm and T = 300°K. Use the bisection method to solve the following equivalent cubic equation for v: $pv^3 - (\beta p + RT)v^2 + \alpha v - \alpha\beta = 0$.

3. The Cawker City Construction Company can purchase a new piece of equipment for $4,440 or by paying $141.19 per month for the next 36 months. You are to determine what annual interest rate is being charged in the monthly payment plan. The equation that governs this calculation is the *annuity formula*

$$A = P \cdot \left(\frac{(1+R)^N - 1}{R(1+R)^N} \right)$$

where A is the amount borrowed, P is the monthly payment, R is the monthly interest rate (annual rate/12), and N is the number of payments. Use the bisection method to solve this equation for R.

4. In level flight, the total drag on the Cawker City Construction Company jet is equal to the sum of parasite drag (D_P) and the drag due to lift (D_L), which are given by

$$D_P = \frac{\sigma f\, V^2}{391}$$

and

$$D_L = \frac{1245}{\sigma e} \left(\frac{W}{b}\right)^2 \frac{1}{V^2}$$

where V is velocity (mph), W is weight (15,000 lb), b is the span (40 ft), e is the wing efficiency rating (0.800), f = parasite drag area (4 ft²), and σ = (air density at altitude)/(air density at sea level) = 0.533 at 20,000 ft (for standard atmosphere). Use the bisection method to find the constant velocity V needed to fly at minimum drag (level flight), which occurs when $D_P = D_L$.

5. The following figure shows a mass M attached to a slender steel rod of mass m:

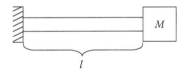

The frequency equation for the free undamped longitudinal vibration is

$$\beta \tan \beta = \frac{m}{M}$$

where

$$\beta = \frac{\omega l}{c}$$

Here,

$$c = \sqrt{\frac{E}{\rho}},$$

l = length of the rod (115 in), E = Young's modulus (= 3 × 10⁷ psi), ρ = mass per unit volume (= 7.2 × 10⁻⁴ lb × sec²/in⁴). Use the bisection method to find the smallest root of the frequency equation if m/M = 0.40.

6. Flexible cables have many applications in engineering, such as suspension bridges and transmission lines. Cables used as transmission lines carry their own uniformly distributed weight and assume the shape of a *catenary* shown in the following figure.

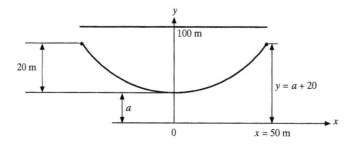

These curves have equations of the form

$$y = a \cosh\left(\frac{x}{a}\right)$$

Assume that the cable has a span of 100 m and maximum deflection of 20 m, and that the weight of the cable per unit length is $w = 50$ N/m. The minimum and maximum tensions occur in the middle (when $y = a$) and at the ends (when $y = a + 20$) and can be computed as $T_{min} = w \times a$ and $T_{max} = w \times (a + 20)$. Find the extreme tension values by using the bisection method to solve the equation

$$a + 20 = a \cosh\left(\frac{50}{a}\right)$$

to find the value of a and then substituting this value into the equations for T_{min} and T_{max}.

7. The cross section of a trough with length L is a semicircle with radius $r = 1$ m. Assume that the trough is filled with water to within a distance h from the top. The volume V of the water is given by

$$V(h) = L\left(\frac{1}{2}\pi r^2 - r^2 \arcsin\left(\frac{h}{r}\right) - h\sqrt{r^2 - h^2}\right)$$

where the three terms represent the area of the semicircle and areas $2A_1$ and $2A_2$.

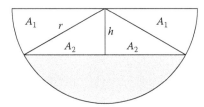

Assume that $L = 10$ m and $V = 10$ m³. Find the depth of the water (which is $r - h = 1 - h$). Note that $V(0) = \pi L/2 \approx 15.7$, that $V(r) = 0$, and that V decreases as h increases; thus, there is a unique solution for h. Use the bisection method to find this solution.

8–14. The *Newton-Raphson method* is another commonly used method for finding an approximate solution of an equation $f(x) = 0$. It consists of taking an initial approximation x_1 to the root and constructing the tangent line to the graph of $f()$ at point $P_1(x_1, f(x_1))$. The point x_2 at which this tangent line crosses the x-axis is the second approximation to the root. Another tangent line may be constructed at point $P_2(x_2, f(x_2))$, and the point x_3 at which this tangent line crosses the x-axis is the third approximation. For many functions, the sequence of approximations $x_1, x_2, x_3, \ldots$ converges to the root, provided that the first approximation is sufficiently close. The following diagram illustrates the Newton-Raphson method:

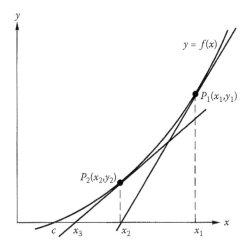

If x_n is an approximation to the zero of $f()$, then the formula for obtaining the next approximation x_{n+1} is

$$x_{n+1} = x_n - \frac{f(x_n)}{f'(x_n)}$$

where $f'()$ is the derivative of $f()$. Proceed as in Problems 1–7, but use the Newton-Raphson method in place of the bisection method.

Numerical Integration

For Problems 15–21, use the trapezoidal method to find the approximate values of the integrals.

15. The current i passing through a capacitor is given by

$$i(t) = 10\sin^2\left(\frac{t}{\pi}\right)$$

amps, where t denotes the time in seconds. Assume that the capacitance C is 5 F (farads). The voltage across the capacitor is given by

$$v(T) = \frac{1}{C}\int_0^T i(t)\ dt$$

volts. Find the value of $v(T)$ for $T = 1, 2, 3, 4,$ and 5 seconds.

16. The circumference C of an ellipse with major axis $2a$ and minor axis $2b$ is given by

$$C = 4a\int_0^{\pi/2}\sqrt{1-\left(\frac{a^2-b^2}{a^2}\right)\sin^2\Phi}\ d\Phi$$

Assume that a room has the shape of an ellipse with $a = 20$ m, $b = 10$ m, and height $h = 5$ m. Find the total area $A = hC$ of the wall.

17. The fraction f of certain fission neutrons having energies above a certain threshold energy E^* can be determined by the formula

$$f = 1-0.484\int_0^{E^*} e^{-E}\sin(h\sqrt{2E})\ dE$$

Find the value of f for $E^* = 0.5, 1.0, 1.5, 2.0, 2.5,$ and 3.0.

18. A particle with mass $m = 10$ kg is moving through a fluid and is subjected to a viscous resistance $R(v) = -v^{3/2}$, where v is its velocity. The relation between time t, velocity v, and resistance R is given by

$$t = \int_{v_0}^{v(t)} \frac{m}{R(v)}dv$$

seconds, where v_0 is the initial velocity. Assuming that $v_0 = 15$ m/sec, find the time T required for the particle to slow down to $v(T) = 7.5$ m/sec.

19. Suppose that a spring has been compressed a distance x_c:

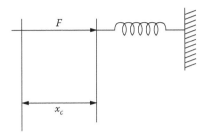

If $F(x)$ is the external force, then the absorbed energy can be expressed as

$$E = \int_0^{x_c} F(x)\,dx$$

Assume that

$$F(x) = \frac{1}{2}e^{x^2}\sin^2(3x^2)$$

newtons and $x_c = 1$ cm. Compute the absorbed energy.

20. The work done (in joules) by a force (in newtons) that is applied at angle θ as it moves an object from $x = a$ to $x = b$ on the x-axis (with meters as units) is given by

$$W = \int_a^b F(x)\cos(\theta(x))\ dx$$

where $F(x)$ is the force applied at point x at angle $\theta(x)$ (radians). Compute the work done for $a = 20.0$ m, $b = 50.0$ m, and the following pairs of values for $F(x)$ and $\theta(x)$:

F(x)	θ(x)
0.0	0.60
4.3	0.87
7.8	1.02
9.9	0.99
12.5	1.20
16.3	0.98
18.4	0.86
21.7	0.43
25.4	0.23
22.3	0.14
20.9	0.15
18.7	0.08

21. Write a program that uses the trapezoidal method to find the cross-sectional area of the hill in the road-construction problem where the height of the hill has been measured in increments of 100 feet and tabulated as follows:

Distance	Height
0	0
100	6
200	10
300	13
400	17
500	22
600	25
700	20
800	13
900	5
1000	0

Then use this value to find the volume of dirt to be removed.

22.–28. Another method of numerical integration that generally produces better approximations than the trapezoidal method is based on the use of parabolas and is known as *Simpson's rule*. In this method, the interval [a, b] is divided into an even number n of subintervals, each of length Δx, and the sum

$$\frac{\Delta x}{3}(y_0 + 4y_1 + 2y_2 + 4y_3 + 2y_4 + \cdots + 2y_{n-2} + 4y_{n-1} + y_n)$$

is used to find the area under the graph of f over the interval [a, b]. Write a program to implement Simpson's method and use it to solve Problems 15–21.

Differential Equations

For Problems 29–33, write a program that uses Euler's method with the specified step size to obtain approximate solutions.

29. The linear-lag behavior of the components of control systems usually is modeled by the differential equation

$$y' + \frac{1}{\tau}y = \frac{Au(t)}{\tau}$$

where

$y =$ time-dependent output of the component

$A =$ gain factor

$u =$ time-dependent input of the component

$\tau =$ time constant.

Select $\tau = 1$, $A = 8$, and assume that $y(0) = 0$ and that the input is given by $u(t) = e^{t/2} \sin\ t$. Find an approximate solution to this initial-value problem for $y(t)$ in the interval $[0, 1]$.

30. Consider a spherical water tank with radius R drained through a circular orifice with radius r at the bottom of the tank:

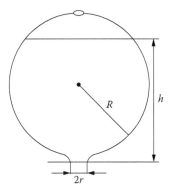

Because there is an air hole at the top of the tank, an atmospheric pressure can be found in the empty portion of the tank. In order to determine the time when the tank should be drained from any level to any other level, the water height h as a function of time should be determined. For any height, the volume of the tank is known to be

$$V = \frac{1}{3}\pi h^2 (3R - h)$$

If the area of the orifice is A and the velocity of the water flowing through the orifice is V, then

$$\frac{dV}{dt} = -\pi r^2 \sqrt{2gh}$$

where $g = 115.8$ ft/min^2 is the gravitation constant. Differentiation of the equation for V gives

$$\frac{dV}{dt} = (2\pi h R - \pi h^2)\frac{dh}{dt},$$

so

$$\frac{dh}{dt} = \frac{-r^2 \sqrt{2gh}}{2hR - h^2}$$

Assume that $R = 15$ ft and $r = 0.2$ ft and that the initial condition is $h(0) = 28$ ft. Solve this initial-value problem for h, using a step size of $\Delta t = 1$ min and continuing the calculations until the water height becomes less than 0.2 ft.

31. A new gas well is estimated to contain 9000 tons of natural gas at a pressure of $p_0 = 900$ psia. The distribution system is connected to the well at the discharge pressure $p^* = 400$ psia. The discharge Q in tons per day through the outlet pipe from the well is approximated by the equation $Q = \alpha(p^2 - p^{*2})^\beta$, where $\alpha = 1.115 \times 10^{-4}$, $\beta = 0.8$, and p = initial pressure of gas in the well in psia. It is also assumed that the gas pressure is directly proportional to the discharge Q, which is modeled by the differential equation

$$\frac{dp}{dt} = -kQ$$

with

$$k = \frac{900 \text{ psia}}{9000 \text{ tons}} = 0.1 \frac{\text{psia}}{\text{tons}}$$

Combining the two equations leads to the initial-value problem

$$\frac{dp}{dt} = -k\alpha(p^2 - p^{*2})^\beta$$

$$p(0) = p_0$$

Solve this initial-value problem using a step size of $\Delta t = 0.01$ and performing 10 steps.

32. The following figure shows a circuit consisting of a coil wound around an iron core, a resistance, a switch, and a voltage source:

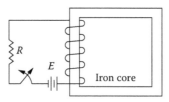

The magnetization curve is given by the equation

$$Ni = \frac{1}{2}\Phi + \frac{3}{1000}\Phi^3$$

where N = number of turns of coil, i = current (amperes), and Φ = flux in core (kilolines). Kirchoff's law gives the differential equation

$$E = Ri + L\frac{di}{dt} = Ri + 10^{-5}\frac{d\Phi}{dt}$$

where L = self-inductance (henrys), R = resistance (ohms), and t = time (sec). Assume $N = 100$ and $R = 500$. Then

$$\frac{d\Phi}{dt} = E - 2.5\Phi - 0.015\Phi^3, \quad \Phi(0) = 0$$

where t is now measured in milliseconds. For $E = 20$ V, solve this initial-value problem in the interval $[0, 2]$ using a step size $\Delta t = 0.01$.

33. The number of individuals in a population is measured each year. Let $P(t)$ denote the population at year t. Let α denote the birthrate, and assume that the death rate β is proportional to the size of the population, that is, $\beta = \gamma P(t)$, where γ is a constant. Hence, the growth rate of the population is given by the logistic equation $P'(t) = \alpha P(t) - \gamma [P(t)]^2$. Assume that $P(0) = 50,000$, $\alpha = 3 \times 10^{-2}$, and $\gamma = 1.5 \times 10^{-7}$. Solve this initial-value problem with step size $\Delta t = 0.01$ to find the population after 5 years.

34.–38. The *Runge-Kutta* method is a popular and one of the most accurate numerical methods for solving a first-order differential equation $y' = f(x, y)$ with initial condition $y(x_0) = y_0$:

1. Select an x-increment Δx.

2. The approximate solution y_{n+1} at $x_{n+1} = x_0 + (n + 1)\Delta x$ for $n = 0, 1, 2, \ldots$ is given by

$$y_{n+1} = y_n + \frac{1}{6}\left(K_1 + 2K_2 + 2K_3 + K_4\right)$$

where

$$K_1 = \Delta x \cdot f(x_n, y_n)$$

$$K_2 = \Delta x \cdot f(x_n + \frac{\Delta x}{2}, y_n + \frac{K_1}{2})$$

$$K_3 = \Delta x \cdot f(x_n + \frac{\Delta x}{2}, y_n + \frac{K_2}{2})$$

$$K_4 = \Delta x \cdot f(x_n + \Delta x, y_n + K_3)$$

Write a program to implement the Runge-Kutta method and use it to solve Problems 29–33.

Using Classes

CONTENTS

There are three social classes in America: upper middle class, middle class, and lower middle class.

JUDITH MARTIN ("MISS MANNERS")

A program is a spell cast over a computer, turning input into error messages.

ANONYMOUS

For every complex problem there is an answer that is clear, simple, and wrong.

H. L. MENCKEN

Carry out a random act of kindness, with no expectation of reward, safe in the knowledge that one day someone might do the same for you.

DIANA, PRINCESS OF WALES

THE WORD *CLASS* IS often used to describe a group or category of objects that have a set of attributes in common. For example, the high school sports teams in one state are described as *class A*, *class AA*, *class AAA*, or *class AAAA*, depending on the number of students in the school. The usual classifications of secondary and college students are *freshman, sophomore, junior,* and *senior. Metals, ceramics, polymers, semiconductors,* and *composites* are classes of engineering materials, based on their chemical composition. The U.S. Navy uses *Skipjack class, Thresher class,* and *Sturgeon class* to characterize different kinds of submarines. Economists describe families as *lower class, middle class,* or *upper class,* based on their annual income. In these examples, *class* is used as a synonym for *type* because it provides a name for a group of related objects.

In Chapter 3, we studied the **fundamental types** provided in C++ to model basic objects such as numbers and characters: `int`, `unsigned`, `double`, `char`, `bool`, and variations of these such as `short`, `long`, and `float`. These types, however, cannot model more complex objects such as names of chemical elements, units of measurement, names and addresses of persons, and so on. None of the fundamental types, `char` type in particular, can adequately model these names because they consist of several characters and `char` values can only be single characters.

For situations like this, C++ provides libraries of standard **classes** to represent more complex objects. Also, programmers can build their own classes, thereby *extending* the C++ language. In this chapter, we will examine four standard C++ classes:

- the `istream` class to model input from a keyboard;

- the `ostream` class for modeling output to a screen;

- the `string` class for modeling sequences of characters;

- the `complex` class template to model complex numbers.

In the last section we will use

- a programmer-defined `RandomInt` class to generate random integers in a simulation program.

In Chapter 14 we will see how we can create our own classes.

7.1 INTRODUCTORY EXAMPLE: INTERNET ADDRESSES

IP (Internet Protocol) addresses are used to uniquely identify computers in the Internet; for example, `titan.ksc.nasa.gov` is the symbolic IP address of a site at the NASA Kennedy Space Center. Such an address is made up of four fields that represent specific parts of the Internet,

`host.subdomain.subdomain.rootdomain`

and is translated into a unique numeric IP address. This address is a 32-bit value, but it is usually represented in a dotted-decimal notation by separating the 32 bits into four 8-bit fields, expressing each field as a decimal integer, and separating the fields with a period; for example, at the time of this writing, 163.205.10.1 was the IP address for the above site at the NASA Kennedy Space Center. Each of the four parts is an 8-bit integer, so the entire address can be stored in 32 bits.

7.1.1 Problem

A *gateway* is a device used to interconnect two different computer networks. Suppose that a gateway connects an engineering firm to the Internet and that the company's network administrator needs to monitor connections through this gateway. Each time a connection is made, the IP address of the employee's computer is transmitted to the network administrator, who analyzes the components that make up the address and records them.

7.1.2 Object-Centered Design

7.1.2.1 Behavior

The program will display on the screen a prompt for an IP address. The user will enter a string of characters from the keyboard, which the program will read. The program will then compute and output its four network/host information blocks on the screen along with descriptive text, or display an error message if the user did not enter a valid address.

7.1.2.2 Objects

From the statement of the desired behavior, we can identify the following objects:

| | Software Objects | | |
Problem Objects	Type	Kind	Name
screen	output device	variable	cout
prompt	text string	constant	none
IP address	text string	variable	*address*
keyboard	input device	variable	cin
network/host information blocks	text string	variable	*block1, block2, block3, block4*
descriptive text error message	text string	constant	none

7.1.2.3 Operations

Again, from the statement of the desired behavior, we can identify the following operations:

i. Display a string on the screen.

ii. Read a string from the keyboard.

iii. Find each network/host information block and extract it from the string.

iv. Display each information block on the screen.

7.1.2.4 Algorithm

We now organize these operations into the following algorithm, which also shows how operation (iii) requires some refinement into simpler operations that are provided in C++.

1. Output a prompt for an IP address to cout.

2. Input *address* from cin.

3. Fill *block1* with appropriate substring of *address* or halt if it can't be found.

4. Fill *block2* with appropriate substring of *address* or halt if it can't be found.

5. Fill *block3* with appropriate substring of *address* or halt if it can't be found.

6. Fill *block4* with appropriate substring of *address* or halt if it can't be found.

7. Output *block1*, *block2*, *block3*, and *block4* to cout.

7.1.2.5 Coding and Testing

The program in Example 7.1 implements this algorithm. It uses the C++ string type, which is a *class* for processing strings of characters.

Example 7.1 Processing IP Addresses

```cpp
/* This program finds the four network/host information blocks in an
   IP address.

   Input:   an IP address
   Output:  four network/host information blocks, or an error message
   -------------------------------------------------------------------*/

#include <iostream>        // cin, cout
#include <string>          // string class
#include <cassert>         // assert()
using namespace std;

int main()
{
  string address, block1, block2, block3, block4;

  cout << "Enter an IP address: ";
  cin >> address;

  // Search address to find first period -- start at position 0
  int dot1 = address.find(".", 0);
  assert(dot1 != string::npos);
  // Period found; set block1 = substring preceding it
  block1 = address.substr(0, dot1);

  // Search address for second period -- start at position dot1 + 1
  int dot2 = address.find(".", dot1 + 1);
```

```
assert(dot2 != string::npos);
// 2nd period found; set block2 = substring between 1st and 2nd
block2 = address.substr(dot1 + 1, dot2 - dot1 - 1);

// Search address for third period -- start at position dot2 + 1
int dot3 = address.find(".", dot2 + 1);
assert(dot3 != string::npos);
// 3rd period found; set block3 = substring between 2nd and 3rd
block3 = address.substr(dot2 + 1, dot3 - dot2 - 1);

// Check for more periods
assert(address.find(".", dot3 + 1) == string::npos);
// No more periods; set block4 = rest of address after 3rd one
block4 = address.substr(dot3 + 1, address.size() - dot3 - 1);

cout << "The network/host blocks are:\n"
     << block1 << endl << block2 << endl
     << block3 << endl << block4 << endl;
}
```

SAMPLE RUNS:

```
Enter an IP address: titan.ksc.nasa.gov
The network/host blocks are:
titan
ksc
nasa
gov

Enter an IP address: 163.205.10.1
The network/host blocks are:
163
205
10
1
```

7.2 INTRODUCTION TO CLASSES

In Chapter 2, we saw how items in a computation can be modeled by variables and constants whose types are the fundamental C++ types—int, double, bool, char, and their variations. For example, we used

```
double celsius;
double fahrenheit = 1.8 * celsius + 32;
```

to declare double variables celsius and fahrenheit that modeled temperature readings. In Chapter 6, we saw how we can use functions to effectively create new operations, such as the temperature-conversion function celsiusToFahrenheit():

```
double celsiusToFahrenheit(double tempCels);
```

These C++ features have, thus far, been adequate for programming tasks in which the items and operations being modeled are sufficiently simple.

The problem, however, is that most real-world objects are not this simple. For example, suppose that the network administrator in the preceding section must also create login accounts for the employees at the engineering firm. Each account will contain information about that employee such as the following:

1. Name

2. User ID

3. Password

4. Employee ID number

5. Limit on minutes of access

6. Current minutes used

.

.

.

Suppose, for simplicity, that we use only the six items of information listed. One way to proceed would be to declare a separate variable for each one:

```
string name,
       userId,
       password,
int    idNumber,
       accessLimit;
double timeUsed;
```

Although this approach will work, it can be quite clumsy. For example, suppose we need a function to display the computer information about an employee:

```
void printUserInfo(string name, string userId, string password,
                   int idNumber, int accessLimit, double timeUsed)

{
   cout << name << endl
        << "User Id: " << userId << endl
        << "Password: " << password<< endl
        << "Id Number: " << idNumber<< endl
        << "Access Limit: " << accessLimit<< endl
        << "Time Used: " << timeUsed<< endl
}
```

Given such a function and the variables defined earlier, we can call it with a statement like

```
printUserInfo(userName, userId, userPassword,
              empNumber, resourceLimit, resourcesUsed);
```

A function to input information about an employee's computer usage would be similar and would be called in much the same way:

```
readUserInfo(userName, userId, userPassword,
             empNumber, resourceLimit, resourcesUsed);
```

The major problem with this approach is that every function we define for processing information about computer users will need a separate parameter for each item of information and calls to these functions must pass each argument to the correct parameter. With only six parameters, as in our example, this may not seem problematic, but for more complex objects, more information must be passed and this approach becomes more cumbersome and prone to error. Just imagine having to enter the source code for a program that had 50 function calls, each having 100 or more arguments!

The basic difficulty with this approach is that there is *one* kind of object (a computer user) that we want to model, and yet we must pass *more than one* piece of information to the operations. To alleviate this problem, C++ provides the **struct** construct (from C) and its extension to a **class**. When programmers create a struct or class, they create a *new type*, with space for the characteristics of objects of that type. For example, the type string used in Example 7.1 is actually a class that was created by some programmer.

In the discussion that follows, we will only describe the major features of structs and classes, leaving most of the details to a later chapter. However, by the end of this section, you should know what they are and how to use them.

7.2.1 Data Encapsulation

Structs and classes provide a way to encapsulate an object's characteristics—*both attributes and behavior*—within a single "wrapper." For example, to create a new type named ComputerUser to model computer users described earlier, we could use either of the following:

```
struct ComputerUser                 class ComputerUser
{                                   {
  string name,                        public:
         userId,                        string name,
         password;                             userId,
  int    idNumber,                             password;
         accessLimit;                  int     idnumber,
  double timeUsed;                             accessLimit;
  // ...                               double timeUsed;
};                                     // ...
                                     };
```

The identifiers name, userId, password, idNumber, accessLimit, and timeUsed are called the **data members** (or **instance variables** or **attribute variables**) of the struct and class.

The name ComputerUser is treated by C++ as *the name of a new type*, and can therefore be used to declare objects:

```
ComputerUser design17,
             management8;
```

These declarations create distinct ComputerUser objects, each of which can store its own characteristics:

design17			management8	
?	name		?	name
?	userId		?	userId
?	password		?	password
?	idNumber		?	idNumber
?	accessLimit		?	accessLimit
?	timeUsed		?	timeUsed

This encapsulation is important, because a single object like design17 now stores all of its own data and solves our earlier problem, because it allows us to pass a complicated object to a function by using just one argument (and declaring just one parameter); for example,

```
void printUserInfo(ComputerUser aUser)
{
  cout << aUser.name << endl
       << "User Id: " << aUser.userId << endl
       << "Password: " << aUser.password << endl
       << "Id Number: " << aUser.idNumber << endl
       << "Access Limit: " << aUser.accessLimit << endl
       << "Time Used: " << aUser.timeUsed << endl;
  // ... and possibly more ...
}
```

Note that the members of a ComputerUser object are accessed using **dot notation**. The expression

```
aUser.name
```

will access the name member of the ComputerUser object associated with parameter aUser. If we call this function with

```
printUserInfo(design17);
```

the values of the data members of design17 will be displayed, producing something like

```
Joe Blow
User Id: joey00
Password: windbag
Id Number: 1238
Access Limit: 200
Time Used: 68.5
```

If we call the same function with a different argument,

```
printUserInfo(management8);
```

the values of `management8`'s data members will be displayed; for example,

```
Betty Boss
User Id: daBoss
Password: denied
Id Number: 1
Access Limit: 10000
Time Used: 0
```

For a complicated object, structs and classes thus provide a convenient way to package the data items needed to describe that object in one container. This is what is meant by **data encapsulation**.

7.2.2 Encapsulating Operations

The major addition when C++ was developed from C was the addition of classes that can have not only data members but also **member functions** (also called **methods**).[1] The data members are placed in a private section of the class, and member functions are placed in a public section. This means that the member functions can be accessed from outside using dot notation, but the data members cannot unless a member function is provided to do this.

For example, to declare member functions `read()` to perform input and `display()` for output in our `ComputerUser` class, we could modify it by adding prototypes for these functions as follows:

```
class ComputerUser
{
  public:
    void read();
    void display();
    // ... function prototypes for other operations
  private:
    string  name,
            userId,
            password;
```

---

[1] In C++, structs may also have member functions, but classes are preferred for encapsulating both data and operations.

```
    int     idNumber,
            accessLimit;
    double  timeUsed;
    // ... declarations of other data members
};
```

The corresponding function definitions, which are usually outside of the class declaration, are modified to indicate where their prototypes can be found. For example, `display()` could be defined by

```
void ComputerUser::display(ComputerUser aUser)
{
  cout << name << endl
       << "User Id: " << userID << endl
       << "Password: " << password << endl
       << "Id Number: " << idNumber << endl
       << "Access Limit: " << accessLimit << endl
       << "Time Used: " << timeUsed << endl;
  // ... and possibly more ...
}
```

Attaching `ComputerUser::` to the function name indicates that this is a member function of the `ComputerUser` class and thus can access the (private) data members.

The data members in our `ComputerUser` class are designated as `private` to prevent unauthorized operations on the data such as assigning a negative value to a data member that should always be nonnegative. All access to the data is controlled by the member functions, of which there would usually be a fairly large collection. However, because this is only an introduction to classes, we will keep our example simple and use only the two member functions `read()` and `display()`.

When class objects are declared, as in

```
ComputerUser design17,
             management8;
```

each has its own data and member functions. When we wish to access one of the member functions, we *send a message* to an object via the dot operator. That object responds by executing the corresponding member function, which can access the data members. For example, to display the information stored in `design17`, we can send it a `display()` message:

```
design17.display();
```

and to display the information stored in `management8`, we would send it a `display()` message:

```
management8.display();
```

We can use the statement

```
management8.read();
```

to send management8 a read() message, which fills the data members in management8 with values extracted from cin; for example,

management8

Betty Boss	name
daBoss	userId
denied	password
1	idNumber
10000	accessLimit
0	timeUsed

Similarly, the statement

```
design17.read();
```

sends design17 a read() message to fill its data members with input values such as

design17

Joe Blow	name
joey00	userId
windbag	password
1238	idNumber
200	accessLimit
68.5	timeUsed

A member function is thus what gets executed when we send an object a message.

Classes also make it possible to redefine C++ operators. For example, the string class redefines + so that when it is given two strings, it concatenates those strings. For example, the declaration

```
string name = "Popeye" + " the " + "Sailor";
```

initializes the object name with the string value "Popeye the Sailor". We can also redefine operators for classes we build. For example, we could redefine the input and output operators for our ComputerUser class, which would enable us to write statements like

```
cout << "Enter computer usage for " << management8 << "\n";
cin >> management8;
```

Redefining a function with a new definition is called **overloading** that function, a topic discussed in Chapter 10.

Once a class has been built, it is usually stored in a library, with its declaration stored in the library's header file and the definitions of its nontrivial member functions in an implementation file. To use the class, a program must include the library's header file.

This brief introduction to classes should indicate their importance in C++. Classes are the single biggest difference between C++ and its parent language C. In fact, prior to 1983, the C++ language was called "C with Classes." An important part of learning to program in C++ is learning how to use the standard classes that are part of the language, to avoid reinventing the wheel. In the sections that follow, we will examine four of these classes: `istream`, `ostream`, `string`, and `complex`.

In summary, C++ provides the class mechanism for building types to represent complicated objects. This mechanism allows us to create a single structure that encapsulates (i) *data members* defining the characteristics of the object and (ii) *member functions* defining the operations on the object. Class members that are designated public within the class can be accessed using dot notation.

7.3 THE `istream` AND `ostream` CLASSES

When C++ was first developed in the early 1980s by Bjarne Stroustrup, it used the I/O system of its parent language C. Later in that decade, Jerry Schwarz, one of the early users of C++ at AT&T, used classes to develop a better I/O system. After revisions and improvements over the years, this set of classes became today's `iostream` library. In this section we will consider two of these classes: the `istream` class for input and the `ostream` class for output.[2]

7.3.1 The `istream` Class

One problem in designing a general purpose model for input is how information entered from a keyboard or other input device gets from there to the computer. The keyboard may actually be a part of the computer as in a laptop or a smart phone, or the information may have to travel from it to an entirely different computer across a network.

Because programmers should not have to concern themselves with the messy details of how data actually gets from the keyboard to the program, an *abstraction* for input is preferred that hides these details—a model that can capture the basic idea of *every* input system, regardless of the low-level details.

7.3.1.1 Streams

Dennis Ritchie, creator of the C programming language, and Ken Thompson were key members of a team of AT&T employees at Bell Labs who developed the Unix operating system. The I/O system they developed for Unix was based on such a model called a **stream**. They envisioned streams of characters flowing from an input device to a program and from a program to an output device, like a stream of water flowing from one place to another. Schwarz used this idea to develop the `iostream` library, which contains the `istream` and `ostream` classes for carrying out I/O in C++.

---

[2] `istream` and `ostream` are really specializations of class templates `basic_istream` and `basic_ostream` to type `char`, which are derived from the `basic_ios` class that handles the low-level details of formatting, buffers, and so on. Using `wchar_t` instead of `char` produces wide-character streams `wistream` and `wostream`.

The istream class represents a flow of characters from an arbitrary input device to an executing program.

an istream

Schwarz then defined an istream object named cin in the iostream library, so that any C++ program that included the library's header file would automatically have access to the input stream flowing from the keyboard to the program.

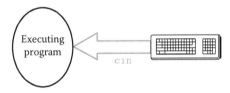

7.3.1.2 The >> Operator

Schwarz also defined a binary input operator >> whose first operand is an istream object such as cin and whose second operand is a variable:

```
istream_object >> variable
```

When this input expression is executed in a program, the >> operator tries to extract a sequence of characters corresponding to a value of the type of *variable* from *istream_object*. If there are no characters, it *blocks execution* until characters are entered.

To illustrate, suppose execution of a program has just begun so that cin is empty. When the following statements are executed,

```
int minutes;
cin >> minutes;
```

and the >> operator attempts to read an integer value for minutes but finds that cin is empty, execution cannot proceed. If the user now enters 40 from the keyboard, the characters 4 and 0 enter into cin:

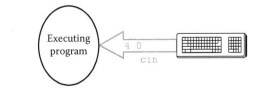

Because cin is no longer empty, execution resumes and >> extracts the characters 4 and 0, converts them to the integer 40, and stores this value into its right operand minutes. Because >> extracts values from an istream, it is often referred to as the *extraction operator*.

7.3.1.3 Status Indicators

So far we have assumed that the user enters appropriate values. Suppose, however, that in the preceding example, 40 was mistyped as f0, causing cin to contain the characters f and 0:

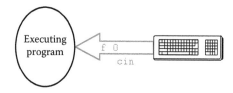

What happens now when >> tries to extract an integer and encounters the letter f?

The condition of an istream is known as its **state** and there are three possible states:

- *good* state: all is well with the stream.

- *bad* state: something has gone wrong with the stream.

- *fail* state: the last operation on the stream did not succeed.

The istream class maintains a **flag** for each of these states, each of which is essentially a boolean variable with the good flag initialized to true and the bad and fail flags initialized to false. In our scenario, if the >> operator encounters the letter f while trying to read an integer, it will set the stream's *good* flag to false, and its *bad* and *fail* flags to true.

For each of these flags, the istream class provides a boolean member function having the same name as its flag, that reports on the value of that flag.[3]

Message	Returns True If and Only If
cin.good()	all is well in the istream
cin.bad()	something is wrong with the istream
cin.fail()	the last operation could not be completed

We can use good() to check that an input step has succeeded:

```
assert(cin.good());
```

or, alternatively, we might use one of the other status indicators; for example,

```
assert(!cin.fail());
```

---

[3] There is a fourth *end-of-file* state that occurs when the last input operation encounters an end-of-file mark before finding any data. This state can be checked by sending an istream the eof() message. It is described in more detail in Chapter 11.

Combined with the `assert()` mechanism, these status indicators provide an easy way to guard against data-entry errors.

7.3.1.4 The `clear( )` and `ignore( )` Member Functions

Once the *good* flag of an `istream` has been set to false, all subsequent input operations on that stream are blocked until its state is cleared. This is accomplished by using the `clear()` member function,

```
cin.clear();
```

which resets the good flag to true and the other flags to false.

Resetting the status flags, however, does not remove the offending input from the `istream`. For this, we can use `ignore()`,

```
cin.ignore();
```

which, as its name implies, skips the next character in `cin`. More generally, `ignore()` can be called with arguments to skip more than one character:

```
cin.ignore(skip, stop_char);
```

where *skip* is an integer expression and *stop_char* is a character. This statement will skip the next *skip* characters in `cin`, unless *stop_char* is encountered. The default value of *skip* is 1, and the default value of *stop_char* is the end-of-file mark. For example, the message

```
cin.ignore(100, ' ');
```

will skip all characters up to the next space (assuming it is within the next 100 characters), and

```
cin.ignore(100, '\n');
```

will skip all characters remaining on a given line of input (assuming that the end of the line is within 100 characters). This `clear()` and `ignore()` combination is useful for detecting bad input characters and removing them from the input stream, as in the following program segment:

```
cin >> minutes;
if (cin.fail())          // e.g., invalid input character
{
 cin.ignore(100, '\n');
 cin.clear();
}
```

7.3.1.5 White Space

One of the nice features of the >> operator is that, by default, it skips leading white space—spaces, tabs, and returns. To illustrate, suppose that in response to the statements

```
double length, weight;
cout << "Enter the length (inches) and weight (pounds): ";
cin >> length >> weight;
```

the user enters 70.5 and 180. Then cin contains the characters pictured, where ⌴ represents a space and ↵ represents the return character:

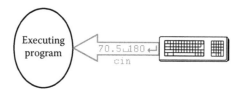

When the input statement

```
cin >> height >> weight;
```

is executed, the first >> reads 70.5 and stores it in length, leaving the space (⌴) unread:

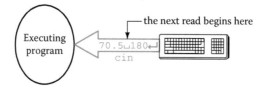

The second >> begins reading where the previous one left off and skips the space:

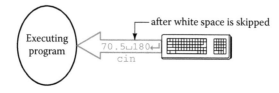

It then reads 180 and stores it in weight:

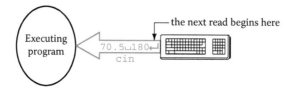

The return character remains in the stream, but because it is a white-space character, >> will skip it in any subsequent input expression.

White space will also be skipped even when reading a single character, as in

```
char ch;
cin >> ch;
```

The input statement will skip all leading white-space characters and read the first non-white-space character from `cin` into `ch`.

Suppose, however, that we want to read all characters, including white-space characters, as, for example, spaces within names of persons, chemical compounds, pieces of equipment, and so on. One way to do this is with an **input manipulator**, which is a keyword in an input statement that changes some property of the `istream`. For example, if we use the **noskipws manipulator**,

```
cin >> ... >> noskipws ...
```

then in all subsequent input, white-space characters will not be skipped. The **skipws manipulator** can be used to reactivate white-space skipping.

Alternatively, the `istream` class has a member function named **get()** that reads a single character without skipping white space. For example, if `ch` is of type `char`, the statement

```
cin.get(ch);
```

will read the next character from `cin` into `ch`, even if it is a white-space character.

7.3.2 The ostream Class

In the same way that the `iostream` library provides an `istream` class for processing streams of characters from an input device to a program, it provides an `ostream` class to represent the "flow" of characters from an executing program to an output device, thereby hiding the low-level details from the user:

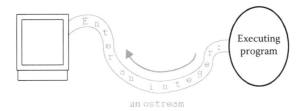

an ostream

Any C++ program that includes the `iostream` library will automatically have access to three output streams from the program to whatever device the user is using for output:

`cout`: the standard buffered output stream for displaying normal output

cerr and clog: standard output streams intended for displaying error or diagnostic messages (clog is buffered; cerr is not)

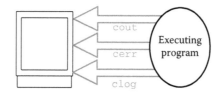

The assert() function typically writes its diagnostic messages to cerr.

7.3.2.1 The << Operator

The << operator is applied to an ostream object and an expression,

```
ostream_object << expression
```

The expression will be evaluated, its value converted into the corresponding sequence of characters, and those characters inserted into the output stream. Thus, if the constant PI is defined by

```
const double PI = 3.1416;
```

and we output it with

```
cout << PI;
```

the << operator will convert the real number 3.1416 into the sequence of characters 3, ., 1, 4, 1, and 6, and insert them into cout:

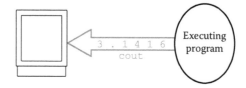

Because cout and clog, unlike cerr, are buffered, characters actually remain in them until they are *flushed* to empty their contents onto the screen. This can sometimes be confusing because we know that an output statement has been executed but its output doesn't appear on the screen. One common way to flush an ostream is to use an identifier called an **output manipulator** that causes something to happen on the ostream immediately when it is encountered in an output statement. The most commonly used manipulator to flush an output stream is **endl**. It inserts a newline character ('\n') into the ostream and then flushes it, thus ending a line of output.[4]

---

[4] A less commonly used alternative is **flush**, which simply flushes the ostream without inserting anything, which means that subsequent output will appear on the same line.

Whenever >> is used to input a value from cin, cout *is automatically flushed.* Thus, when the statements

```
double radius;
cout << "Enter radius: ";
cin >> radius;
```

are executed, << inserts the prompt "Enter radius: " into cout:

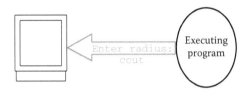

but it is the >> operator in the next statement that moves the prompt "Enter radius: " out of the ostream and onto the screen:

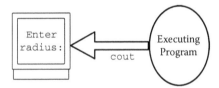

7.3.3 Format Control

Various features of the form in which a value is displayed or is entered can be specified by using **format manipulators**. For example, in the program in Example 4.2, we used the statement

```
cout << showpoint << fixed << setprecision(2)
     << "For Route #" << route << ":\n"
     << "\n\tMiles per gallon: " << setw(8) << milesPerGallon
     << "\n\tTotal cost:      $" << setw(8) << totalTripCost
     << "\n\tCost per mile:   $" << setw(8) << costPerMile
     << endl;
```

Here, fixed is a format manipulator that causes real values to be displayed in fixed-point notation instead of in scientific notation; showpoint ensures that the decimal point will be displayed; and setprecision(2) sets the number of decimal places of precision to 2. The first two manipulators are provided by the iostream library and setprecision() by the iomanip library.

As another example of how format manipulators affect a stream, consider the following code:

```
int i = 26;
```

```
cout << showbase
     << oct << i << endl
     << dec << i << endl
     << hex << i << endl;
```

The output

```
032
26
0x1A
```

will be produced by these statements because $32_8 = 26_{10} = 1A_{16}$.

The following table lists the format manipulators provided by iostream. They and other manipulators are described in more detail in Chapter 11.

Format Manipulator	Description
fixed	Use fixed-point notation for real values
scientific	Use scientific notation for real values
showpoint	Show decimal point and trailing zeros for whole real numbers
noshowpoint	Hide decimal point and trailing zeros for whole real numbers
dec	Use base-10 notation for integer input or output
hex	Use base-16 (hexadecimal) notation for integer input or output
oct	Use base-8 (octal) notation for integer input or output
showbase	Display integer values indicating their base (e.g., 0x for hex)
noshowbase	Display integer values without indicating their base
showpos	Display + sign for positive values
noshowpos	Do not display + sign for positive values
boolalpha	Read or display bool values as true or false
noboolalpha	Do not read or display bool values as true or false
uppercase	In scientific, use E; in hexadecimal, use symbols A–F
nouppercase	In scientific, use e; in hexadecimal, use symbols a–f
flush	Write contents of stream to screen (or file)
endl	Insert newline character into output stream and flush the stream
left	Left-justify displayed values, pad with fill character on right
right	Right-justify displayed values, pad with fill character on left
internal	Pad with fill character between sign or base and value
skipws	Skip white space on input
noskipws	Do not skip white space on input

Some format manipulators require arguments, and to use them, the **iomanip library** must be included. For example, in the program in Example 4.2 we wanted to specify the number of decimal places, or *precision*, to use in displaying real values so we used the **setprecision(n)** manipulator provided in iomanip with $n = 2$ to set the number of decimal places of precision in subsequent reals to 2.

Another kind of format has to do with the space or *field* in which a data value is displayed. The default width of an output field is 0 and automatically grows to accommodate the value being displayed, which is usually what the programmer wants. In situations where a wider field is needed, we can use the **setw(n)** manipulator, which sets the width of the *next field* to n. However, the field width for subsequent outputs is automatically set back to zero, and so setw() *must be used for each value for which we want to override the default output format.*

To illustrate the use of **setw(),** suppose we are creating a report of expenditures for items used in constructing a piece of equipment. If we use output statements like the following

```
cout << showpoint << fixed
     << "Electrical:     $" << elecCost<< endl
     << "Sheet metal:    $" << metalCost << endl
     << "Wages:          $" << wages << endl
     << "Miscellaneous: $" << miscCost << endl;;
```

the output produced might be

```
Electrical:     $ 1013.
Sheet metal:    $  549.2297
Wages:          $  728.5
Miscellaneous:  $   25.
```

This would look much better if the amounts were rounded to two decimal points and were right-aligned. One way to do this is to use the setprecision(), setw(), and right manipulators as in the following:

```
cout << showpoint << fixed
     << setprecision(2) << right
     << "Electrical:     $" << setw(7) << elecCost<< endl
     << "Sheet metal:    $" << setw(7) << metalCost << endl
     << "Wages:          $" << setw(7) << wages << endl
     << "Miscellaneous: $" << setw(7) << miscCost << endl;
```

which would generate output with the following format:

```
Electrical:     $ 1013.00
Sheet metal:    $  549.23
Wages:          $  728.50
Miscellaneous:  $   25.00
```

The character used to fill the empty part of a field is called the **fill character**. As the preceding example illustrates, it is a *space* by default, but we can change it to another character ch by using the **setfill(ch)** manipulator. For example, if we change the preceding output statement to

```
cout << showpoint << fixed
     << setprecision(2) << right
     << setfill('*')
     << "Electrical:     $" << setw(7) << elecCost<< endl
     << "Sheet metal:    $" << setw(7) << metalCost << endl
     << "Wages:          $" << setw(7) << wages << endl
     << "Miscellaneous: $" << setw(7) << miscCost << endl;
```

the resulting output will be

```
Electrical:     $ 1013.00
Sheet metal:    $ *549.23
Wages:          $ *728.50
Miscellaneous: $ **25.00
```

7.4 THE string CLASS

The introductory example in Section 7.1 used C++'s string class to process IP addresses. In this section we will describe some of the other features of this class.

7.4.1 Declaring string Objects

As we saw in Section 7.1, the string class is declared in the string library, which must be included in each program that uses it:

```
#include <string>        // string class
```

Because string is the name of a class and the name of a class is a type, string objects can be declared and initialized in the same manner as other objects we have studied. For example, the declaration

```
string address, block1, block2, block3, block4;
```

in Example 7.1 creates five string objects named address, block1, block2, block3, and block4 and initializes each of them to an **empty string**, which contains no characters. A literal for the empty string can be written as two consecutive double quotes ("").

A declaration of a string object can also initialize that object by providing a string expression for its initial value. For example, the declarations

```
string address = "aaa.bbb.ccc.ddd",
       block1 = block2 = block3 = block4 = "";
```

create the string object address, initializing it to contain 15 characters,

address	a	a	a	.	b	b	b	.	c	c	c	.	d	d	d

and the `string` objects `block1`, `block2`, `block3`, and `block4`, initializing each of them to be empty strings.

These `string` objects are *variables*, and they can be assigned new values. For example, the first sample run in Example 7.1 would replace the 15-character string used to initialize `address` with an 18-character string.

address	t	i	t	a	n	.	k	s	c	.	n	a	s	a	.	g	o	v

Constant `string` objects are defined by preceding a normal string declaration with the keyword `const`; for example,

```
const string UNITS = "centimeters";
```

As with other constants, the `string` value of such objects cannot be changed later in the program.

7.4.2 String Operations

7.4.2.1 Input, Output, and Assignment

The `string` class has extended `<<`, `>>`, and `=` to perform these basic I/O and assignment operations on `string` objects. For example, we have used the output operator `<<` to display string literals as input prompts in most of our programs, as in Example 7.1:

```
cout << "Enter an IP address: ";
```

In that same program we used `<<` to display the values of `string` variables:

```
cout << "The network/host blocks are:\n"
     << block1 << endl << block2 << endl
     << block3 << endl << block4 << endl;
```

The assignment operator `=` was used to assign values to the four `string` variables `block1`, `block2`, `block3`, and `block4`.

Also, when the statements

```
cout << "Enter an IP address: ";
cin >> address;
```

were executed and the user entered `titan.ksc.nasa.gov`,

```
Enter an IP : titan.ksc.nasa.gov
```

the string `titan.ksc.nasa.gov` was assigned to `address`. If, however, the user entered some spaces between some of the blocks,

```
Enter an IP address: titan. ksc. nasa.gov
```

the string `titan.` would be assigned to `address`, but what happens to the remainder of the input?

Input of strings proceeds in the same way as we have described for other types: `>>` skips leading white space and stops at the next white-space character, which we might describe as getting the next "word" from the input stream. Thus, in the preceding example, the unread characters `ksc. nasa.gov` (including the space before `ksc`) remain in the `istream` and will be read in subsequent input statements if there are any. For the program in Example 7.1, however, execution will terminate with an error message something like the following:

```
Enter an IP address: titan. ksc. nasa.gov
Assertion failed: (dot2 != string::npos), function main, file fig7-1.cpp,
line 28.
```

We will explain the reason for this later in this section.

The `string` library does provide a function **`getline()`** that can be used to read an entire line of text into a `string` variable including white-space characters. For example, if we rewrite the earlier code segment as

```
cout << "Enter an IP address: ";
getline(cin, address);
```

and enter `titan. ksc. nasa.gov` as input, the function `getline()` will read the entire line as input and assign it to the variable `address` sent to it. The output produced by the program will be

```
titan
 ksc
 nasa
gov
```

Note the leading spaces before `ksc` and `nasa`. They were read as part of the string that was entered and were assigned to `block2` and `block3`. In general and more precisely, `getline()` extracts characters from an `istream` (e.g., `cin`) and transfers them into a `string` variable until a newline character is encountered, which is removed from the `istream` but is not stored in the `string` variable.

In summary, to read an entire line of input, the `getline()` function should be used. To skip leading white space and then read all characters up to the next white-space character, the input operator `>>` should be used. But one must be careful when using both `>>` and `getline()` because a newline character that terminates input of characters via `>>` will be left in the `istream`, and if the next input statement uses `getline()`, it will not read any characters because encountering this newline character will cause it to terminate immediately.

7.4.2.2 The Subscript and `size()` *Operations*

Associated with each character in a string is an integer called an **index** that is 0 for the first character, 1 for the second, and so on. For example, for a string `units` declared by

```
string units = "Foot Lbs";
```

a sequential container in memory will be associated with `units` with the elements numbered 0, 1, 2, . . ., and the individual characters that make up the string, F, o, o, t, a space, L, b, and s, will be stored in these elements:

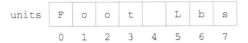

The **subscript operator []** can be used to access individual characters of a string. In our example, `units[0]` is a variable of type `char` whose value is `'F'`, `units[1]` is a variable of type `char` whose value is `'o'`, and so on; in general, an expression of the form *string_object[index]* accesses the character at the specified *index* of *string_object*. For example, the four assignment statements

```
units[0] = 'I';
units[1] = 'n';
units[2] = 'c';
units[3] = 'h';
```

would change the value of `units` to `"Inch Lbs"`. If we wish to output these first four characters, we can use a `for` loop that varies the index over the element positions:

```
for (int i = 0; i < 4; i++)
   cout << units[i] << endl;
```

The size of a string can be found by sending it the **`size()`** message, which returns the number of characters in that string. For example, `units.size()` will return the value 7. Note that *the index of the last character in a string is always* `size()` − 1 because indices start at zero. Thus, if we want to process the characters in reverse order, we must start at this index as in the following example:

```
for (int i = units.size() - 1; i >= 0; i--)
   cout << units[i] << end;;
```

7.4.2.3 Relational Operators

The relational operators `<`, `>`, `==`, `<=`, `>=`, and `!=` have all been defined for the `string` class and can be used to compare `string` objects. For each operation, the elements of the string operands are compared character by character until a mismatch occurs or the end of one (or both) of the strings is reached.

To illustrate, consider the declarations

```
string units1 = "Milligrams",
       units2 = "Millimeters";
```

The boolean expression units1 < units2 is true because the first element where these two strings differ is in position 5 and 'g' < 'm' is true. For the same reason, units1 <= units2 and units1 != units2 are true, whereas units1 > units2, units1 >= units2 and units1 == units2 are false.

If units2 were changed to

```
units2 = "MILLIMETERS";
```

however, units1 < units2 would be false because the first element where these two strings differ is in position 1 and 'i' < 'I' is false (because the numeric code for 'i' is 105 and that for 'I' is 73). This would also be the case if units2 were changed to

```
units2 = "Mill Output";
```

because the first element where units1 and units2 differ is in position 4 and 'i' < ' ' is false (because the numeric code for 'i' is 105 and that for a space is 32).

7.4.2.4 Other String Operations

The string class provides several other operations. For example, the program in Example 7.1 used find() to locate the periods in an address and substr() to extract substrings that were the blocks in the address. These and several other commonly used string operations are described in the following table. In this table, s, s1, and s2 are of type string; pos and n are nonnegative integers. Several variations of these operations as well as others are described in Appendix D.

SOME USEFUL STRING FEATURES

Output: <<	out << s outputs the characters of s to ostream out
Input: >>, getline()	in >> s inputs characters from istream in into s until a whitespace character is encountered getline(in, s) inputs an entire line from istream in into s
Assignment: =	s = s1 assigns the string expression s1 to s
Relational Operators: <, >, ==, <=, >=, !=	s1 < s2, s1 > s2, s1 == s2, s1 <= s2, s1 >= s2, and s1 != s2 compare strings
Subscript: []	s[i] is the character in position i of s
+	s1 + s2 is the string obtained by concatenating s1 with s2 by appending s2 to s1
string::npos	some integer that is either negative or greater than the maximum length of strings—used to signal unsuccessful string operations (e.g., see find())

`append()`	`s1.append(s2)` appends `s2` onto `s1` as in `s1 + s2`
`empty()`	`s.empty()` is true if `s` contains no characters and is false otherwise
`erase()`	`s.erase(pos, n)` erases `n` characters from `s`, starting at position `pos`; if `n` is too large or is omitted, characters are erased to the end of `s`
`find()`	`s.find(s1, pos)` returns the first position `pos` at which `s1` occurs in `s` or `string::npos` if `s1` is not found
`find_first_of()`	`s.find_first_of(s1, pos)` returns the first position $\geq$ `pos` of a character in `s1` that matches *any character* in `s` or `string::npos` if none is found
`insert()`	`s.insert(pos, s1) )` inserts the string `s1` into `s` at position `pos`
`length(), size()`	`s.length()` and `s.size()` return the number of characters in `s`
`replace()`	`s.replace(pos, n, s1)` replaces the substring of `s` of length `n`, starting at position `pos`, with `s1`; if `n` is too large, all the characters to the end of `s` are replaced
`substr()`	`s.substr(pos, n)` returns a copy of the substring consisting of `n` characters from `s`, starting at position `pos`; if `n` is too large or is omitted, characters are copied only to the end of `s`

7.5 THE C++ complex TYPE

Although the complex number system has many important applications, it is perhaps not as familiar as the real number system. Thus, in this section, we review the basic properties of and operations on complex numbers and describe how they are represented and used in C++ programs.

7.5.1 Representation of Complex Numbers

Because complex numbers have two parts, a *real part* and an *imaginary part*, they can be plotted in a coordinatized plane by taking the horizontal axis to be the real axis and the vertical axis to be the imaginary axis, so that the complex number $a + bi$ can be represented as the point $P(a, b)$ or, alternatively, as the vector $\overrightarrow{OP}$ from the origin to the point $P(a, b)$:

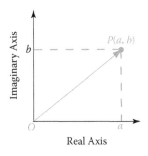

Real Axis

In C++, the type of a complex object is complex<T>, where T is one of the real types float, double, or long double. Each of the types **complex<float>**, **complex<double>**, and **complex<long double>** is a class defined in the **complex library** that must be included in any program that uses one of these types.[5] A **complex literal** has the form complex<T>(a, b), where a and b are of type T or can be promoted to that type and represent the real part and the imaginary part of the complex number. The following are some examples:

C++ Representation	Mathematical Representation
complex<double>(-6.283, 17.1)	$-6.283 + 17.1i$
complex<double>(1, 1)	$1 + i$
complex(<double>(0, 1)	i

7.5.2 Operations on Complex Numbers

The **sum**, **difference**, and **product** of two complex numbers $z = a + bi$ and $w = c + di$ are defined by:

$$z + w = (a + c) + (b + d)i$$

$$z - w = (a - c) + (b - d)i$$

$$z \cdot w = (ac - bd) + (ad + bc)i$$

$$z / w = \frac{ac + bd}{c^2 + d^2} + \frac{bc - ad}{c^2 + d^2}i \text{ (provided } c^2 + d^2 \neq 0)$$

The sum and difference correspond to the sum and difference of the vector representations of z and w.

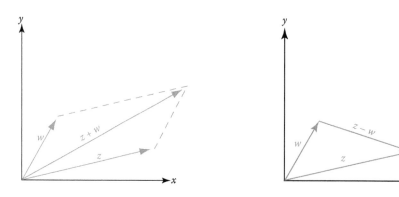

---

[5] The complex library actually contains a *class template* complex<T>, which is a pattern for a class for complex numbers. When T is replaced by one of float, double, or long double, a class for complex numbers whose real and imaginary parts have that type is created.

The product can be represented by a vector whose length is the product of the lengths of the vectors representing z and w and whose angle from the positive x-axis is the sum of the angles for z and w.

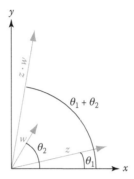

To see why this geometric representation of the product of two complex numbers is correct, it is helpful to consider the **polar representation** of complex numbers. To describe this representation, consider a vector $\overrightarrow{OP}$ from the origin to point P, and suppose that r is the length of $\overrightarrow{OP}$ and that θ is the angle from the positive x-axis to $\overrightarrow{OP}$, so that the polar coordinates of point P are (r, θ):

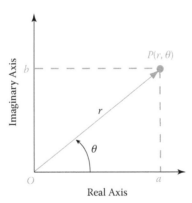

It is clear from this diagram that the relation between the rectangular coordinates (a, b) of point P and its polar coordinates (r, θ) is

$$a = r \cos \theta$$

$$b = r \sin \theta$$

It follows that the complex number represented by $\overrightarrow{OP}$ can be written in the polar form as

$$r \cos \theta + ir \sin \theta = r(\cos \theta + i \sin \theta)$$

A basic property of complex numbers is that

$$e^{i\theta} = \cos\theta + i\sin\theta$$

and thus an alternative form of the polar representation of a complex number is

$$re^{i\theta}$$

Now consider two complex numbers

$$z = r_1 e^{i\theta_1}$$

$$w = r_2 e^{i\theta_2}$$

The familiar properties of exponents then give

$$z \cdot w = (r_1 e^{i\theta_1})(r_2 e^{i\theta_2}) = r_1 r_2 e^{i(\theta_1 + \theta_2)}$$

which agrees with the geometric representation of the product of two complex numbers shown earlier. In the same manner, we can use polar representation and properties of exponents to show that the quotient of two complex numbers corresponds to a vector whose magnitude is the quotient of the magnitudes of the vectors representing the numbers and whose angle of inclination is the difference of the vectors' angles of inclination:

$$z / w = (r_1 e^{i\theta_1}) / (r_2 e^{i\theta_2}) = (r_1 / r_2) e^{i(\theta_1 - \theta_2)}$$

These basic arithmetic operations for complex numbers are defined in C++'s complex number classes and are denoted by the usual arithmetic operators +, -, *, and /; + and − may also be used as unary operators. Also defined are the assignment operator = along with assignment shortcuts +=, -=, *=, and /= and the logical operations == and !=.[6]

The input and output operators << and >> have also been overloaded for the three complex types. Complex values are output in the form (a, b), where a is the real part and b is the imaginary part. They can be input in the form a or (a) if the imaginary part is 0 and (a, b) otherwise.

In addition to these operations, C++ defines several functions for its complex classes. The following table describes some of the more useful ones. In the descriptions, z is a complex number $a + bi$.

COMPLEX FUNCTIONS

Real part of z	`real(z)` returns the real part of z
	`z.real(a)` assigns a to the real part of z
Imaginary part of z	`imag(z)` returns the imaginary part of z
	`z.imag(b)` assigns b to the imaginary part of z

---

[6] <, >, ≤, and ≥ are not defined for complex numbers.

Phase angle of z: θ	`arg(z)`
Norm of z: $a^2 + b^2$	`norm(z)`
Absolute value (or magnitude) of z:	`abs(z)`
$\|z\| = \sqrt{a^2 + b^2}$	
Conjugate of z: $\bar{z} = a - bi$	`conj(z)`
Power: z^r, where r is a real value	`pow(z, r)`
Square root	`sqrt(z)`
Trig Functions	`cos(z), sin(z), tan(z),`
Hyperbolic Functions	`cosh(z), sinh(z), tanh(z),`
Exponential Function: e^z	`exp(z)`
Logarithms—natural, base-10	`log(z), log10(z)`

7.5.3 Example: Solving Quadratic Equations

In Section 5.4 we considered the problem of solving quadratic equations, using the `if` statement to check if there were real roots and if not, display a message. The program in Example 7.2 is a variation of the program in Example 5.4 that finds the roots—possibly complex—of any quadratic equation.

Example 7.2 Quadratic Equation Solver—Complex Roots

```
/* This program solves quadratic equations using the quadratic formula.

   Input:  the three coefficients of a quadratic equation
   Output: the complex roots of the equation
 ------------------------------------------------------------------*/

#include <iostream>      // cout, cin, <<, >>
#include <complex>       // complex types
using namespace std;

int main()
{
  complex<double> a, b, c;
  cout << "Enter the coefficients of a quadratic equation: ";
  cin >> a >> b >> c;

  complex<double> discriminant = b*b - 4.0*a*c,
                  root1, root2;
  root1 = (-b + sqrt(discriminant)) / (2.0*a);
  root2 = (-b - sqrt(discriminant)) / (2.0*a);
  cout << "Roots are " << root1 << " and " << root2 << endl;
}
```

SAMPLE RUNS:

```
Enter the coefficients of a quadratic equation: 1 4 3
Roots are (-1,0) and (-3,0)
```

```
Enter the coefficients of a quadratic equation: 2 0 -8
Roots are (2,0) and (-2,0)

Enter the coefficients of a quadratic equation: 2 0 8
Roots are (0,2) and (-0,-2)

Enter the coefficients of a quadratic equation: 1 2 3
Roots are (-1,1.41421) and (-1,-1.41421)

Enter the coefficients of a quadratic equation: (1,2) (3,4), (5,6)
Roots are (-0.22822,0.63589) and (-1.97178,-0.23589)
```

7.5.4 Application: A-C Circuits

The following a-c circuit contains a capacitor, an inductor, and a resistor in series:

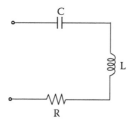

The impedance Z_R for a resistor is simply the resistance R (in ohms), but for inductors and capacitors, it is a function of the frequency. The impedance Z_L of an inductor is the complex value given by

$$Z_L = \omega L i$$

where ω is the frequency (in radians per second) of the a-c source, and L is the self-inductance (in henrys). For a capacitor, the impedance is

$$Z_C = \frac{-i}{\omega C}$$

where C is the capacitance (in farads). The total impedance Z is then given by

$$Z = Z_R + Z_L + Z_C$$

and the current I (in amperes) by

$$I = \frac{V}{Z}$$

The program in Example 7.3 inputs the resistance, inductance, capacitance, the frequency of the a-c source, and the voltage, and then uses the preceding equations to calculate the current and output it.

Example 7.3 Current in an A-C Circuit

```
/* This program computes the current in an a-c circuit containing a
   capacitor, an inductor, and a resistor in series.

   Input:  resistance, inductance, capacitance, frequency, voltage
   Output: current
   ------------------------------------------------------------------*/

#include <iostream>           // cin, cout, <<, >>
#include <complex>            // complex<double> type
#include <iomanip>            // format output
using namespace std;

int main()
{
  const complex<double> i = complex<double>(0, 1);

  double R,              // resistance (ohms)
         L,              // inductance (henrys)
         C,              // capacitance (farads)
         omega;          // frequency (radians per second)
  complex<double> V,     // voltage (volts)
                  Z,     // total impedance
                  I;     // current (amperes)

  cout << "\nEnter resistance (ohms), inductance (henrys),\n"
       << "and capacitance (farads): ";
  cin >> R >> L >> C;

  cout << "\nEnter frequency (radians/second): ";
  cin >> omega;

  cout << "\nEnter voltage as a complex number in the form (x, y): ";
  cin >> V;

  // Calculate resistance using complex arithmetic
  Z = R + omega * L * i - i / (omega * C);

  // Calculate and display current using complex arithmetic
  I = V / Z;
  cout << fixed << setprecision(2)
       << "\nCurrent = " << real(I) << " + " << imag(I) << "I"
       << "\nwith magnitude = " << abs(I) << endl;
}
```

SAMPLE RUN:

```
Enter resistance (ohms), inductance (henrys),
and capacitance (farads): 5000 .03 .02

Enter frequency (radians/second): 377

Enter voltage as a complex number in the form (x, y): (60000, 134)

Current = 12.00 + -0.00I
with magnitude = 12.00
```

7.6 SIMULATION WITH RANDOM NUMBERS: SHIELDING A NUCLEAR REACTOR

The term **simulation** refers to modeling a dynamic process and using this model to study the behavior of the process. The behavior of some *deterministic* processes can be modeled with an equation or a set of equations. For example, processes that involve exponential growth or decay are commonly modeled with an equation of the form $A(t) = A_0 e^{kt}$, where $A(t)$ is the amount of some substance A present at time t, A_0 is the initial amount of the substance, and k is a rate constant. In many problems, however, the process being studied can be modeled using *randomness*; for example, Brownian motion, the arrival of airplanes at an airport, the number of defective parts manufactured by a machine, and so on. Computer programs that simulate such processes use random number generators to introduce randomness into the values produced during execution.

7.6.1 Random Number Generators—The Randomint Class

A **random number generator** is a function that produces a number selected at *random* from some fixed range in such a way that a sequence of these numbers tends to be uniformly distributed over the given range. Although it is not possible to develop an algorithm that produces truly random numbers, there are some methods that produce sequences of *pseudorandom numbers* that are adequate for most purposes. Most of these algorithms have two properties:

1. Some initial value called a *seed* is required to begin the process of generating random numbers. Different seeds will produce different sequences of random numbers.

2. Each random number produced is used in the computation of the next random number.

Although C++ does not provide a random number generator, its parent language C does. This can be used to construct a class RandomInt for data objects whose values are pseudorandom integers. (See the text's website for more information about this class.)[7]

---

[7] This RandomInt class was developed by Professor Joel Adams of Calvin College who was a coauthor of the text *C++: An Introducton to Computing*, 3/e, Pearson Education Inc., 2003.

Two basic operations on a `RandomInt` object are provided that correspond directly to the properties of a (pseudo-)random number generator given above:

1. *Construction* (i.e., declaration), which initializes the object to a random number

2. *Generation*, which provides the object with a new random number

Additional operations, such as assignment (=) and output (<<) are also provided. The class is also designed so that any of the operations (numeric, relational, etc.) that can be applied to integers can be applied to class objects, although the result is an `int`, as opposed to a `RandomInt`.

7.6.1.1 Construction

A `RandomInt` object can be constructed by writing a declaration of the form:

```
RandomInt object_name(lower_bound, upper_bound);
```

Such a declaration constructs a `RandomInt` object named *object_name* whose value is a random number in the range *lower_bound* to *upper_bound*. If *lower_bound* and *upper_bound* are omitted, a random number is generated in the range 0 to some (large) upper limit defined in the `RandomInt` class.

7.6.1.2 Generation

The second operation is to generate a new random number. Given a `RandomInt` object constructed as just described, this can be done by sending it the `generate()` message:

```
object_name.generate();
```

When this statement is executed, the value of *object_name* is changed to another random integer from the range specified when *object_name* was constructed.[8] In addition to changing the value of *object_name*, `generate()` returns the randomly generated value.

7.6.2 Problem

When the enriched uranium fuel of a nuclear reactor is burned, high-energy neutrons are produced. Some of these are retained in the reactor core, but most of them escape. Because this radiation is dangerous, the reactor must be shielded. We wish to develop a program that simulates neutrons entering the shield and determines what percentage of them get through it.

In our model of the shielding, we will make the simplifying assumption that neutrons entering the shield follow random paths by moving forward, backward, left, or right with equal likelihood, in jumps of one unit. We will also assume that losses of energy occur only when there is a collision that causes a change of direction, and that after a certain number

---

[8] Optional arguments *lower_bound* and *upper_bound* can be used with `generate()` to alter the range of the random numbers to be generated.

of such collisions, the neutron's energy is dissipated and it dies within the shield, provided that it has not already passed back inside the reactor core or outside through the shield.

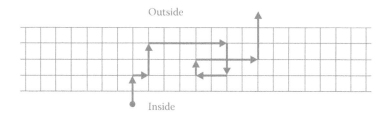

7.6.3 Solution[9]

The program in Example 7.4 inputs the thickness of the shield, the limit on the number of collisions before energy is dissipated, and the number of neutrons simulated. To simulate the path a neutron takes, it repeatedly generates random integers 1, 2, 3, or 4, corresponding to movement forward, backward, to the left, or to the right, respectively. It continues to do this until the net movement in the forward direction equals the shield thickness, indicating that the neutron escapes; or it becomes 0, indicating that it returns back inside the reactor; or the number of collisions reaches the specified limit, indicating that the neutron dies within the shield. This is repeated for each neutron simulated, and the percentage of neutrons that escape is calculated and displayed.

Example 7.4 Simulate Shielding of a Nuclear Reactor

```
/* This program simulates particles entering the shield described in
   the text and determines what percentage of them reaches the
   outside.

   Input:  thickness of the shield, limit on the number of direction
           changes, number of neutrons, current direction a neutron
           traveled
   Output: the percentage of neutrons reaching the outside
   ------------------------------------------------------------------*/

#include <iostream>           // cin, cout, <<, >>
#include "RandomInt.h"        // random integer generator
using namespace std;

int main()
{
   int thickness,
       collisionLimit,
       neutrons;
```

```
cout << "\nEnter the thickness of the shield, the limit on the \n"
     << "number of collisions, and the number of neutrons:\n";
cin >> thickness >> collisionLimit >> neutrons;

RandomInt direction(1,4);

int forward,
    collisions,
    oldDirection,
    escaped = 0;

for (int i = 1; i <= neutrons; i++)
{
  // Next neutron
   forward = oldDirection = collisions = 0;

    while (forward < thickness && forward >= 0 &&
           collisions < collisionLimit)
    {
      direction.generate();

      if (direction != oldDirection)
        collisions++;

      oldDirection = direction;

      if (direction == 1)
        forward++;
      else if (direction == 2)
        forward--;
    }
    if (forward >= thickness)
      escaped++;
}

cout << '\n' << 100 * double(escaped) / double(neutrons)
     << "% of the particles escaped.\n";
}
```

SAMPLE RUNS:

```
Enter the thickness of the shield, the limit on the
number of collisions, and the number of neutrons:
1 1 100

26% of the particles escaped

Enter the thickness of the shield, the limit on the
number of collisions, and the number of neutrons:
100 5 1000
```

```
0% of the particles escaped

Enter the thickness of the shield, the limit on the
number of collisions, and the number of neutrons:
4 5 100

3% of the particles escaped

Enter the thickness of the shield, the limit on the
number of collisions, and the number of neutrons:
8 10 500

0.2% of the particles escaped
```

In the four sample runs of the program, the first two are test cases. In the first test case, the neutron will move only once because each first move is interpreted as a collision and the limit on the number of collisions is 1. Because each of the four possible moves is equally likely, we would expect 25 percent of the neutrons to escape on the average, and the result produced by the program is consistent with this value. In the second test case, the shielding is 100 units thick and the limit on the number of collisions is small, so we expect almost none of the neutrons to escape through the shield.

7.6.4 Normal Distributions

Most random number generators generate random numbers having a **uniform distribution**, but they can also be used to generate random numbers having other distributions. The **normal distribution** is especially important because it models many physical processes. For example, the heights and weights of people, the lifetime of lightbulbs, the tensile strength of steel produced by a machine, and, in general, the variations in parts produced in almost any manufacturing process, have normal distributions. The normal distribution has the familiar bell-shaped curve,

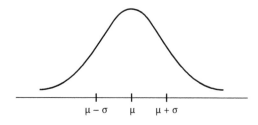

where μ is the mean of the distribution, σ is the standard deviation, and approximately two-thirds of the area under the curve lies between $\mu - \sigma$ and $\mu + \sigma$.

A normal distribution having $\mu = 0$ and $\sigma = 1$ is called a **standard normal distribution**, and random numbers having approximately this distribution can be generated quite easily from a uniform distribution with the following algorithm:

Algorithm for the Standard Normal Distribution

> 1. Set *sum* equal to 0.
>
> 2. Do the following 12 times:
>
>> a. Generate a random number *x* from a uniform distribution.
>>
>> b. Add *x* to *sum*.
>
> 3. Calculate $z = sum - 6$.

The numbers *z* generated by this algorithm have an approximate standard normal distribution. To generate random numbers *y* having a normal distribution with mean μ and standard deviation σ, we simply add the following step to the algorithm:

> 4. Calculate $y = \mu + \sigma \times z$.

Implementing this algorithm as a program is left as an exercise.

CHAPTER SUMMARY

Key Terms

<< operator	input manipulator
>> operator	insertion operator
Bjarne Stroustrup	`istream`
class	Jerry Schwarz
`complex`	member functions
data encapsulation	normal distribution
data members	`ostream`
dot notation	output manipulator
empty string	overloading
`endl` manipulator	polar representation
extraction operator	precision
fill character	random number generator
format manipulator	randomness
fundamental types	simulation
`getline()`	state
index	stream

string   uniform distribution

struct   white space

subscript operator ([])

NOTES

- Classes provide a way to encapsulate the characteristics of an object. This makes it possible for a single object to contain all of its own information, which in turn makes it possible to pass a complicated object to a function by using just one argument.

- Public members of a class object can be accessed using dot notation.

- Data members are designated private to prevent unauthorized operations from being performed on them.

- Member functions get executed when messages are sent to objects.

- The state of an istream (good, bad, fail) can be checked by sending a message (good(), bad(), fail()) to the istream object (e.g., cin). The clear() message can be used to reset states and the ignore() message to skip characters in the stream.

- The input operator >> skips leading white space (space, tab, or newline). The noskipws and skipws manipulators can be used to turn this feature off or on. The get() method provides a useful alternative for reading any character, whether white space or not.

- The endl manipulator inserts a newline character ('\n') into an ostream to end a line of output and then flushes it. The ostream cout is flushed automatically whenever the istream cin is used.

- The format of output is controlled by inserting format manipulators into output lists. Except for setw(), which affects only the next output value, these manipulators control the format of all subsequent output in the program unless changed by some other manipulator.

- When the input operator >> is applied to a string object, it extracts characters from an istream and transfers them into a string object until a white-space character is encountered.

- The getline() function stops extracting and transferring characters when a newline character is encountered. In can thus be used to read an entire line of input. To read characters only up to the next white-space character, the input operator >> should be used.

- The characters in a string object are indexed, beginning with 0. An expression of the form *string_object*[*index*] accesses the element at the specified *index* of *string_object*. The index of the last character is always *string_object*.size() – 1.

- The `complex` library provides data types `complex<T>`, where `T` is `float`, `double`, or `long double` and is the type of the real and imaginary parts of complex values; it also provides the customary complex operations and functions.

- Random numbers generators are used in simulations to model processes that involve randomness.

Style and Design Tips

- *If any of the objects needed to solve a problem have been identified but cannot be defined using the fundamental types, examine the predefined classes of C++ to see if any of them is appropriate for representing such objects.*

- *If any of the operations needed to solve a problem involve a class object, study the available documentation for that class to see if any of its methods perform that operation.* In the worst case, this may involve looking at the header file for that class and experimenting with the methods.

Warnings

1. *When a value is read for a* `char` *variable using* `>>`, *white-space characters are skipped, and the first nonwhite-space character is removed from the* `istream` *and assigned. For other types, characters are read and removed until white space or some other character that cannot belong to a value of that type is encountered and is left in the stream for the next input operation.* The characters that were read are then converted to the type of the variable and assigned.

2. *In the preceding warning, if no character in an* `istream` *is found that can belong to a value of the required type, the input operation is said to fail.* No characters will be read and removed and the value of the variable remains unchanged.

3. *The first character in a string object has index 0.*

4. *Be careful when using both* `>>` *and* `getline()` *for input.* A newline character that terminates input via `>>` will be left in the `istream` and a subsequent attempt to read a string using `getline()` will not read any characters because it terminates as soon as this newline is encountered. An infinite loop may result.

5. *The* `string` *subscript operation uses square brackets containing an index,*

   ```
   stringVariable[index]
   ```

   whereas the `string` *substring operation uses parentheses containing an index and a size:*

   ```
   string_object.substr(index, num_chars)
   ```

6. *Run-time bounds checking is performed on most* `string` *functions that use an index and causes an out-of-bounds exception if the index is negative or greater than the size of the string, which terminates program execution.*

7. *Run-time bounds checking is not performed on the subscript operation.* Special care must be taken when accessing the characters of a string with the subscript operator, to ensure that all accesses fall within the boundaries of the string.

8. *To fill a* string *object with a word from an* istream, *use the input operator* >>:

```
cin >> someString;
```

to fill a string *object with a line from an* istream, *use the* getline() *function:*

```
getline(cin, someString);
```

TEST YOURSELF

Section 7.2

1. Packaging data items that are needed to describe an object in one container is known as data _____.

2. Classes have two kinds of members: _____ members and _____ members.

3. Redefining a function with a new definition is called _____ that function.

4. Public members of a class can be accessed using _____ notation.

Questions 5–7 assume the following declarations:

```
class Date
{
  public:
    void display();
  private:
    string month;
    int day;
    int year;
};
Date launch;
```

5. List the data members of Date.

6. List the function members of Date.

7. Write a statement to call the display() function in launch.

Section 7.3

1. Who designed the C++ language?

2. Who developed the classes that constitute the iostream library?

3. What are two of the main classes in the iostream library?

4. A _____ is an abstraction that models how input gets from the keyboard to a program, or from a program to the screen.

5. _____ is a class for inputting characters from an arbitrary input device to an executing program.

6. _____ is an `istream` object defined in the `iostream` library, which is a stream from the keyboard to a program.

7. Three states of an `istream` are _____ , _____ , and _____ .

8. The statement `assert(cin.` _____ `);` will stop program execution if there is a data-entry error.

9. The method _____ in the `istream` class is used to reset the states of an input stream.

10. The method _____ in the `istream` class is used to skip characters in an input stream.

11. (True or false) By default, `>>` skips white space in an input stream.

12. (True or false) The method `get()` skips white space in an input stream.

13. _____ is a class for outputting characters from an executing program to an arbitrary output device.

14. _____ and _____ are `ostream` objects defined in the `iostream` library, which are streams from a program to an output device.

15. Two manipulators that can be used to flush an output stream are _____ and _____ .

16. (True or false) When the statement

```
cout << setw(10) << 1.234 << 5.678 << endl;
```

is executed, the real numbers 1.234 and 5.678 will be displayed in 10-space fields.

17. (True or false) When the statement

```
cout << setprecision(1) << 1.234 << 5.678 << endl;
```

is executed, the real numbers 1.2 and 5.7 will be displayed.

Section 7.4

1. A string that contains no characters is called a(n) _____string.

2. Write a declaration that initializes a `string` variable `label` to an empty string.

3. Write a declaration that initializes a `string` constant UNITS to "meters."

4. If the input for the statement `cin >> s1 >> s2;`, where s1 and s2 are string variables, is

    ```
    ABC
    DEF GHI
    ```

    then s1 will be assigned the value _____ and s2 the value _____ .

Questions 5–22 assume the following declarations:

```
string s1 = "shell",
s2 = "seashore",
s3 = "She sells seashells by the seashore.",
s4;
```

For Questions 5–19 find the value of the expression, and for Questions 20–22 give the new value of the variable (s1, s2, or s3).

5. `s1[2]`

6. `s2.size()`

7. `s4.size()`

8. `s3.empty()`

9. `s4.empty()`

10. `s1 > s2`

11. `s1 < s3`

12. `s2 + s1`

13. `s1.substr(0, 3)`

14. `s3.find("sea", 0)`

15. `s3.rfind("sea", 35)`

16. `s3.find_first_of("abc",0)`

17. `s3.find_last_of("abc",35)`

18. `s3.find_first_not_of("abc",0)`

19. `s3.find_last_not_of("abc",35)`

20. `s1.replace(0, 2, 'b');`

21. `s2.insert(3, "l on the ");`

22. `s3.erase(9, 13);`

Section 7.5

For Questions 1–6, calculate the given expression given that $z = 8 + 3i$ and $w = 7 + 2i$.

1. $z + w$

2. $z - w$

3. $z \cdot w$

4. z / w

5. z^2

6. $|z|$

Questions 7–11 assume that the following declarations have been executed:

```
double r1 = 1.5, r2 = 2.1, r3;
complex<double> c1 = complex<double>(1, 3),
                c2 = complex<double>(2, 1),
                c3;
```

What value, if any, will be assigned to the given variable in each of the assignment statements?

7. `c3 = c1 * c2;`   10. `r3 = real(c2) + imag(c2);`

8. `c3 = conj(c2);`   11. `c3 = r1;`

9. `c3 = complex<double>(r1, r2);`

12. For the variable `c3` declared in Questions 7–11, how should the data be entered so that the value assigned to `c3` by the input statement `cin >> c3` is the complex constant $1.5 + 2.5i$?

EXERCISES

Section 7.3

For Exercises 1–11, assume that `i1`, `i2`, and `i3` are `int` variables, `r1`, `r2`, and `r3` are `double` variables, and `c1`, `c2`, and `c3` are `char` variables. Tell what value, if any, will be assigned to each of these variables or explain why an error occurs, when the input statements are executed with the given input data.

1. ```
   cin >> i1 >> i2 >> i3
       >> r1 >> r2 >> r3;
   ```
            Input: 1   2   3
                             4   5.5   6.6

2. ```
   cin >> i1 >> i2 >> i3;
   cin >> r1 >> r2 >> r3;
   ```
            Input: 1
                             2
                             3
                             4
                             5
                             6

3. ```
   cin >> i1 >> r1;
   cin >> i2 >> r2;
   cin >> i3 >> r3;
   ```
            Input: 1 2.2
                             3 4.4
                             5 6.6

4. ```
   cin >> i1 >> i2 >> i3
       >> r1 >> r2 >> r3;
   ```
            Input: 1 2.2
                             3 4.4
                             5 6.6

5. ```
   cin >> i1 >> c1 >> r1
       >> i2 >> c2 >> r2
       >> i3 >> c3 >> r3;
   ```
            Input: 1 2.2
                             3 4.4
                             5 6.6

6. ```
   cin >> noskipws
       >> i1 >> c1 >> r1
       >> i2 >> c2 >> r2
       >> i3 >> c3 >> r3;
   ```
            Input: 1 2.2
                             3 4.4
                             5 6.6

```
 7. cin >> i1 >> c1 >> i2                    Input:  1A 2B 3C
         >> c2 >> i3 >> c3;

 8. cin >> i1 >> i2 >> i3;                    Input:  012 345 678

 9. cin >> dec >> i1 >> i2 >> i3;             Input:  012 345 678

10. cin >> oct >> i1 >> i2 >> i3;             Input:  012 345 678

11. cin >> hex >> i1 >> i2 >> i3;             Input:  12 3A BC
```

For Exercises 12–20, assume that alpha and beta are real variables with values 2567.392 and 0.004, respectively, and that num1 and num2 are integer variables with values 12 and 436, respectively. Show precisely the output that each of the sets of statements produces, or explain why an error occurs.

```
12. cout << num2 << num2 + 1 << num2 + 2;

13. cout << num2 << setw(4) << num2 + 1 << num2 + 2;

14. cout << num1 << num1 + 1 << num1 + 2;

15. cout << oct << num1 << num1 + 1 << num1 + 2;

16. cout << showbase << oct << num1 << num1 + 1 << num1 + 2;

17. cout << hex << num1 << num1 + 1 << num1 + 2;

18. cout << showbase << hex << num1 << num1 + 1 << num1 + 2;

19. cout << fixed << showpoint << right
         << setw(9)  << setprecision(3) << alpha << endl
         << setw(10) << setprecision(5) << beta << endl
         << setw(7)  << setprecision(4) << beta << endl;

20. cout << scientific << showpoint << left
         << setprecision(1) << setw(10) << alpha << endl
         << setw(5) << beta << endl
         << "Tolerance:"
         << setw(12) << setprecision(3) << beta << endl;
```

For Exercises 21 and 22, assume that i and j are integer variables with values 15 and 8, respectively; that ch is a character variable whose value is 'c'; and that x and y are real variables with values 2559.50 and 8.015, respectively. Show precisely the output produced, or explain why an error occurs.

```
21. cout << setw(j) << setprecision(2)
         << fixed << showpoint << right
```

```
        << "New balance =" << x << ' ' << setw(i % 10) << ch
        << setw(j) << setprecision(j-6) << y << endl;

22. cout << "i =" << setw(i) << i
        << fixed << showpoint << right
        << "j =" << setw(j) << setprecision(j) << j << endl
        << setw(j) << i << ' '
        << setw(i) << j << endl;
```

For Exercises 23–26, assume that n1 and n2 are integer variables with values 987 and –6789, respectively; that r1 and r2 are real variables with values 12.3456 and –0.00246, respectively; and that ch is a character variable with the value 'T'. For each, write an output statement that uses these variables and format manipulators to produce the given output. (The underlining dashes are shown here only to help you determine the spacing.)

23.   12.3 T   987

   -6789TTT-0.00246
    ----

24.   12.35 -0.0025

    12.345600T***987
    ----

25. Values: 12.34560 and -0.00246
    ----

26. Observations: 12.3 and -0.0

    Locations: 987 and 6789

Section 7.4

Exercises 1–26 assume the following declarations:

```
string s1,
       s2 = "row, row, row your boat",
       s3 = "row",
       s4 = "boat.";
```

For Exercises 1–21 find the value of the expression or explain why an error occurs.

1. s2[3]

2. s1.size() + s4.size()

3. s3 < s4

4. s3 <= s2

5. s3 + s4

6. "fl" + s4.substr(1,3) + " a " + s4

7. s2.substr(1,3)

8. s3.substr(1,3)

9. s2.find("ow", 1)

10. s2.find("ow", 2)

11. s2.find("ow", 12)

12. s2.rfind("ow", 12)

13. s2.rfind("ow", 11)

14. s2.rfind("ow", 1)

15. s2.find_first_of(s4,0)

16. s2.find_first_of(s4,2)

17. s2.find_first_not_of(s4,0)

18. s2.find_first_not_of(s4,1)

19. s2.find_last_of(s3, 22)

20. s2.find_last_of(s3, 19)

21. s2.find_last_not_of(s3, s2.size() − 1)

For Exercises 22–26, give the output produced, or explain why an error occurs.

```
22. int i = s2.find_last_of(s3, s2.size() - 1);
    cout << s2.substr(i, s2.size() - i) << endl;
```

```
23. s1 = s3 + s4;
    s1[0] = 'g';
    s1.replace(2, 2, " fl");
    cout << s1 << endl;
```

```
24. s1 = s2;
    s1.erase(3, 10);
    s1.insert(0, "Bor");
    cout << s1 + "?\n";
```

```
25. s1 = s2;
    s1[s1.find("b", 0)] = 'g';
    for (int i = 0; i <= 10; i += 5)
       s1[i] = 'm';
    cout << s1 << endl;
```

```
26. s1 = s2;
    int i;
    i = s1.find_first_of("aeiou", 0);
    while (i != string::npos)
    {
      s1.replace(i, 1, "xx");
      i = s1.find_first_of("aeiou", i + 1);
```

```
}
cout << s1 << endl;
```

27. Given that `string` variable `last_first` has the value `"Smith, Bill"`, write statements to extract the first and last names from `last_first` and then combine them so that `"Bill Smith"` is assigned to the `string` variable `first_last`.

28. Write a function that accepts a single `string` object consisting of a first name, a middle name or initial, and a last name, and returns a single `string` object consisting of the last name, followed by a comma, and then the first name and the middle initial. For example, for `"John Quincy Doe"`, the function should return the string `"Doe, John Q."`.

29. Write a function that accepts the number of a month and returns the name of the month.

30. Write a function that accepts the name of a month and returns the number of the month.

31. Write a function that, given a string of lowercase and uppercase letters, returns a copy of the string in all lowercase letters; and another function that, given a string of lowercase and uppercase letters, returns a copy of the string in all uppercase letters. (Hint: Use a `for` loop and the functions provided in `cctype`.)

32. A `string` is said to be a palindrome if it does not change when the order of its characters is reversed. For example:

    madam

    463364

    ABLE WAS I ERE I SAW ELBA

    are palindromes. Write a function that, given a string, returns true if that string is a palindrome and returns false otherwise.

Section 7.5

For Exercises 1–12, calculate each expression, given that $z = 1 + 2i$ and $w = 3 - 4i$.

1. $z + w$

2. $z - w$

3. $z \cdot w$

4. z / w

5. z^2

6. $\bar{z}$

7. $\bar{w}$

8. $\dfrac{z + \bar{z}}{2}$

9. $\dfrac{z - \bar{z}}{2}$

10. $z \cdot \bar{z}$

11. $\dfrac{1}{z}$

12. $\dfrac{z+w}{z-w}$

13.–24. Repeat Exercises 1–12 for $z = 6 - 5i$ and $w = 5 + 12i$.

25.–36. Repeat Exercises 1–12 for $z = 1 + i$ and $w = 1 - i$.

For Questions 37–41, assume that the following declarations have been executed:

```
double r1 = 2.5, r2 = 6.0, r3;
complex<double> c1 = complex<double>(1, -1),
                c2 = complex<double>(-3, 4),
                c3;
```

What value, if any, will be assigned to the given variable in each of the assignment statements?

37. `c3 = c1 * c2;`

38. `c3 = conj(c2);`

39. `c3 = complex<double>(r1, r2);`

40. `r3 = real(c2) + imag(c2);`

41. `c3 = r1;`

PROGRAMMING PROBLEMS

Sections 7.2–7.4

1. Write a driver program to test the name-conversion function of Exercise 27.

2. Write a driver program to test the name-conversion function of Exercise 28.

3. Write a program to print personalized contest letters like those frequently received in the mail. They might have a format like that of the following sample. The user should enter the three strings in the first three lines, and the program then prints the letter with the underlined locations filled in.

Mr. John Q. Doe
123 SomeStreet
AnyTown, AnyState 12345

Dear Mr. Doe:

How would you like to see a brand new BMW parked in front of 123 SomeStreet in AnyTown, AnyState? Impossible, you say? No, it isn't, Mr. Doe. Simply keep the enclosed raffle ticket and validate it by sending a $100.00 tax-deductible political contribution and 10 from Shyster & Sons chewing tobacco. Not only will you become eligible for the drawing to be conducted

> on February 29 by the independent firm of G. Y. P. Shyster, but you will also be helping to reelect Sam Shyster. That's all there is to it, John. You may be a winner!!!

4. Write a driver program to test the name-of-a-month function of Exercise 29.

5. Write a driver program to test the number-of-a-month function of Exercise 30.

6. Write a driver program to test the case-conversion functions of Exercise 31.

7. Write a driver program to test the palindrome-checker function of Exercise 33.

8. There are 3 teaspoons in a tablespoon, 4 tablespoons in a quarter of a cup, 2 cups in a pint, and 2 pints in a quart. Write a program to convert units in cooking. The program should call for the input of the amount, the units, and the new units desired.

9. Write a function to count occurrences of a string in another string. Then write a driver program to input a string and then input several lines of text, using the function to count occurrences of the string in the lines of text.

10. Reverend Zeller developed a formula for computing the day of the week on which a given date fell or will fall. Suppose that we let a, b, c, and d be integers defined as follows:

a = the number of a month of the year, with March = 1, April = 2, and so on, with January and February being counted as months 11 and 12 of the preceding year

b = the day of the month

c = the year of the century

d = the century

For example, July 31, 1929, gives $a = 5$, $b = 31$, $c = 29$, $d = 19$; January 3, 1988, gives $a = 11$, $b = 3$, $c = 87$, $d = 19$. Now calculate the following integer quantities:

w = the integer quotient $(13a - 1)/5$

x = the integer quotient $c/4$

y = the integer quotient $d/4$

$z = w + x + y + b + c - 2d$

$r = z$ reduced modulo 7; that is, r is the remainder of z divided by 7: $r = 0$ represents Sunday; $r = 1$ represents Monday, and so on.

Write a function day_of_the_Week() that receives the name of a month, the day of the month, and a year, and returns the name of the day of the week on which that date fell or will fall. Write a program that inputs several strings representing dates, calls the function day_of_the_Week(), and displays the day returned by the function.

a. Verify that December 12, 1960, fell on a Monday, and that January 1, 1991, fell on a Tuesday.

b. On what day of the week did January 25, 1963, fall?

c. On what day of the week did June 2, 1964, fall?

d. On what day of the week did July 4, 1776, fall?

e. On what day of the week were you born?

Section 7.5

1. Write a program that reads three complex numbers and then determines whether the triangle whose vertices are the points in the plane that represent those three numbers is (a) equilateral, (b) isosceles, (c) a right triangle.

2. Write a program that inputs complex numbers, converts them to their polar representation, and displays each number in both forms.

3. Extend the program in Problem 2 to input a positive integer n and then compute and display the *nth roots* of each complex number. The *nth* roots of $z = re^{i\theta}$ are given by

$$z^{1/n} = r^{1/n}\left[\cos\left(\frac{\theta + 2k\pi}{n}\right) + i\sin\left(\frac{\theta + 2k\pi}{n}\right)\right], \text{ for } k = 0, 1, \ldots, n-1$$

4. In a circuit containing a resistor and an inductor in series, the voltage is given by $V = (R + i\omega L)I$, where V is the voltage (in volts), R is the resistance (in ohms), L is the inductance (in henrys), and ω is the angular velocity (in radians per second). Write a program that can be used to compute the voltage (complex) given the current (complex), or to find the current given the voltage. Execute the program with $R = 1.3\omega$, $L = 0.55$ mH, and $\omega = 365.0$ rad/sec.

Section 7.6

1. A coin is tossed repeatedly, and a payoff of 2^n dollars is made, where n is the number of the toss on which the first head appears. For example, TTH pays $8, TH pays $4, and H pays $2. Write a program to simulate playing the game several times and to print the average payoff for these games.

2. Suppose that a gambler places a wager of $5 on the following game: a pair of dice is tossed, and if the result is odd, the gambler loses his wager. If the result is even, a card is drawn from a standard deck of 52 playing cards. If the card drawn is an ace, 3, 5, 7, or 9, the gambler wins the value of the card (with aces counting as 1, Jacks as 11, Queens as 12, and Kings as 13); otherwise, he loses. What will be the average winnings for this game? Write a program to simulate the game.

3. Johann VanDerDoe, centerfielder for the Klavin Klodhoppers, has the following lifetime hitting percentages:

Out	63.4%
Walk	10.3%
Single	19.0%
Double	4.9%
Triple	1.1%
Home run	1.3%

Write a program to simulate a large number of times at bat, for example, 1000, for Johann, counting the number of outs, walks, singles, and so on, calculating his

$$\text{batting average} = \frac{\text{number of hits}}{\text{number of times at bat} - \text{number of walks}}$$

4. Consider a quarter circle inscribed in a square whose sides have length 1:

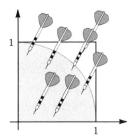

Imagine throwing q darts at this square and counting the total number p that hit within the quarter circle. For a large number of throws, we would expect

$$\frac{p}{q} \approx \frac{\text{area of quarter circle}}{\text{area of square}} = \frac{\pi}{4}$$

Write a program to approximate π using this method. To simulate throwing the darts, generate two random numbers x and y and consider point (x, y) as being where the dart hits.

5. The simulation in Problem 4 can be generalized to find the area under the graph of any function and is known as a *Monte Carlo* method of approximating integrals. To illustrate it, consider a rectangle that has base $[a, b]$ and height m, where $m \geq f(x)$ for all x in $[a, b]$:

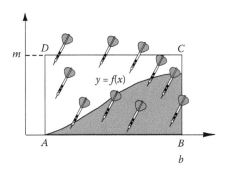

Imagine throwing q darts at rectangle $ABCD$ and counting the total number p that hit the shaded region. For a large number of throws, we would expect

$$\frac{P}{q} \approx \frac{\text{area of shaded region}}{\text{area of rectangle } ABCD}$$

Write a program to calculate areas using this Monte Carlo method. To simulate throwing the darts, generate two random numbers x from $[a, b]$ and y from $[0, m]$, and consider the point (x, y) as being where the dart hits. Use your program to compute the current passing through a capacitor with capacitance $C = 5$ farads for 6 seconds, which can be calculated as $1/C$ times the area under the curve

$$f(t) = 10\sin^2\left(\frac{t}{\pi}\right)$$

from $t = 0$ to $t = 6$.

6. The famous **Buffon Needle problem** is as follows: A board is ruled with equidistant parallel lines, and a needle of length equal to the distance between these lines is dropped at random on the board. Write a program to simulate this experiment and estimate the probability p that the needle crosses one of these lines. Display the values of p and $2/p$. (The value of $2/p$ should be approximately equal to a well-known constant. What constant is it?)

7. The tensile strength of a certain metal component has an approximate normal distribution with a mean of 10,000 pounds per square inch and a standard deviation of 100 pounds per square inch. Specifications require that all components have a tensile strength greater than 9800; all others must be scrapped. Write a program that uses the algorithm described in this section to generate 1000 normally distributed random numbers representing the tensile strength of these components, and determine how many must be rejected.

More Selection Control Structures

If you don't know where you're going, you'll wind up somewhere else.

YOGI BERRA

'Would you tell me, please, which way I ought to go from here?'
'That depends a good deal on where you want to get to', said the Cat.

LEWIS CARROLL, ALICE'S ADVENTURES IN WONDERLAND

We are all special cases.

ALBERT CAMUS

IN CHAPTER 5, we saw that the logical flow of execution in a function or program is governed by three basic control mechanisms: **sequence**, **selection**, and **repetition**. We looked at examples where an `if` statement was used to perform selective execution and `for` and `while` statements to perform repetitive execution. Our introductions to these statements were quite brief, however, so we will take a closer look at them in this chapter and the next. We will begin by looking at the `if` statement in more detail and then introduce the `switch` statement that sometimes provides a more efficient way to perform selection. In Chapter 9, we will reexamine the `for` statement and see some of the other ways it can be used to control repetition. We will also introduce the `do-while` statement, and we will compare it with the `while` statement we studied in Chapter 5.

8.1 INTRODUCTORY EXAMPLE: AIR QUALITY INDEXES REVISITED

The Air Quality Index (AQI) in the introductory example in Chapter 5 involved only two categories: safe and hazardous. In reality, however, there are usually several more. For example, the United States AQI has six categories and a color for each:

AQI	Level of Health Concern	Color
0 to 50	Good	Green
51 to 100	Moderate	Yellow
101 to 150	Unhealthy for sensitive groups	Orange
151 to 200	Unhealthy	Red
201 to 300	Very unhealthy	Purple
301 and above	Hazardous	Maroon

The program in Example 8.1 is a modification of that in Example 5.1 that classifies the index using these six categories. It uses the `if-else if` construct that we introduced in Chapter 5 and will consider in more detail in the next section.

Example 8.1 Air Quality Indexes

```
/* This program reads thee pollution levels, calculates an air
   quality index (AQI) as their integer average, and then displays
   an appropriate air-quality message.

   Input:    the three pollution levels
   Constant: the cutoff values (parts per million)
   Output:   the air quality index, the corresponding level of health
             concern using the six U.S. categories, and its color
-----------------------------------------------------------------*/

#include <iostream>                   // cin, cout, <<, >>
using namespace std;
```

```cpp
int main()
{
  const int CUTOFF1 = 50,
            CUTOFF2 = 100,
            CUTOFF3 = 150,
            CUTOFF4 = 200,
            CUTOFF5 = 300;

  int level1, level2, level3;
  cout << "Enter 3 pollution readings (parts per million): ";
  cin >> level1 >> level2 >> level3;    // get pollution levels

                                        // compute AQI
  int index = (level1 + level2 + level3) / 3;
                                        // display AQI
  cout << "AQI: " << index << " -- ";
                                        // display condition
  if (index < CUTOFF1)
    cout << "Good -- Green\n";
  else if (index < CUTOFF2)
    cout << "Moderate -- Yellow\n";
  else if (index < CUTOFF3)
    cout << "Unhealthy for Sensitive Groups -- Orange\n";
  else if (index < CUTOFF4)
    cout << "Unhealthy -- Red\n";
  else if (index < CUTOFF5)
    cout << "Very Unhealthy -- Purple\n";
  else
    cout << "Hazardous -- Maroon\n";
}
```

SAMPLE RUNS:

```
Enter 3 pollution readings (parts per million): 30 40 50
AQI:  40 -- Good -- Green


Enter 3 pollution readings (parts per million): 40 50 60
AQI:  50 -- Moderate -- Yellow


Enter 3 pollution readings (parts per million): 95 111 98
AQI:  101 -- Unhealthy for Sensitive Groups -- Orange


Enter 3 pollution readings (parts per million): 144 159 148
AQI:  150 -- Unhealthy -- Red


Enter 3 pollution readings (parts per million): 299 298 311
AQI:  302 -- Hazardous -- Maroon
```

8.2 MORE ABOUT THE if STATEMENT

In Section 5.4 we saw three different forms of the if statement:

The **single-branch** or **simple if** form:

```
if (condition)
    statement
```

The **dual-branch** or **if-else** form:

```
if (condition)
    statement₁
else
    statement₂
```

The **multibranch** or **if-else-if** form:

```
if (condition₁)
    statement₁
else if (condition₂)
    statement₂
...
else if (conditionₙ)
    statementₙ
else
    statementₙ₊₁
```

where each *condition* is a boolean expression.

Although these may look like three distinct statements, the first and last are actually special forms of the second if statement. The first form is a simpler way of writing an if statement in which the else part is empty:

```
if (condition)
    statement₁
else;
```

The multibranch form is simply a series of **nested** if statements written as one. To illustrate this, if we were to write a 5-branch if-else-if and we were to start each new if statement on a new line with each else aligned with its corresponding if, our code would appear as follows:

```
if (condition₁)
   statement₁
else
   if (condition₂)
     statement₂
   else
```

```
if (condition₃)
    statement₃
else
    if (condition₄)
        statement₄
    else
        statement₅
```

However, the free-form nature of C++ allows us to begin each nested `if` on the same line as the preceding `else` and align the `else-if` combinations:

```
if (condition₁)
    statement₁
else if (condition₂)
    statement₂
else if (condition₃)
    statement₃
else if (condition₄)
    statement₄
else
    statement₅
```

This latter style reflects more clearly the multibranch nature and is therefore more readable. It is important, however, to understand that each `else` is really a continuation of the nearest preceding `if`:

```
if (condition₁)
↑   statement₁
else if (condition₂)
    statement₂
else if (condition₃)
    statement₃
else if (condition₄)
    statement₄
else
        statement₅
```

8.2.1 Pitfall: The Dangling-Else Problem

Each *statement*$_i$ in an `if` statement can be any C++ statement and, in particular, may be another `if` statement; for example,

```
if (x > 0)
    if (y > 0)
        z = sqrt(x) + sqrt(y);
```

When such *nested* `if` statements are followed by an `else`, it is not evident with which `if` the `else` corresponds. Does the `else` match the outer `if`?

```
if (x > 0)
    if (y > 0)
        z = sqrt(x) + sqrt(y);
else
    cerr << "\n*** Unable to compute z!" << endl;
```

Or does it match the inner `if`?

```
if (x > 0)
    if (y > 0)
        z = sqrt(x) + sqrt(y);
    else
        cerr << "\n*** Unable to compute z!" << endl;
```

This ambiguity is known as the **dangling else problem**, and C++ resolves it by stipulating that

In a nested `if` statement, an `else` is matched with the nearest preceding unmatched `if`.

Thus, for the preceding `if` statement, the second matching is used; that is, the `else` is associated with the inner `if` (whose condition is $y > 0$). Consequently, the output statement is executed only in the case that x is positive and y is nonpositive. If we wish to associate this `else` with the outer `if`, we can force the association by enclosing the inner `if` in curly braces as follows:

```
if (x > 0)
{
    if (y > 0)
        z = sqrt(x) + sqrt(y);
}
else
    cerr < "*** Unable to compute z!" << endl;
```

Putting the inner `if` inside a **block** makes it a complete statement, so that the `else` must associate with the outer `if`. Thus, the output statement is executed whenever x is nonpositive.

8.2.2 Pitfall: Confusing = and ==

We turn now to what is one of the most common errors made by beginning C++ programmers in constructing boolean expressions, and, as a consequence, in writing `if` statements. We begin by looking at two features of C++ that are relevant to this problem.

1. True and false in C++. To maintain compatibility with its parent language C, in any boolean context, C++ interprets the value 0 as equivalent to the boolean value `false` and any nonzero value as equivalent to the boolean value `true`.[1] Thus, the statement

```
if (0)
    cout << "T\n";
```

---

[1] C has no `bool` type. In its place, C uses the zero/nonzero mechanism described here.

```
    else
       cout << "F\n";
```

will always display F because the condition controlling the selection is zero, which C++ treats as false. Similarly, the statement

```
    if (23)
       cout << "T\n";
    else
       cout << "F\n";
```

will always display the value T because the condition controlling the selection is non-zero, which is interpreted as true in C++.

2. *Assignments are expressions.* We saw in Chapter 3 that in C++, assignment (=) is an operator that, in addition to assigning a value to a variable, returns the value being assigned as its result. For example, the assignment expression

```
    x = 23
```

does two things: It assigns x the value 23 and it produces the value 23 as its result. Similarly,

```
    x = 0
```

both assigns x the value 0 and produces the value 0 as its result.

By themselves, neither of these C++ features is particularly troublesome, but when coupled with the similarity of the assignment and equality operators, they make it easy to write if statements that contain logical errors. To illustrate, suppose that a programmer encodes the instruction

> *If x is equal to zero, then*
> > *Display the character string "Zero"*
> *Otherwise*
> > *Display the character string "Nonzero"*

as

```
  if (x = 0)
     cout << "Zero\n";
  else
     cout << "Nonzero\n";
```

Because the assignment operator

```
  x = 0
```

is used instead of an equality comparison

```
x == 0
```

and the value returned by the assignment operator is the value that was assigned, this statement is equivalent to

```
if (0)
   cout << "Zero\n";
else
   cout << "Nonzero\n";
```

Because 0 is treated as false, the statement to output Nonzero will *always* be selected, regardless of the value of x.

Similarly, if a programmer writes

```
cout << MENU;            // display menu of choices: A, B, C
cin >> choice;
if (choice = 'A')
   statement₁            // do something when choice is A
else if (choice = 'B')
   statement₂            // do something else when choice is B
else if (choice = 'C')
   statement₃            // do something else when choice is C
else
   cout << choice << " must be A, B, or C.\n";
```

then the statement associated with choice A will *always* be selected, regardless of the value entered by the user. The reason is that instead of the first condition testing whether choice is equal to A,

```
choice == 'A'
```

it assigns choice the numeric code of A (65 in ASCII):

```
choice = 'A'
```

The result produced by the assignment operator is the value assigned (65), and so this if-else-if form is equivalent to

```
cout << MENU;            // display menu of choices: A, B, C
cin >> choice;
if (65)
   statement₁            // do something when choice is A
else if (choice = 'B')
   statement₂            // do something else when choice is B
else if (choice = 'C')
```

```
    statement₃              // do something else when choice is C
  else
    cout << ch       oice << " must be A, B, or C.\n";
```

Because nonzero values are treated as true, the value 65 is treated as true, so $statement_1$ will be executed, and $statement_2$, $statement_3$, and the output statement will be bypassed, regardless of the value of choice.

This kind of mistake is easy to make in constructing boolean expressions. Unfortunately, these errors can be fiendishly difficult to find, because the equality operator (==) and the assignment operator (=) are similar in appearance. *Any time an algorithm calls for an equality comparison, the code that implements the algorithm should be carefully checked to ensure that an assignment operator has not been inadvertently used instead of the equality operator.*

8.3 THE switch STATEMENT

An if statement can be used to implement a multialternative selection statement in which exactly one of several alternative actions is selected and performed. In the if-else-if form described in the preceding section, a selection is made by evaluating one or more boolean expressions. Because selection conditions can usually be formulated as boolean expressions, an if-else-if form can be used to implement virtually any multialternative selection.

In this section, another multialternative selection statement called the switch statement is described. Although it is not as general as the if statement, it is more efficient for implementing certain forms of selection. As usual, we begin with an example that illustrates the use of the statement.

8.3.1 Example: Temperature Conversions

In the early sections of Chapter 6, we considered the problem of converting temperatures from the Fahrenheit scale to the Celsius scale. In Section 6.2 we constructed a function to do this conversion, and in Section 6.4 we showed how a library Heat could be created to store various functions for converting temperatures between the Fahrenheit, Celsius, and Kelvin scales.

Now that we know about selective execution, we can consider a more general version of the problem—writing a program that allows the user to choose which conversion to be performed: Fahrenheit to Celsius (or vice versa), Fahrenheit to Kelvin (or vice versa), or Celsius to Kelvin (or vice versa).

8.3.1.1 Behavior

Our program should display on the screen a menu of the six possible conversion options and then read the desired conversion from the keyboard. Next, it should display on the screen a prompt for a temperature, which it should read from the keyboard. The program should then display the result of converting the input temperature, as determined by the specified conversion.

8.3.1.2 Objects

From our behavioral description, we have the following objects in this problem:[2]

	Software Objects		
Problem Objects	**Type**	**Kind**	**Name**
menu	string	constant	*MENU*
conversion	char	variable	*conversion*
temperature	double	variable	*temperature*
result	double	variable	*result*

We can thus specify the problem as follows:

Input: *temperature*, a double; and *conversion*, a char

Precondition: *conversion* is in the range A–F (or a–f)

Output: *MENU*, prompts for input, and the result of converting the temperature

8.3.1.3 Operations

Our behavioral description leads to the following list of operations:

   i. Display prompts and *MENU* (strings) on the screen.

  ii. Read *temperature* (a double) from the keyboard.

 iii. Read *conversion* (a char) from the keyboard.

 iv. Select the conversion function corresponding to *conversion* and apply it to *temperature*.

All of these are provided in C++. For operation (iv), we must compare *conversion* to each of the valid menu choices, and based on that comparison, select an appropriate conversion function.

8.3.1.4 Algorithm

The following algorithm applies this strategy.

Algorithm to Convert Arbitrary Temperatures

    1. Display *MENU* via cout.

    2. Read *conversion* from cin.

    3. Display a prompt for a temperature via cout.

    4. Read *temperature* from cin.

---

[2] Because almost every problem uses the screen, keyboard, and prompt objects, we will omit them from our lists of objects and simply assume that they are needed. This will save space and will also allow us to focus our attention on user-defined objects.

5. If *conversion* is 'A' or 'a'

>>>Convert *temperature* from Fahrenheit to Celsius and store in *result*.

>>Otherwise, if *conversion* is 'B' or 'b'

>>>Convert *temperature* from Celsius to Fahrenheit and store in *result*.

>>Otherwise, if *conversion* is 'C' or 'c'

>>>Convert *temperature* from Celsius to Kelvin and store in *result*.

>>Otherwise, if *conversion* is 'D' or 'd'

>>>Convert *temperature* from Kelvin to Celsius and store in *result*.

>>Otherwise, if *conversion* is 'E' or 'e'

>>>Convert *temperature* from Fahrenheit to Kelvin and store in *result*.

>>Otherwise, if *conversion* is 'F' or 'f'

>>>Convert *temperature* from Kelvin to Fahrenheit and store in *result*.

>>Otherwise

>>>Display an error message.

8.3.1.5 Coding and Testing

We could implement the algorithm using an `if-else-if` construct:

```
if (conversion == 'A' || conversion == 'a')
  result = fahrToCelsius(temperature)
else if (conversion == 'B' || conversion == 'b')
  result = celsiusToFahr(temperature);
  ...
else
{
  cerr << "\n*** Invalid conversion: " << conversion << endl;
  result = 0.0;
}
```

The C++ `switch` statement, however, provides a more convenient way to do this, as shown in the program in Example 8.2.

Example 8.2 Arbitrary Temperature Conversions

```
/* This program converts temperatures from one scale to another.

   Input:  menu choices, temperatures
   Output: menu, temperatures on other scale
-----------------------------------------------------------------*/
```

```cpp
#include <iostream>        // cin, cout, <<, >>
#include <string>          // string class
using namespace std;
#include "Heat.h"          // fahrToCelsius(), celsiusToFahr(), ...

int main()
{
  const string MENU = "To convert arbitrary temperatures, enter:\n"
                      " A - to convert Fahrenheit to Celsius;\n"
                      " B - to convert Celsius to Fahrenheit;\n"
                      " C - to convert Celsius to Kelvin;\n"
                      " D - to convert Kelvin to Celsius;\n"
                      " E - to convert Fahrenheit to Kelvin; or\n"
                      " F - to convert Kelvin to Fahrenheit.\n"
                      " --> ";

  cout << MENU;
  char conversion;
  cin >> conversion;

  cout << "\nEnter the temperature to be converted: ";
  double temperature;
  cin >> temperature;
  double result;
  switch (conversion)
  {
    case 'A': case 'a':
            result = fahrToCelsius(temperature);
            break;
    case 'B': case 'b':
            result = celsiusToFahr(temperature);
            break;
    case 'C': case 'c':
            result = celsiusToKelvin(temperature);
            break;
    case 'D': case 'd':
            result = kelvinToCelsius(temperature);
            break;
    case 'E': case 'e':
            result = fahrToKelvin(temperature);
            break;
    case 'F': case 'f':
            result = kelvinToFahr(temperature);
            break;
    default:
            cerr << "\n*** Invalid conversion: "
                 << conversion << endl;
            result = 0.0;
  }
  cout << "The converted temperature is " << result << endl;
}
```

SAMPLE RUN:

```
To convert arbitrary temperatures, enter:
    A - to convert Fahrenheit to Celsius;
    B - to convert Celsius to Fahrenheit;
    C - to convert Celsius to Kelvin;
    D - to convert Kelvin to Celsius;
    E - to convert Fahrenheit to Kelvin; or
    F - to convert Kelvin to Fahrenheit.
--> B


Enter the temperature to be converted: 100
The converted temperature is 212
```

Note the convenience of the `switch` statement: by allowing us to specify any number of cases for a given alternative, we can test the value of `conversion` quite conveniently, regardless of whether it is uppercase or lowercase:

```
switch (conversion)
{
  case 'A': case 'a':
            result = fahrToCelsius(temperature);
            break;

  . . .

}
```

The equivalent `if-else-if` version seems clumsy by comparison, and many people find it to be more work and less readable. In addition, using a `switch` statement to select from among several alternatives is typically *more time-efficient* than using an `if` statement, as discussed at the end of this section.

8.3.2 Form of the switch Statement

The C++ `switch` statement has the following general form:

THE switch STATEMENT

FORM:

```
switch (expression)
{
  case_list₁ :
            statement_list₁;
  case_list₂ :
            statement_list₂;
            .
            .
            .
```

```
    case_list_n :
                statement_list_n
  default :
                statement_list_{n+1}
  };
```

where
  switch and default are C++ keywords;
  expression is an integer (or integer-compatible) expression
  each `case_list_i` is a sequence of cases of the form

```
    case constant_value :
```

  the default clause is optional; and
  each *statement_list_i* is a sequence of statements.

PURPOSE:

When the switch statement is executed, *expression* is evaluated. If the value of *expression* is in `case_list_i`, then execution begins in *statement_list_i* and continues until one of the following is reached:

  A break statement
  A return statement
  The end of the switch statement

If the value of *expression* is not in any `case_list_i`, then *statement_list_{n+1}* in the default clause is executed. If the default clause is omitted and the value of *expression* is not in any `case_list_i`, then execution "falls through" the switch statement.
  Note that *expression* must be an integer-compatible expression (in particular, it may not evaluate to a real or a string value).

8.3.3 The break Statement

As illustrated in the program in Example 8.2, each of the statement lists in a switch statement usually ends with a **break statement** of the form

```
    break;
```

When it is executed, this statement transfers control to the first statement following the switch statement. As we will see in the next chapter, the break statement can also be used to terminate repetition of a loop. In both situations break has the same behavior: execution jumps to the first statement following the switch statement or loop in which it appears.

8.3.4 Drop-Through Behavior

An important feature to remember when using the switch statement is its drop-through behavior. To illustrate it, suppose we had written the switch statement in Example 8.2 without the break statements:

```
switch (conversion)
{
  case 'A': case 'a':
            result = fahrToCelsius(temperature);
  case 'B': case 'b':
            result = celsiusToFahr(temperature);
  case 'C': case 'c':
            result = celsiusToKelvin(temperature);
  case 'D': case 'd':
            result = kelvinToCelsius(temperature);
  case 'E': case 'e':
            result = fahrToKelvin(temperature);
  case 'F': case 'f':
            result = kelvinToFahr(temperature);
  default:
            cerr << "\n*** Invalid conversion: "
                 << conversion << endl;
            result = 0.0;
}
```

The output produced when this modified version is run may be rather unexpected. Here is
one example:

```
To convert arbitrary temperatures, enter:
    A - to convert Fahrenheit to Celsius;
    B - to convert Celsius to Fahrenheit;
    C - to convert Celsius to Kelvin;
    D - to convert Kelvin to Celsius;
    E - to convert Fahrenheit to Kelvin; or
    F - to convert Kelvin to Fahrenheit.
--> B

Enter the temperature to be converted: 100
*** Invalid conversion: B
The converted temperature is 0
```

As in the sample run in Example 8.2, the value of `conversion` is B, so control is transferred to the statement:

```
result = celsiusToFahr(temperature);
```

However, there is no `break` following this statement to transfer control past the other statements, and so execution drops through to the statement in the next case, which resets `result`:

```
result = celsiusToKelvin(temperature);
```

Again, there is no `break` statement, so execution drops through to the statement in the
next case, which resets `result` again:

```
result = kelvinToCelsius(temperature);
```

This drop-through behavior continues until a `break`, a `return`, or the end of the `switch` statement is reached. Because there are no `break` or `return` statements here, execution proceeds through the next two cases and reaches the `default` case, which displays

```
*** Invalid conversion: B
```

and sets result to zero. The output statement following the `switch` then displays

```
The converted temperature is 0
```

To avoid this behavior, we must remember to end each statement list in a `switch` statement with a `break` (or `return` statement), except for the final statement list, where it is not necessary.

The program in Example 8.2 uses the `switch` statement in a `main` function, but it is perhaps more commonly used to control selection in functions other than `main`. In this case, the function is probably using the `switch` to select its return value, and so a `return` statement can be used instead of a `break` statement. The following example illustrates this.

8.3.5 Example: Converting Engineering Program Codes to Names

Numeric or letter codes are commonly used to represent information about someone or something; for example, a student's class level at a university, an employee's salary bracket at a company, the stage in some manufacturing processes, a student's major program, and so on. Here we will use the last example and develop a function that accepts the code for an engineering student's program and returns the name of the program: C for civil, E for electrical, I for industrial, M for mechanical, and U for undecided.

Suppose we start with the following *stub* for our function:

```
/*-------------------------------------------------------------------
   This function returns the name of an engineering program
   corresponding to a given program code.

   Receive: a character
   Return:  the appropriate (string) name of an engineering program
            (Civil, Electrical, Industrial, Mechanical, Undecided)
            or an empty string for a nonvalid code
   Output:  An error message in case of a nonvalid code
   --------------------------------------------------------------*/
string engrProgram(int progCode)
{
}
```

Here, the key word in the function's documentation is *corresponding*. Because we must return the name of the program corresponding to `progCode`, we must compare `progCode` to each of the possible codes, and then select an appropriate `return` statement.

The following algorithm applies this strategy.

If progCode is 'C'
    Return "Civil".
Otherwise, if progCode is 'E'
    Return "Electrical".
Otherwise, if progCode is 'I':
    Return "Industrial".
Otherwise, if progCode is 'M'
    Return "Mechanical".
Otherwise, if progCode is 'U'
    Return "Undecided".
Otherwise
    Display an error message.
    Return the empty string.

Although we clearly could implement this algorithm using an `if` statement, the function in Example 8.3 solves this problem using a `switch` statement. Note that no `break` statements are required in the cases of this `switch`. The function uses the `switch` to select a `return` statement, and a `return` statement causes execution of the function to terminate.

Example 8.3 Converting Engineering Program Codes to Names

```
/* This is a driver program to test the function engrProgram.

   Input:  a character
   Output: names of programs (Civil, Electrical, ... )
-------------------------------------------------------------------*/

#include <iostream>          // cin, cout, <<, >>
#include <string>            // string
using namespace std;

string engrProgram(char progCode);

int main()
{
  char code;
  cout << "Enter the code of an engineering program: ";
  cin >> code;
  cout << engrProgram(code) << endl;
}

/*-------------------------------------------------------------
   This function returns the name of an engineering program
   corresponding to a given program code.

   Receive: a character
```

```
        Return:   the appropriate (string) name of an engineering program
                  (Civil, Electrical, Industrial, Mechanical, Undecided)
                  or an empty string for a nonvalid code
        Output:   An error message in case of a nonvalid code
    ----------------------------------------------------------------*/
    string engrProgram(char progCode)
    {
      switch (progCode)
      {
        case 'C': return "Civil";
        case 'E': return "Electrical";
        case 'I': return "Industrial";
        case 'M': return "Mechanical";
        case 'U': return "Undecided";
        default:  cerr << "*** code error: " << progCode << " ***\n";
                  return "";
      }
    }
```

SAMPLE RUNS:

```
Enter the code of an engineering program: C
Civil


Enter the code of an engineering program: E
Electrical


Enter the code of an engineering program: I
Industrial


Enter the code of an engineering program: M
Mechanical


Enter the code of an engineering program: U
Undecided


Enter the code of an engineering program: a
*** code error: a ***
```

8.4 CONDITIONAL EXPRESSIONS

The selection statements (if and switch) we have considered thus far are similar to statements provided by other languages. However, C++ has inherited a third selection mechanism from its parent language C, an expression that produces either of two values, based on the value of a boolean expression (also called a *condition*).

To illustrate it, consider the simplified form of the Air Quality Index computation problem from Chapter 5 in which we wish to determine whether the air quality is safe or hazardous, based on the value of `index`, the Air Quality Index. There we used the following `if` statement:

```
if (index < CUTOFF)
  cout << "Safe condition\n";
else
  cout << "Hazardous condition\n";
```

An alternative is to use the following output statement

```
cout << ((index < CUTOFF) ? "Safe" : "Hazardous")
       << " condition\n";
```

which, like the preceding `if` statement, will display `Safe  condition` if the condition `index < CUTOFF` is true, but it will display `Hazardous  condition` if it is false.

Because the value produced by such expressions depends on the value of their condition, they are called **conditional expressions**,[3] and have the following general form:

THE CONDITIONAL EXPRESSION

FORM:

> `condition ? expression₁ : expression₂`

where
> `condition` is a boolean expression; and
> `expression₁` and `expression₂` are type-compatible expressions.

BEHAVIOR:

> `condition` is evaluated.
> If the value of `condition` is true (i.e., nonzero), the value of `expression₁` is returned as the result.
> If the value of `condition` is false (i.e., zero), the value of `expression₂` is returned as the result.

Note that in a conditional expression, only one of `expression₁` and `expression₂` is evaluated. Thus, an assignment such as

```
reciprocal = ( (x == 0) ? 0 : 1 / x);
```

is safe because if the value of `x` is zero, the expression `1 / x` will not be evaluated, and so no division-by-zero error results. A conditional expression can thus sometimes be used in

---

[3] A conditional expression has the form `C ? A : B` and is actually a *ternary* (three-operand) operation, in which C, A, and B are the three operands and `? :` is the operator.

place of an `if` statement to guard a potentially unsafe operation. When it is used as a sub-expression in another expression, the conditional expression should be enclosed in parentheses, because its precedence is lower than most of the other operators (see Appendix C).

This mechanism has many different uses, because it can be used anywhere that an expression can appear. In fact, the conditional expression can be used in place of most `if`-`else` statements. To illustrate, suppose that we wanted to write a function `largerOf()` to find the maximum of two `int` values. Although we could do this with an `if` statement,

```cpp
int largerOf(int value1, int value2)
{
  if (value1 > value2)
    return value1;
  else
    return value2;
}
```

a conditional expression provides a simpler alternative:

```cpp
int largerOf(int value1, int value2)
{
  return ( (value1 > value2) ? value1 : value2);
}
```

Using such a function, we can write

```cpp
max = largerOf(x, y);
```

and `max` will be assigned the larger of the two values `x` and `y`.

As a final example, suppose that `numCourses` is an `int` variable containing the number of courses a student is taking in the current semester. Then the output statement

```cpp
cout << "\nYou are taking " << numCourses << " course"
     << ( (numCourses == 1) ? "" : "s" )
     << " this semester.\n";
```

will display the *singular* message

```
You are taking 1 course this semester.
```

if `numCourses` is equal to 1, and will display a *plural* message if `numCourses` has a value other than 1:

```
You are taking 3 courses this semester.
```

CHAPTER SUMMARY

Key Terms

block	dual-branch `if` or `if-else` form
`break` statement	multialternative selection
condition	multibranch `if` or `if-else-if` form
conditional expression	single-branch or simple `if` form
dangling else problem	`switch` statement
drop-through behavior	ternary operation

NOTES

- The multibranch `if` is a series of nested `if` statements written as one.

- In a nested `if`, each `else` is matched with the nearest preceding unmatched `if`.

- Be sure to use the `==` operator for equality comparisons, and not `=` (assignment). Any time an algorithm calls for an equality comparison, the code that implements the algorithm should be carefully checked to ensure that `=` (assignment) has not been inadvertently used instead of the `==` (equality operator).

- Remember that the type of the expression of a `switch` statement and the constants in its case lists must be integer-compatible. Note that they may not be real or string expressions.

- To prevent drop-through behavior in a `switch` statement, remember to end the statement list in each case with a `break` or `return` statement (except for the final statement list, where it is not needed).

- In deciding which statement to use to implement a selection, use a `switch` if all of the following hold and an `if` otherwise:

  1. An equality (`==`) comparison is being performed.

  2. The same expression is being compared in each condition.

  3. The expression being compared is integer-compatible.

- Conditional expressions can be used in place of many `if-else` statements and sometimes provide a simpler alternative.

Style and Design Tips

In this text, we use the following conventions for formatting the selection statements considered in this chapter.

- *For an* if *statement,* if (boolean_expression) *is on one line, with its statement indented on the next line. If there is an* else *clause,* else *is on a separate line, aligned with* if, *and its statement is indented on the next line. If the statements are compound, the curly braces are aligned with the* if *and* else *and the statements inside the block are indented.*

```
if (boolean_expression)
   statement₁
else
   statement₂

if (boolean_expression)
{
   statement₁
      . . .
   statementₖ
}
else
{
   statementₖ₊₁
      . . .
   statementₙ
}
```

An exception is made when the if-else-if *form is used to implement a multialternative selection structure. In this case the format used is*

```
if (boolean_expression₁)
   statement₁
else if (boolean_expression₂)
   statement₂
      . . .
else if (boolean_expressionₙ)
   statementₙ
else
   statementₙ₊₁
```

- *For a* switch *statement,* switch (expression) *is on one line, with its curly braces aligned and on separate lines; each case list is indented within the curly braces, and each statement list and* break *or* return *statement is indented past its particular case list.*

```
switch (expression)
{
   case_list₁:
            statement_list₁;
            break;              // or return
```

```
    case_list₂:
            statement_list₂;
            break;                // or return
    . . .
    case_listₙ:
            statement_listₙ;
            break;                // or return
    default:
            statement_listₙ₊₁
}
```

Alternatively, each $statement\_list_i$ may begin on the same line as $case\_list_i$.

- *Program defensively by using the* if *statement to test for illegal values.* This provides an alternative to the assert() mechanism that does not terminate the program on a failed precondition.

- *Multialternative selection constructs can be implemented more efficiently with an* if-else-if *construct than with a sequence of separate* if *statements.*

- *Multialternative selection statements of the form*

```
if (variable == constant₁)
  statement₁
else if (variable == constant₂)
  statement₂
  . . .
else if (variable == constantₙ)
  statementₙ
else
  statementₙ₊₁
```

where variable *and each* $constant_i$ *are* int-*compatible, are usually implemented more efficiently using a* switch *statement. A second advantage of the* switch *statement is that a problem solution implemented with a* switch *is often more readable than an equivalent solution implemented using an* if *statement.*

Warnings

1. *One of the most common errors in an* if *statement is using an assignment operator* (=) *when an equality operator* (==) *is intended.*

2. *When real quantities that are algebraically equal are compared with* ==, *the result may be a false boolean expression, because most real numbers are not stored exactly.*

3. *In a nested* if *statement, each* else *clause is matched with the nearest preceding unmatched* if. *Indentation and alignment of each* else *with its corresponding* if *should be used to make these associations clear.*

4. *Each* switch *statement and block must contain matching curly braces. A missing* } *can be difficult to locate. In certain situations, the compiler may not find that a* { *is unmatched until it reaches the end of the file. In such cases, an error message such as*

```
Error...: Compound statement missing } in function ...
```

will be generated.

5. *The selector in a* switch *statement must be integer-compatible. In particular, the values of the selector in case lists*

- *may not be real constants, and*

- *may not be string constants.*

6. *A* switch *statement has drop-through behavior. Execution of the statement list in a particular case will continue on into subsequent cases until a* break, *a* return, *or the end of the* switch *statement is reached. To avoid this behavior, you must remember to end each statement list in a* switch *statement with a* break *or* return *statement (except for the final statement list, where it is not necessary).*

TEST YOURSELF

Section 8.3

For the following questions, assume that number is an int variable, code is a char variable, and x is a double variable.

1. If number has the value 99, tell what output is produced by the following switch statement, or indicate why an error occurs:

```
switch(number)
{
   case 99:
            cout << number + 99 << endl;
            break;
   case -1:
            cout << number - 1 << endl;
            break;
   default:
            cout << "default\n";
}
```

2. Proceed as in Question 1, but suppose the break statements are omitted.

3. Proceed as in Question 1, but suppose number has the value 50.

4. Proceed as in Question 2, but suppose number has the value 50.

5. Proceed as in Question 1, but suppose number has the value –1.

6. Proceed as in Question 2, but suppose `number` has the value –1.

7. If the value of `code` is the letter B, tell what output is produced by the following `switch` statement, or indicate why an error occurs:

```
switch (code)
{
  case 'A': case 'B':
                              cout << 123 << endl;
                              break;
  case 'P': case 'R': case 'X':
                              cout << 456 << cndl;
                              break;
}
```

8. Proceed as in Question 7, but suppose the value of `code` is the letter X.

9. Proceed as in Question 7, but suppose the value of `code` is the letter M.

10. If the value of `x` is 2.0, tell what output is produced by the following `switch` statement, or indicate why an error occurs:

```
switch (x)
{
  case 1.0:
            cout << x + 1.0 << endl;
            break;
  case 2.0:
            cout << x + 2.0 << endl;
            break;
}
```

EXERCISES

Section 8.2

1. Describe the output produced by the following poorly indented program segment:

```
int number = 4;
double alpha = -1.0;
if (number > 0)
  if (alpha > 0)
    cout << "first\n";
else
  cout << "second\n";
  cout << "third\n";
```

Exercises 2 and 3 refer to the following `if` statement, where `honors`, `awards`, and `goodStudent` are of type `bool`:

```
if (honors)
   if (awards)
     goodStudent = true;
   else
     goodStudent = false;
else if (!honors)
   goodStudent = false;
```

2. Write a simpler `if` statement that is equivalent to this one.

3. Write a single assignment statement that is equivalent to this `if` statement.

4. What output (if any) will be produced by the statements

```
int x = 0,
    y;
if ((x = 15) > y)
   cout << "Yes\n"
else
   cout << "No\n";
```

if y has the value (a) 10? (b) 20?

For Exercises 5–12, you are asked to write functions. To test these functions, you should write driver programs as instructed in Programming Problems 1–8 for Section 8.2 at the end of this chapter.

5. In a certain region, pesticides can be sprayed from an airplane only if the temperature is at least 70°, the relative humidity is between 15 and 35 percent, and the wind speed is at most 10 miles per hour. Write a boolean-valued function `okToSpray()` that receives three numbers representing temperature, relative humidity, and wind speed, and returns true if the conditions allow spraying and false otherwise.

6. A certain credit company will approve a loan application if the applicant's income is at least $25,000 or the applicant's assets are at least $100,000; in addition, the applicant's total liabilities must be less than $50,000. Write a boolean-valued function `creditApproved()` that receives three numbers representing income, assets, and liabilities, and returns true if the criteria for loan approval are satisfied and false otherwise.

7. Write a function that returns true if the value of an `int` parameter `year` is the number of a leap year and return false otherwise. (A leap year is a multiple of 4; and if it is a multiple of 100, it must also be a multiple of 400.)

8. Write a function that returns the number of days in a given `int` parameter `month` (1, 2, . . . , 12) of a given parameter `year`. Use Exercise 7 to determine the number of days if the value of `month` is 2.

9. Proceed as in Exercise 8, but assume that `month` is a `string` variable whose value is the name of a month.

10. Write a function that checks if *t* is in the range shown in the graph

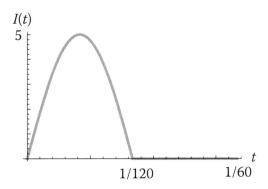

and if so, returns the value of the rectified half-wave function $I(t)$; the curve is a sine function for half the cycle and zero for the other half. (Maximum current is 5 amps.)

11. Proceed as in Exercise 10, but for a sawtooth graph

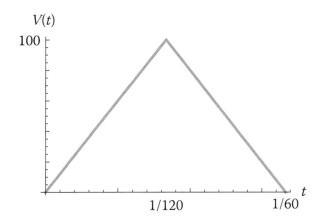

consisting of two straight lines. The maximum voltage of 100 V occurs at the middle of the cycle.

12. Proceed as in Exercise 10, but for a piecewise graph

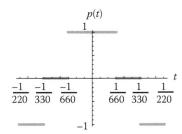

of the excess pressure $p(t)$ in a sound wave.

1. Write a `switch` statement that increases `balance` by adding `amount` to it if the value of the character variable `transCode` is `'D'`; decreases `balance` by subtracting `amount` from it if `transCode` is `'W'`; displays the value of `balance` if `transCode` is `'P'`; and displays an illegal-transaction message otherwise.

2. Write a `switch` statement that, for two given integers `a` and `b`, and a given character `operation`, computes and displays `a + b`, `a - b`, `a * b`, or `a / b` according to whether the `operation` is `'+'`, `'-'`, `'*'`, or `'/'`, and displays an illegal-operator message if it is not one of these.

For Exercises 3–6, write functions that use `switch` statements to compute what is required. To test these functions, you should write driver programs as instructed in Programming Problems 1–4 for Section 8.3 at the end of this chapter.

3. Given a distance less than 1000, return a shipping cost as determined by the following table:

Distance	Cost
0 through 99	$5.00
At least 100 but less than 300	$8.00
At least 300 but less than 600	$10.00
At least 600 but less than 1000	$12.00

4. Given a number representing a TV channel, return the call letters of the station that corresponds to that number, or some message indicating that the channel is not used. Use the following channel numbers and call letters (or use those that are available in your locale):

    2: WCBS

    4: WNBC

    5: WNEW

    7: WABC

    9: WOR

    11: WPIX

    13: WNET

5. Given the number of a month, return the name of the month (or an error message indicating an illegal month number).

6. Proceed as in Exercise 5, but return the number of days in the month. (A leap year is a multiple of 4; but if it is a multiple of 100, it must also be a multiple of 400.)

Section 8.4

1. Describe the operation that the following function performs:

```
int doSomething1(int value)
{
   return ( (value >= 0) ? value : -value );
}
```

2. Describe the operation that the following function performs:

```
char doSomething2(char ch)
{
   return ( ( ('A' <= ch) && (ch <= 'Z') ) ? ch+32 : ch );
}
```

3. Write conditional expressions that can replace the blanks in the output statement:

```
cout << _____ << month << '/'
     << _____ << day << '/'
     << _____ << year % 100<< endl;
```

so that the output produced will be as follows:

month	day	year	Output
12	25	1999	12/25/99
10	1	1980	10/01/80
7	4	1976	07/04/76
2	2	2002	02/02/02

4. Write a conditional expression that can replace the blank in the output statement

```
cout << _____ << number<< endl;
```

so that the output produced will be as follows:

number	Output
123	123
23	023
3	003

5. Write a function `smallerOf()` that returns the smaller of two given integer values.

6. Using nested conditional expressions, write a function:

   a. `largestOf()` that, given three int values, returns the largest of the three.

   b. `smallestOf()` that, given three int values, returns the smallest of the three.

7. The mathematician Carl Friedrich Gauss discovered that the sum of the integers from 1 through n is given by the form

$$\cfrac{1}{\dfrac{1}{R_1}+\dfrac{1}{R_2}+\dfrac{1}{R_3}}$$

Using a conditional expression, construct a function sum() that returns the value according to Gauss's formula if the value of its parameter is positive and zero otherwise.

PROGRAMMING PROBLEMS

Section 8.2

1. Write a driver program to test the function okToSpray() of Exercise 5.

2. Write a driver program to test the function creditApproved() of Exercise 6.

3. Write a driver program to test the leap-year function of Exercise 7.

4. Write a driver program to test the days-in-a-month function of Exercise 8.

5. Write a driver program to test the days-in-a-month function of Exercise 9.

6. Write a driver program to test the rectified half-wave function of Exercise 10.

7. Write a driver program to test the sawtooth function of Exercise 11.

8. Write a driver program to test the piecewise pressure function of Exercise 12.

9. Suppose the following formulas give the safe loading L in pounds per square inch for a column with slimness ratio S:

Write a program that reads a slimness ratio and then calculates the safe loading.

10. Suppose that charges by a gas company are based on consumption according to the following table:

Gas Used	Rate
First 70 cubic meters	$5.00 minimum cost
Next 100 cubic meters	5.0¢ per cubic meter
Next 230 cubic meters	2.5¢ per cubic meter
Over 400 cubic meters	1.5¢ per cubic meter

Write a function that computes the charges for a given amount of gas usage. Use this function in a program in which the meter reading for the previous month and the current meter reading are entered, each a four-digit number and each representing cubic meters, and that then calculates and displays the amount of the bill. Note: Because of rollover, the current reading may be less than the previous one; for example, the previous reading may be 9897, and the current one may be 0103. Execute the program with the following meter readings: 3450, 3495; 8810, 8900; 9950, 0190; 1275, 1982; 9872, 0444.

11. Write a program that reads values for the coefficients A, B, C, D, E, and F of the equations

$$Ax + By = C$$

$$Dx + Ey = F$$

of two straight lines, and then determine whether the lines are parallel (their slopes are equal) or the lines intersect. If they intersect, determine whether the lines are perpendicular (the product of their slopes is equal to –1).

12. Write a program that reads the coordinates of three points and then determines whether they are collinear.

Section 8.3

1. Write a driver program to test the distance-cost function of Exercise 3.

2. Write a driver program to test the TV-channel function of Exercise 4.

3. Write a driver program to test the month-name function of Exercise 5.

4. Write a driver program to test the days-in-month function of Exercise 6.

5. Modify the program in Example 8.3 to allow codes CH for chemical, CI for civil, CO for computer, EL for electrical, EN for environmental, I for industrial, and M for mechanical. Use a nested `switch` statement to process codes that begin with the same letter.

6. Locating avenues' addresses in mid-Manhattan is not easy; for example, the nearest cross street to 866 Third Avenue is 53rd Street, whereas the nearest cross street to 866 Second Avenue is 46th Street. To locate approximately the nearest numbered cross street for a given avenue address, the following algorithm can be used:

Cancel the last digit of the address, divide by 2, and add or subtract the number given in the following abbreviated table:

1st Ave.	Add 3
2nd Ave.	Add 3
3rd Ave.	Add 10
4th Ave.	Add 8
5th Ave. up to 200	Add 13
5th Ave. up to 400	Add 16
6th Ave. (Ave. of the Americas)	Subtract 12
7th Ave.	Add 12
8th Ave.	Add 10
10th Ave.	Add 14

Write a function that uses a `switch` statement to determine the number of the nearest cross street for a given address and avenue number according to the preceding algorithm. Then write a program to test your function.

7. A wholesale lab equipment company discounts the price of each of its products depending on the number of units bought and the price per unit. The discount increases as the numbers of units bought and/or the unit price increases. These discounts are given in the following table:

Number Bought	Unit Price (dollars)		
	0–10.00	10.01–100.00	100.01+
1–9	0%	2%	5%
10–19	5%	7%	9%
20–49	9%	15%	21%
50–99	14%	23%	32%
100+	21%	32%	43%

Write a function that calculates the percentage discount for a specified number of units and unit price. Use this function in a program that reads the number of units bought and the unit price, and then calculates and prints the total full cost, the total amount of the discount, and the total discounted cost.

8. An airline vice president in charge of operations needs to determine whether the current estimates of flight times are accurate. Because there is a larger possibility of variations due to weather and air traffic in the longer flights, he allows a larger error in the time estimates for them. He compares an actual flight time with the estimated flight time and considers the estimate to be too large, acceptable, or too small, depending on the following table of acceptable error margins:

Estimated Flight Time in Minutes	Acceptable Error Margin in Minutes
0–29	1
30–59	2
60–89	3
90–119	4
120–179	6
180–239	8
240–359	13
360 or more	17

For example, if an estimated flight time is 106 minutes, the acceptable error margin is 4 minutes. Thus, the estimated flight time is too large if the actual flight time is less than 102 minutes, or the estimated flight time is too small if the actual flight time is greater than 110 minutes; otherwise, the estimate is acceptable. Write a function that uses a `switch` statement to determine the acceptable error for a given estimated flight time, according to this table. Use your function in a program that reads an estimated flight time and an actual flight time, and then determines whether the estimated time is too large, acceptable, or too small. If the estimated flight time is too large or too small, the program should also print the amount of the overestimate or underestimate.

9. Write a function `convertLength()` that receives a real value and two strings `inUnits` and `outUnits`, then converts the value given in `inUnits` to the equivalent metric value in `outUnits` and displays this value. The function should carry out the following conversions:

inUnits	outUnits	
I	C	(inches to centimeters; 1 in = 2.54001 cm)
F	C	(feet to centimeters; 1 ft = 30.4801 cm)
F	M	(feet to meters; 1 ft = 0.304801 m)
Y	M	(yards to meters; 1 yd = 0.914402 m)
M	K	(miles to kilometers; 1 mi = 1.60935 km)

Also, write a driver program to test your function. What happens if you enter units other than those listed?

10. Proceed as in Problem 9, but write a function `convertWeight()` that carries out the following conversions:

inUnits	outUnits	
O	G	(ounces to grams; 1 oz = 28.349527 g)
P	K	(pounds to kilograms; 1 lb = 0.453592 kg)

11. Proceed as in Problem 9, but write a function `convertVolume()` that carries out the following conversions:

inUnits	outUnits	
P	L	(pints to liters; 1 pt = 0.473167 L)
Q	L	(quarts to liters; 1 qt = 0.94633 L)
G	L	(gallons to liters; 1 gal = 3.78541 L)

12. Write a menu-driven program to test the three functions `convertLength()`, `convertWeight()`, and `convertVolume()` of Problems 9–11. It should allow the user to select one of three options according to whether lengths, weights, or volumes are to be converted; read the value to be converted and the units; and then call the appropriate function to carry out the conversion.

More Repetition Control Structures

We are what we repeatedly do. Excellence then, is not an act, but a habit.

ARISTOTLE

Reader, suppose you were an idiot. And suppose you were a member of Congress. But I repeat myself.

MARK TWAIN

It's déjà vu all over again.

YOGI BERRA

A rose is a rose is a rose.

GERTRUDE STEIN

A s we saw in Chapter 5, the three basic control behaviors used in programming are **sequence**, **selection**, and **repetition**. We have now considered all of the C++ selection structures, but we have only introduced the two basic types of repetition:

1. *Repetition controlled by a counter,* in which the body of the loop is executed once for each value of some control variable in a specified range of values.

2. *Repetition controlled by a logical expression,* in which the decision to continue or to terminate repetition is determined by the value of some logical expression.

In Chapter 5, we introduced C++'s `for` loop, which implements repetition of the first kind, and the `while` loop, which is one of C++'s implementations of repetition of the second kind. In this chapter we will take a closer look at these repetition structures and also introduce some others.

9.1 TWO INTRODUCTORY EXAMPLES: SUMMATION AND CALCULATING DEPRECIATION

In Section 5.5 we introduced the basic `for` loop with the example of calculating factorials. In this section, we begin with a similar example of a function to calculate sums of integers, and then use this function in a program that calculates and displays depreciation tables.

9.1.1 Example 1: Gauss's Punishment—Calculating Sums

We begin with an incident in the life of Carl Friedrich Gauss, one of the greatest mathematicians of all time. When Gauss was young, he attended a school in Brunswick, Germany, and one day when the students were being particularly mischievous, the teacher asked them to sum the integers from 1 to 100, expecting that this would keep them busy for a while. (As described later, Gauss discovered a very easy way to do this.)

Although calculating the sum of the integers from 1 to 100 is not a particularly important computation, a function that computes and returns the sum $1 + 2 + ... + n$ for any positive integer n is useful. The program in Example 9.1 shows a function that uses a `for` loop to calculate this sum.

Example 9.1 Summation Problem

```
/* This is a driver program to test function sum().

   Input:  an integer n
   Output: the sum of the integers from 1 through n
   ---------------------------------------------------------------------*/

#include <iostream>    // cout, cin, <<, >>
using namespace std;

int sum(int n);        // prototype for sum()
```

```
int main()
{
   cout << "This program computes the sum 1 + 2 + ... + n.\n";
   cout << "Enter a value for n: ";
   int n;
   cin >> n;
   cout << "--> 1 + ... + " << n << " = " << sum(n) << endl;
}


/* Function to compute the sum of the integers from 1 to n.

   Receive: n, an integer
   Return:  the sum 1 + 2 + ... + n
---------------------------------------------------------------------*/

int sum(int n)
{
   int runningTotal = 0;
   for (int count = 1; count <= n; count++)
      runningTotal += count;
   return runningTotal;
}
```

SAMPLE RUNS:

```
This program computes the sum 1 + 2 + ... + n.
Enter a value for n: 5
--> 1 + ... + 5 = 15

This program computes the sum 1 + 2 + ... + n.
Enter a value for n: 100
--> 1 + ... + 100 = 5050
```

We noted earlier that Gauss's teacher expected that computing the sum $1 + 2 + ... + 100$ would keep the students busy for some time. However, Gauss responded with the correct answer (5050) almost immediately, using a particularly clever approach. The simplicity and efficiency of his algorithm compared to the repetitive algorithm we used is an indication of his genius.

To compute the sum of the integers from 1 through 100, Gauss perhaps observed that writing the sum forward,

$$sum = 1 + 2 + 3 + ...+ 98 + 99 + 100$$

and then backward,

$$sum = 100 + 99 + 98 + ...+ 3 + 2 + 1$$

and then adding corresponding terms in these two equations gives

$$2 \times sum = 101 + 101 + \ldots 101 + 101$$

$$= 100 \times 101$$

Thus, the sum is equal to

$$\frac{1}{\rule{4cm}{0.4pt}}$$

Applying his algorithm to the more general summation problem, we begin with the sum

$$sum = 1 + 2 + 3 + \ldots + (n - 2) + (n - 1) + n$$

reverse it,

$$sum = n + (n - 1) + (n - 2) + \ldots + 3 + 2 + 1$$

and then add these two equations to get

$$2 \times sum = (n + 1) + (n + 1) + \ldots + (n + 1) + (n + 1)$$

$$= n \times (n + 1)$$

Dividing by 2 gives

◀

This formula, known as *Gauss's formula*, implies that function sum() can be written without using a loop at all, as shown in Example 9.2.

Example 9.2 Function sum()**—No-Loop Version**

```
/* Function to compute the sum of the integers from 1 to n.

   Receive: n, an integer
   Return:  the sum 1 + 2 + ... + n (0 if n < 0)
   ----------------------------------------------------------------*/

int sum (int n)
{
   return n*(n + 1) / 2;
}
```

This solution obviously is better than one that uses a loop, because it solves the same problem in less time. For example, to compute the sum of the integers from 1 through 1000, the first version of `sum()` must perform 1000 additions of `count` to `runningTotal`, 1000 assignments of that result to `runningTotal`, 1000 increments of `count`, and 1000 comparisons of `count` to `n`, for a total of 4000 operations. For an arbitrary value of `n`, each of these operations would be performed n times for a total of 4n operations. We say that the number of operations performed by the loop version of `sum()` **grows linearly** with the value of its parameter n.

By contrast, the no-loop version of `sum()` always does 1 addition, 1 multiplication, and 1 division, for a total of 3 operations, regardless of the value of n. Thus, the time taken by the last version of `sum()` is **constant**, no matter what the value of its parameter n.

This is a simple example from an area of computer science called **analysis of algorithms**. If there are several algorithms that solve the same problem, we analyze the number of operations required by each. The algorithm using Gauss's formula solves the summation problem in constant time, while the algorithm using a loop solves the problem in time proportional to n; consequently, Gauss's algorithm is to be preferred.

9.1.2 Example 2: Calculating Depreciation

Depreciation is a decrease in the value over time of some asset due to wear and tear, decay, declining price, and so on. For example, suppose that a company purchases a new robot for $200,000 that will serve its needs for 5 years. After that time, called the *useful life* of the robot, it can be sold at an estimated price of $50,000, which is the robot's salvage value. Thus, the value of the robot will have depreciated $150,000 over the 5-year period. The calculation of the value lost in each of several years is an important accounting problem, and there are several ways of calculating this quantity. We want to write functions that use some of these methods to calculate depreciation for each year of an item's useful life and display tables that show these annual depreciations. Each function will receive the amount to be depreciated and the number of years in an item's useful life, and then output to the screen a depreciation table that displays the depreciation for each year.

There are several different methods of calculating depreciation. One standard method is the **straight-line method**, in which the amount to be depreciated is divided evenly over the specified number of years. For example, straight-line depreciation of $150,000 over a 5-year period gives an annual depreciation of $150,000/5 = $30,000:

Year	Depreciation
1	$30,000
2	$30,000
3	$30,000
4	$30,000
5	$30,000

With this method, the value of an asset decreases a fixed amount each year.

Another common method of calculating depreciation is called the **sum-of-the-years'-digits method**. To illustrate it, consider again depreciating $150,000 over a 5-year period. We first calculate the "sum of the years' digits," $1 + 2 + 3 + 4 + 5 = 15$. In the first year,

5/15 of $150,000 ($50,000) is depreciated; in the second year, 4/15 of $150,000 ($40,000) is depreciated; and so on, giving the following depreciation table:

Year	Depreciation
1	$50,000
2	$40,000
3	$30,000
4	$20,000
5	$10,000

The program in Example 9.3 outputs depreciation tables using these two methods of depreciation. It uses the function sum() from Example 9.2 to calculate the "sum of the years' digits."

Example 9.3 Calculating Depreciation

```cpp
/* This program computes depreciation tables using straight-line and
   sum-of-the-years'-digits methods.

   Input:  purchase price, salvage value, and useful life of an item
   Output: depreciation tables
   ----------------------------------------------------------------*/

#include <iostream>              // <<, >>, cout, cin
#include <iomanip>               // output formatters
using namespace std;

int sum(int n);
void straightLine(double amount, int numYears);
void sumOfYears(double amount, int numYears);

int main()
{
   cout << "This program computes depreciation tables using\n"
        << "straight-line and sum-of-the-years'-digits methods.\n\n";

   double purchasePrice,    // item's purchase price,
          salvageValue,     // salvage value,
          amount;           // amount to depreciate, and
   int usefulLife;          // useful life in years

   cout << "What is the item's purchase price? ";
   cin >> purchasePrice;
   cout << " salvage value? ";
   cin >> salvageValue;
   cout << " useful life? ";
   cin >> usefulLife;
   amount = purchasePrice - salvageValue;
```

```
   straightLine(amount, usefulLife);

   sumOfYears(amount, usefulLife);
}

/* Function to compute the sum of the integers from 1 to n.

   Receive: n, an integer
   Return:  the sum 1 + 2 + ... + n (0 if n < 0)
-------------------------------------------------------------------*/
int sum(int n)
{
   return n*(n + 1) / 2;
}

/* Function to output a straight-line depreciation table.

   Receive: amount to depreciate and number of years
   Output:  depreciation table
-------------------------------------------------------------------*/
void straightLine(double amount, int numYears)
{
   double depreciation = amount / numYears;

   cout << "\nYear - Depreciation"
        << "\n-------------------\n";

   cout << fixed << showpoint << right      // set up format for $$
        << setprecision(2);

   for (int year = 1; year <= numYears; year++)
      cout << setw(3) << year
           << setw(13) << depreciation << endl;
}

/* Function to output a sum-of-the-years'-digits depreciation table.

   Receive: amount to depreciate and number of years
   Return:  depreciation table
-------------------------------------------------------------------*/
void sumOfYears(double amount, int numYears)
{
   cout << "\nYear - Depreciation"
        << "\n-------------------\n";

   double yearSum = sum(numYears);
   double depreciation;

   cout << fixed << showpoint << right      // set up format for $$
        << setprecision(2);
```

```
for (int year = 1; year <= numYears; year++)
{
    depreciation = (numYears - year + 1) * amount / yearSum;
    cout << setw(3) << year
         << setw(13) << depreciation << endl;
}
}
```

SAMPLE RUN:

```
This program computes depreciation tables using
straight-line and sum-of-the-years'-digits methods.

What is the item's purchase price? 200000
   salvage value? 50000
   useful life? 5

Year - Depreciation
--------------------
  1      30000.00
  2      30000.00
  3      30000.00
  4      30000.00
  5      30000.00

Year - Depreciation
--------------------
  1      50000.00
  2      40000.00
  3      30000.00
  4      20000.00
  5      10000.00
```

9.2 THE for LOOP

Counting loops, or **counter-controlled loops,** are loops in which a set of statements is executed once for each value in a specified range:

for each value of a *counter_variable* in a specified range:

    statement

For example, our solution to the summation problem of the preceding section used a counting loop to execute the statement

```
runningTotal += count;
```

once for each value of count in the range 1 through n.

Because counting loops are used so often, nearly all programming languages provide a special statement to implement them. In C++ this is the **for statement** or **for loop**. The four components of a counting for loop were introduced in Section 5.5:

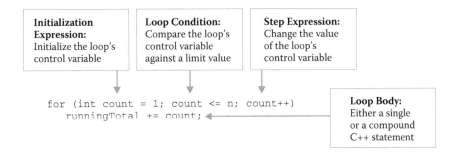

As noted in Chapter 5, a **trace table** is a useful tool to trace the action of a loop (especially in debugging). For example, in the first sample run of Example 9.1 where the value 5 is entered for n, the loop counts through the values 1 through 5, so that the body of the for loop is executed five times. The following table shows the value of the various variables and loop condition as the function sum() executes:

count	n	count <= n	Action	runningTotal
1	5	true	Execute loop body	1
2	5	true	Execute loop body	3
3	5	true	Execute loop body	6
4	5	true	Execute loop body	10
5	5	true	Execute loop body	15
6	5	false	Terminate repetition	15

A similar trace table for the second sample run where n has the value 100 would show that the loop counts through the values 1 through 100, so that the loop body is executed 100 times.

There are two forms of the for statement that are commonly used to implement counting loops: an ascending form, in which the loop control variable is incremented,

```
for (int control_variable = initial_value;
         control_variable <= limit_value;
         increment_expression)
    statement
```

and a descending form, in which the loop control variable is decremented:

```
for (int control_variable = initial_value;
         control_variable >= limit_value;
         decrement_expression)
    statement
```

The first form counts through an *ascending range*, and the second counts through a *descending range*.

To illustrate the first form, consider the following for statement:

```
for (int number = 1; number <= 10; number++)
    cout << number << '\t' << number * number << endl;
```

Here, number is the control variable, the initial value is 1, the limit value is 10, and the increment expression is number++. This for loop will execute the statement

```
cout << number << '\t' << number * number << endl;
```

once for each value of number in the ascending range 1 through 10. On the first pass through the loop, number will have the value 1; on the second pass it will have the value 2; and so on until the final pass when number will have the value 10. Thus, the output produced will be

```
1    1
2    4
3    9
4    16
5    25
6    36
7    49
8    64
9    81
10   100
```

By using an appropriate increment expression, a for statement can be used to step through a range of values in increments of size other than 1. For example, the for statement

```
for (int number = 0; number <= 100; number += 20)
    cout << number << '\t' << number * number << endl;
```

uses the expression

```
number += 20
```

to count upwards in increments of 20, producing the output

```
0    0
20   400
40   1600
60   3600
80   6400
100  10000
```

The second form of a `for` loop performs a decrement operation following each execution of the loop body. For example, the following loop

```
for (int number = 10; number >= 6; number--)
    cout << number << '\t' << number * number << endl;
```

will count downward from 10 to 6, producing the output

```
10    100
9     81
8     64
7     49
6     36
```

Note that whereas the ascending form continues the repetition as long as the control variable is less than or equal to the limit value, the descending form is counting downward, and so must continue the repetition as long as the control variable is *greater than or equal to* the limit value.

Although the preceding examples used `int` variables to control the repetition, this is not a requirement. For example, the following `for` statement

```
for (double x = 0; x <= 1; x += 0.1)
    cout << x << '\t' << exp(x) * sin(x) << endl;
```

will display a table of points on the graph of $\frac{total\ resistance = \frac{1}{\frac{1}{resistor\ 1} + \frac{1}{resistor\ 2} + \frac{1}{resistor\ 3}}}{}$ for x ranging from 0 to 1 in increments of 0.1:

```
0     0
0.1   0.110333
0.2   0.242655
0.3   0.398911
0.4   0.580944
0.5   0.790439
0.6   1.02885
0.7   1.2973
0.8   1.59651
0.9   1.92667
1     2.28736
```

The loop conditions in the preceding examples of `for` loops have used <= or >= to test if the value of the control variable has reached its limit value. In some problems, however, it may be more natural to use < instead of <= or > instead of >=. For example, in a loop that counts through the (integer) number of degrees in a circle, it seems more intuitive to write

```
for (int degrees = 0; degrees < 360; degrees++)
```

than to use `degrees <= 359` for the loop condition.

In fact, in some problems, we may not even know the limiting value of the control variable. For example, to calculate and print all sums of consecutive integers, 1, 1 + 2, 1 + 2 + 3, . . ., until this sum exceeds 1000, we could use the following `for` loop:

```
int sum = 0;
for (int count = 1; sum <= 1000; count++)
{
    sum += count;
    cout << sum << endl;
}
```

9.2.1 Nested for Loops

The statement that appears within a `for` statement may itself be a `for` statement; that is, one `for` loop may be nested within another `for` loop. When this happens, the two loops behave something like the hands of a clock:

```
for (int hours = 1; hours <= 12; hours++)
    for (int minutes = 0; minutes < 60; minutes++)
        cout << hours << ':'
             << setw(2) << setfill('0') << minutes << endl;
```

The *inner* "minutes" loop executes sixty times for each execution of the *outer* "hours" loop:

```
1:00
1:01
1:02
  .
  .
  .
1:59
2:00
2:01
2:02
  .
  .
  .
2:59
3:00
  .
  .
  .
12:59
```

The inner loop thus acts like a clock's minutes hand, and the outer loop acts like the clock's hours hand.

To see how nested loops can be useful, consider the problem of printing a multiplication table by calculating and displaying products of the form x * y for each x in the range 1 through lastX and each y in the range 1 through lastY (where lastX and lastY are arbitrary integers). Such a multiplication table can be easily generated using nested `for` statements as shown in Example 9.4.

Example 9.4 Multiplication Tables

```cpp
/* This program calculates and displays a multiplication table.

   Input:  lastX and lastY, the largest numbers to be multiplied
   Output: a list of products: 1*1 ... lastX * lastY
--------------------------------------------------------------------*/

#include <iostream>        // cout, cin, <<, >>
#include <iomanip>         // right, setw()
using namespace std;

int main()
{
  cout << "This program constructs a multiplication table\n"
          "for the values 1*1 through lastX*lastY.\n\n";

  int lastX,               // the largest numbers being multiplied
      lastY,
      product;             // the product of the two numbers

  cout << "Enter two integer limit values (lastX and lastY): ";
  cin >> lastX >> lastY;

  cout << "|";
  for (int y = 1; y <= lastY; y++)
    cout << right << setw(5) << y;
  cout << "\n---|";
  for (int y = 1; y <= lastY; y++)
    cout << "-----";
  cout << endl;

  for (int x = 1; x <= lastX; x++)
  {
    cout << setw(3) << x << "|";
    for (int y = 1; y <= lastY; y++)
    {
    product = x * y;
    cout << setw(5) << product;
    }
    cout << "\n";
    if (x < lastX) cout << "  |\n";
  }
}
```

SAMPLE RUN:

```
This program constructs a multiplication table
for the values 1*1 through lastX*lastY.

Enter two integer limit values (lastX and lastY): 4 6

    | 1      2      3      4      5      6
 ---|------------------------------------
  1 | 1      2      3      4      5      6
    |
  2 | 2      4      6      8      10     12
    |
  3 | 3      6      9      12     15     18
    |
  4 | 4      8      12     16     20     24
```

In the sample run, lastX is given the value 4 and lastY the value 6. The first for loop displays the y values as a heading for the table and the next for loop displays some underlining dashes. Control then goes on to the nested for loops, where the control variable x of the outer loop is assigned its initial value 1. The statement it controls first displays the value of x as a label for that row, and the inner loop is then executed, which counts through the values 1 through 6 for y and thus calculates and displays the first six products: 1 * 1, 1 * 2, ... , 1 * 6. Control then passes from the inner loop to the increment expression of the outer loop, where the value of x is incremented to 2. It is displayed as a label for that row and inner loop is then executed again. It again counts through the values 1 through 6 for y, but because the value of x is now 2, this pass calculates and displays the next four products: 2 * 1, 2 * 2, ... , 2 * 6. This continues until the control variable x increments to 5, making the loop condition x <= lastX false, so that repetition stops. The compound statement

```
{
   product = x * y;
   cout << setw(5) << product;
}
```

was executed a total of 24 times, because the inner loop was executed 6 times for each of the 4 executions of the outer loop.

9.2.2 Words of Warning

A for loop must be constructed carefully to ensure that its initialization expression, loop condition, and increment expression will eventually cause the loop condition to become false so that repetition terminates. To illustrate this, suppose the for loop in our earlier example to compute values of $2v^2 \sin a \cos a$ were changed to

```
for (double x = 0; x != 1; x += 0.1)
     cout << x << '\t' << exp(x) * sin(x) << endl;
```

with the termination condition x <= 1 changed to x != 1. In the following execution of this statement, the `for` loop never terminated:

```
0          0
0.1        0.110333
0.2        0.242655
0.3        0.398911
0.4        0.580944
0.5        0.790439
0.6        1.02885
0.7        1.2973
0.8        1.59651
0.9        1.92667
1          2.28736
1.1        2.67733
1.2        3.09448
1.3        3.53558
1.4        3.9962
           .
           .
           .

709.6     -5.83067e+307
709.7     -4.88986e+307
709.8     -inf
709.9     -inf
           .
           .
           .
```

The reason for this is that, as we saw in Section 3.3, most real values cannot be stored exactly. In particular, 0.1 is such a real number, and the value stored for it in memory is not exactly 0.1. Consequently, when this value was used repeatedly to increment x, the value of x was no longer exact and did not ever equal the termination value 1, which can be stored exactly.

9.2.3 The Forever Loop

The `for` statement is used primarily to implement counting loops where the number of repetitions is known (or can be computed) in advance. However, as we noted earlier, there are many problems in which it isn't possible to determine how many repetitions will be needed. What is needed for these situations is a statement that provides for *indefinite repetition*, that is, one that allows for any number of repetitions without specifying this in advance.

Although some programming languages provide a separate statement for indefinite loops, C++ does not, but instead allows the programmer to construct such loops from other loops. One way this can be done is by removing the initialization expression, the loop condition, and the step expression from a `for` loop, as illustrated in the following general form:[1]

THE FOREVER LOOP

FORM:

```
for ( ; ; )  // forever loop
    statement
```

where
    statement is usually a compound statement.

BEHAVIOR:

The specified statement will be executed an unspecified number of times until it is terminated, usually by an `if-break` or `if-return` combination that it contains (or by user intervention).

If a `break` statement is encountered, execution of the loop will terminate and will continue with the statement following the loop.

If a `return` statement is encountered, the loop and the function containing it will terminate and control returns to the calling function.

We will call such a loop a **forever loop** because it contains no built-in loop condition that specifies when repetition should terminate. It is an **indefinite loop** that executes the statements in its body until one of them causes it to stop.[2]

To illustrate, consider the following :

```
for ( ; ; )  // forever loop
    cout << "Help ! I'm stuck in a loop!\n";
```

This statement will produce the output

```
Help ! I'm stuck in a loop!
Help ! I'm stuck in a loop!
Help ! I'm stuck in a loop!
Help ! I'm stuck in a loop!
        .
        .
        .
```

an unlimited number of times, unless the user *interrupts* execution (usually by pressing the `Control` and `C` keys).

To prevent such infinite looping, the body of a forever loop is usually a compound statement that contains, in addition to the usual kinds of statements we've been using that are

---

[1] Alternatively, we can achieve the same effect with either of these forms:
    `while (true)` *statement* or do *statement* `while (true);`
[2] This will still be an indefinite loop if the initialization expression and the step expression are retained. Sometimes it is convenient to use one or both of them.

executed repeatedly, a statement that will *terminate* execution of the loop when some condition is satisfied. This terminating statement is usually an **if-break combination**—an if statement containing a break statement,

```
if (condition) break;
```

When *condition* becomes true, the break statement causes repetition to stop and control is transferred to the statement following the loop. To distinguish this condition from the loop conditions of the other loops we will call it a **termination condition** instead of a *loop condition*.

Most forever loops, therefore, have the following form:

```
for (;;)                              // loop:
{
   statement_list₁
   if (termination_condition) break;
   statement_list₂
}                                     // end loop
```

where either *statement_list₁* or *statement_list₂* can be empty.

To illustrate forever loops, here is a useful utility function called getMenuChoice() that receives a menu and the characters that denote the first and last choices on the menu. (A precondition of this function is that the menu choices are a closed range such as A–D.) It repeatedly displays the menu and reads the user's choice until that choice is in the range of valid choices:

```
char getMenuChoice(string MENU,
                   char firstChoice, char lastChoice)
{
   char choice;                         // what the user enters
   for (;;)                             // loop:
   {
      cout << MENU;                     // statement_list₁
      cin >> choice;
                                        // if break combination
      if ((choice >= firstChoice) && (choice <= lastChoice))
         break;
                                        // statement_list₂
      cerr << "\nI'm sorry, but " << choice
           << " is not a valid menu choice.\n";
   }                                    // end loop
   return choice;
}
```

The effect here is to "lock" the user inside the forever loop until a valid menu choice is entered. For each invalid menu choice, the termination condition is false, so the break statement is bypassed, the output statement displays an error message, and control returns to the beginning of the loop for the next repetition and gives the user another chance.

When a valid choice is entered, the termination condition becomes true, so the break statement is executed and transfers control to the return statement following the loop.

A statement related to break that is sometimes useful for modifying (but not terminating) execution of a loop is a **continue statement**. When it is executed, the current iteration is terminated and a new one begins. The continue statement is useful when one wants to skip to the bottom part of a loop if a certain condition is true:

```
for (...)
{
   ...
   if (condition) continue;
   //- Skip from here to end of loop body if
   //- condition is true and begin a new iteration
}
```

9.2.3.1 Returning From a Loop

An alternative and somewhat clearer termination condition for the forever loop in get-MenuChoice() is to have the function return immediately when the condition is true instead of breaking out of the loop to a return statement:

```
char getMenuChoice(string MENU,
                   char firstChoice, char lastChoice)
{
   char choice;                          // what the user enters
   for (;;)                              // loop:
   {
      cout << MENU;                      // statement_list₁
      cin >> choice;
                                         // if break combination
      if ((choice >= firstChoice) && (choice <= lastChoice))
         return choice;
                                         // statement_list₂
      cerr << "\nI'm sorry, but " << choice
           << " is not a valid menu choice.\n";
   }                                     // end loop
}
```

As before, an invalid choice produces an error message and the body of the loop is repeated. A valid choice causes the return statement to be executed, terminating not only the loop but also the function.

9.3 THE while LOOP

A loop of the form

```
loop
      if (termination_condition) exit the loop.
      other statements
end loop
```

in which the termination test occurs before the loop's statements are executed is called a
pretest or **test-at-the-top** loop. Although such loops can be implemented in C++ using a
forever loop,

```
for (;;)                // loop:
{
  if (termination_condition) break;
  statement_list
}                       // end loop
```

the **while loop** that we introduced briefly in Section 5.5 provides a simpler alternative. We
will use a while loop to solve the following problem.

9.3.1 Example: Follow the Bouncing Ball

Suppose that when a ball is dropped and bounces from the pavement, it reaches a height
one-half of its previous height. The program in Example 9.5 simulates the behavior of the
ball when it is dropped from a given height that is input to the program. It displays the
number of each bounce and the height of that bounce, repeating this until the height is less
than a constant SMALL_NUMBER set to 1 millimeter.[3]

Example 9.5 Bouncing Ball

```
/* This program calculates and displays the rebound heights of a
   dropped ball.

   Input:   a height in meters from which a ball is dropped.
   Output:  for each rebound of the ball from the pavement below:
            the number and the height of that rebound
            assuming that the height of each rebound is
            one-half the previous height
------------------------------------------------------------------*/

#include <iostream>                        //   <<, >>, cout, cin
using namespace std;

int main()
{
  const double SMALL_NUMBER = 1.0e−3;  //  1 millimeter
```

---

[3] A for loop could also be used:
```
for (int bounce = 0; height >= SMALL_NUMBER; bounce++)
{
  height /= 2;
  cout << "Rebound # " << bounce << ": "
       << height << " meters" << endl;
}
```

```cpp
cout << "This program computes the number and height\n"
    << "of the rebounds of a dropped ball.\n";

cout << "\nEnter the starting height (in meters): "; double height;
cin >> height;
cout << "\nStarting height: " << height << " meters\n";

int bounce = 0;
while (height >= SMALL_NUMBER)
{
  height /= 2.0;
  bounce++;
  cout << "Rebound # " << bounce << ": "
      << height << " meters" << endl;
}
}
```

SAMPLE RUN:

```
This program computes the number and height
of the rebounds of a dropped ball.

Enter the starting height (in meters): 15

Starting height: 15 meters
Rebound # 1: 7.5 meters
Rebound # 2: 3.75 meters
Rebound # 3: 1.875 meters
Rebound # 4: 0.9375 meters
Rebound # 5: 0.46875 meters
Rebound # 6: 0.234375 meters
Rebound # 7: 0.117188 meters
Rebound # 8: 0.0585938 meters
Rebound # 9: 0.0292969 meters
Rebound # 10: 0.0146484 meters
Rebound # 11: 0.00732422 meters
Rebound # 12: 0.00366211 meters
Rebound # 13: 0.00183105 meters
Rebound # 14: 0.000915527 meters
```

9.3.2 The while Statement

While loops are implemented in C++ using a **while statement** of the following form:

THE while STATEMENT
FORM:

```cpp
while (loop_condition)
    statement
```

where
   while is a C++ keyword;
   loop_condition is a boolean expression; and
   statement is a simple or a compound statement.

BEHAVIOR:

When execution reaches a while statement:

   1. *loop_condition* is evaluated.
   2. If *loop_condition* is true:
        a. The specified *statement*, called the **body** of the loop, is executed.
        b. Control returns to Step 1.
      Otherwise
        Control is transferred to the statement following the while statement.

As in a for loop, the loop condition in a while loop is placed before the body of the loop, which means that a while loop is a pretest loop:

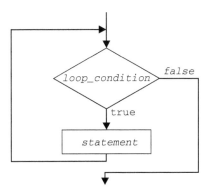

As this diagram indicates, this condition is evaluated before the body of the loop is executed, and execution of the specified statement is repeated as long as the loop condition remains true and terminates when it becomes false. It also indicates that if this condition is initially false, the body of the loop will not be executed. This is sometimes given the name *zero-trip behavior* because the body of a pretest loop will be executed zero or more times. Thus, in the program in Example 9.5, if we input a value for height that is less than SMALL_NUMBER, the loop condition height >= SMALL_NUMBER will be false the first time it is evaluated, and the statements in the while loop will not be executed. It is important to keep this zero-trip behavior in mind when trying to decide which loop to use in designing a solution for a problem.

9.4 THE do STATEMENT

In the last section, we considered *pretest* (or *test-at-the-top*) loops of the form

    loop
        if (termination_condition) exit the loop.
        other statements
    end loop

and saw that `while` loops have this structure. Such loops are said to have *zero-trip behavior* because they will make zero "trips" through the loop if the termination condition is false initially.

There are some problems, however, for which zero-trip behavior is not appropriate and at least one pass through the loop is mandatory. For such problems, we need a *posttest* loop of the form

```
loop
    other statements
    if (termination_condition) exit the loop.
end loop
```

and in this section we will look at C++'s implementation of posttest loops.

9.4.1 Example: How Many Digits?

It is easy for us to determine the number of digits in an integer, whether it is 214 or 2147483647, simply by scanning it and counting. But how do we construct a program that can "scan and count" the digits?

Our approach will be to repeatedly divide the number by 10, counting how many times it takes to produce 0. For example, 214 has three digits, and three integer divisions by 10 are required to reach 0:

The `int` value	Operation I Divide by 10	Operation II Add 1 to Digit Count
214	$214/10 = 21$	$0 + 1 = 1$
21	$21/10 = 2$	$1 + 1 = 2$
2	$2/10 = 0$	$2 + 1 = 3$

The key observation in the preceding computation is that we must perform the two operations *at least once* because every integer has at least one digit. Thus, the loop that we use must make at least one pass through its body—*one-trip behavior.* Putting the loop condition at the "bottom" of the loop, as in the following algorithm, will accomplish this.

Algorithm to Count Digits

1. Initialize *digitCount* to 0.

2. Loop:

    a. Increment *digitCount.*

    b. Divide *intValue* by 10, storing the result in *intValue.*

    c. If *intValue* is 0, terminate the repetition.

  End loop

3. Return *numDigits.*

The loop in this algorithm can be implemented with a forever loop:

```
for (;;)
{
  numDigits++;
  intValue /= 10;
  if (intValue == 0) break;
}
```

But as with the test-at-the-top loops in the preceding section, C++ provides a clearer alternative—the do statement. Example 9.6 illustrates its use.

Example 9.6 Digit Counter

```
/* This program uses the function digitsIn() to find the number of
   digits in an integer value.

   Input:  an integer
   Output: the number of digits in the integer
-------------------------------------------------------------------*/

#include <iostream>                          // <<, >>, cout, cin
using namespace std;

int digitsIn(int intValue);

int main()
{
  int theValue;

  cout << "Enter an integer value: ";
  cin >> theValue;
  cout << theValue << " contains "
       << digitsIn(theValue) << " digit(s).\n";
}

/* Function to count the digits in an integer value.

   Receive: intValue, an integer value
   Return:  intDigits, the number of digits in intValue
-------------------------------------------------------------------*/

int digitsIn(int intValue)
{
  int numDigits = 0;
```

```
do
{
   numDigits++;
   intValue /= 10;
}
while (intValue != 0);

return numDigits;
}
```

SAMPLE RUNS:

```
Enter an integer value: 5
5 contains 1 digit(s).

Enter an integer value: 12345678
12345678 contains 8 digit(s).
```

9.4.2 A Posttest Loop

The function in Example 9.6 uses a new C++ repetition statement, the **do statement**, which has the following form:

THE do STATEMENT

FORM:

```
do
   statement
while (loop_condition);
```

where
> do and while are C++ keywords;
> *loop_condition* is a boolean expression;
> *statement* is a simple or a compound statement; and
> a semicolon must follow the expression at the end of the statement.

BEHAVIOR:

When execution reaches a do loop:

1. *statement* is executed.
2. *loop_condition* is evaluated.
3. If *loop_condition* is true, then
      Control returns to Step 1.
    Otherwise
      Control is transferred to the statement following the do loop.

Note that the loop condition in a do statement appears after the body of the loop, as pictured in the following diagram:

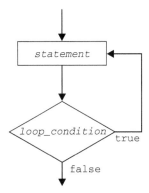

This means that a do loop is a **posttest** or **test-at-the-bottom loop** and has **one-trip behavior**; that is, the body of the loop will be executed at least once.

9.5 INPUT LOOPS

One important use of loops is to input a collection of values into a program. Because the number of data values may not be known before repetition begins, we need some method to signal that the end of the data has been reached. In this section, we look at three different approaches: *counting*, *sentinels*, and *queries*. To illustrate how these three techniques are used, we will use each of them in solving the problem of calculating the average of a set of failure times.

9.5.1 Running Example: Mean Time to Failure

One important statistic that is used in measuring the reliability of a component in a circuit is the mean time to failure, which can be used to predict the circuit's lifetime. This is especially important in situations in which repair is difficult or even impractical, such as a computer circuit in a satellite. Suppose that an engineering laboratory has been awarded a contract by NASA to evaluate the reliability of a particular component for a space probe to Mars. As part of this evaluation, several of these circuits have been tested, and the times at which each failed have been recorded. We are to develop a program to process this data and determine the mean time to failure.

We will solve this problem in three different ways to illustrate the three kinds of input loops—a counting approach, a query-based approach, and a sentinel-based approach. Each will involve performing the following tasks (but not necessarily in the order listed):

- Determine the number of components (*numComponents*)

- Repeatedly input a failure time (*failureTime*)

- Find their sum (*failureTimeSum*)

- If *numComponents* is not zero, divide the sum by *numComponents* to find the mean failure time (*meanFailureTime*); otherwise, display an appropriate "No Data" message.

9.5.2 Input Loops: The Counting Approach

Probably the easiest way to process a set of input values is to first input the number of values in the data set and then use a counting loop to read and process that many values. The program in Example 9.7 uses this approach.

Example 9.7 The Counting Approach

```
/* This program uses a counting loop to process a collection of
   failure times and find the mean time to failure.

   Input:   the number of component failure times and the collection
            of failure times
   Output: prompts and the average of the failure times
 ----------------------------------------------------------------*/

#include <iostream>                    // <<, >>, cout, cin
using namespace std;

int main()
{
  cout << "Computing Component Mean Time to Failure\n\n";

  int numComponents;
  double failureTime,
         failureTimeSum = 0.0;

  cout << "How many failure times will be entered? ";
  cin >> numComponents;

  for (int count = 1; count <= numComponents; count++)
  {
    cout << "Enter failure time #" << count << ": ";
    cin >> failureTime;
    failureTimeSum += failureTime;
  }

  if (numComponents > 0)
    cout << "\nThe mean failure time of the "
         << numComponents << " components is "
         << failureTimeSum / numComponents << endl;
  else
    cerr << "\nNo failure times to process!\n";
}
```

SAMPLE RUN:

```
Computing Component Mean Time to Failure

How many failure times will be entered? 5
Enter failure time #1: 2.3
```

```
Enter failure time #2: 2.5
Enter failure time #3: 2.4
Enter failure time #4: 2.6
Enter failure time #5: 2.1

The mean failure time of the 5 components is 2.38
```

Conceptually, this approach is quite simple. We ask the user how many values are to be entered and then use a `for` loop to read and process that many values. The general pattern is as follows:

PATTERN FOR COUNTING INPUT LOOP

Display a prompt for the number of values to be processed.
Read *numberOfValues* to be processed.
`for (int var = 1; var <= numberOfValues; var++)`
{
    Display a prompt for a data value.
    Read *theValue* to be processed.
    Process *theValue*.
}

The main disadvantage of this approach is that it requires knowing in advance how many values will be entered. This may be difficult and even impractical to determine for large data sets. The two other types of input loops circumvent this difficulty by having the computer, rather than the user, do the counting.

9.5.3 Input Loops: The Sentinel Approach

The second approach is the **sentinel approach** introduced in Section 5.5. It uses a special value called an **end-of-data flag** or **sentinel** to signal the end of the data values to be processed.

9.5.3.1 Forever Loops and Sentinels

The program in Example 9.8 uses a forever loop to implement the sentinel approach.

Example 9.8 The Sentinel Approach in a Forever Loop

```
/* This failure-time program uses a forever loop and a sentinel to
   process a collection of failure times and find the mean time to
   failure.

   Input:         a collection of component failure times
   Precondition:  failure times >= 0
   Output:        prompts and the average of the failure times
-------------------------------------------------------------------*/
```

```cpp
#include <iostream>                      // <<, >>, cout, cin
using namespace std;

int main()
{
  cout << "Computing Component Mean Time to Failure\n\n";
  int numComponents = 0;
  double failureTime,
         failureTimeSum = 0;
  for (;;)               // or while (true)
  {
    cout << "Enter a failure time (-1 to quit): ";
    cin >> failureTime;

    if (failureTime < 0) break;

    failureTimeSum += failureTime;
    numComponents++;
  }

  if (numComponents != 0)
    cout << "\nThe mean failure time of the "
         << numComponents << " components is "
         << failureTimeSum / numComponents << endl;
  else
    cerr << "\nNo failure times to process!\n";
}
```

SAMPLE RUN:

```
Computing Component Mean Time to Failure

Enter a failure time (-1 to quit): 2.3
Enter a failure time (-1 to quit): 2.4
Enter a failure time (-1 to quit): 2.5
Enter a failure time (-1 to quit): 2.6
Enter a failure time (-1 to quit): 2.1
Enter a failure time (-1 to quit): -1

The mean failure time of the 5 components is 2.38
```

As the comment at the beginning of the forever loop indicates, a `while` statement could also be used to implement the input loop:

```cpp
while (true)
{
  cout << "Enter a failure time (-1 to quit): ";
  cin >> failureTime;

  if (failureTime < 0) break;
```

```
    failureTimeSum += failureTime;
    numComponents++;
}
```

If we examine this and earlier examples for similarities, we see the following pattern:

PATTERN FOR SENTINEL INPUT LOOP (FOREVER LOOP VERSION)

```
for (;;)                    // or while(true)
{
    Display a prompt (for a data value).
    Read theValue to be processed.
    If (theValue is the sentinel) terminate the repetition.
    Process theValue.
}
```

This pattern provides a succinct and intuitive way to input any data set for which a sentinel value exists.

9.5.3.2 While Loops and Sentinels

Some programmers prefer not to use forever loops and choose instead "standard" while loops as described in Section 9.3. We gave an example of using a while loop to input coefficients of a quadratic equation and output the solution (or an error message if there was none) in Section 5.5. The general pattern for such while-based input as follows:

PATTERN FOR SENTINEL INPUT LOOP (WHILE LOOP VERSION)

```
Display a prompt (for a data value).
Read theValue to be processed.
while (theValue is not the sentinel)
{
    Process theValue.
    Display a prompt (for a data value).
    Read theValue to be processed.
}
```

For example, a while loop version of the sentinel loop used in the mean-time-to-failure program in Example 9.8 would be written as follows:

```
cout << "Enter a failure time (-1 to quit): ";
cin >> failureTime;
while (failureTime >= 0)
{
    failureTimeSum += failureTime;
    numComponents++;
```

```
    cout << "Enter a failure time (-1 to quit): ";
    cin >> failureTime;
}
```

Note the two input steps, one before the loop and one at the bottom of the loop. The input statement before the loop is needed because the `while` loop is a pretest loop that tests its condition upon entry into the loop. The input statement at the bottom of the loop is needed to input a new value that can be compared with the sentinel when execution returns to the beginning of the loop.

9.5.3.3 End-of-File as a Sentinel Value

For some problems there may not be a suitable sentinel value because all values are valid inputs. In such cases, one could use the query approach described next. But there is an alternative. In Chapter 7, some of the status flags for `istream` objects were described such as `cin` that monitor the state of a stream. In addition to those, there is an *eof flag* that is set when a special character called the **end-of-file** (or **eof**) **mark** is read and a function named `eof()` that monitors this flag, returning `true` if the eof flag is set and `false` otherwise.[4]

This function can be used like a sentinel test to check whether the eof mark was entered as in Example 9.9, signaling the end of input. For the particular (Unix) system used in the sample run, the eof mark is `Control-d`, so entering it sets the eof flag in `cin`, causing the function member `cin.eof()` to return `true`, which terminates the repetition. (For input from files as described in Chapter 11, the end-of-file mark is placed automatically at the end of the file when it is created.)

Example 9.9 Using End-of-File as a Sentinel

```
/* This failure-time program uses a forever loop and the eof
   character as a sentinel to process a collection of failure
   times and find the mean time to failure.

   Input:  a collection of component failure times
   Output: prompts and the average of the failure times
   ----------------------------------------------------------------*/

#include <iostream>            // <<, >>, cout, cin, eof()
using namespace std;

int main()
{
   cout << "Computing Component Mean Time to Failure\n\n";
   int   numComponents = 0;
   double failureTime,
          failureTimeSum = 0;
```

---

[4] In the Windows environment, the eof mark can be entered by typing `Control-z` followed by the `Enter` key. In the Unix and Macintosh environments, `Control-d` is used.

```
for (;;)                    // or while (true)
{
   cout << "Enter a failure time (eof char to quit): ";
   cin >> failureTime;

   if (cin.eof()) break;

   failureTimeSum += failureTime;
   numComponents++;
}

if (numComponents != 0)
   cout << "\nThe mean failure time of the "
        << numComponents << " components is "
        << failureTimeSum / numComponents << endl;
else
   cerr << "\nNo failure times to process!\n";
}
```

SAMPLE RUN:

```
Computing Component Mean Time to Failure

Enter a failure time (eof char to quit): 2.3
Enter a failure time (eof char to quit): 2.4
Enter a failure time (eof char to quit): 2.5
Enter a failure time (eof char to quit): 2.6
Enter a failure time (eof char to quit): 2.1
Enter a failure time (eof char to quit): ^d
The mean failure time of the 5 components is 2.38
```

There are two drawbacks to using the eof mark as a sentinel. The first is that it is *platform dependent*, that is, it depends on the particular operating system being used. The second is that, as discussed in Section 7.3, if we wish to read any values after the eof mark has been read, the `istream` function member `clear()` must be called to reset the status flags.

9.5.4 Input Loops: The Query Approach

Each of the preceding kinds of input loops has its disadvantages. The counting loop approach requires knowing in advance the number of values to be entered. The sentinel approach can only be used in problems where there is a suitable value to use as the sentinel. The eof mark is platform dependent. The final approach we consider is to **query the user** at the end of each repetition to determine whether there is more data to process. Although not without its disadvantages, this approach is the most broadly applicable.

The program in Example 9.10 is a modification of the previous mean-time-to-failure programs that use this query approach. After a data value is entered by the user, the program asks

```
Do you have more data to enter (y or n)?
```

and then reads the user's response from the keyboard. If it is y or Y, another data value is read. This continues until the user answers something other than y or Y, and the loop condition

```
response == 'y' || response == 'Y'
```

then terminates the repetition.

Example 9.10 The Query Approach

```cpp
/* This program uses a query-controlled loop to process a collection
   of failure times and find the mean time to failure.

   Input:   a collection of component failure times and user's
            response to "more data?" query
   Output: prompts and the average of the failure times
   ----------------------------------------------------------------*/

#include <iostream>                 // <<, >>, cout, cin
using namespace std;

int main()
{
  cout << "Computing Component Mean Time to Failure\n\n";

  int numComponents;
  double failureTime,
         failureTimeSum = 0.0;

  char response;
  do
  {
     cout << "\nEnter a failure time: ";
     cin >> failureTime;

     failureTimeSum += failureTime;
     numComponents++;

     cout << "Do you have more data to enter (y or n)? ";
     cin >> response;
  }
  while (response == 'y' || response == 'Y');

  cout << "\nThe mean failure time of the "
       << numComponents << " components is "
       << failureTimeSum / numComponents << endl;
}
```

SAMPLE RUN:

```
Computing Component Mean Time to Failure
```

```
Enter a failure time: 2.3
Do you have more data to enter (y or n)? y

Enter a failure time: 2.4
Do you have more data to enter (y or n)? y

Enter a failure time: 2.5
Do you have more data to enter (y or n)? y

Enter a failure time: 2.6
Do you have more data to enter (y or n)? y

Enter a failure time: 2.1
Do you have more data to enter (y or n)? n

The mean failure time of the 5 components is 2.38
```

In problems where it is reasonable to assume that there is at least one data value to be entered and processed, the query and the corresponding loop condition are placed at the bottom of the loop, making it a posttest loop.[5] This suggests the following *pattern* for a query-controlled input loop:

PATTERN FOR QUERY-CONTROLLED INPUT LOOP

```
char response;
do
{
    Display a prompt for a data value.
    Read theValue to be processed.
    Process theValue.
    Display a query that asks if there is more data.
    Input the user's response (y or n)
}
while ((response != 'n' && (response != 'N'));
```

---

[5] To allow no values to be entered and processed, a forever loop can be used:
```
for (;;)
{
  cout << "\nDo you wish to continue (y or n)? ";
  char response;
  cin >> response;
  if (response == 'n' || response == 'N') break;
  cout << "\nEnter a value: ";
  double value;
  cin >> value;
  // process value
}
```

9.5.4.1 Query Functions

The code to perform a query tends to clutter a loop, which may obscure the program's structure, especially if the query is lengthy, perhaps due to special instructions to the user. One way to avoid this is to use a **query function** like the following to perform the query, read the user's response, and return true or false based on the response, thus hiding the actual code that does the querying:

```cpp
bool moreData()
{
    char answer;
    cout << "Do you have more values to enter (y or n)? ";
    cin >> answer;
    return (answer == 'y') || (answer == 'Y');
}
```

Because it returns a boolean value, a call to such a **query function** can be used as a loop condition. For example, we could modify the mean-time-to-failure program of Example 9.10 to use the query function moreData() to control the input loop:

```cpp
int main()
{
    .
    .
    .
    do
    {
        cout << "\nEnter a failure time: ";
        cin >> failureTime;
        failureTimeSum += failureTime;
        numComponents++;
    }
    while ( moreData() );
    .
    .
    .
}
```

Each time execution reaches the loop condition, the function moreData() is called. It queries the user, reads the response, and returns true if the response was either y or Y and returns false otherwise. If moreData() returns true, the body of the do loop is repeated, but if it returns false, repetition is terminated.

The general pattern of a loop controlled by a query function is as follows:

PATTERN FOR INPUT LOOP CONTROLLED BY A QUERY FUNCTION

```
bool queryFunction();

     .
     .
     .
do
{
    Display a prompt for a data value.
    Read theValue to be processed.
    Process theValue.
}
while ( queryFunction() );
```

9.5.4.2 The Disadvantage of the Query Approach

The counting and sentinel approaches require one interaction by the user to enter each data value, but the query approach requires two—one for the data value, and one for the response to the query. This doubling of user effort may make the query approach too cumbersome for large data sets.

9.6 CHOOSING THE RIGHT LOOP

With several different kinds of loops, it can be somewhat difficult to decide which to use. Asking the following questions as the problem is analyzed and a design plan developed may help.

The first question is

Does the algorithm require counting through some fixed range of values?

If the answer is yes, then a counting loop is needed, and a `for` loop is the appropriate choice. Otherwise, one of the more general loops—`while`, `do`, or `forever`—is a better choice.

In the second case, the next question is:

Which of the general loops should I use?

If the answer isn't apparent, you might begin in the algorithm with a *generic loop* of the form

```
Loop
    body-of-the-loop
End loop
```

As you continue developing the algorithm, add any necessary initialization statements before the loop and the statements that make up the body of the loop:

initialization statements
Loop
    statement$_1$
       ·
       ·
       ·
    statement$_n$
End loop

Finally, formulate an appropriate termination condition. Then, determining where it should be placed in the loop will lead to which kind of loop to use:

If the termination condition appears

- at the beginning of the loop, it's a pretest loop; choose a `while` loop;
- at the bottom of the loop, it's a posttest loop; choose a `do` loop;
- within the list of statements, it's a test-in-the-middle loop; choose a forever loop with an `if-break` (or `if-return`) combination.

To demonstrate this procedure, let's look again at the bouncing-ball problem in Section 9.3. Using a generic loop, we might use the following as a first version of the algorithm:

1. Initialize *bounce* to 0.

2. Enter a value for *height* and display it.

3. Loop

    a. Replace *height* with *height* divided by 2.

    b. Increment *bounce* by 1.

    c. Display *bounce* and *height*.

    End loop

As a termination condition for the loop, we can use

$$height < SMALL\_NUMBER$$

because repetition is to stop when *height* is less than some *SMALL_NUMBER*. However, the user could have entered zero or a negative value for height, in which case the body of the loop should not be executed. Thus, we should evaluate this condition immediately upon entering the loop:

1. Initialize *bounce* to 0.

2. Enter a value for *height*.

3. Display original *height* value with appropriate label.

4. Loop

    a. If *height* < *SMALL_NUMBER*, terminate the repetition.

    b. Replace *height* with *height* divided by 2.

    c. Add 1 to *bounce.*

    d. Display *bounce* and *height.*

  End loop

This is a pretest loop, and we should therefore use a `while` loop to implement it.

By contrast, if we reconsider the sentinel approach to reading a collection of values, we begin by constructing the generic loop

Loop

    Display a prompt for input.

    Input *theValue.*

    Process *theValue.*

  End loop

Because we are using the sentinel approach, an appropriate termination condition is

    theValue is the sentinel

Before this termination condition can be evaluated, *theValue* must have been read, which means that the termination condition must appear after the input statement. Also, a sentinel value must not be processed, which means that the termination condition should be placed before the processing statements:

Loop

    a. Display a prompt for input.

    b. Input *theValue.*

    c. If *theValue* is the sentinel, terminate repetition.

    d. Process *theValue.*

  End loop

This is a test-in-the-middle loop, and we can use a forever loop to implement it.

CHAPTER SUMMARY

Key Terms

analysis of algorithms	initialization list
`break` statement	nested loops
`continue` statement	one-trip behavior
counting (or counter-controlled) loops	posttest loop
do loop	pretest loop
`do` statement	query-controlled loop
end-of-data flag	query function
end-of-file mark	sentinel
for loop	test-at-the-bottom loop
forever loop	termination condition
grows linearly	trace table
`if`-`break` combination	while loop
indefinite loop	zero-trip behavior

NOTES

- A trace table is a useful tool (especially in debugging) to trace the action of a loop.

- A `for` statement can be used to step through a range of values with any increment by using an appropriate increment expression.

- An indefinite loop can be implemented in C++ with a `for` statement with no loop condition—`for(;;){ ... }`—or a `while` statement of the form `while(true){ ... }` can be used.

- A forever loop continues repetition when its termination condition is false and terminates repetition when that condition is true.

- A `break` statement can be used to terminate execution of an enclosing loop. It terminates the innermost enclosing loop.

- In some cases, a `return` statement is a useful alternative to `break` for terminating execution of a function and returning to the calling function.

- A `continue` statement is useful for skipping the rest of the current iteration and beginning a new one.

- A *while* loop is a *pretest* (or test-at-the-top) loop. It continues repetition so long as its condition is true, and terminates repetition when that condition is false. A while loop has *zero-trip behavior.*

- A *do* loop is a *posttest* (or test-at-the-bottom) loop. Because its loop body will always be executed at least once before the loop condition is tested, it is said to have *one-trip behavior.*

- Three common kinds of input loops are:

  - *counting* loops in which the number of input items is known in advance;

  - *sentinel-controlled* loops in which an end-of-data flag (or sentinel) signals the end of data;

  - *query-controlled* loops in which the user is asked whether there is more data.

- The following guidelines may help with deciding which kind of loop to use:

  1. If the algorithm requires counting through some fixed range of values, use a counting loop.

  2. If a general loop is required, use a generic indefinite loop to formulate the algorithm. Then determine where the termination condition should go. If at the beginning, use a while loop; if at the end, use a do loop; if somewhere else, use a forever loop with an `if-break` or `if-return`.

Style and Design Tips

In this text, we use the following conventions for formatting the repetition statements considered in this chapter.

- *The statement in a* `for`, `while`, `do`, *and forever loop is indented. If the statement is compound, the curly braces are aligned with the* `for`, `while`, *or* `do`, *and the statements inside the block are indented. In a* `do` *loop,* `do` *is aligned with its corresponding* `while`.

```
for (...)
    statement
```

```
for (...)
{
    statement₁
    ...
    statementₙ
}
```

```
while(loop_condition)
    statement
```

```
while(loop_condition)
{
    statement₁
    ...
    statementₙ
}
```

```
do                                          do
    statement                               {
while (loop_condition);                         statement₁
                                                ...
                                                statementₙ
                                            }
                                            while (loop_condition);
```

Warnings

1. *Care must be taken to avoid infinite looping.*

   - *The loop condition of a* for *loop must eventually become false; the body of a* while *loop or a* do *loop must contain statements that eventually cause its loop condition to become false.* **For example, the code fragment**

     ```
     x = 0.0;
     do
     {
        cout << x << endl;
        x += 0.3;
     }
     while (x != 1.0);
     ```

     produces an infinite loop because the value of x is never equal to 1.0:

     ```
     0.0
     0.3
     0.6
     0.9
     1.2
     1.5
     1.8
       .
       .
       .
     ```

   - *The body of a forever loop should always contain an* if-break *or* if-return *combination and statements that ensure that the termination condition of the loop will eventually become true.*

2. *The body of a* while *loop will not be executed if the loop condition is false initially, but the statements in a* do *loop will always be executed at least once.*

3. *Be sure to enclose multiple statements in the body of a* for, while, do, *or forever loop within curly braces because these loops control only a single statement.* **For example, the output statement in the following segment is outside the body of the loop**

   ```
   for (int i = 1; i <= 10; i++)
      j = i*i;
      cout << j << endl;
   ```

and thus will display only a single value,

```
100
```

Likewise, the statement that increments count in the following is outside the body of the loop:

```
int count = 1;
while (count <= 10)
    cout << count << " " << count*count << endl;
    count++;
```

and an infinite loop results:

```
1 1
1 1
1 1
 .
 .
 .
```

4. *In a do loop, the closing* while (loop_condition) *must be followed by a semicolon.*

5. *In a* for *loop, the control variable as well as variables in the loop condition should not usually be modified within the body of the loop, because it is intended to run through a specified range of consecutive values. Strange and undesirable results may be produced otherwise.*

6. *Each use of the equality operator in a loop condition should be double-checked to make certain that the assignment operator is not being used.* Using = instead of == is one of the easiest errors to make. This error is illustrated by the following:

```
do
{
    // ... do some processing ...

    cout << "Do you wish to continue (y or n)? ";
    cin >> answer;
}
while (answer = 'y');
```

This loop will be executed infinitely many times, regardless of what the user enters, because

a. The loop condition is an *assignment* that sets answer to y (it is not a *comparison*);

b. The assignment operator (=) produces the value that was assigned as its result;

c. This assignment thus produces the value 121 (the ASCII value of character y); and

d. C++ treats any nonzero value as true.

Similarly, the forever loop

```
for
{
    cout << "\nPlease enter an integer value (0 to quit): ";
    cin >> value;

    if (value = 0) break;

    // ... do something with value ...
}
```

is an infinite loop, because its termination condition is an assignment, not a comparison. Because the result of that assignment is zero, which represents the value false in C++, this termination condition will always be false, and so the break statement will never be executed.

TEST YOURSELF

Section 9.4

1. _____ loops execute a set of statements once for each value in a specified range and they are implemented in C++ by a _____ statement.

2. What are the four components of a counting for loop?

3. The terminating statement in a forever loop is usually a(n) _____ combination.

Answer Questions 4–7 using "pretest" or "posttest."

4. A while loop is a _____ loop.

5. A do loop is a _____ loop.

6. The body of a _____ loop is always executed at least once.

7. A _____ loop has zero-trip behavior.

For Questions 8–15, describe the output produced. For Exercises 11–15, assume that i, j, and k are of type int.

8.
```
for (int i = 0; i < 10; i++)
    cout << "2*" << i << " = " << 2*i << endl;
```

9.
```
for (int i = 0; i <= 5; i++)
    cout << 2*i + 1 << " ";
cout << endl;
```

10.
```
for (int i = 1; i < 4; i++
{
    cout << i;
```

```
      for (int j = i; j >= 1; j--)
         cout << j << endl;
   }
```

11.
```
   i = 0;
   j = 0;
   for (;;)
   {
      k = 2 * i * j;
      if (k > 10) break;
      cout << i << j << k << endl;
      i++;
      j++;
   }
   cout << k << endl;
```

12.
```
   k = 5;
   i = -2;
   while (i <= k)
   {
      i += 2;
      k--;
      cout << i + k << endl;
   }
```

13.
```
   i = 4;
   while (i >= 0)
   {
      i--;
      cout << i << endl;
   }
   cout << "\n*****\n";
```

14.
```
   i = 0;
   do
   {
      k = i * i + 1;
      cout << i << ' ' << k << endl;
      i++;
   }
   while (k <= 10);
```

15.
```
i = 4;
do
{
   k = i * i - 4;
   cout << i << ' ' << k << endl;
   i--;
}
while (k >= 0);
```

Section 9.6

1. Name the three kinds of input loops.

2. A special value used to signal the end of data is called a(n) _____ or _____.

3. (True or false) A disadvantage of using a while loop instead of a forever loop for sentinel-based input is that duplicate input steps are required.

4. The eof flag is set when a special character called the _____ mark is read.

5. The method _____ from class istream is used to check the eof flag.

6. (True or false) One advantage of using the eof mark is that it is platform independent.

7. (True or false) The counting method of input is one of the most flexible methods.

8. A _____ is a question asked of the user to determine whether there are more data values.

EXERCISES

Section 9.4

For Exercises 1–14, describe the output produced. For Exercises 7–14, assume that i, j, and k are of type int.

1.
```
for (int i = 10; i > 0; i--)
    cout << i << " cubed = " << i*i*i << endl;
```

2.
```
for (int i = 10; i > 0; i -= 2)
    cout << i << " squared = " << i*i << endl;
```

3.
```
for (int i = 1; i <= 5; i++)
{
    cout << i << endl;
    for (int j = i; j >= 1; j -= 2)
        cout << j << endl;
}
```

```
4. int k = 5;
   for (int i = -2; i < 5; i += 2)
   {
      cout << i + k << endl;
      k = 1;
   }

5. for (int i = 3; i > 0; i--)
      for (int j = 1; j <= i; j++)
         for (int k = i; k >= j; k--)
            cout << i << j << k << endl;

6. for (int i = 1; i <= 3; i++)
      for (int j = 1; j <= 3; j++)
      {
         for (int k = i; k <= j; k++)
            cout << i << j << k << endl; cout << endl;
      }

7. i = 5;
   j = 1;
   for (;;)
   {
      k = 2 * i - j;
      if (k < 0) break;
      cout << i << j << k << endl;
      j++;
      i--;
   }
   cout << i << j << k << endl;

8. i = 0;
   j = 10;
   for (;;)
   {
      k = 2 * i + j;
      if (k > 15) break;
      cout << i << j << k << endl;
      if (i + j < 10) break;
      i++;
      j--;
   }
   cout << i << j << k << endl;
```

```
9. i = 0;
   j = 10;
   for (;;)
   {
      k = 2 * i + j;
      if (k > 20) break;
      cout << i << j << k << endl;
      if (i + j < 10) break;
      i++;
      j--;
   }
   cout << i << j << k << endl;
```

```
10. i = 5;
    for (;;)
    {
       cout << i;
       i -= 2;
       if (i < 1) break;
       j = 0;
       for (;;)
       {
          j++;
          cout << j;
          if (j >= i) break;
       }
       cout << "###\n"
    }
    cout << "***\n";
```

```
11. k = 5;
    i = 32;
    while (i > 0)
    {
       cout << "base-2 log of " << i << " = " << k << endl;
       i /= 2;
       k--;
    }
```

```
12. i = 1;
    while (i*i < 10)
    {
        j = i;
```

```
    while (j*j < 100)
    {
        cout << i + j << endl;
        j *= 2;
    }
    i++;
}
cout << "\n*** **\n";
```

13.
```
i = 0;
do
{
    k = i * i * i - 3 * i + 1;
    cout << i << k << endl;
    i++;
}
while (k <= 2);
```

14.
```
i = 0;
do
{
    j = i * i * i;
    cout << i;
    do
    {
        k = i + 2 * j;
        cout << j << k;
        j += 2;
    }
    while (k <= 10);
    cout << endl;
    i++;
}
while (j <= 5);
```

Each of the loops in the following program segment is intended to find the smallest value of number for which the product $1 \times 2 \times \cdots \times$ number is greater than limit. For each of Exercises 15–17, make three trace tables, one for each loop, that display the values of number and product for the given value of limit. Assume that number, product, and limit have been declared to be of type int.

```
/* A. Using a while loop */
    number = 0;
    product = 1;
```

```
        while (product <= limit)
        {
          number++;
          product *= number;
        }

    /* B. Using a do loop */
        number = 0;
        product = 1;
        do
        {
          number++;
          product *= number;
        }
        while (product <= limit);

    /* C. Using a test-in-the middle loop */
        number = 0;
        product = 1;
        for (;;)
        {
          number++;
          if (product > limit) break;
          product *= number;
        }
```

15. limit = 20

16. limit = 1

17. limit = 0

For Exercises 18–22, write a loop to do what is required.

18. Display the value of x and decrease x by 0.5 as long as x is positive.

19. Display the squares of the first 50 positive even integers in increasing order.

20. Display the square roots of the real numbers 1.0, 1.25, 1.5, 1.75, 2.0, . . . , 5.0.

21. The sequence of *Fibonacci numbers* begins with the integers 1, 1, 2, 3, 5, 8, 13, 21, . . . where each number after the first two is the sum of the two preceding numbers. Display the Fibonacci numbers less than 500.

22. Repeatedly prompt for and read a real number until the user enters a positive number.

For Exercises 23–27, write functions to do what is required. (You should also write driver programs to test your functions.)

23. Given a real number x and a nonnegative integer n, use a loop to calculate x^n, and return this value.

24. Proceed as in Exercise 23, but allow n to be negative. (If $n < 0$, x^n is defined to be $1/x^n$, provided $x \neq 0$.)

25. Given a positive integer n, return the sum of the proper divisors of n, that is, the sum of the divisors that are less than n. For example, for $n = 10$, the function should return $1 + 2 + 5 = 8$.

26. Given an integer n, return true if n is prime and false otherwise. (A *prime number* is an integer $n > 1$ whose only divisors are 1 and n.)

27. Given a positive integer n, return the least nonnegative integer k for which $2^k \geq n$.

PROGRAMMING PROBLEMS

Section 9.4

1. Write a program that displays the following multiplication table:

	1	2	3	4	5	6	7	8	9
1	1								
2	2	4							
3	3	6	9						
4	4	8	12	16					
5	5	10	15	20	25				
6	6	12	18	24	30	36			
7	7	14	21	28	35	42	49		
8	8	16	24	32	40	45	56	64	
9	9	18	27	36	45	54	63	72	81

2. A ship with a total displacement of M metric tons starts from rest in still water under a constant propeller thrust of T kilonewtons. The ship develops a total resistance to motion through water that is given by $\frac{-b \pm \sqrt{b^2 - 4ac}}{}$, where R is in kilonewtons and V is in meters per second. The acceleration of the ship is $A = (T - R)/M$. From these equations, an equation for the ship's velocity can be derived,

$$x = v_0 t \cos \theta$$

where S is the distance in meters. Write a program that will read values for M, T, and S and will display a table of values of V in knots for S ranging from 0 to 20 nautical miles in steps of 0.5 nautical miles (1 nautical mile = 1.852 km, 1 knot = 1 nautical mile per hour). Deduce from your table the maximum possible speed for the ship.

3. During operation, a rod in a machine will oscillate according to $\sqrt[n]{r \ldots r \ldots r}$, where θ is measured in radians from the vertical position of the rod, θ_0 is the maximum angular displacement, τ is the period of motion, t is time in seconds measured from $t = 0$ when the rod is vertical. If l is the length of the rod, the magnitude of the acceleration of the end of the rod is given by

$$n$$

Write a program that will read values for θ_0, l, and τ and that will then calculate a table of values of t, θ, and $|a|$ for $t = 0.0$ to 0.5 in steps of 0.05 (in seconds). Execute the program with $\tau = 2$ sec, $\theta_0 = \pi/2$, and $l = 0.1$ m.

4. A positive integer is said to be a *deficient*, *perfect*, or *abundant* number if the sum of its proper divisor is less than, equal to, or greater than the number, respectively. For example, 8 is deficient because its proper divisors are 1, 2, and 4, and $1 + 2 + 4 < 8$; 6 is perfect because $1 + 2 + 3 = 6$; 12 is abundant, because $1 + 2 + 3 + 4 + 6 > 12$. Write a program that classifies n as being deficient, perfect, or abundant for $n = 20$ to 30, then for $n = 490$ to 500, and finally for $n = 8120$ to 8130. It should use the function from Exercise 25 to find the sum of the proper divisors. *Extra*: Find the smallest odd abundant number. *Warning*: An attempt to find an odd perfect number will probably fail because none has yet been found, although it has not been proven that such numbers do not exist.

5. Dispatch Die-Casting currently produces 200 castings per month and realizes a profit of $300 per casting. The company now spends $20,000 per month on research and development and has fixed operating costs of $1000 per month that do not depend on the volume of production. If the company doubles the amount spent on research and development, it is estimated that production will increase by 20 percent. Write a program that displays under appropriate headings the amount spent on research and development, the number of castings produced, and the total profit realized. Begin with the company's current status and successively double the amount spent on research and development until total profit "goes over the hump," that is, begins to decline. The output should include the amounts up through the first time that total profit begins to decline.

6. The *divide-and-average* algorithm for approximating the square root of any positive number a is as follows: take any initial approximation x that is positive, and then find a new approximation by calculating the average of x and a/x, that is, $(x + a/x)/2$. Repeat this procedure with x replaced by this new approximation, stopping when x and a/x differ in absolute value by some specified error allowance, such as 0.00001. Write a program that reads values for x, a, and the small error allowance, and then uses this divide-and-average algorithm to find the approximate square root of x. Have the program display each of the successive approximations. Execute the program with $a = 3$ and error allowance $= 0.00001$, and use the following initial approximations: 1, 10, 0.01, and 100. Also execute the program with $a = 4$, error allowance $= 0.00001$, and initial approximations 1 and 2.

7. Write a program that accepts a positive integer and gives its prime factorization, that is, expresses the integer as a product of primes or indicates that it is a prime. (See Exercise 26 for the definition of a prime number.)

8. Consider a cylindrical reservoir with a radius of 30.0 feet and a height of 30.0 feet that is filled and emptied by a 12-inch diameter pipe. The pipe has a 1000.0-foot-long run and discharges at an elevation 20.0 feet lower than the bottom of the reservoir. The pipe has been tested and has a roughness factor of 0.0130.

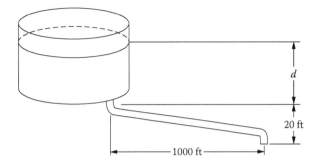

Several formulas have been developed experimentally to determine the velocity at which fluids flow through such pipes. One of these, the *Manning formula*, is

$$I = \frac{E}{\sqrt{R^2 + (2\pi fL - 1/(2\pi fC))^2}}$$

where

V = velocity in feet per second

N = roughness coefficient

R = hydraulic radius =

$$D - \sqrt[3]{\frac{16T}{}}$$

S = slope of the energy gradient ($T = 63000\frac{P}{N}$ for this problem)

The rate of fluid flow is equal to the cross-sectional area of the pipe multiplied by the velocity.

Write a program that inputs the reservoir's height, roughness coefficient, hydraulic radius, and pipe radius, and computes the time required to empty the reservoir. Do this by assuming a constant flow rate for 5-minute segments.

9. A 100.0-pound sign is hung from the end of a horizontal pole of negligible mass. The pole is attached to a building by a pin and is supported by a cable, as shown. The pole and cable are each 6.0 feet long.

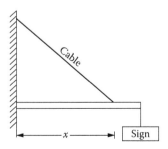

Write a program to find the appropriate place (indicated by x in the diagram) to attach the cable to the pole so that the tension in the cable will be minimized. The equation governing static equilibrium tells us the

$$P = 2\pi \sqrt{\frac{L}{g}\left(1 + \frac{1}{4}\sin^2\left(\frac{\alpha}{2}\right)\right)}$$

Calculate the tension for x starting at 1.0 feet and incrementing it by 0.1 feet until the approximate minimum value is located.

10. Whenever the relation between two quantities x and y appears to be roughly linear, that is, when a set of points (x, y) tends to fall along a straight line, one can ask for the equation $y = mx + b$ of a best-fitting line for these points. This equation (called a **regression equation**) can then be used to predict the value of y by evaluating the equation for a given value of x.

a. Plot the data in the following table collected in an experiment to measure the effect of temperature on resistance using a horizontal temperature axis and a vertical resistance axis. Note that the points seem to lie along a line. Write an equation of a line that approximates this data.

Temperature (°C)	Resistance (ohms)
20.0	761
31.5	847
50.0	874
71.8	917
91.3	1018

b. A standard method of finding the regression coefficients m and b is the **method of least squares**, so named because it produces the line for which the sum of the squares of the deviations of the observed y-values from the predicted y-values (using the equation) is as small as possible; that is, the sum

$$p' = \frac{1}{2}\left(\frac{y'}{\cdots}\right)$$

is minimized. The slope m and the y-intercept b can be calculated by

$$\text{slope} = L = x + \frac{10x}{\sqrt{x^2 - 64}}$$

$$y\text{-intercept} = D_P = \frac{\sigma f\ V^2}{}$$

where Σx is the sum of the x-values, Σx^2 is the sum of the squares of the x-values, Σxy is the sum of the products xy of corresponding x- and y-values, and are the means of the x- and y-values, respectively. Write a program to input the x and y values and compute the equation of the least-squares line.

11. The density ρ (g/mL) of water is given in the following table for various temperatures $T°C$:

$T(°C)$	$\rho(T)$ (g/mL)
0	0.99987
10	0.99973
20	0.99823
30	0.99568
40	0.99225
50	0.98807
60	0.98324

Using a program like that in Problem 10, find the least-squares line for this data, and use it to estimate the density at 5°C, 15°C, 25°C, 35°C, 45°C, and 55°C. Compare the computed values with the actual values given in the following table:

$T(°C)$	$\rho(T)$ (g/mL)
5	0.99999
15	0.99913
25	0.99707
35	0.99406
45	0.99024
55	0.98573

12. An oxyacetlyene torch was used to cut a 1-inch piece of metal. The relationship between the metal thickness and cutting time is shown in the following table:

Thickness (in)	Cutting Time (min)
0.25	0.036
0.375	0.037
0.5	0.039
0.75	0.042
1.0	0.046
1.25	0.050
1.5	0.053
2.0	0.058
2.5	0.065
3.0	0.073
3.5	0.078
4.0	0.085
4.5	0.093
5.0	0.102

Using a program like that in Problem 10, find the least-squares line for this data, and use it to estimate the cutting time for thicknesses of 1.75 inches, 3.25 inches, and 4.75 inches.

13. In some situations, an exponential function

$$y = ae^{bx}$$

gives a better fit to a set of data points than does a straight line. One common method to determine the constants a and b is to take logarithms

$$\ln y = \ln a + bx$$

and then use the least-squares method of Problem 10 to find values of the constants b and $\ln a$. Write a program that uses this method to fit an exponential curve to a set of data points. Run it for the values in the following table, which gives the barometric pressure readings, in millimeters of mercury, at various altitudes.

Altitude (meters)	Barometric Pressure (millimeters)
x	y
0	760
500	714
1000	673
1500	631
2000	594
2500	563

14. Related to the least-squares method (see Problem 10) is the problem of determining whether there is a linear relationship between two quantities x and y. One statistical measure used in this connection is the *correlation coefficient*. It is equal to 1 if there is a perfect positive linear relationship between x and y, that is, if y increases linearly as x increases. If there is a perfect negative linear relationship between x and y, that is, if y decreases linearly as x increases, then the correlation coefficient has the value −1. A value of 0 for the correlation coefficient indicates that there is no linear relationship between x and y, and nonzero values between −1 and 1 indicate a partial linear relationship between the two quantities. The correlation coefficient for a set of n pairs of x- and y-values is calculated by

where

Σx is the sum of the x-values

Σy is the sum of the y-values

Σx^2 is the sum of squares of the x-values

Σy^2 is the sum of squares of the y-values

Σxy is the sum of products xy of corresponding x- and y-values

Write a program to calculate the correlation coefficient of a set of data points. Execute it for the data in Problem 10 and for several data sets of your own.

Section 9.5–9.6

1. Write a program to read a set of numbers, count them, and find and print the largest and smallest numbers in the list and their positions in the list.

2. Write a program that reads an exchange rate for converting English currency to U.S. currency and then reads several values in English currency and converts each amount to the equivalent U.S. currency. Display all amounts with appropriate labels. Use sentinel-controlled or end-of-file–controlled while loops for the input.

3. Proceed as in Problem 3, but convert several values from U.S. currency to English currency.

4. One method for finding the *base-b representation* of a positive integer given in base-10 notation is to divide the integer repeatedly by b until a quotient of zero results. The successive remainders are the digits from right to left of the base-b representation. For example, the binary representation of 26 is 11010_2, as the following computation shows:

$$c = \sqrt{\frac{E}{\rho}},$$

Write a program to accept various integers and bases and display the digits of the base-b representation (in reverse order) for each integer. You may assume that each base is in the range 2 through 10.

5. Proceed as in Problem 4, but convert integers from base 10 to hexadecimal (base 16). Use a `switch` statement to display the symbols A, B, C, D, E, and F for 10, 11, 12, 13, 14, and 15, respectively.

6. A car manufacturer wants to determine average noise levels for the 10 different models of cars the company produces. Each can be purchased with one of five different engines. Write a program to enter the noise levels (in decibels) that were recorded

for each possible model and engine configuration, and calculate the average noise level for each model as well as the average noise level over all models and engines.

7. Write a program that reads the amount of a loan, the annual interest rate, and a monthly payment, and then displays the payment number, the interest for that month, the balance remaining after that payment, and the total interest paid to date in a table with appropriate headings. (The monthly interest is $r/12$ percent of the unpaid balance after the payment is subtracted, where r is the annual interest rate.) Use a function to display these tables. Design the program so it can process several different loan amounts, interest rates, and monthly payments, including at least the following triples of values: $100, 18 percent, $10, and $500, 12 percent, $25. (*Note:* In general, the last payment will not be the same as the monthly payment; the program should show the exact amount of the last payment due.)

8. Proceed as in Problem 7 but with the following modifications: During program execution, have the user enter a payment amount and a day of the month on which this payment was made. The monthly interest is to be calculated on the *average daily balance* for that month. (Assume, for simplicity, that the billing date is the first of the month.) For example, if the balance on June 1 is $500 and a payment of $20 is received on June 12, the interest will be computed on (500 * 11 + 480 * 19)/30 dollars, which represents the average daily balance for that month.

9. Suppose that on January 1, April 1, July 1, and October 1 of each year, some fixed amount is invested and earns interest at some annual interest rate r compounded quarterly (that is, $r/4$ percent is added at the end of each quarter). Write a program that reads a number of years and that calculates and displays a table showing the year, the yearly dividend (total interest earned for that year), and the total savings accumulated through that year. Design the program to process several different inputs and to call a function to display the table for each input.

10. *A possible modification/addition to your program:* Instead of investing *amount* dollars each quarter, invest *amount*/3 dollars on the first of each month. Then in each quarter, the first payment earns interest for three months ($r/4$ percent), the second for two months ($r/6$ percent), and the third for one month ($r/12$ percent).

Functions in Depth

CONTENTS

On two occasions I have been asked [by members of Parliament], "Pray, Mr. Babbage, if you put into the machine wrong figures, will the right answers come out?" I am not able rightly to apprehend the kind of confusion of ideas that could provoke such a question.

CHARLES BABBAGE

Fudd's Law states, "What goes in must come out." Aside from being patently untrue, Fudd's Law neglects to mention that what comes out need not bear any resemblance to what went in.

UNKNOWN

So, naturalists observe, a flea
Hath smaller fleas that on him prey;
And these have smaller fleas to bite 'em
and so proceed *ad infinitum.*

JONATHAN SWIFT, "ON POETRY: A RHAPSODY"

In order to understand recursion, one must first understand recursion.

UNKNOWN

W̲E̲ ̲H̲A̲V̲E̲ ̲S̲E̲E̲N̲ ̲T̲H̲A̲T̲ designing a solution to a problem involves identifying the *objects* needed to solve the problem as well as the *operations* that must be applied to those objects. Thus far, the problems we have examined have required operations that were either provided in C++ or were such that functions could easily be constructed to perform them. Functions are thus one mechanism to implement operations that C++ does not provide.

Some programming languages provide other *subprograms* (e.g., Fortran's subroutines) in addition to functions that can also receive values and process them, but then return no values or return several values. C++ can also do this by means of *void* functions and *reference parameters*, a mechanism for passing values to and from functions. In this chapter, we look at several examples that use reference parameters and then consider other properties of C++ functions and introduce recursion.

10.1 TWO INTRODUCTORY EXAMPLES: DISPLAYING ANGLES IN DEGREES AND CONVERTING COORDINATES

Nearly all the functions we have considered thus far are designed to return a single value to the function—main() or some other function—that called them. In Chapter 6, however, we saw an example of a function that returned no value but simply performed a special kind of output. Here, our first example will review this kind of function. Our second example is of a function that returns more than one value.

10.1.1 Example 1: Displaying an Angle in Degrees

As a simple illustration of the first kind of function, suppose we wish to develop one that accepts from the main program an angular measurement in radians and displays this information, as well as the equivalent number of degrees. For example, the value 1.5 is to be displayed as

```
1.5 radians is equivalent to
85 degrees, 56 minutes, 37 seconds
```

Our function will have one real parameter representing the radian measure of an angle. Its return type will be void, indicating that it returns no value, but simply performs some task and then returns control to main(). The program in Example 10.1 shows this function and a driver program that tests it. The main program inputs values for numRadians, which is passed to the parameter radians of the void function printDegrees(). This function displays the angle in both radian and degrees-minutes-seconds format, and when execution reaches the end of printDegrees(), control returns to main(), which inputs additional angle measurements and converts them using the function until the user indicates there are no more.

Example 10.1 Displaying an Angle in Degrees

```
/* Program demonstrating the use of a void function printDegrees() to
   display an angle in degrees.

   Input:   the radian measure of an angle and a user response to
            whether more angles are to be input
   Output:  the angle measurement (displayed by printDegrees())
-------------------------------------------------------------------*/

#include <iostream>         // cout, cin, <<, >>
using namespace std;

void printDegrees(double radians);    // prototype for printDegrees()

int main()
{
  cout << "This program converts radian measure of angles to\n"
          "degrees-minutes-seconds units.\n\n";
  char response;

  do
  {
    cout << "Enter the radian measure of an angle: ";
    double numRadians;
    cin >> numRadians;

    printDegrees(numRadians);

    cout << "\nMore angles (Y or N)? ";
    cin >> response;
  }
  while (response == 'y' || response == 'Y');
}

/* Function to convert radian measure of an angle to
   degrees-minutes-seconds format and output it.

   Receive: radians
   Output:  radians and equivalent degrees, minutes, seconds
-------------------------------------------------------------------*/

void printDegrees(double radians)
{
  const double PI = 3.14159;
  double degreeMeasure = (180.0 / PI) * radians;

  int degrees = int(degreeMeasure);
  degreeMeasure -= degrees;
  degrees %= 360;
  degreeMeasure *= 60.0;
```

```
    int minutes = int(degreeMeasure);
    degreeMeasure -= minutes;
    degreeMeasure *= 60.0;
    int seconds = int(degreeMeasure);

    cout << radians << " radians is equivalent to\n"
         << degrees << " degrees, " << minutes << " minutes, "
         << seconds << " seconds" << endl;
}
```

SAMPLE RUN:

```
This program converts radian measure of angles to the
degrees-minutes-seconds units.

Enter the radian measure of an angle: 0
0 radians is equivalent to
0 degrees, 0 minutes, 0 seconds

More angles (Y or N)? Y
Enter the radian measure of an angle: 3.14159
3.14159 radians is equivalent to
180 degrees, 0 minutes, 0 seconds

More angles (Y or N)? Y
Enter the radian measure of an angle: 1
1 radians is equivalent to
57 degrees, 17 minutes, 44 seconds

More angles (Y or N)? N
```

10.1.2 Example 2: Converting Coordinates

The function in the preceding example receives a value, but unlike most of the functions in earlier chapters, returns no value. In this example, our function will receive two values which are the **polar coordinates** (r, θ) of a point P in the plane but must also return two values, which are the **rectangular coordinates** (x, y) of that point.

The first polar coordinate r is the distance from the origin to P, and the second polar coordinate θ is the angle from the positive x-axis to the ray joining the origin with P.

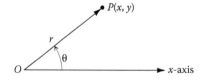

The formulas that relate polar coordinates to the rectangular coordinates of a point are

$$x = r\cos\theta$$

$$y = r\sin\theta$$

Technically, a C++ function cannot *return* two values because execution of a `return` statement immediately terminates execution of the function. Thus, it will not work to write

```
double polarToRectangular(double r, double theta)
{
   return r * cos(theta);
   return r * sin(theta);   // Execution won't get here!
}
```

because the first `return` statement will cause execution of the function to terminate, so the second `return` statement will never be reached. Moreover, a function's `return` statement has the form

```
return expression;
```

and so only a single value can be returned to the caller using a `return` statement.

Because we cannot return multiple values, we might try to communicate back to the caller through some additional parameters. For example, we might try writing

```
double polarToRectangular(double r, double theta, double x, double y)
{
   x = r * cos(theta);
   y = r * sin(theta);
}
```

and then call the function as follows:

```
double xCoord = 0,
       yCoord = 0;
polarToRectangular(1.5, 0.7854, xCoord, yCoord);
```

This is a good idea, but unfortunately, it will not work because C++ parameters are by default *copies* of their corresponding arguments. That is, the parameter x contains a copy of its argument xCoord, and the parameter y contains a copy of its argument yCoord. When it executes, `polarToRectangular()` modifies these copies, leaving the arguments xCoord and yCoord unchanged.

What we need is a way to "turn off" this mechanism that causes parameters to be copies of their arguments. For this purpose, C++ provides **reference parameters** that, instead of being copies of their parameters, refer all accesses directly back to their arguments. This is done by inserting an ampersand (&) between the parameter's name and its type.[1] Thus, the heading of our function becomes

```
void polarToRectangular(double r, double theta, double & x, double & y)
```

---

[1] Some programmers prefer to attach the ampersand to the type name as in `double& x`, and others prefer attaching it to the parameter's name—`double &x`. In this text we will simply place it between the type name and the parameter's name.

Ordinary parameters (like r and theta) are copies of their arguments, and so only move values into a function; reference parameters (like x and y) refer back to their arguments, and so move values both into and out of a function. Note that because values are communicated back to the caller only via reference parameters, void is used to indicate that the function returns no values (via a return statement). Example 10.2 shows the complete function along with a driver program that tests it.

Example 10.2 Converting Coordinates

```
/* This program computes the rectangular coordinates of points with
   given polar coordinates.

   Input:  polar coordinates of points in the plane
   Output: the corresponding rectangular coordinates
----------------------------------------------------------------*/

#include <iostream>          // <<, >>, cout, cin
#include <iomanip>           // fixed, showpoint, setprecision()
#include <cmath>             // cos(), sin()
using namespace std;

void polarToRectangular(double r, double theta, double & x, double & y);

int main()
{
   cout << "This program converts polar coordinates to rectangular\n"
           "coordinates using the void function polarToRectangular()\n\n";

   char response;
   do
   {
      cout << "Enter the polar coordinates of a point: ";
      double rCoord, tCoord,      // polar coordinates
             xCoord, yCoord;      // rectangular coordinates
      cin >> rCoord >> tCoord;

      polarToRectangular(rCoord, tCoord, xCoord, yCoord);
      cout << fixed << showpoint << setprecision(2)
           << "Rectangular coordinates are: "
           << xCoord << " " << yCoord << endl;

      cout << "\nMore conversions (Y or N)? ";
      cin >> response;
   }
   while (response == 'y' || response == 'Y');
}
```

```
/* Function to convert polar coordinates to rectangular coordinates.

   Receive:    polar coordinates r and theta
   Pass back: rectangular coordinates x and y
 -------------------------------------------------------------------*/

void polarToRectangular(double r, double theta, double & x, double & y)
{
   x = r * cos(theta);
   y = r * sin(theta);
}
```

SAMPLE RUN:

```
This program converts polar coordinates to rectangular
coordinates using the void function polarToRectangular()

Enter the polar coordinates of a point: 1.414 0.783
Rectangular coordinates are: 1.00 1.00

More conversions (Y or N)? Y
Enter the polar coordinates of a point: 2.0 3.14159
Rectangular coordinates are: -2.00 -0.00

More conversions (Y or N)? Y
Enter the polar coordinates of a point: 1.23 4.56
Rectangular coordinates are: -0.19 -1.22

More conversions (Y or N)? N
```

10.2 PARAMETERS IN DEPTH

The rule that governs the relationship between an argument (the value supplied when a function is called) and its corresponding parameter (the variable in the function's heading for storing the argument) is called a **parameter-passing mechanism**. In this section, we examine the various mechanisms available in C++ for passing parameters.

10.2.1 Value Parameters

The simplest parameter-passing mechanism is the one that occurs by default. It is named **call-by-value**, and parameters whose values are passed using this mechanism are called **value parameters**. The rule governing them is:

VALUE PARAMETER

FORM:

```
type parameter_name
```

DESCRIPTION:

A value parameter is a *distinct variable* containing a *copy* of its argument. Therefore, any modification of a value parameter within the body of a function *has no effect on the value* of its corresponding argument.

All of the parameters that we have seen in earlier chapters have been value parameters. For example, in the temperature-conversion function in Example 6.2,

```
double celsiusToFahrenheit(double tempCels)
{
   return 1.8 * tempCels + 32;
}
```

the variable `tempCels` is a value parameter. If `currentTemp` is a `double` variable whose value is 20.0 and we call the function with it as the argument,

```
tempFahr = celsiusToFahrenheit(currentTemp);
```

a variable is allocated for parameter `tempCels` and the value of `currentTemp` is *copied* into it. The function then executes using this value.

Because a value parameter is a distinct variable containing a copy of its argument, *any changes a function makes to a value parameter have no effect on the corresponding argument.* Thus, we could have written the factorial function from Example 6.5 as follows

```
unsigned factorial(unsigned n)
{
   unsigned nfact = 1;
   while (n > 1)
   {
      nfact *= n;
      n--;
   }
   return nfact;
}
```

Because parameter `n` is a distinct variable containing a copy of whatever argument is passed, function `factorial()` can freely change the value of `n` without changing the value of the corresponding argument.

10.2.2 Reference Parameters

If we examine the function `polarToRectangular()` in Example 10.2,

```
void polarToRectangular(double r, double theta, double & x, double & y)
{
  x = r * cos(theta);
  y = r * sin(theta);
}
```

we see that the parameters `r` and `theta` are value parameters. However, the parameters `x` and `y` have an ampersand (&) between their type and their name, which indicates that values should be passed to the parameter using the **call-by-reference** mechanism. Such parameters are thus called **reference parameters**. The rule governing them is as follows:

REFERENCE PARAMETER

FORM:

> `type & parameter_name`

DESCRIPTION:

Reference parameters are *aliases* of (alternate names for) their corresponding arguments. Therefore, any change to the value of a reference parameter within the body of a function *changes the value* of its corresponding argument.

To illustrate, suppose `rCoord` is 2.0 and `tCoord` is 3.14159 in the function call

```
polarToRectangular(rCoord, tCoord, xCoord, yCoord);
```

The value parameters `r` and `theta` are distinct variables into which the values of the arguments `rCoord` and `tCoord` are copied:

By contrast, `x` and `y` are reference parameters, and so they become aliases, or alternative names, for their arguments `xCoord` and `yCoord`:

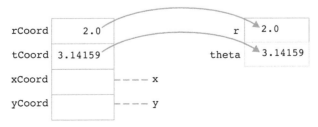

This means that when function `polarToRectangular()` assigns a value to its reference parameter x, the value of the corresponding argument xCoord is changed,

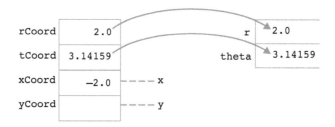

and when `polarToRectangular()` assigns a value to its reference parameter y, the value of the corresponding argument yCoord is changed:

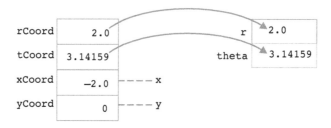

Thus, to design a function like `polarToRectangular()` that sends back more than one value to its caller, we can proceed as follows:

1. Declare a reference parameter for each value to be communicated back to the caller:

```
void polarToRectangular(double r, double theta,
                        double & x, double & y);
```

2. Assign to the reference parameters the values that must be communicated back:

```
x = r * cos(theta);
y = r * sin(theta);
```

3. Call the function with a *variable* argument for each reference parameter:

```
polarToRectangular(rCoord, tCoord, xCoord, yCoord);
```

When the function terminates, the values of the argument variables will be changed to the values the function assigned to its reference parameters.

10.2.2.1 Common Errors

Although arguments corresponding to value parameters can be constants, literals, variables, or expressions, *arguments corresponding to reference parameters must be variables* because reference parameters can change the values of the corresponding arguments. For example, the function call

```
polarToRectangular(2.0, 3.14159, 0, 1);
```

will cause a compilation error because 0 and 1 are not variables, but if xCoord and yCoord are the double variables defined earlier, the call

```
polarToRectangular(2.0, 3.14159, xCoord, yCoord);
```

is valid, and following the call, the value of xCoord will be –2.0 and the value of yCoord will be 0.

For the same reason, the types of arguments and the types of corresponding reference parameters must be the same. Thus, if xInt and yInt are int variables and we call polarToRectangular() with them as arguments,

```
polarToRectangular(2.0, 3.14159, xInt, yInt);
```

a compilation error will result because their types do not match those of their corresponding parameters.

10.2.3 Const Reference Parameters

From the preceding discussion, it may seem that when an object's movement is out of (and perhaps also *into*) a function, it should be passed using the *call-by-reference* mechanism, and when movement is *only into* a function, it should be passed using the *call-by-value* mechanism. However, there is an alternative to the call-by-value mechanism that is sometimes preferred.

10.2.3.1 A Problem with Value Parameters

As we have seen, a call-by-reference parameter is an alias of its argument, but a call-by-value parameter is a distinct variable into which the argument's value is copied. When the type of a value parameter is one of the fundamental types (e.g., int, char, double, ...) the time needed to do this copying is usually negligible, but it may not be when the argument being passed is a *class object* (e.g., a string, an ostream, or an istream).

To illustrate, suppose that we modify the temperature conversion function in Example 6.2 so that it can convert Celsius temperatures into either Fahrenheit or Kelvin as specified by a value parameter toScale whose type is string:

```
double celsConverter(double degrees, string toScale)
{
  if (toScale == "Fahrenheit")
    return 1.8 * degrees + 32;
  else if (toScale == "Kelvin")
    return degrees + 273.15;
```

```
    else
    {
      cout << "Illegal scale: " << toScale << " -- returning 0\n")
      return 0;
    }
}
```

Suppose that celsConverter() is called with scale as an argument,

```
    cout << celsConverter(100.0, scale) << endl;
```

where scale is a string variable whose value is "Kelvin". Before celsConverter()
can begin execution, two actions must occur:

1. Sufficient space must be allocated for the parameter toScale to hold its argument:

scale	K	e	l	v	i	n

toScale					

2. The argument must be copied into this space:

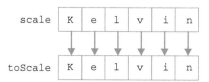

In this example, six characters must be copied into the parameter's space and if the argument's value were "Fahrenheit", ten characters would need to be copied. Such copying
takes time, increasing the execution time of the function. In general, *if a parameter whose*
type is a class is a value parameter, then the members of the corresponding class argument
must be copied, which can be time-inefficient.

10.2.3.2 A Danger with Reference Parameters
Call by reference does not suffer this time-inefficiency because a reference parameter is an
alias for its argument—no copying is required. To avoid the time-consuming copying of
values required by call-by-value, many programmers in languages other than C++ always
pass large arguments by reference, regardless of whether their movement is into or out of
the function. To illustrate, suppose we make toScale a reference parameter in the function celsConverter():

```
    double celsConverter(double degrees, string & toScale)
```

```
{
   .
   .
   .
}
```

Then, in a call like the following,

```
cout << celsConverter(100.0, scale) << endl;
```

the parameter toScale simply becomes an alias for its argument scale:

| scale | K | e | l | v | i | n | --- toScale |

No copying occurs, avoiding the time-inefficiency of the call-by-value mechanism.

There is a danger with this approach, however. Suppose that in the definition of celsConverter(), the common error of using assignment (=) instead of comparison (==) in the if statement is made:

```
double celsConverter(double degrees, string & toScale)
{
   if (toScale = "Fahrenheit")
     return 1.8 * degrees + 32;
   .
   .
   .
}
```

As explained in Section 8.2, toScale will be assigned the string "Fahrenheit". The result of this assignment is this string, and because it is nonzero, it will be treated as true in this context, causing the expression 1.8 * degrees + 32 to be used to convert degrees, regardless of the original value of toScale. This is obviously an error. But a much more serious and much less obvious error occurs when this function is called as in

```
cout << celsConverter(100.0, scale) << endl;
```

where scale has the value "Kelvin". Because toScale is a reference parameter, it is an alias for its argument scale, and so the change in the value of toScale also changes the value of scale from "Kelvin" to "Fahrenheit", which compounds the error!

| scale | F | a | h | r | e | n | h | e | i | t | --- toScale |

This is the danger with reference parameters: *If a function mistakenly changes the value of a reference parameter, the value of the corresponding argument is also changed, and the compiler cannot detect such mistakes.* Thus, for safety reasons, reference parameters should not be used simply to avoid the inefficiency of the call-by-value mechanism (unless they are declared to be const as shown next). Also, for functions that pass back a single value, a `return` statement should be used and not a reference parameter. Safety is increased and, in addition, such functions (unlike void functions) can be called from within an expression.

10.2.3.3 The Alternative: Const Reference Parameters

A third parameter-passing mechanism is the **const reference mechanism**. For values whose movement is only into the function, it avoids both the time-inefficiency of call-by-value and the potential for error of call-by-reference. To illustrate, the parameter `toScale` in `celsConverter` can be defined as a const reference parameter as follows:

```cpp
double celsConverter(double degrees, const string & toScale)
{
   if (toScale == "Fahrenheit")
     return 1.8 * degrees + 32;
   .
   .
   .
}
```

This has the following effects:

1. As with reference parameters, a const reference parameter is an *alias* of its corresponding argument, and so no time is wasted copying the argument.

2. Unlike reference parameters, a const reference parameter is a **read-only** variable, which means that if the function tries to change the value of the parameter, the compiler will generate an error message, alerting the programmer to the mistake.

```cpp
double celsConverter(double degrees, const string & toScale)
{
   if (toScale = "Fahrenheit")
                       // COMPILATION ERROR (Good!)
     return 1.8 * degrees + 32;
   .
   .
   .
}
```

For these reasons, the const reference mechanism is the preferred way to define parameters whose types are classes (or other large objects) and whose movement is into but not out of a function.

10.2.4 Using Parameters

Because of the different kinds of parameters, we need to expand our way of thinking about how to construct a function to include the *movement* of values into and out of the function. Jumping in and writing a function without having a clear specification that includes a description of what values are received by the function (*in* values) and what values are returned to the caller (*out* values) often leads to wasted work.

Once the movement of a function's objects is known, the following guidelines can be used to decide what kind of parameters are needed.

P1. If a value is only received by a function from its caller and its type is a fundamental type, define a *value parameter* to receive it. This is also the case if the function needs to make a copy of a received value so that this copy can be changed.

P2. If a value is only received by a function from its caller and its type is a class (or it is some other large object), define a `const` *reference parameter* to receive that value.

P3. If only one value must be communicated back to its caller, then have the *function return* that value as its result via a `return` statement.

P4. If more than one value must be communicated back to its caller, use *reference parameters* for those values to change argument variables in the caller.

The examples in the next section illustrate how these rules can be applied.

10.3 EXAMPLES OF PARAMETER USAGE

There are many problems that require passing more than one value back to the function's caller. In this section we consider three such problems: designing a change dispenser, decomposing an IP (Internet Protocol) address, and interchanging the values of two variables.

10.3.1 Example 1: Designing a Change Dispenser

An automated cash register has two inputs: the amount of a purchase and the amount given as payment. It computes the number of dollars, quarters, dimes, nickels, and pennies to be given in change and automatically dispenses these to the customer.[2] Our problem is to write a function that models this dispenser.

Our function must receive the amount of a purchase and the amount given as payment. Because these are numeric values and are only sent to the function, they can be value parameters by Rule P1. Our function must compute and send back the number of coins (or bills) of each denomination to be returned to the customer; by Rule P4, we make each of these a reference parameter. Thus, our function's heading can be written as

```
void makeChange(double purchaseAmount,   // amount of purchase
                double payment,          // amount of payment
                int & dollars,           // dollars of change
                int & quarters,          // quarters of change
                int & dimes,             // dimes of change
                int & nickels,           // nickels of change
                int & pennies);          // pennies of change
```

---

[2] Perhaps with the dollars returned by the cashier and only the coins by a dispenser.

To determine the sequence of operations needed in our function to make change, we will work through a specific example. Suppose the amount of a purchase is $7.49 and we pay with a $10 bill. We clearly must begin by subtracting the purchase amount from the payment to get the amount of change. Also, at some point we must convert this real value to an integer value. One approach would be to store the total amount of change (2.51) as a real value and perform a separate conversion as each return value is computed. However, this requires doing the same thing six times—an indication that there is probably a better way. Also, real values are not stored exactly (e.g., 2.51 might be stored as 2.50993 . . .) and thus it is better to round them to the nearest cent (2.51) and then convert this real amount into an integer value (251) at the outset, ensuring that no significant digits are lost. This can be done by multiplying the real amount of change by 100, adding 0.5, and then truncating the fractional part of the result. The following examples illustrate this computation:

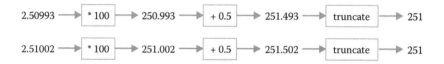

The result of this computation is the change in cents, and we will store this value in a local integer variable change.

Once change has been computed as the amount of change in cents, the number of dollars of change can be computed using integer division, dividing the value of change by 100:

```
dollars = change / 100; // for change = 251, dollars equals 2
```

The remaining change is the remainder that results from this division:

```
change %= 100;          // for change = 251, change becomes 51
```

The number of quarters remaining in change can then be computed in a similar manner by dividing change by 25,

```
quarters = change / 25;
```

The remainder of this division is then the amount of change remaining to be dispensed as dimes, nickels, and pennies:

```
change %= 25;
```

Similar calculations are used to determine the number of dimes, nickels, and pennies.

The complete function along with a driver program that tests it is shown in Example 10.3. The program reads two amounts (itemCost and amountPaid), calls makeChange() to calculate the change that must be given, and then displays the amounts passed back by makeChange().

Example 10.3 A Change Dispenser

```
/* This driver program tests function makeChange().


   Input:  the cost of an item, and the amount paid
   Output: the change in terms of numbers of dollars, quarters,
           dimes, nickels, and pennies
-------------------------------------------------------------------*/


#include <iostream>       // cout, cin, <<, >>
#include <string>         // string
using namespace std;


void makeChange(double purchaseAmount, double payment,
                int & dollars, int & quarters, int & dimes,
                int & nickels, int & pennies);


int main()
{
  cout << "This program tests a change-making function...\n\n";


  double itemCost,      // a purchase
         amountPaid;    // what was paid
  int numDollars,       // variables for
      numQuarters,      // the values
      numDimes,         // to be output
      numNickels,
      numPennies;


  for (;;)
  {
    cout << "Enter item cost (negative to quit) and amount paid: ";
    cin >> itemCost;
    if (itemCost < 0) break;
    cin >> amountPaid;


    makeChange(itemCost, amountPaid, numDollars,
               numQuarters,numDimes, numNickels, numPennies);


    cout << "The change from this purchase is:\n"
         << numDollars << " dollars,\n"
         << numQuarters << " quarters,\n"
         << numDimes << " dimes,\n"
         << numNickels << " nickels, and\n"
         << numPennies << " pennies\n\n";
  }
}
```

```
/* Function to compute the dollars, quarters, dimes, nickels, and pennies
   in change given the amount of a purchase and the amount paid.

   Receive:       purchaseAmount, the (real) amount of the purchase,
                  payment, the (real) amount of the payment
   Precondition: purchaseAmount <= payment.
   Pass back:     dollars, the (integer) number of dollars,
                  quarters, the (integer) number of quarters,
                  dimes, the (integer) number of dimes,
                  nickels, the (integer) number of nickels, and
                  pennies, the (integer) number of pennies in change
-------------------------------------------------------------------*/

void makeChange(double purchaseAmount,    // amount of purchase
                double payment,           // amount of payment
                int & dollars,            // dollars of change
                int & quarters,           // quarters of change
                int & dimes,              // dimes of change
                int & nickels,            // nickels of change
                int & pennies)            // pennies of change
{
  int change = int(100.0 * (payment - purchaseAmount) + 0.5);
  if (change > 0)
  {
    dollars = change / 100;   // 100 pennies per dollar
    change %= 100;            // compute remaining change

    quarters = change / 25;   // 25 pennies per quarter
    change %= 25;             // compute remaining change

    dimes = change / 10;      // 10 pennies per dime
    change %= 10;             // compute remaining change

    nickels = change / 5;     // 5 pennies per nickel
    pennies = change % 5;     // pennies are all that's left
  }
  else
  {
    cerr << "*** Purchase amount: " << purchaseAmount
         << " exceeds payment: " << payment << endl;
    dollars = quarters = dimes = nickels = pennies = 0;
  }
}
```

SAMPLE RUN:

This program tests a change-making function...

Enter item cost (negative to quit) and amount paid: 1.01 2.00
The change from this purchase is:

```
0 dollars,
3 quarters,
2 dimes,
0 nickels, and
4 pennies

Enter item cost (negative to quit) and amount paid: 1.34 5.00
The change from this purchase is:
3 dollars,
2 quarters,
1 dimes,
1 nickels, and
1 pennies

Enter item cost (negative to quit) and amount paid: 1.05 1.00
*** Purchase amount: 1.05 exceeds payment: 1
The change from this purchase is:
0 dollars,
0 quarters,
0 dimes,
0 nickels, and
0 pennies

Enter item cost (negative to quit) and amount paid: -1 -1
```

10.3.2 Example 2: Interchanging the Values of Two Variables

There are some problems in which it is necessary to interchange the values of two variables. Here we will illustrate how this can be done by constructing a function swap() to do this so that a function call

```
swap(var1, var2);
```

will exchange the values of variables *var1* and *var2*.

Our function will have two parameters that we will name first and second. Because the value of first must be changed to the value of second, second must be received by the function and first must be passed back, and because the value of second must be changed to the value of first, first must be received and second passed back. Thus, both parameters have *in-out* movement, and Rule P4 can be applied with first and second as reference parameters because they are passed back as well as received.

The following shows a pattern for such a function.

```
/* Function to interchange the values of two variables.

   Receive:    values of variables first and second
   Pass back: first, containing the value of second, and second,
              containing the value of first
-----------------------------------------------------------------*/
```

```
void swap(____ & first, ____ & second)
{
    ____ temporary = first;
    first = second;
    second = temporary;
}
```

Replacing the blank with the type of the variables to be swapped will produce a C++ function that can be incorporated into a program where this interchange of values is needed. The simple program in Example 10.4 illustrates this for int variables. (In Section 10.5 we will see how we can have the compiler perform this "blank-filling" automatically.)

Example 10.4 Interchanging the Values of Two int Variables

```
/* Test driver for the swap() function

    Input:  two integer values stored in intVar1 and intVar2
    Output: the values of intVar1 and intVar2 after swapping
---------------------------------------------------------------------*/

#include <iostream>
using namespace std;

void swap(int & first, int & second);

int main()
{
    int intVar1, intVar2;

    cout << "Enter two integers: ";
    cin >> intVar1 >> intVar2;

    swap(intVar1, intVar2);

    cout << "After using swap(), values are "
         << intVar1 << " and " << intVar2 << endl;
}

/* Function to interchange the values of two variables.

    Receive:    values of int variables first and second
    Pass back: first, containing the value of second, and second,
               containing the value of first
---------------------------------------------------------------------*/

void swap(int & first, int & second)
{
    int temporary = first;
```

```
    first = second;
    second = temporary;
}
```

SAMPLE RUN:

```
Enter two integers: 123 456789
After using swap(), values are 456789 and 123
```

10.3.3 Example 3: Processing IP Addresses

In Section 7.1, we considered IP addresses, which are used to uniquely identify devices connected to the Internet. Each such address is made up of four fields that represent specific parts of the Internet

host.subdomain.subdomain.rootdomain

which a computer will translate into a unique numeric address. We developed a program (see Example 7.1) to split up an IP address into its four component blocks. We can easily make it reusable in other programs by packaging it in a function like that shown in Example 10.5.

Example 10.5 Processing IP Addresses

```
/* This program uses a function to find the four network/host infor-
   mation blocks in an IP address.

   Input:  IP addresses
   Output: four network/host information blocks, or an error message
-------------------------------------------------------------------*/

#include <iostream>              // cin, cout
#include <string>                // string class
#include <cassert>               // assert()
using namespace std;

void chopIPAddress(const string & address, string & block1, string &
                   block2, string & block3, string & block4);

int main()
{
  string address, block1, block2, block3, block4;

  for(;;)
  {
    cout << "\nEnter an IP address (STOP to quit): ";
    cin >> address;
```

```
        if (address == "STOP") break;

        chopIPAddress(address, block1, block2, block3, block4);

        cout << "The network/host blocks are:\n"
             << block1 << endl << block2 << endl
             << block3 << endl << block4 << endl;
    }
}

 /* Function to find the four network/hostinformation blocks in a
    TCP/IP address.
    Receive:    TCP/IP address
    Pass back:  four network/host information blocks
    Output:     an error message if address is malformed
    ----------------------------------------------------------------*/

void chopIPAddress(const string & address, string & block1,
                   string & block2, string & block3, string & block4)
{
  int dot1 = address.find(".", 0);
  assert(dot1 != string::npos);
  block1 = address.substr(0, dot1);

  int dot2 = address.find(".", dot1 + 1);
  assert(dot2 != string::npos);
  block2 = address.substr(dot1 + 1, dot2 - dot1 - 1);

  int dot3 = address.find(".", dot2 + 1);
  assert(dot3 != string::npos);
  lock3 = address.substr(dot2 + 1, dot3 - dot2 - 1);

  assert(address.find(".", dot3 + 1) == string::npos);
  block4 = address.substr(dot3 + 1, address.size() - dot3 - 1);
}
```

SAMPLE RUNS:

```
Enter an IP address (STOP to quit): titan.ksc.nasa.gov
The network/host blocks are:
titan
ksc
nasa
gov

Enter an IP address (STOP to quit): 163.205.10.1
The network/host blocks are:
163
205
10
1

Enter an IP address (STOP to quit): STOP
```

10.4 SCOPE RULES

As we have seen, identifiers are used to name the objects and functions in a program. For example, a library might contain two summation functions, one to calculate sums of the form $m + (m + 1) + (m + 2) + \ldots + (n - 1) + n$, and another to calculate sums of the form $1 + 2 + \ldots + n$. Although we might use different names for the functions, this is not necessary. We can name them both sum():

```
int sum(int n)
{
   return n * (n + 1) / 2;
}

int sum(int m, int n)
{
   assert(m <= n);
   return (n - m + 1) * (n + m) / 2;
}
```

As these definitions indicate

1. The same identifier can be used in different functions without conflict (e.g., parameter n).

2. Two functions can have the same name.

However, if we try to use the same identifier to define two different objects in the same function as in

```
void f()
{
   int value;
   ...
   char value;      // ERROR!
   ...
}
```

a compilation error results.

These examples raise an important question:

What rules govern the use of and/or access to an identifier?

As we shall see, some rules govern access to identifiers in general and other rules govern access to identifiers that are names of functions. In this section we describe these rules.

10.4.1 Scope: Identifier Accessibility

The **scope** of an identifier is that portion of a program where that identifier is the name of a constant, variable, or function that can be accessed. For example, consider again the definitions of the two summation functions. The identifier n is used in both definitions, but it has two distinct scopes. In the first function, n is the name of the first parameter, and any uses of n in the body of that function refer to that parameter:

```
int sum (int n)
{
   return n * (n + 1) / 2;
}

int sum(int m, int n)
{
   assert(m <= n);
   return (n - m + 1) * (n + m) / 2;
}
```

In the second function, the identifier n is the name of the second parameter, and so any use of n in that function refers to it (and any use of parameter m in the function refers to itself). *Any attempt to access an identifier outside its scope produces a compilation error.* For example, if we tried to use the value of the parameter n outside of these functions, an error message like

```
Identifier 'n' is undeclared
```

would be generated unless n is declared somewhere else and its scope includes this use of it. Understanding scope is important for understanding certain compilation errors.

Stated simply, the scope of an identifier depends on *where the identifier is declared*. The scopes of the identifiers we have seen thus far have been determined by four simple rules:

S1. If an identifier is declared within a block, then its scope runs from its declaration to the end of the block.

S2. If an identifier is a parameter of a function, then its scope is the body of the function.

S3. If an identifier is declared in the initialization expression of a for loop, then its scope runs from its declaration to the end of the loop.

S4. If an identifier is a declared outside all blocks and is not a parameter, then its scope runs from its declaration to the end of the file.

Note that an identifier's scope always begins at its declaration. One basic rule that follows from this is that *an identifier must be declared before it can be used.*

10.4.1.1 An Example

To illustrate how the compiler uses these scope rules, suppose that we were to rewrite the function getMenuChoice() from Section 9.2 as follows:

```
char getMenuChoice(const string & MENU, char firstChoice, char lastChoice)
{
   for (;;)
   {
      cout << MENU;
```

```
   char choice;
   cin >> choice;

   if (choice >= firstChoice && choice <= lastChoice)
     break;

   cerr << "\nI'm sorry, but " << choice
        << " is not a valid menu choice.\n";
 }
 return choice;            // ERROR!
}
```

When this function is compiled, the return statement generates an error message like

```
Identifier 'choice' is undeclared
```

(Before continuing, try to find the cause of the error using the scope rules.)

The problem is that by Scope Rule S1, the scope of choice only runs through the last statement of the block that is the body of the loop. Because the return statement appears after the block, it lies outside the scope of choice, and so the compiler generates an error when it attempts to access choice. The problem can be avoided by moving the declaration of choice to the outermost block of the function, thus ensuring that all statements that refer to choice lie within its scope:

```
char getMenuChoice(const string& MENU, char firstChoice, char lastChoice)
{
   char choice;
   for (;;)
   {
     cout << MENU;
     cin >> choice;

     if (choice >= firstChoice && choice <= lastChoice)
       break;

     cerr << "\nI'm sorry, but " << choice
          << " is not a valid menu choice.\n";
   }
   return choice;              // OK!
}
```

10.4.1.2 Objects Declared Outside All Blocks

Although we have not put any declarations of variables or constants outside all blocks, we have done so with prototypes of functions and with libraries that are included in a program. For example, in Example 10.5, the prototype of the function chopIPaddress() precedes the main function, so that the identifier chopIPaddress is declared outside all blocks. By Scope Rule S4, its scope thus extends from its declaration to the end of the file.

This is the reason why a function prototype *must* precede calls to that function. Because a prototype is a declaration of the function, it begins the scope of the function. Calls to that function later in the program thus lie within its scope.

Identifiers that are declared in a library's header file, such as cin, cout, cerr, and clog in iostream and the class string in the string library, are also governed by Scope Rule S4. When the compiler processes a #include directive outside all blocks as in

```cpp
#include <iostream>
using namespace std;

int main()
{
  ...
}
```

all objects in that file are inserted outside all blocks and thus fall under Scope Rule S4. The scope of such declarations thus extends from their declarations (i.e., from the #include directive) to the end of the file.

10.4.1.3 Scopes of for-Loop Control Variables

Scope Rule S3 governs the scope of an identifier declared in the initialization expression of a for loop. To illustrate it, suppose that a program contains the statements

```cpp
for (int i = 1; i <= someLimit; i++)
{
  ...
}

cout << "i = " << i << endl;          // ERROR
```

When the program is compiled, an error message like

```
Identifier 'i' is undeclared
```

results because Scope Rule S3 asserts that the scope of i ends at the bottom of the loop, so the output statement lies outside the scope of i. But this also means that i could be redeclared and used after the loop has ended. For example, it could be used in a later for loop,

```cpp
for (int i = 1; i <= 100; i++)
{
  ...
}

...

for (int i = 50; i > 0; i--)
{
  ...
}
```

because the second loop lies outside the scope of the first i. Alternatively, we could move the declaration of i *before* the first loop so that it is governed by Scope Rule S1 instead of S3, and both loops fall within its scope:

```
int i;
for (i = 1; i <= 100; i++)
{
   ...
}
   ...
for (i = 50; i > 0; i--)
{
   ...
}
...
```

10.4.2 Name Conflicts

Our discussion of scope would not be complete without mentioning the rule:

> *Within the scope of an identifier, no redeclaration of that identifier that results in an ambiguity for the compiler is permitted.*

As a simple illustration, suppose that we write

```
int main()
{
   int value = 1;
   ...
   double value = 2.0;          // ERROR!
   ...
   cout << value << endl;
}
```

The redeclaration of value creates an *ambiguity* for the compiler, because in the output statement, it is unclear which value is to be displayed. From the standpoint of the compiler, value is declared as the name of an int object, starting the scope of the identifier value. When the compiler encounters a second declaration of the same name that it cannot distinguish from the first declaration, a **name conflict** arises,

```
int main()
{
   int value = 1;

   double value = 2.0;                    // ERROR!

   cout << value << endl;

}
```

and so the compiler generates an error. By contrast, there is no name conflict in the following, because the scopes of the two versions of `value` do not overlap:

```
int main()
{
    ...
    if (x > 0)
    {
        int value = 1;
        cout << value << endl;

    }
    else
    {
        double value = 2.0;
        cout << value << endl;

    }
}
```

The compiler will eliminate ambiguity between names whenever it can. To illustrate, consider the following example:

```
int main()
{
    int value = 1;
    {
        double value = 2.3;
        cout << value << endl;
    }
    cout << value << endl;

}
```

The effect here is that the outer (`int`) version of `value` will be hidden from the statements within the scope of the nested version of `value`. Because `value` is redeclared in its own *nested block*, the compiler will assume that accesses within the scope of the nested declaration are to the local (`double`) version of `value`, and accesses outside of its scope are to the outer (`int`) version of `value`, and so the compiler will proceed to translate the code. The output produced will be

```
2.3
1
```

Because the outer variable `value` is accessible before and after this nested scope, but is not accessible within it, the nested variable `value` is sometimes described as creating a **hole in the scope** of the outer `value`. Some compilers may generate a warning message like

```
Warning: redeclaration of 'value' hides previous declaration
```

to alert us to this hole.

10.4.3 Namespaces

If name is declared or defined in a namespace declaration of the form

```
namespace Something
{
   // declaration or definition of name
   // other declarations and definitions
}
```

then the scope of name is this namespace block. It can be accessed outside of the namespace only in the following ways:

1. By its fully qualified name Something::name;[3]

2. By its unqualified name name, if a **using declaration** of the form

   ```
   using namespace Something::name;
   ```

   or a **using directive** of the form

   ```
   using namespace Something;
   ```

   has already been given, provided that no other item has the same name (in which case qualification is required as described in 1).

A using directive makes all names in the namespace available. This is why we add the directive using namespace std; after #includes of standard I/O libraries—so that the names declared and defined in them can be used without qualification.

10.5 AN INTRODUCTION TO RECURSION

All the examples of function calls considered thus far have involved one function f() calling a different function g() (with the calling function f() often being the main function). However, a function f() may also *call itself*, a phenomenon known as **recursion**, and in this section, we show how recursion is implemented in C++. Many examples of recursion could be given, but we will consider only two of the classic ones: calculating factorials and analyzing a street network. Others are given in the exercises and in Chapter 16 where we revisit recursion.

10.5.1 Example 1: The Factorial Problem Revisited

To illustrate the basic idea of recursion, we look again at the problem of calculating the factorial function that we introduced in Section 6.3 and revisited in Section 10.2. Although the first definition of the factorial *n*! of an integer *n* that one usually learns (and the one we used earlier) is

$$n! = \begin{cases} 1 \text{ if } n = 0 \\ 1 \times 2 \times \dots n \text{ if } n > 0 \end{cases}$$

---

[3] This is the reason for using string::npos in Chapter 7 to access the constant named npos in the string class.

it would be foolish to use it to calculate a sequence of consecutive factorials, that is, to multiply together the numbers from 1 through n each time:

$$0! = 1$$

$$1! = 1$$

$$2! = 1 \times 2 = 2$$

$$3! = 1 \times 2 \times 3 = 6$$

$$4! = 1 \times 2 \times 3 \times 4 = 24$$

$$5! = 1 \times 2 \times 3 \times 4 \times 5 = 120$$

A great deal of the effort would be redundant, because it is clear that once a factorial has been calculated, it can be used to calculate the next factorial. For example, given the value of 4!, we can then compute 5! simply by multiplying the value of 4! by 5:

$$5! = 5 \times 4! = 5 \times 24 = 120$$

This value of 5! can in turn be used to calculate 6!,

$$6! = 6 \times 5! = 6 \times 120 = 720$$

and so on. Indeed, to calculate $n!$ for any positive integer n, we need only know the value of 0!, and the fundamental relation between one factorial and the next:

$$n! = n \times (n - 1)!$$

In general, a function is said to be **defined recursively** if its definition consists of two parts:

1. An **anchor** or **base case**, in which the value of the function is specified for one or more values of the parameter(s).

2. An **inductive** or **recursive step**, in which the function's value for the current value of the parameter(s) is defined in terms of previously defined function values and/or parameter values.

For the factorial function we have:

$$0! = 1 \qquad \text{(the anchor or base case)}$$

$$\text{For } n > 0, n! = n \times (n - 1)! \qquad \text{(the inductive or recursive step)}$$

The first statement specifies a particular value of the function, and the second statement defines its value for n in terms of its value for $n - 1$.

This recursive definition can be used as an algorithm for calculating factorials. To see how it works, consider using it to calculate 5!. We must first calculate 4! because 5! is defined as the product of 5 and 4!. But to calculate 4! we must calculate 3! because 4! is defined as 4 × 3!. And to calculate 3!, we must apply the inductive step of the definition again, 3! = 3 × 2!, then again to find 2!, which is defined as 2! = 2 × 1!, and once again to find 1! = 1 × 0!. Now we have finally reached the anchor case and because the value of 0! is given, we can backtrack to find the value of 1!,

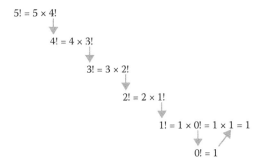

then again to find the value of 2!,

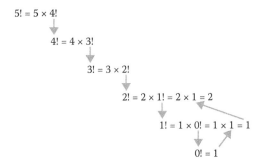

and so on, until we eventually obtain the value 120 for 5!:

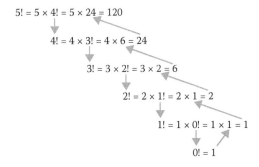

As this example demonstrates, a recursive definition may require considerable book-keeping to record information at the various levels of the recursion, because information from the anchor case is used to backtrack from one level to the preceding one. Fortunately, most modern high-level languages (including C++) support recursion by automatically performing all of the necessary bookkeeping and backtracking.

Example 10.6 shows a definition of `factorial()` that implements this algorithm.

Example 10.6 Computing n! Recursively

```
/* factorial computes n! recursively.

   Receive:       n, a nonnegative integer
   Precondition: n >= 0
   Return:        n!
   ------------------------------------------------------------------*/

int factorial(int n)
{
  assert(n >= 0);
  if (n == 0)
    return 1;                          // anchor case
  else
    return n * factorial(n-1);         // inductive step
}
```

When this function is called with an argument greater than zero, the inductive step

```
else
   return n * factorial(n-1);
```

causes the function to call itself repeatedly, each time with a smaller parameter, until the anchor case

```
if (n == 0)
   return 1;
```

is reached. To illustrate, consider the statement

```
int fact = factorial(4);
```

which calls the function `factorial()` to calculate 4!. Because the value of n (4) is not 0, the inductive step executes

```
return n * factorial(n-1);
```

which calls `factorial(3)`. Before control is transferred to `factorial(3)`, the current value (4) of the parameter n is saved so that the value of n can be restored when control returns. This might be pictured as follows:

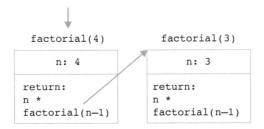

Because the value of n (3) in this function call is not 0, the inductive step in this second call to `factorial()` generates another call `factorial(n-1)` passing it the argument 2. Once again, the value of n (3) is saved so that it can be restored later:

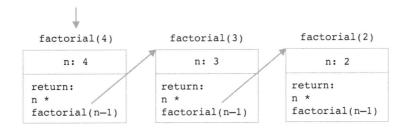

Because the value of n (2) in this function call is not 0, the inductive step in this third call to `factorial()` generates another call `factorial(n-1)` passing it the argument 1. Once again, the value of n (2) is saved so that it can be restored later. The call `factorial(1)` in turn generates another call, `factorial(0)`:

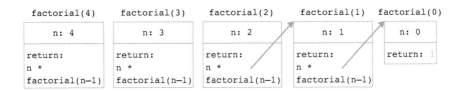

Because the anchor condition

```
if (n == 0)
    return 1;
```

is now satisfied in this last function call, no additional recursive calls are generated. Instead, the value 1 is returned as the value for `factorial(0)`:

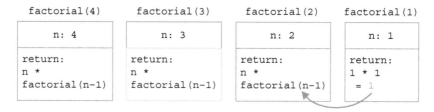

Now that `factorial(0)` has completed its computation, execution resumes in `factorial(1)` where this returned value can now be used to complete the evaluation of

```
n * factorial(n - 1) = 1 * factorial(0) = 1 * 1 = 1
```

giving 1 as the return value for `factorial(1)`:

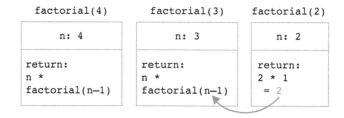

Once `factorial(1)` has completed its computation, execution resumes in `factorial(2)` where the return value of `factorial(1)` can now be used to complete the evaluation of

```
n * factorial(n - 1) = 2 * factorial(1) = 2 * 1 = 2
```

giving 2 as the return value for `factorial(2)`:

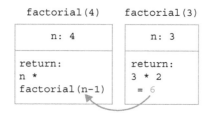

Because `factorial(2)` has completed its computation, execution resumes in `factorial(3)` where the return value of `factorial(2)` is used to complete the evaluation of

```
n * factorial(n - 1) = 3 * factorial(2) = 3 * 2 = 6
```

giving 6 as the return value for `factorial(3)`:

This completes the function call to `factorial(3)`, so execution resumes in the call to `factorial(4)`, which computes and returns the value

```
n * factorial(n - 1) = 4 * factorial(3) = 4 * 6 = 24
```

giving 24 as the return value for `factorial(4)`:

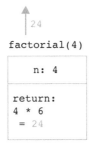

Note that in the definition of `factorial()` we used `assert()` to test that the parameter n is not negative. To see the reason for this, consider what would happen if we had not tested this precondition and the function were called with a negative integer, as in

```
int fact = factorial(-1);
```

Because –1 is not equal to 0, the inductive step

```
else
   return n * factorial(n - 1);
```

would be performed, recursively calling `factorial(-2)`. Execution of this call would begin, and since –2 is not equal to 0, the inductive step

```
else
   return n * factorial(n - 1);
```

would be performed, recursively calling `factorial(-3)`. This behavior would continue until memory was exhausted, at which point the program would terminate abnormally, possibly producing an error message like

```
Stack overruns Heap.
```

Such behavior is described as **infinite recursion** and is obviously undesirable. To avoid it we programmed defensively by including the parameter-validity check.

10.5.2 Example 2: Street Network

Consider a network of streets laid out in a rectangular grid, for example,

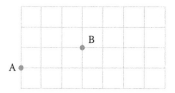

In a *northeast path* from one point in the grid to another, one may go only to the north (up) and to the east (right). For example, there are four northeast paths from A to B in the preceding grid:

To count the paths, we might proceed (recursively) as follows: To get from A to B, there are two ways to begin:

Case 1: Go one block north (up); call this point A_1

Case 2: Go one block east (right); call this point A_2

In either case, we must count paths on a smaller grid. In the first case, there are the same number of columns separating A_1 and B, but there is one less row. In the second case, there are the same number of rows separating A_2 and B, but there is one less column. This suggests the following inductive step in a recursive solution to the problem:

Number of paths from A to B

= (Number of paths from A_1 to B) + (Number of paths from A_2 to B)

The function `numberOfPaths()` in Example 10.7 uses this recursive approach to count the number of northeast paths from one point to another in a rectangular grid. The anchor case is that in which points A and B coincide.

Example 10.7 Counting Paths in a Street Network

```
/* Program to count "northeasterly" paths from a given starting
   point in a network of streets to a specified ending point.

   Input:  coordinates of starting and ending points

   Output: number of paths from start to end
   --------------------------------------------------------------*/
```

```cpp
#include <iostream>
using namespace std;

int numberOfPaths(int numRows, int numColumns);
int main()
{
  int startRow, startColumn, endRow, endColumn;

  cout << "Enter starting coordinates (row then column): ";
  cin >> startRow >> startColumn;
  cout << "Enter ending coordinates (row then column): ";
  cin >> endRow >> endColumn;

  cout << "\nThere are "
       << numberOfPaths(endRow - startRow, endColumn  - startColumn)
       << " paths\n";
}

int numberOfPaths(int numRows, int numColumns)
{
  if (numRows == 0 || numColumns == 0)
    return 1;
  else
    return numberOfPaths(numRows - 1, numColumns)
         + numberOfPaths(numRows, numColumns - 1);
}
```

SAMPLE RUNS:

```
Enter starting coordinates (row then column): 0 1
Enter ending coordinates (row then column): 3 2


Number of paths from start to end: 4


Enter starting coordinates (row then column): 1 1
Enter ending coordinates (row then column): 1 1


Number of paths from start to end: 1


Enter starting coordinates (row then column): 0 1
Enter ending coordinates (row then column): 7 4


Number of paths from start to end: 120


Enter starting coordinates (row then column): 0 0
Enter ending coordinates (row then column): 7 4


Number of paths from start to end: 330
```

10.6 INLINING, OVERLOADING, AND TEMPLATES

In this section we will consider three additional properties of functions:

1. Inlining: In applications where function calls and returns are too time-consuming, inlining can be used to improve speed of execution.

2. Overloading: Functions may have the same name, provided the compiler is able to distinguish between them; this is known as overloading the functions.

3. Templates: A function template is a pattern for a function that can be used to generate a collection of functions that have the same name but process different kinds of data.

10.6.1 Inline Functions

We have seen that when one function f() calls another function g(), execution is transferred from f() to g(). When g() terminates, a second transfer of control is necessary to return execution back to f(). *Each function call significantly increases the amount of time required for a program to execute*—a considerable amount of time in fact, compared to the speed at which a computer normally operates. In many situations, the *overhead* associated with function calls is acceptable, but there are other occasions in which execution time is so important that this overhead cannot be tolerated. For these problems, C++ provides **inlining** of functions.

One of the details we have omitted (for simplicity) in the discussion of functions is that a definition can be preceded by a function specifier, which provides the compiler with special instructions about the function:

```
specifier return_type function_name(parameter_declaration_list)
{
   statement_list
}
```

One such specifier is the **inline specifier** that instructs the compiler to use the C++ inline mechanism to avoid the overhead of normal function calls.

To illustrate, suppose we want to inline the temperature-conversion function celsiusToFahrenheit() from Section 6.2. We can do so by preceding its prototype and its definition with the inline specifier:

```
inline double celsiusToFahrenheit(double tempCelsius);
   .
   .
   .
inline double celsiusToFahrenheit(double tempCelsius)
{
   return 1.8 * tempCelsius + 32;
}
```

When the function prototype and definition occur in the same file, both must be labeled as `inline`. Otherwise, there will be a mismatch between the function's prototype and its definition, and a compiler error results.

The `inline` specifier suggests to the C++ compiler that it *replace each call to this function with the body of the function, with the arguments for that function call substituted for the function's parameters.*[4] For example, the compiler is being asked to effectively replace the function call in the statement

```
tempFahr = celsiusToFahrenheit(currentTemp);
```

with the body of the function, but with the argument `currentTemp` substituted for the parameter `tempCelsius`:

```
tempFahr = 1.8 * currentTemp + 32.0;
```

Thus, the effect of the `inline` specifier is to ask the compiler to take additional *compile time* to perform this substitution *for each call* to the function. This saves *run time*, because the elimination of each function call means that no transfers of execution need to be performed, eliminating the overhead incurred by those function calls.

10.6.1.1 Inline Functions and Libraries

The normal procedure for a function stored in a library is to put its prototype in the library's header file and its definition in the implementation file. This separation is not allowed, however, if we want to inline a function. An inline library function must be defined in the header file.

To illustrate, suppose that a library `MyMath` contains, among other mathematical functions, the function `sum()` from Example 9.2, with the prototype of `sum()` in `MyMath.h` and its definition in `MyMath.cpp`. To designate `sum()` as an inline function, we would remove its definition from the implementation file and replace its prototype in the header file with this definition, preceding it with the `inline` specifier, as shown in Example 10.8.

Example 10.8 Inlined Library Functions

```
/* This header file provides an interface for library MyMath.
   It contains prototypes and inline definitions of non-predefined math
   functions.
   ... remaining opening documentation omitted ...
   ----------------------------------------------------------------*/

int factorial(int n);
```

---

[4] This *suggestion* may be ignored by the compiler if it judges that the function definition is too complicated to be inlined efficiently.

```
inline int sum(int n)
{
  return n * (n + 1) / 2;
}

// ... other function prototypes and/or inline definitions
```

These steps are necessary because the only part of the library that the compiler sees when it compiles a program is the header file, which is inserted when it processes the #include directive. If the *compiler* (and not the *linker*) is to replace calls to a function with the function's definition, that function's entire definition must be visible to the compiler, which means that it must be moved from the implementation file to the header file.

10.6.1.2 To Inline or Not to Inline: A Space-Time Trade-Off

Because inlined functions eliminate the overhead of function calls and make programs run faster, it may be tempting to make all functions inline. There is, however, a *space-time trade-off*: a program that uses inlined functions may indeed run *faster* than its noninlined equivalent, but it may also be much *larger* and thus require more memory.

To see why this happens, we must realize that, unlike the simple examples we have been using, functions written for real-world software projects typically (1) contain many statements, and (2) may be called at many different places in a program. When such functions are inlined, *each substitution of a function's body for its call replaces a single statement (the call) with many statements (the body)*, which increases the overall size of the program. Thus, if the definition of a function consists of a large number of statements and is called many times, then substituting the body of the function definition at each call will significantly increase the overall size of the program, a phenomenon known as *code bloat*.

To avoid code bloat, inline *should be used with restraint*. If a function uses just a few operations (such as the celsiusToFahrenheit() and sum() functions), then it may be designated as inline to eliminate the overhead associated with calling the function. But if a function uses many operations (e.g., more than five or six), it should probably not be inlined.

10.6.2 Overloading Functions

Earlier in this section, we considered two summation functions from a MyMath library both named sum(). It would seem that using the same name for two different functions in the same file should generate an error. However, no name conflict occurs when these definitions are compiled, even though there are two definitions for the same identifier! As we shall see, this is consistent with what we stated earlier:

> *Within the scope of an identifier, no redeclaration of that identifier that results in an ambiguity for the compiler is permitted.*

The key here is the phrase *an ambiguity for the compiler.* What is necessary for the C++ compiler to be able to distinguish functions having the same name from one another without ambiguity will now be described.

10.6.2.1 Signatures

The **signature** of a function is a list of the types of its parameters, including any `const` and reference parameter indicators (`&`). For example, the signature of the `polarToRect-angular()` function in Example 10.2 is

```
(double, double, double &, double &)
```

and the signature of the `makeChange()` function in Example 10.3 is

```
(double, double, int &, int &, int &, int &, int &)
```

while the signature of the `chopIPAddress` function in Example 10.5 is

```
(const string &, string &, string &, string &, string &)
```

Signatures are important, because the compiler essentially considers a function's signature to be part of its name.

10.6.2.2 Function Overloading

If two different functions have the same name, that name is said to be **overloaded**. For example, when we use the same name for both of the summation functions,

```
int sum(int n)
{
    return n * (n + 1) / 2;
}

int sum(int m, int n)
{
    assert(m < n);
    return (n - m + 1) * (n + m) / 2;
}
```

we are overloading the name `sum` with two different definitions. The compiler is able to distinguish them from one another, because they have different signatures. If we call `sum()` with

```
cout << sum(100) << endl;
```

the call has one `int` argument, and so the compiler associates this call with the first definition of `sum()` whose signature has one `int` type. However, if we write

```
cout << sum(20, 40) << endl;
```

the function call has two `int` arguments, and so the compiler associates this call with the second definition of `sum()` whose signature has two `int` types.

Signatures thus allow the C++ compiler to distinguish calls to different functions with the same name. As a result, we have the following rule governing overloading: *The name of a function can be overloaded, provided no two versions of the function have the same signature.* Note that the return type of a function is not a part of its signature. Thus, two functions with identical signatures but with different return types cannot have the same name.

Names should be overloaded only when it is appropriate. Otherwise the code you write may be difficult to read and understand. Different functions that perform the same operation (e.g., summation, finding the minimum operation, finding the maximum) on different data types are prime candidates for overloading. But giving the same name to operations that have nothing to do with each other simply because the language allows you to do so is an abuse of the overloading mechanism and is not good programming style.

It should be evident that overloading has been used since Chapter 3. For example, the operators +, -, *, and / are all overloaded so that they can be applied to any of the numeric types. In the expression

$$2.0 / 5.0$$

the C++ compiler uses the real division operation (which produces the value 0.4), but in the integer expression

$$2 / 5$$

it uses the integer division operation (which produces the value 0). Many of the other operators, including <<, >>, =, +=, -=, *=, /=, <, >, ==, <=, >=, and !=, have also been overloaded with multiple definitions. For an expression of the form

operand₁ operator operand₂

the compiler simply matches the types of *operand₁* and *operand₂* against the signatures of the available definitions of *operator* to determine which definition to apply.

10.6.3 Function Templates

In Section 10.3, we considered a pattern for a function to interchange the values of two variables:

```
void swap(____ & first, ____ & second)
{
    ____ temporary = first;
    first = second;
    second = temporary;
}
```

We saw with an example how we could replace the blanks with the type `char` to obtain an actual function that could be incorporated into a program that processes individual characters. But the swap operation is needed in many other problems such as sorting a list of strings, interchanging the smaller and larger of two integers, and interchanging two rows in a table of doubles.

One possibility would be to make a library of overloaded `swap()` functions, one for each type of values we may ever want to interchange, that we could #include in any program where one of these functions is needed. However, each definition of `swap()` in this library would be doing exactly the same thing (on a different type of data). This should raise red flags for us, because we have said before that *any time we find ourselves repeating the same work, there is probably a better way.*

10.6.3.1 Parameters for Types

Here, the better way is to recognize that the only differences in any of these definitions are the three places where a type is specified. It would be nice if we could define the function and leave these types "blank" as in our earlier pattern and somehow *pass the type* to the function when we call it. Then we could replace all of these definitions with one.

This is effectively what C++ allows us to do, as shown in Example 10.9.

Example 10.9 The `swap()` Template

```
/* This header file provides a template that generates functions
   for interchanging the values of two variables.
   ... remaining opening documentation omitted ...
------------------------------------------------------------------*/

template<typename Item>
inline void swap(Item & first,  Item & second)
{
  Item temporary = first;
  first = second;
  second = temporary;
}
```

Rather than specify that the function is to exchange two values of a particular type such as char, int, and so on, this definition uses the identifier Item as a placeholder for the type of the value to be exchanged. More precisely, the line informs the compiler of two things:

1. This definition is a **template**: *a pattern from which the compiler can create a function.*

2. The identifier Item is the name of a **type parameter** for this definition that will be given a value when the function is called.[5]

The rest of the definition simply specifies the behavior of the function, using the type parameter Item in place of any specific type.

Using this version of the swap library, we can now write

```
#include "swap.h"
#include <iostream>
#include <string>
using namespace std;

int main()
{
  int i1 = 11,
      i2 = 22;
```

---

[5] C++ also allows the keyword class to be used to specify a type parameter.

```
    swap(i1, i2);
    cout << i1 << ' ' << i2 << endl;

    double d1 = 33.3,
    d2 = 44.4;
    swap(d1, d2);
    cout << d1 << ' ' << d2 << endl;

    string s1 = "Hi",
    s2 = "Ho";
    swap(s1, s2);
    cout << s1 << ' ' << s2 << endl;
}
```

When the compiler encounters the first call to swap(),

```
    swap(i1, i2);
```

in which the two arguments i1 and i2 are of type int, it uses the pattern given by our template to generate a new definition of swap() in which the type parameter Item is replaced by int:

```
inline void swap(int & first, int & second)
{
    int temporary = first;
    first = second;
    second = temporary;
}
```

When it reaches the second call,

```
    swap(d1, d2);
```

where the two arguments d1 and d2 are of type double, the compiler will use the same pattern to generate a second definition of swap() in which the type parameter Item is replaced by double:

```
inline void swap(double & first, double & second)
{
    double temporary = first;
    first = second;
    second = temporary;
}
```

When the compiler reaches the final call,

```
    swap(s1, s2);
```

in which the two arguments `s1` and `s2` are of type `string`, it will use the same pattern to generate a third definition of `swap()` in which the type parameter `Item` is replaced by `string`:

```
inline void swap(string & first, string & second)
{
   string temporary = first;
   first = second;
   second = temporary;
}
```

We are spared from all of the redundant coding of the earlier approach because the compiler is providing multiple versions of the swap operation as they are needed.

10.6.3.2 Templates vs. Overloading

If there are several versions of the same operation to be encoded as functions, it may not be clear whether one should write several functions that overload the same name or write one function template. The following guideline helps with making this decision:

If each version of the operation behaves in exactly the same way,

regardless of the type of data being used,

then define a **function template** to perform the operation.

Otherwise, define a separate function for each operation

and use **overloading** to give them the same name.

Thus, because each version of `swap()` uses the three-way swap and behaves in exactly the same manner regardless of the type of values being interchanged, a function template is appropriate.

By contrast, the two summation functions we considered earlier use different formulas, and thus behave differently. These operations are therefore best performed by two separate functions that overload the name `sum()`.

The reasoning behind this guideline should be clear. When the compiler creates a function definition from a template, it blindly replaces each occurrence of the type parameter with the type of the arguments. As a result, each definition created from a template must behave in exactly the same way, except for the type of data being operated upon. Overloading has no such restriction and so can be used for a wider variety of operations.

CHAPTER SUMMARY

Key Terms

alias	call by reference
anchor	call by value
argument	code bloat
base case	`const` reference parameter

hole in the scope

inductive step

`inline` specifier

namespace

overload

parameter

parameter-passing mechanism

recursion

recursive definition of a function

reference parameter

scope

signature

template

type parameter

`using` directive

value parameter

NOTES

- By default, parameters are value parameters; their values are *copies* of the corresponding arguments. Thus, changing the value of a value parameter has no effect on the value of the corresponding argument.

- Reference parameters are aliases for their corresponding arguments; they refer to the same memory locations. Thus, changing the value of a reference parameter also changes the value of the corresponding argument, which must be a variable. To specify that a parameter is a reference parameter, insert an ampersand (&) between its type and its name in the function heading.

- A `const` reference parameter, specified by inserting the keyword `const` before a reference parameter's type, is a preferred alternative to a value parameter whose type is a class (or that is some other large object) because it does not require copying the value of the corresponding argument.

- Use a `return` statement to send back a single value from a function; use reference parameters to send back multiple values.

- An identifier's scope is the portion of a program where the object or function that it names can be accessed. Basic scope rules for an identifier are:

  - *Declared in a block*: from declaration to end of block

  - *Parameter*: body of the function

  - *In initialization of for loop*: from declaration to end of loop body

  - *Declared outside all blocks*: from declaration to end of file

- A `using` clause makes names in a namespace available without qualification.

- A recursive function must have an anchor that will eventually be executed and cause a return from the function and an inductive step that specifies the current action of the function in terms of previously defined actions.

- A simple function can be inlined to avoid the overhead of calling that function and thereby reduce execution time. Prepend the keyword `inline` to the function heading to do this.

- The signature of a function is the list of its parameter types. A function may overload another function (i.e., have the same name) provided they have different signatures.

- Preceding the prototype and the definition of a function with a template clause of the form `template<typename T>` converts the function into a function template, which is a type-independent pattern for a function. When the function is called, the compiler will use the type of the arguments to determine what type to substitute for *T* and generate a type-specific function.

Style and Design Tips

- *Functions should be documented in the same way that programs are.* **The documentation should include specifications and descriptions of:**

  - The purpose of the function

  - Any items that must be received by the function from its caller

  - Any items that must be input to the function

  - Any items that are returned or passed back by the function to its caller

  - Any items that are output by the function

- *Follow the same stylistic standards for functions that are used for main programs.*

- *Declaring variable objects as close as possible to their first use increases the readability of a program and also aids with debugging.*

- *Value parameters receive copies of the corresponding arguments and are used for simple/small arguments. For arguments such as classes (and arrays studied in the next chapter) that are stored in large amounts of memory, use const reference parameters.*

- *Use a* `return` *statement to return a single value from a function; use reference parameters when several values are to be returned.*

- *All variables used within a function should be defined within that function, either as parameters or as local variables.* **This keeps functions self-contained and increases their generality and reusability.**

- *Recursive functions should be clearly marked as such. For clarity and readability, the anchor case and inductive steps of a recursive function should be marked with comments.*

- *Only simple functions should be specified as being inline functions.* **Substitution of an inline function's body for each of its calls can increase a program's size considerably if the function is nontrivial and/or it is called at several places in the program.**

Warnings

1. *When a function is called, the number of arguments should be the same as the number of parameters in the function heading.*[6]

2. *The type of an argument corresponding to a value parameter should be compatible with the type of that parameter.*

3. *An argument that corresponds to a reference parameter must be a variable whose type matches the type of that parameter; it may not be a constant or an expression.*

4. *Check carefully the prototype of each function that is used to see if it has any reference parameters;* changes in their values may lead to unexpected or unintended changes in the values of arguments.

5. *Any attempt to use a function's parameters outside the function will generate an error because they are allocated memory only during execution of that function; there is no memory associated with them either before or after execution of that function.*

6. *An identifier must be declared before it can be used.* For example, if a function f() calls a function g(), then a prototype (or definition) of g() must precede the definition of f().

7. *Inlined functions cannot be split with prototypes in a header (.h) file and definitions in an implementation (.cpp) file.*

8. *Use restraint with inlining functions, because code bloat may result otherwise.*

TEST YOURSELF

Section 10.3

1. The parameter-passing mechanism that occurs by default is call-by-_____.

2. A _____ parameter contains a copy of the corresponding argument.

3. If a _____ parameter is changed in the body of the function, then the value of the corresponding argument is not changed.

4. If a _____ parameter is changed in the body of the function, then the value of the corresponding argument is also changed.

5. Placing a(n) _____ between a parameter's type and its name indicates that the parameter is a reference parameter.

---

[6] An exception is when one or more parameters at the end of the parameter list have *default arguments*; for example,

```
void f(int x, int y = 0, double z = 3.5);
```

As the following examples show, this function may be called with 1, 2, or 3 arguments:
f(1, 2, 4.9); Within f, x will have the value 1, y the value 2, and z the value 4.9
f(1, 2); Within f, x will have the value 1, y the value 2, and z the value 3.5
f(1); Within f, x will have the value 1, y the value 0, and z the value 3.5

Questions 6–11 assume the following function definition

```
void f(int x, int & y, int & z)
{
   z = y = x * x + 1;
}
```

and the following declarations in the calling function:

```
int a = 0, b = 1, c = 2, d = 3;
const int E = 4;
```

6. (True or false) After d = f(a, b, c); is executed, d will have the value 1.

7. (True or false) After f(a, b, c); is executed, b and c will both have the value 1.

8. (True or false) After f(c, d, E); is executed, d will have the value 5.

9. (True or false) After f(c+1, c-1, d); is executed, d will have the value 10.

10. (True or false) The function call f(c, d, E); makes z a const reference parameter.

11. Rewrite the definition of f() so that x is a const reference parameter.

12. Given the function definition

```
void change(int number, string & a, string & b, string & c)
{
   const string BAT = "bat";

   if (number < 3)
   {
      a = BAT;
      b = BAT;
   }
   else
      c = BAT;
}
```

what output will the following program segment produce?

```
string str1 = "cat", str2 = "dog", str3 = "elk";
change(2, str1, str2, str3);
cout << "String = " << str1 << str2 < str3 << endl;
```

Section 10.4

1. The part of a program where an identifier refers to a particular object or function is called the _____ of that identifier.

2. (True or false). A compilation error results if an identifier is accessed within its scope.

3. The scope of an identifier declared within a block runs from its declaration to the _____.

4. The scope of a parameter is the _____.

5. What will the following statements produce?

```
for (int i = 0; i < 3; i++)
   cout << i << endl;
cout << "i is now " << i << endl;
```

Section 10.5

1. _____ is the phenomenon of a function referencing itself.

2. Name and describe the two parts of a recursive definition of a function.

3. (True or false) A nonrecursive function for computing some value may execute more rapidly than a recursive function that computes the same value.

4. For the following recursive function, find `f(5)`.

```
int f(int n)
{
   if (n == 0)
      return 0;
   else
      return n + f(n - 1);
}
```

5. For the function in Question 4, find `f(0)`.

6. For the function in Question 4, suppose + is changed to * in the inductive step. Find `f(5)`.

7. For the function in Question 4, what happens with the function call `f(-1)`?

Section 10.6

1. A function's _____ is a list of the types of its parameters.

2. Two functions are said to be overloaded if they have the same _____.

3. A function's name can be overloaded provided no two definitions of the function have the same _____.

4. A _____ is a pattern from which a function can be constructed.

5. Templates may have _____ parameters, but ordinary functions may not.

6. Given the template definition

```
template<typename something>
void print(something x)
{ cout << "***" << x << "***\n"; }
```

describe what the compiler will do when it encounters the statements

```
int number = 123;
print(number);
```

EXERCISES

Exercises 1–11 assume the following program skeleton:

```
#include <cmath>
using namespace std;

void calculate(double a, double & b,
               int m, int & k, int & n, char & c);

int main()
{
  const double PI = 3.14159;
  const int TWO = 2;
  const char INITIAL = 'N';

  int month, day, year, p, q;
  double hours, rate, amount, u, v;
  char code, dept;
}
```

Determine whether the given statement can be used to call the function `calculate()`. If it cannot be used, explain why.

1. `calculate(u, v, TWO, p, q, code);`

2. `calculate(PI, u, TWO, p, v, dept);`

3. `calculate(hours, PI, TWO, day, year, dept);`

4. `calculate(13, hours, PI, 13, year, dept);`

5. `calculate(PI * hours, PI, TWO, day, year, dept);`

6. `calculate(PI, PI * hours, TWO, day, year, dept);`

7. `while (u > 0)`
    `calculate(u, v, TWO, p, q, code);`

8. `calculate(0, hours, (p + 1) / 2, day, year, code)`

9. `calculate(sqrt(amount), rate, 7, p, q, INITIAL);`

10. `while (amount > 0)`
    `calculate(TWO, amount, day, p + q, day, dept);`

11. `cout << calculate(u, v, TWO, p, q, code);`

The following exercises ask you to write functions to compute various quantities. To test these functions, you should write driver programs as instructed in Programming Problems 1–4.

12. The change-dispenser function in Example 1 rounds a real amount to the nearest hundredth before converting it to cents. Write a function that receives a real value and a nonnegative integer, and returns the real value rounded to the number of decimal places specified by that integer—the nearest integer if the integer received is 0, the nearest tenth if it is 1, the nearest hundredth if it is 2, and so on.

13. Write a function that receives a weight in pounds and ounces and returns the corresponding weight in grams. (1 oz = 28.349527 g)

14. Write a function that receives a weight in grams and returns the corresponding weight in pounds and ounces.

15. Write a function that receives a length in yards, feet, and inches and returns the corresponding measurement in centimeters. (1 in = 2.54001 cm)

16. Write a function that receives a length in centimeters and returns the corresponding measurement in yards, feet, and inches.

17. Write a function that receives a time in the usual representation of hours, minutes, and a character value that indicates whether this is A.M. ("A") or P.M. ("P") and returns the corresponding military time.

18. Write a function that receives a time in military format and returns the corresponding time in the usual representation in hours, minutes, and A.M./P.M. For example, a time of 0100 should be returned as 1 hour, 0 minutes, and "A" to indicate A.M.; and a time of 1545 should be returned as 3 hours, 45 minutes, and "P" to indicate P.M.

Section 10.5

Exercises 1–12 assume the following function f():

```
void f(int num)
{
    if ((1 <= num) && (num <= 8))
    {
        f(num - 1);
        cout << num;
    }
    else
        cout << endl;
}
```

For Exercises 1–3, tell what output is produced by the function call.

1. f(3);

2. f(7);

3. `f(10);`

4–6. Tell what output is produced by the function calls in Exercises 1–3 if num – 1 is replaced by num + 1 in the function definition.

7–9. Tell what output is produced by the function calls in Exercises 1–3 if the cout << num; statement and the recursive call to f() are interchanged.

10–12. Tell what output is produced by the function calls in Exercises 1–3 if a copy of the statement cout << num; is inserted before the recursive call to f().

13. Given the following function f(), use the method illustrated in this section to trace the sequence of function calls and returns in evaluating f(1, 5).

```
int f(int num1, int num2)
{
  if (num1 > num2)
    return 0;
  else if (num2 == num1 + 1)
    return 1;
  else
    return f(num1 + 1, num2 - 1) + 2;
}
```

14. Proceed as in Exercise 13, but for f(8, 3).

Exercises 15–17 assume the following function g():

```
void g(int num1, int num2)
{
  if (num2 <= 0)
    cout << endl;
  else
  {
    g(num1 - 1, num2 - 1);
    cout << num1;
    g(num1 + 1, num2 - 1);
  }
}
```

15. What output is produced by the function call g(14, 4)? (*Hint:* First try g(14, 2), then g(14, 3)).

16. How many numbers are output by the call g(14, 10)?

17. If the cout << num1; statement is moved before the first recursive call to g(), what output will be produced by g(14, 4)?

For Exercises 18–22, determine what is calculated by the given recursive function.

18.
```cpp
void f(unsigned n)
{
    if (n == 0)
        return 0;
    else
        return n * f(n - 1);
}
```

19.
```cpp
double f(double x, unsigned n)
{
    if (n == 0)
        return 0;
    else
        return x + f(x, n - 1);
}
```

20.
```cpp
double f(double x, unsigned n)
{
    if (n == 0)
        return 1;
    else
        return x * f(x, n - 1);
}
```

21.
```cpp
unsigned f(unsigned n)
{
    if (n < 2)
        return 0;
    else
        return 1 + f(n / 2);
}
```

22.
```cpp
unsigned f(unsigned n)
{
    if (n == 0)
        return 0;
    else
        return f(n / 10) + n % 10;
}
```

The following exercises ask you to write functions to compute various quantities. To test these functions, you should write driver programs as instructed in Programming Problems 8–11.

23. Write a recursive power function that calculates x^n, where x is a real value and n is a nonnegative integer.

24. Write a recursive function that returns the number of digits in a nonnegative integer.

25. Write a recursive function `printReverse()` that displays an integer's digits in reverse order.

26. Modify the recursive power function in Exercise 23 so that it also works for negative exponents. One approach is to modify the recursive definition of x^n so that for negative values of n, division is used instead of multiplication and n is incremented rather than decremented:

$$x^n = \begin{cases} 1 & \text{if } n \text{ is } 0 \\ x^{n-1} * x & \text{if } n \text{ is greater than } 0 \\ x^{n+1} / x & \text{otherwise} \end{cases}$$

PROGRAMMING PROBLEMS

Section 10.3

1. Write a driver program to test the rounding function of Exercise 12.

2. Write a driver program to test the weight-conversion functions of Exercise 13 and 14.

3. Write a driver program to test the length-conversion functions of Exercises 15 and 16.

4. Write a driver program to test the time-conversion functions of Exercises 17 and 18.

5. Write a program that reads a positive integer and then calls a function that displays its prime factorization, that is, a function that expresses the positive integer as a product of primes or indicates that the number is a prime.

6. The *greatest common divisor* of two integers a and b, GCD(a, b), not both of which are zero, is the largest positive integer that divides both a and b. The *Euclidean algorithm* for finding this greatest common divisor of a and b was described in Exercise 15 of Section 6.3. The *least common multiple* of a and b, LCM(a, b), is the smallest nonnegative integer that is a multiple of both a and b, and can be calculated using

$$\text{LCM}(a, b) = \frac{|a \times b|}{\text{GCD}(a, b)}$$

Write a program that reads two integers and then calls a function that calculates and passes back the greatest common divisor and the least common multiple of the integers. The program should then display the two integers together with their greatest common divisor and their least common multiple.

7. Consider a simply supported beam to which a single concentrated load is applied:

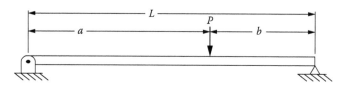

For $a \geq b$, the maximum deflection is given by

$$max\_deflection = \frac{-Pa(L^2 - a^2)^{3/2}}{9\sqrt{3}EIL}$$

the deflection at the load by

$$deflection\_at\_load = \frac{-Pa^2b^2}{3EIL}$$

and the deflection at the center of the beam by

$$deflection\_at\_center = \frac{-Pa(3L^2 - 4a^2)}{48EI}$$

where P is the load, E is the modulus of elasticity, I is the moment of inertia, a is the distance from the left end of the beam to the load, and $b = L - a$ is the distance from the right end of the beam. For $a \leq b$, simply replace a with b and b with a in the preceding equations.

Write a main program that inputs values for L, P, E, I, and an *increment* by which to move the load, and then outputs a table of values for the *load position*, *max_deflec-tion*, *deflection_at_load*, and *deflection_at_center* as the load position moves along the beam from the left end to the right end in the specified increment. The main program must call a single function that receives L, P, E, I, and the current load position (from the left end), and that calculates and passes back the three deflections to main(), which outputs these to the table. Execute your program with the following values: L = 360 inches, P = 24,000 pounds, E = 30 x 10^6 psi, I = 795.5 in⁴, 6 in. increment. Use I/O-manipulators to produce nice even columns in the table. Your table should also have column headings to label the output.

8. The graph of a person's emotional cycle $y = f(x)$ is a sine curve having an amplitude of 1 and a period of 28 days. On a given day, the person's emotional index is *f(age)*, where *age* is his or her age in days. Similarly, the physical and intellectual cycles are sine curves having an amplitude of 1 and periods of 23 and 33 days, respectively. Write a function that receives a person's age and returns his or her physical, intellectual, and emotional indices for the current day. Write another function to compute a person's biorhythm index, which is the sum of the physical, intellectual, and emotional cycles. Write a driver program to test your functions.

1. Write a driver program to test the power function of Exercise 23.

2. Write a driver program to test the digit-counting function of Exercise 24.

3. Write a driver program to test the reverse-printing function of Exercise 25.

4. Write a driver program to test the modified power function of Exercise 26.

5. Write a test driver for one of the functions in Exercises 18–22. Add output statements to the function to trace its actions as it executes. For example, the trace displayed for f(21) for the function f in Exercise 21 should have a form like

```
f(21)  =  1  +  f(10)
    f(10)  =  1  +  f(5)
        f(5)  =  1  +  f(2)
            f(2)  =  1  +  f(1)
                f(1)  returns  0
            f(2)  returns  1
        f(5)  returns  2
    f(10)  returns  3
f(21)  returns  4
```

where the indentation level reflects the depth of the recursion. (*Hint*: This can be accomplished by declaring a variable level outside all blocks, initially zero, that is incremented when the function is entered and decremented just before exiting the function.)

6. Write a recursive function that prints a nonnegative integer with commas in the correct locations. For example, it should print 20131 as 20,131. Write a driver program to test your function.

7. The sequence of *Fibonacci numbers*, 1, 1, 2, 3, 5, 8, 13, 21, . . . (see Exercise 21 in Section 9.4) can be defined recursively by:

$$f_1 = f_2 = 1 \qquad \text{(anchor)}$$

$$\text{For } n \geq 3, f_i = f_{i-1} + f_{i-2} \qquad \text{(inductive step)}$$

A recursive function seems like a natural way to calculate these numbers. Write such a function and then write a driver program to test your function. (*Note:* You will probably find that this function is very inefficient. See if you can figure out why by tracing some function calls as was done in the text.)

Files and Streams

CONTENTS

I can only assume that a "Do Not File" document is filed in a "Do Not File" file.

SENATOR FRANK CHURCH

The Internet has no such organization—files are made available at random locations. To search through this chaos, we need smart tools, programs that find resources for us.

CLIFFORD STOLL

. . . it became increasingly apparent to me that, over the years, Federal agencies have amassed vast amounts of information about virtually every American citizen. This fact, coupled with technological advances in data collection and dissemination, raised the possibility that information about individuals conceivably could be used for other than legitimate purposes and without the prior knowledge or consent of the individuals involved.

PRESIDENT GERALD R. FORD

MANY COMPUTER USERS HAVE had the unfortunate experience of an unexpected hardware failure (e.g., an interruption of power) or a software "crash" before they have saved a document they are preparing or editing. This happens because while the information is being entered or edited, it is stored in the computer's main memory and when this failure occurs, this memory is deallocated and its contents are lost. Most users soon learn to follow the "save often" advice.

Saving a document stores it in a *stable* location in secondary memory (e.g., hard drives and CDs) so that information is not lost, unless, of course, that device fails. Such information must be stored in such a way that:

- it is kept separate from other documents, programs, etc., that are saved; and

- it can be retrieved when needed.

Organizing secondary memory into **files** that store separate items of information accomplishes the first requirement. When information is saved to secondary memory, it is *written* to a file. When it must be retrieved so that it can be edited, compiled, or processed in some other way, it is *read* from that file and *loaded* from secondary memory to main memory.

In computing, files are usually either **text files** or **binary files**, depending on how information is stored in them. A text file stores information using a standard coding scheme such as ASCII (American Standard Code for Information Interchange) or Unicode, which makes it possible to view and process it as regular text by a text editor, for example. Binary files may contain any kind of data encoded in binary form such as images, sound, and object code generated by a compiler.

In this chapter we will focus our attention on those C++ features that enable us to input data from or write data to text files. When there is a large set of data values that need to be processed, they can be stored in a file and a program can read these values directly from that file and process them. Similarly, if a program produces a large amount of output, we may want to have it written to disk for later printing or for processing in some other way. We begin with a simple example that illustrates reading environmental data from a file that is then processed and the results written to another file.

11.1 INTRODUCTORY EXAMPLE: ENVIRONMENTAL DATA ANALYSIS

In our programs up to now, the user has entered data directly from the keyboard in response to prompts displayed on the screen. There are many problems, however, in which the sheer volume of data to be processed makes it impractical to enter it from the keyboard. For such problems, the data can be stored in a file and the program designed to read data values from that file. In this section we look at one such problem.

11.1.1 Problem: Processing a File of Environmental Data

A team of environmental scientists has been collecting various kinds of data for several months. This data includes atmospheric pressure readings that were recorded at the same time each day and saved in a file. One part of the analysis of this data is to calculate the average of the readings and also the maximum and minimum readings along with the days on which they occurred. A program is needed that will read this data, calculate these statistics, and write the results to an output file that can be accessed by the team personnel.

11.1.2 Object-Centered Design

11.1.2.1 Behavior

For maximum flexibility, our program should display on the screen a prompt for the name of the input file containing the pressure readings to be processed and then read this file name from the keyboard. It should then read an arbitrary number of values from that file and compute the average along with the minimum and maximum of those values and when they occurred. The program should display on the screen a prompt for the name of the output file, which is entered from the keyboard. It must then write the computed values to that output file.

11.1.2.2 Objects

In addition to the objects identified in the preceding behavior, the program must maintain a count of how many values have been read and their sum, because these values are needed to compute the average of a collection of values. This gives the following list of objects needed to solve the problem:

	Software Objects		
Problem Objects	**Type**	**Kind**	**Name**
The name of the file in which the pressure readings are stored	string	variable	*inputFileName*
A pressure reading	double	variable	*reading*
The number of readings	int	variable	*count*
The sum of the readings	double	variable	*sum*
The average reading	double	variable	none (*sum/count*)
The minimum reading	double	variable	*minimum*
When it occurred	int	variable	*minDay*
The maximum reading	double	variable	*maximum*
When it occurred	int	variable	*maxDay*
The name of the file to which the results are to be written	string	variable	*outputFileName*

From this list we can specify the input and output of our program as follows:

Input(keyboard):	*inputFileName*, a string naming an input file *outputFileName*, a string naming an output file
Input(*inputFileName*):	a sequence of pressure readings
Output(*outputFileName*):	the average of the input values the maximum and minimum values and when they occurred

11.1.2.3 Operations

From our behavioral description, we need the following operations:

    i. Read a string (*inputFileName* and *outputFileName*) from the keyboard.

    ii. Initialize *count, sum, minimum, maximum, minDay,* and *maxDay* to specific values.

    iii. Read a real value (*reading*) stored in a file.

    iv. Increment an integer variable (*count*) by 1.

    v. Add a real value (*reading*) to a real value (*sum*).

    vi. Update *minimum* with *reading* and *minDay* with *count*, if necessary.

    vii. Update *maximum* with *reading* and *maxDay* with *count*, if necessary.

    viii. Repeat operations (iii–vii) until the end of the file is reached.

    ix. Write results (*sum/count, minimum, minDay, maximum, maxDay*) to an output file.

The only operations that have not been discussed in earlier chapters are those that involve files (iii, viii, and ix). In this chapter we will be looking at the basic file-processing features of C++.

Up to now, all input and output for our programs has been accomplished using `cin` and `cout`, which are automatically connected to the necessary input and output devices—the keyboard and the screen. For file I/O, however, we must include instructions in our program to establish these connections:

- A special object called an **ifstream** must be **opened** to connect an input file to our program.

- A special object called an **ofstream** must be **opened** to connect an output file to our program.

- These `ifstream` or `ofstream` connections must be **closed** when execution terminates.

This adds two more objects to our list of objects,

	Software Objects		
Problem Objects	**Type**	**Kind**	**Name**
Stream from program to input file	`ifstream`	variable	*inStream*
Stream from program to output file	`ofstream`	variable	*outStream*

and two more operations to our list of operations:

    x. Open a stream to a file.

    xi. Close a stream to a file.

The preceding operations can be organized into the following algorithm:

1. Read the name of the input file into *inputFileName*.

2. Open an `ifstream` named *inStream* to the file whose name is in *inputFileName*. (If this fails, display an error message and terminate execution.)

3. Initialize *count*, *minDay*, *maxDay* and *sum* to 0; *maximum* to the smallest possible (real) value; and *minimum* to the largest possible (real) value. (These initializations for *maximum* and *minimum* ensure that they will be updated to the first *reading* in 4e and 4f.)

4. Loop:

   a. Read a value for *reading* from *inStream*.

   b. If the end-of-file mark was read, exit the loop.

   c. Increment *count*.

   d. Add *reading* to *sum*.

   e. If *reading* is less than *minimum*

      Set *minimum* to *reading* and *minDay* to *count*.

   f. If *reading* is greater than *maximum*

      Set *maximum* to *reading* and *maxDay* to *count*.

   End loop

5. Close *inStream*.

6. Prompt for and read the name of the output file into *outputFileName*.

7. Open an `ofstream` named *outStream* to the file whose name is in *outputFileName*. (If this fails, display an error message and terminate the algorithm.)

8. Write *count* to *outStream*.

9. If *count* is greater than zero

   Write *sum/count*, *minimum*, *minDay*, *maximum*, and *maxDay* to *outStream*.

10. Close *outStream*.

The program in Example 11.1 encodes the preceding algorithm. The file-processing features that it uses will be described in the following sections.

Example 11.1 Processing Environmental Data

```
/* This program reads environmental data (pressure readings)
   stored in a file, computes the average, minimum, and maximum of
   the readings, the days on which the maximum and minimum occurred,
   and writes these statistics to an output file.

   Input(keyboard): names of the input and output files
   Input(file):     a sequence of atmospheric pressure readings
   Output(screen):  prompts
   Output(file):    the average reading, the minimum and maximum
                    reading, and the days on which these occurred
-----------------------------------------------------------------*/

#include <iostream>                    // cin, cout
#include <fstream>                     // ifstream, ofstream
#include <string>                      // string, getline()
#include <cassert>                     // assert()
#include <cfloat>                      // DBL_MIN and DBL_MAX
using namespace std;

int main()
{
  cout << "This program reads a list of numbers (pressure readings)\n"
          "from an input file, computes their average, minimum, and\n"
          "maximum and the days on which the min. and max. occurred\n"
          "and writes these results to another file.\n\n";

  // ----------- Input Section --------------------------------
  cout << "Enter the name of the input file: ";
  string inputFileName;
  getline(cin, inputFileName);          // get name of input file
                                        // open an input stream
  ifstream inStream;                    // to the input file,
  inStream.open(inputFileName.data());  // establish connection,
  assert( inStream.is_open() );         // and check for success

  int count = 0,                        // number of values
      minDay,                           // day min. reading occurred
      maxDay;                           // day max. reading occurred
  double reading,                       // value being processed
         minimum = DBL_MAX,             // smallest seen so far
         maximum = DBL_MIN,             // largest seen so far
         sum = 0.0;                     // running total
  for (;;)                              // loop:
  {
    inStream >> reading;                //   read a value
    if ( inStream.eof() ) break;        //   if eof, quit
```

```
    count++;                                  //   update: count,
    sum += reading;                           //           sum,
    if (reading < minimum)
    {
      minimum = reading;                      //           minimum,
      minDay = count;                         //           minDay,
    }
    if (reading > maximum)
    {
      maximum = reading;                      //           maximum,
      maxDay = count;                         //           maxDay
    }
  }                                           // end loop
  inStream.close();                           // close the connection

  // ------------ Output Section --------------------------------
  cout << "Enter the name of the output file: ";
  string outputFileName;
  getline(cin, outputFileName);               // get name of input file
                                              // open an output stream
  ofstream outStream;                         // to the output file,
  outStream.open(outputFileName.data());      // establish connection,
  assert( outStream.is_open() );              // and check for success

                                              // write results to file
  if (count == 0)
    outStream << "No values were read\n";
  else
    outStream << "Average pressure reading:   " << sum / count << endl
              << "Minimum pressure reading:   " << minimum
              << " on day " << minDay << endl
              << "Maximum pressure reading:   " << maximum
              << " on day " << maxDay << endl;
  outStream.close();                          // close the stream

  cout << "Processing complete.\n";
}
```

Notice that instead of "hardwiring" the name of the input file into the program, it asks the user to enter it, which makes it easy to use *any* input file. It also simplifies testing our program before executing it with the "real" data so that we can check its correctness. The getline() function from the string library is used to read the file's name so it can process multiword file names; using >> instead would require single-word names.

11.1.2.6 Testing

To test correctness, we can build our own small **test files** with the data in various configurations and for which we can easily check the results by hand. For example, we might place an ascending sequence of numbers

```
11.1  22.2 33.3  44.4
55.5  66.6 77.7
88.8  99.9
```

in a file test1.txt, a descending sequence

```
99 98 97 96 95 94 93 92 91
```

in another file test2.txt, and a set of numbers with some repetitions and some empty lines

```
44 55.5 55.5
11.1 22 33

55 11.1
11
```

in yet another file test3.txt, and so on. The intent is to *exercise* the program, seeing if we can make it fail. Example 11.2 shows sample runs of the program using these three files. Once the program has been thoroughly tested, it can be used with the original data file with confidence.

Example 11.2 Program Testing

Execution #1:

```
This program reads a list of numbers (pressure readings)
from an input file, computes their average, minimum, and
maximum and the days on which the min. and max. occurred
and writes these results to another file.

Enter the name of the input file: test1.txt
Enter the name of the output file: out1.txt
Processing complete.
```

Listing of out1.txt:

```
Average pressure reading: 55.5
Minimum pressure reading: 11.1 on day 1
Maximum pressure reading: 99.9 on day 9
```

Execution #2:

```
This program reads a list of numbers (pressure readings)
from an input file, computes their average, minimum, and
maximum and the days on which the min. and max. occurred
and writes these results to another file.

Enter the name of the input file: test2.txt
Enter the name of the output file: out2.txt
Processing complete.
```

Listing of `out2.txt`:

```
Average pressure reading: 95
Minimum pressure reading: 91 on day 9
Maximum pressure reading: 99 on day 1
Execution #3:

This program reads a list of numbers (pressure readings)
from an input file, computes their average, minimum, and
maximum and the days on which the min. and max. occurred
and writes these results to another file.

Enter the name of the input file: test3.txt
Enter the name of the output file: out3.txt
Processing complete.
```

Listing of `out3.txt`:

```
Average pressure reading: 33.1333
Minimum pressure reading: 11 on day 9
Maximum pressure reading: 55.5 on day 2
```

11.2 THE ifstream AND ofstream CLASSES

In this section, we take a closer look at the types and operations for performing file l/o that are provided by the `fstream` library.

11.2.1 Declaring File Streams

The `iostream` library we have used up to now for interactive I/O establishes connections between programs executing in main memory and I/O devices:

- The `istream` object `cin` connects a program with the keyboard.

- The `ostream` object `cout` (`cerr` and `clog` also) connects a program with the screen.

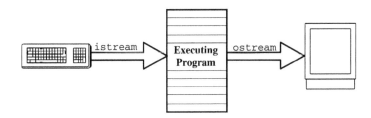

These stream connections are made automatically for interactive programs, but as illustrated by the program in Example 11.1, a program that is to perform input from and/or output to a file must first construct its own streams. This operation of creating a connection between an executing program (in main memory) and a file (stored on some secondary memory device such as a disk drive) is known as **opening** a stream.

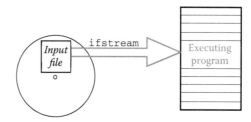

The program in Example 11.1 illustrates this. It first opens an **ifstream** named inStream to a file and then reads data from that file via this stream. When the end of this file is reached, it opens an **ofstream** named outStream to which it outputs the results.[1] These classes are declared in the fstream library, so we must have the #include <fstream> directive in any program where we wish to use them.

To begin, we must first construct an ifstream object to use as a connection from the input file to the program with a declaration of the form

```
ifstream input_stream_name;
```

Similarly, before a program can write values to a file, it must construct an ofstream object to act as a connection from the program to that file with a declaration of the form

```
ofstream output_stream_name;
```

For example, the program in Example 11.1 uses the declaration

```
ifstream inStream;
```

to construct an input stream named inStream and later,

```
ofstream outStream;
```

to construct an output stream named outStream.

11.2.2 The Basic File Stream Operations

The classes ifstream and ofstream are *derived from* the classes istream and ostream, respectively, which means that these classes have been constructed as *extensions* of those classes.[2] As a result, all of the operations on istream objects can also be performed on ifstream objects, and all operations on ostream objects can also be performed on ofstream objects. These include the following commonly used ones:

---

[1] There are also fstream objects that can be used both for input and output, but we will not use them here.
[2] In the language of object-oriented programming, ifstream, ofstream, and fstream are *subclasses* of istream, ostream, and iostream, respectively, and *inherit* their operations.

>>	Inputs a value from a file that has been opened for input
getline()	Reads a line of text from a file opened for input into a `string` object
eof()	Returns true if the last input operation read the end-of-file mark and returns false otherwise
<<	Outputs a value to a file that has been opened for output

There also are several new operations that are limited to file streams, including the following:

open()	Establishes a connection between a program and a file
is_open()	Returns true if a file was opened successfully and returns false otherwise
close()	Terminates a connection between a program and a file

We now examine these operations in more detail.

11.2.2.1 Opening a File Stream

The declarations

```
ifstream inStream;
ofstream outStream;
```

used in Example 11.1 declare that `inStream` is a file stream that can be connected to a file to be used for input and that `outStream` is a file stream that can be connected to a file to be used for output. But these are only *potential* connections. They become *actual* connections when we send them an `open()` message. For example, we could have used the statement

```
inStream.open("test1.txt");
```

in the program in Example 11.1 to open `inStream` as a connection between the program and the file named `test1.txt`. However, such "hardwiring" of a file name into a program means that it cannot be used with other files without modifying the program and recompiling it each time. For more flexibility, a `string` variable `inFileName` was declared and the actual name of the data file read and stored in it. In this case, however, the operating system needs the actual name of that file, not a string object that contains that name as a data member. So a `data()` message is sent to the string variable `inputFileName` to extract the string of characters stored in that `string` object:[3]

```
inStream.open( inputFileName.data() );
```

---

[3] The `c_str()` message can also be used to extract this file name; it appends a terminating null character whereas `data()` does not.

For example, in the first sample execution with test data in Example 11.2, inputFileName contained the string "test1.txt" and the program was connected to the file test1.txt via the file stream inStream:

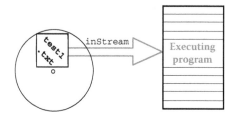

It is important to realize that a **text file** such as test1.txt,

```
11.1 22.2 33.3 44.4
55.5 66.6 77.7
88.8 99.9
```

is simply a *sequence of characters*. If we use the symbol ␣ to represent a blank, the symbol ⏎ to represent a new line, and the symbol ◊ to represent a special **end-of-file mark** automatically placed at the end of the file by the operating system, then after the statement

```
inStream.open( inputFileName.data() );
```

has been executed, inStream may be visualized as

```
11.1␣22.2␣33.3␣44.4⏎55.5␣66.6␣77.7⏎88.8␣99.9⏎◊
```

where the down-arrow (↓) indicates the **read position** at which the next character will be read.

The open() message can also be used to open ofstream objects as connections to output files. For example, the program in Example 11.1 has the declaration

```
ofstream outStream;
```

and opens it with the statement

```
outStream.open( outFileName.data() );
```

This connects our output stream outStream to a file whose name is the string stored in outFileName. If this file does not exist, it will be created, and if there is a file already with this name, its contents will be destroyed. The resulting file will contain only an end-of-file character as does the stream outStream that connects the program to the file:

outStream

The down-arrow (↓) in this diagram represents the **write position**—where the next output will be placed in the stream.

In general, the open() message has the (simplified) form

```
stream_name.open(file_name);
```

where *stream_name* is the name of a file stream and *file_name* is the name of a data file. It initializes *stream_name* as a connection between the program in which this statement occurs and a file named *file_name*.

By default, opening an ofstream *to a file is destructive.* That is, if a file named out-file.txt exists and the statement

```
outStream.open("outfile.txt");
```

is executed, any old contents of outfile.txt will be destroyed. This can be avoided by using open() with one of the following as a second **mode argument**:[4]

Mode	Description
ios::in	*The default mode for* ifstream *objects.* Open a file for input, nondestructively, with the read position at the file's beginning.
ios::trunc	Open a file and delete any contents it contains (i.e., *truncate* it).
ios::out	*The default mode for* ofstream *objects.* Open a file for output, using ios::trunc.
ios::app	Open a file for output, but nondestructively, with the write position at the file's end (i.e., for *appending*).
ios::ate	Open an existing file with the read position (ifstream objects) or write position (ofstream objects) at the end of the file.

For example, to open an ofstream named outStream to a file outfile.txt so that we can add data at the end of it, we can use

```
outStream.open("outfile.txt", ios::app);
```

This second argument makes the open() message *nondestructive* so that the old contents of outfile.txt are preserved and any additional values written to the file will be *appended* to it.[5]

---

[4] ios::binary can also be used to open a file in binary mode.

[5] The file modes can be combined using the bitwise-or (|) operator. For example,

```
fstream inoutStream;
inoutStream.open("ReadWriteFile", ios::in | ios::out);
```
opens ReadWriteFile for both input and output.

In our examples thus far, we declared a file stream and then sent it the `open()` message to connect it to a file, for example,

```
ofStream outStream;
outStream.open( outFileName.data() );
```

We can combine these steps and initialize an `ofstream` object in its declaration:

```
ofStream outStream( outFileName.data() );
```

This statement both declares `outStream` as an `ofstream` and opens it as a connection to the file whose name is stored in `outputFileName`.

11.2.2.2 Programming Defensively—The `is_open()` Message

An attempt to open a stream to a file can fail for a variety of reasons. One obvious one is that the file cannot be found, perhaps because it doesn't exist or we didn't use the correct name or the correct path to the directory where it is located. Obviously, if this happens, any subsequent attempts to read from that file will also fail. One rule when using files for I/O is

> *Always test whether an attempt to open a file was successful before proceeding with any additional operations on the file.*

This testing is easily done using the boolean message `is_open()` whose form is as follows.

```
stream_name.isopen()
```

which will return `true` if *stream_name* was successfully opened and `false` otherwise.

In the program in Example 11.1, we used the `assert()` mechanism to check the value returned by the `is_open()` message:

```
assert( inStream.is_open() );
```

and

```
assert( outStream.is_open() );
```

This provides a succinct way to test each file stream and terminate the program if it failed to open, but it does abort the program when this happens. An alternate and more user-friendly approach is to use an `if` statement inside a loop as in the following function, which might be paired with a similar one for output files and stored in a library for use in other programs:

```
void getInputFile(ifstream & inStream)
{
  char response;
  do
  {
```

```
    cout << "Enter the name of the input file: ";
    string inputFileName;
    getline(cin, inputFileName);

    inStream.open(inputFileName.data());

    if (inStream.is_open()) break;

    cerr << "\n***Unable to open << inputFileName
         << "\nTry again (Y or N)? ";
    cin >> response;
  }
  while (response != 'N' && response != 'n');
}
```

This function gives the user more than one chance to enter a correct file name and is consistent with a design strategy practiced by many programmers: *recover from errors whenever possible and exit gracefully when it isn't.*

11.2.2.3 The Input and Output Operators

One of the attractive features of C++ is its *consistency*—its use of the same operators to perform tasks that are functionally similar. For example, input from a file can be performed in the same manner as input from the keyboard using >> and performs in the same way in both cases—skip leading white space; stop reading characters when a character is encountered that cannot belong to a value of the type being read; convert a string of characters to a number; use of eof() to check for end of input; and so on.

This same consistency is also true of output. Once an ofstream object is connected to a file it is used in the same way as output to the screen—the output operator << and format manipulators such as endl, showpoint, setprecision(), and setw( )can be used to perform output (see Section 7.3).[6]

11.2.2.4 Closing a File Stream

We have seen that initializing a file stream object via the open() function or in its declaration establishes it as a connection between a program and a file. At some point, this file stream must be disconnected so that the file itself can be accessed and used with operating system commands—for example, to print the contents of the file to a printer. Because a file stream object is a variable, this will happen automatically when execution leaves its scope just as is does for any other variable. In particular, a file stream declared within a function will be disconnected when execution leaves that function (if not before).

---

[6] This consistency in the I/O libraries is an example of the power of class *inheritance*; ifstream is derived from istream and ofstream is derived from ostream. (Actually, all are derived from the ios class.) Operations and functions such as eof() defined in a class are automatically inherited by these derived classes.

A file stream can be disconnected explicitly by sending it the `close()` message of the form

```
stream_name.close();
```

where *stream_name* is the name of a file stream of either type. For example, the statement

```
inStream.close();
```

in the program in Example 11.1 severs the connection between the program and the input file and the `istream` variable `inStream` becomes undefined so that any subsequent attempt to read from it before reconnecting it to a file is an error.

For an output stream, the `close()` message performs a bit differently. To see this difference, consider the statement

```
outStream.close();
```

in Example 11.1. Before the connection between the program and the output file is severed, any contents of `outStream` are first copied to the output file. After that, the connection is broken and the `ostream` variable `outStream` becomes undefined. To illustrate, for the first execution in Example 11.2, the contents of `outStream`

are written to the output file `out1.txt`, which we might visualize as follows:

```
Average pressure reading:  55.5
Minimum pressure reading:  11.1 on day 1
Maximum pressure reading:  99.9 on day 9
```

Although a file stream is automatically disconnected when it reaches the end of its scope, there are times when it is necessary to disconnect it earlier. For example, our program may write values to a file and then we want to open that file and read from it. In this case, the output stream to that file must first be disconnected by using `close()` and then it can be reopened. However, even if this is not the case, it is considered good programming practice to use `close()` to disconnect every file stream when it is no longer needed.

One situation where this is important is in programs that use many different files, because an operating system may place a limit on the number of files a program may have open simultaneously. This means that if a program tries to open more files than allowed, the operating system will terminate the program abnormally. This problem can be avoided by always using the `close()` message to sever the connection between a program and a file when the program is done using it. This keeps the number of open files associated with the program from growing beyond the limit allowed by the operating system.

Files and Streams ▪ **411**

11.2.2.5 File Streams as Parameters
An example of a function that allowed a user to enter the name of an input file repeatedly
until it was located and could be opened was given earlier in this section:

```
void getInputFile(ifstream & inStream)
{
  char response;
  do
  {
    cout << "Enter the name of the input file: ";
    string inputFileName;
    getline(cin, inputFileName);

    inStream.open(inputFileName.data());

    if (inStream.is_open()) break;

    cerr << "\n***Unable to open << inputFileName
         << "\nTry again (Y or N)? ";
    cin >> response;
  }
  while (response != 'N' && response != 'n');
}
```

Note that the parameter `inStream` is of type `ifstream` and is a reference parameter,
which must always be the case; *parameters corresponding to file stream arguments must be
reference parameters* because

- Reading from an `ifstream` object alters its read position; and

- Writing to an `ofstream` object alters its write position.

If the file stream parameters were not reference parameters, these changes would not have
been propagated back to `main()` or another function that calls this function.

11.2.3 Summary
The following points summarize some of the important points regarding file I/O in C++.

- A text file stores characters on a secondary memory device.

- Input from or output to a file from a program can only be done indirectly through a
  file stream—an abstract conduit between the program and the file.

- A file stream can be connected to a file by using `open()` or the initialization-
  at-declaration mechanism.

- Opening an `ifstream` to a file initializes the file stream with the contents of that file and the read position at the first character in the file stream.

- Opening an `ofstream` to a file initializes the file stream as empty (containing only the end-of-file mark) with the write position at the end-of-file mark. Any previous contents of the file are destroyed.

- If an `ofstream` to a file is opened using `ios::app`, the stream is initialized with the contents of that file with the write position at the end of the file.

- After skipping leading white space characters, the input operator (`>>`) will extract the first value following the read position in an `ifstream` and advance the read position to the first character after the input value. Numeric values are delimited by nonnumeric characters.

- The `getline()` function extracts the line of input beginning at the read position in an `ifstream` and stores the extracted characters in a `string` variable, stopping when it reaches a newline character. It leaves the read position at the first character beyond that newline. Intermixing calls to `getline()` and the input operator must be done with caution.

- The output operator (`<<`) inserts a value into an `ofstream` at the current write position and advances it to the point immediately following the value.

- A file stream should be disconnected from a file by a call to `close()` when input/output is finished.

11.3 ADDITIONAL STREAM FEATURES

We have looked at the basic file-stream operations of `open()`, `is_open()`, `<<`, `>>`, `get-line()`, `eof()`, and `close()`, but there are many others. Of these, we discussed in earlier chapters: `get()` to read a character from a stream without skipping white space; `good()`, `bad()`, and `fail()` to check a stream's status; `clear()` to reset a stream's status bits; `ignore()` to skip characters; and the various stream manipulators. The following table lists some of the stream messages that can be sent to both `istream` and `ifstream` objects, or to both `ostream` and `ofstream` objects. But this is not an exhaustive list; the `iostream` and `fstream` libraries are extensive and this table indicates only a small part of their capabilities.

Function Member	Description
`good()`	`true` if the *good* bit is set (1) else false
`fail()`	`true` if the *fail* bit is set (1) else false
`bad()`	`true` if the *bad* bit is set (1) else false
`eof()`	`true` if the *eof* bit is set (1) else false
`clear()`	Reset the *good* bit to 1, all other status bits to 0
`setstate(sBit)`	Set the status bit *sBit* to 1, where *sBit* is one or more of `ios::badbit, ios::failbit, ios::eofbit`
`get(ch)`	Read the next character (including white space) into *ch*

`seekg(offset, base)`	Move the read position *offset* bytes from *base*, where *base* is one of `ios::beg`, `ios::cur`, or `ios::end`
`seekp(offset, base)`	Same as `seekg()` but for the write position
`tellg()`	Return the offset of the read position within the stream
`tellp()`	Same as `tellg()` but for the write position
`peek()`	Return the next character in the stream, but leave it unread
`putback(ch)`	Modify the stream so that *ch* will be the next character read
`ignore(n, stopChar)`	Skip *n* characters, or until *stopChar* is encountered; *n* defaults to 1 and *stopChar* to eof

In addition to stream function members, there are a variety of **manipulators** that can be used to affect the formatting of an `ostream` or an `ofstream`. In this section, we will discuss some of the stream operations and manipulators that we have not seen before.

11.3.1 The `seekg()`, `tellg()`, `seekp()`, and `tellp()` Methods

The file-processing programs we have considered thus far have used **sequential access**, which means that items in the file are accessed and processed sequentially, from beginning to end. This obviously means that it takes longer to access an item near the end of the file than one near the beginning. This is in contrast to **direct** or **random access** where an item can be accessed directly by specifying its position in the file.

11.3.1.1 Example: Two-Pass File Processing

Consider the problem of calculating the average of the numbers in the file and then displaying the difference between each number and this average. Two passes must be made through the file. The first pass counts the numbers and calculates their sum. Their average can then be calculated:

```
if (count > 0) average = sum / count;
```

A second pass must now be made to calculate and display the difference between each number and this average. The problem is how to reset the read position back to the beginning of the file for this second pass.

For such situations, the `iostream` library provides the methods `seekg()`, `tellg()`, `seekp()`, and `tellp()`, which make it possible to manipulate a stream's read and write position directly rather than sequentially. The program in Example 11.3 uses the `seekg()` method to solve our problem. As we saw in Section 7.3, `clear()` must be used to clear the *eof* status bit between the two passes because no more input operations on that stream will succeed until this bit is cleared.

Example 11.3 A Two-Pass File-Processing Program

```
/* This program illustrates making two passes through a file:
   one pass to find the average of the numbers in the file, and a
   second pass to find the difference between each number and the
   average.
```

```
   Input:  a series of values from a file
   Output: each value and its difference from the average value
-----------------------------------------------------------------*/

#include <iostream>      // cout, <<, >>
#include <fstream>       // ifstream, is_open(), eof(), clear(), seekg()
#include <iomanip>
#include <string>
#include <cassert>       // assert
using namespace std;

int main()
{
  string inputFileName;
  cout << "Name of data file? ";
  cin >> inputFileName;
  ifstream inStream(inputFileName.data());
  assert(inStream.is_open());

  double newValue, sum = 0.0, average = 0.0;
  int count = 0;

  for (;;)
  {
    inStream >> newValue;
    if (inStream.eof()) break;
    sum += newValue;
    count++;
  }

  if (count > 0) average = sum / count;

  inStream.clear();              // clear eof bit
  inStream.seekg(0, ios::beg);  // reset read position

  cout << endl;

  for (;;)
  {
    inStream >> newValue;
    if (inStream.eof()) break;
    cout << setw(10) << newValue << " : "
         << setw(10) << newValue - average << endl;
  }
  inStream.close();
}
```

SAMPLE RUN:

```
Name of data file? test1.txt

      11.1 :       -44.4
      22.2 :       -33.3
```

```
33.3 :        -22.2
44.4 :        -11.1
55.5 :            0
66.6 :         11.1
77.7 :         22.2
88.8 :         33.3
99.9 :         44.4
```

11.3.1.2 `seekg()`

A message of the form

```
inStream.seekg(offset, base);
```

changes the read position within an `ifstream` named *inStream*.[7] Here `base` is one of the following:

- **ios::beg** move read position *offset* characters from the **beginning** of the stream

- **ios::cur** move read position *offset* characters from its **current position**

- **ios::end** move read position *offset* characters from the **end** of the stream

For example, the statement

```
inStream.seekg(0, ios::beg)
```

in Example 11.3 moves the read position back to the beginning of the file. If we wanted to move it to the end of the file, we could use

```
inStream.seekg(0, ios::end);
```

If the input file has a field containing 10 characters at the beginning of each line that we needed to skip over to get to a numeric value, we could replace each input statement with

```
inStream.seekg(10, ios::cur);
inStream >> newValue;
```

The `seekg()` function is especially useful when all the lines in an input file have the same length, say

---

[7] The 'g' in `seekg()` and `tellg()` refers to *getting* values from the stream being manipulated (i.e., that it is an input stream).

```
const int LINE_LENGTH = 60;
```

If we wanted to move the read position to the beginning of the line numbered `lineNum`, we could use

```
inStream.seekg((lineNum - 1) * LINE_LENGTH, ios::beg);
```

This provides **direct access** to that line in the file in place of *sequentially* processing all of the data items in the file that precede the one we need.

If it happens that the arguments to `seekg()` specify moving the read position before the beginning of the file (e.g., using a negative offset from `ios::beg`) or past the end of the file (e.g., using a positive offset from `ios::end`) the operation fails and the stream status *fail* bit is set. This disables all subsequent operations on that stream until the status bits have been reset using the `clear()` function.

11.3.1.3 tellg()

It is sometimes convenient to think of a stream as a list of characters, in which each position has its own number or index, similar to a string object:

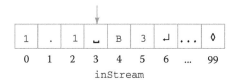

The `tellg()` function can be used to find the index of the read position. For example, the statements

```
inStream.seekg(0, ios::end);
unsigned lastPosition = inStream.tellg();
```

will (1) move the read position to the end-of-file mark

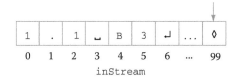

and (2) store the index (99) of the end-of-file mark in `lastPosition`. Note that because the index of the first character is always zero, the index of the end-of-file mark is always the number of characters in the file (not counting the end-of-file mark). A text-processing program could use this fact to determine how many characters are in a file, rather than counting them one at a time.

As with `seekg()`, `tellg()` is especially useful for files that are organized into lines or records of fixed lengths. To illustrate, suppose we need to process such an input file but not all of the data in each line. If each line is made of `LINE_LENGTH` characters as in our earlier example, we can use

```
unsigned charsLeft = inStream.tellg() % LINE_LENGTH;
inStream.seekg(LINE_LENGTH - charsLeft, ios::curr);
```

to move the read position forward to the beginning of the next line.

11.3.1.4 `seekp()` and `tellp()`

Because `seekg()` and `tellg()` only manipulate the read position within a stream, they can only be used with input streams. The write position within an output stream can be manipulated by using the functions `seekp()` and `tellp()`, which behave in the same manner as `seekg()` and `tellg()`, respectively, but must be used with output streams.[8]

11.3.2 The `peek()` and `putback()` Methods

Two other functions allow a programmer to do some useful, if unusual, operations on an input stream. The names of these functions, `peek()` and `putback()`, describe the operations they perform.

11.3.2.1 `peek()`

The `peek()` function allows the programmer to *look ahead* in an input stream and see what the next character is. Thus, it is similar to `get()`, but `peek()` does not advance the read position.

To illustrate, consider what a C++ compiler must do when reading a C++ program from a file that a programmer has created. One of its tasks is to break the source program into a sequence of meaningful groups of characters called **tokens**. For each identifier, each keyword, each operator, and each punctuation mark, there is a distinct token. For example, the compiler might refer to the + operator as PLUS, the ++ operator as INCREMENT, and the += operator as PLUS_ASSIGN. This task of identifying tokens in a program is accomplished by a **lexical analyzer**, which the compiler contacts whenever it needs the next token.

As an example of the problem of identifying tokens, suppose the compiler has just encountered a + character in its input stream `inStream`. We might picture this as follows, where the down arrow, as before, denotes the position from which the next character will be read:

inStream

---

[8] The 'p' in `seekp()` and `tellp()` refers to *putting* values into the stream being manipulated (i.e., that it is an output stream).

The compiler calls for the next token from its lexical analyzer, which retrieves the next character '+' and advances the read position:

The lexical analyzer needs more information in order to determine which operation the programmer specified: if the next character is an equal sign (=), then the operator is +=; if the next character is another +, the operator is the increment operator ++; and if the next character is a white-space character, a letter, a digit, a single quote, or a double quote, then the operator is simply the plus operator +. The peek() function makes it possible to look ahead at the next character without actually moving the read position.[9]

```
nextCh = inStream.peek();          // look ahead at next char
if (nextCh == '=')                 // if it's an =
{
   inStream.get(nextCh);           // get the char
   return PLUS_EQUALS_TOKEN;       // and return +=
}
else if (nextCh == '+')            // else if it's another +
{
   inStream.get(nextCh);           // get the char
   return INCREMENT_TOKEN;         // and return ++
}
else if (isspace(nextCh)           // else if it's whitespace
         || isalnum(nextCh)        // a letter or a digit,
         || nextCh == '\''         // a single quote, or
         || nextCh == '"')         // a double quote,
   return PLUS_TOKEN;              // return +
else                               // else
// ... generate error message      // illegal token
```

Thus, if

```
inStream.peek()
```

returns the character y, then the lexical analyzer can infer that inStream contains

___

[9] The cctype library function isalnum(ch) returns true if and only if its argument ch is an alphanumeric character. Similarly, isspace(ch) returns true if and only if its argument ch is a white-space character.

indicating that the plus operator + was specified, and not ++ or +=. The y is left in the stream where it can be subsequently processed in the normal fashion the next time the compiler asks the lexical analyzer for a token.

11.3.2.2 putback()

An alternative to peek() is to use get() to retrieve the next character,

```
inStream.get(nextCh);
```

and then use putback() to return it to inStream, if necessary:

```
if (nextCh == '=')              // if it's an =
{
   inStream.get(nextCh);        // get the char
   return PLUS_EQUALS_TOKEN;    // and return +=
}
else if (nextCh == '+')         // else if it's another +
{
   inStream.get(nextCh);        // get the char
   return INCREMENT_TOKEN;      // and return ++
}
else if (isspace(nextCh)        // else if it's whitespace
       || isalnum(nextCh)       // a letter or a digit,
       || nextCh == '\''        // a single quote, or
       || nextCh == '"')        // a double quote
{
   inStream.putback(nextCh);    // put it back for now
   return PLUS_TOKEN;           // return +
}
else                            // else
// ... generate error message   // illegal token
```

Thus, if the value of nextCh were 'y', putback(nextCh) would return that character to the stream so that it will be accessed the next time the compiler asks the lexical analyzer for a token.

11.3.3 The setstate() Method

We've seen how the clear() function can be used to reset the status bits of a stream to their default values. Sometimes, however, it is useful to be able to set them to specified values, and this can be done using the setstate() function.

To illustrate, suppose that we have a problem involving five types of beam design: I-beam, C-channel, flitch, cantilever, and hip, and we want to write a readBeam() function to input a beam design, similar to the getline() function provided by the string library. Our function will read a word from a stream into a string variable, which it then passes back to the caller via a reference parameter.

But what do we do if the user enters an invalid value, a word that is not one of the five beam designs? To handle this situation in a way that is consistent with the `iostream` library, our function should set the *fail* status bit for that input stream. Example 11.4 shows how the stream function member `setstate()` can be used to do this.

Example 11.4 Setting Stream-Status Flags

```
void readBeam(istream & theStream, string & beam)
{
  string beamType;
  theStream >> beamType;

  if (beamType == "I-beam" || beamType == "C-channel" ||
      beamType == "flitch" || beamType == "cantilever" ||
      beamType == "hip")
    beam= beamType;
  else
  {
    theStream.seekg(-beam.size(), ios::cur);      // unread the word
    theStream.setstate(ios::failbit);             // indicate failure
  }
}
```

By setting the *fail* status bit in the stream, this function leaves the handling of the error up to its caller. For example, a programmer wishing to treat this as a fatal error can write

```
string aBeamType;
readBeam(cin, aBeamType);
assert( cin.good() );
```

and if the user enters any word other than one of the five beam designs, the call to `assert()` will terminate the program. By contrast, a programmer who wishes to treat this occurrence as a nonfatal error and have the user try again, can write

```
string aBeamType;
for (;;)
{
  readBeam(cin, aBeamType);
  if ( cin.good() ) break;
  cout << "Try again: ";
  cin.clear();
  cin.ignore(80, '\n');
}
```

and the user will be given more chances to enter a valid value. Such flexibility is a trade-mark of good design, because it leaves the decision of how to handle the problem up to users of the function, allowing them to choose the approach they prefer.

The `setstate()` function can be used to set the status bits in a stream, which are referred to by the following names:

Status Bit	Description
`ios::badbit`	The *bad* bit: 1 if an unrecoverable error occurred, 0 otherwise.
`ios::failbit`	The *fail* bit: 1 if a recoverable error occurred, 0 otherwise.
`ios::eofbit`	The *eof* bit: 1 if the end-of-file mark was read, 0 otherwise.

In practice, `setstate()` is rarely passed `ios::eofbit`, because that is set by the input operations.

11.3.3.1 The Formatting Manipulators

As we saw in Chapter 7, the `iostream` and `iomanip` libraries provide manipulators for controlling the format of output streams. Manipulators can be divided into two categories: those without arguments and those that require arguments.

Format manipulators that do not require arguments are available in the `iostream` library. Some of the basic ones are given in the following table:

Manipulator	Description
`boolalpha`	Use strings `false` and `true` for I/O of boolean values
`noboolalpha`	Use integers `0` and `1` for I/O of boolean values
`scientific`	Use floating-point (scientific) notation
`fixed`	Use fixed-point notation
`showpoint`	Show decimal point and trailing zeros for whole real numbers
`noshowpoint`	Hide decimal point and trailing zeros for whole real numbers
`showpos`	Display positive values with a + sign
`noshowpos`	Display positive values without a + sign
`dec`	Display integer values in base 10
`oct`	Display integer values in base 8
`hex`	Display integer values in base 16
`showbase`	Display integer values indicating their base (e.g., `0x` for hex)
`noshowbase`	Display integer values without indicating their base
`uppercase`	In hexadecimal, use symbols A–F; in scientific, use E
`nouppercase`	In hexadecimal, use symbols a–f; in scientific, use e
`skipws`	Skip white space on input
`noskipws`	Don't skip white space on input
`flush`	Write contents of stream to screen (or file)
`endl`	Insert newline character and flush the stream
`left`	Left-justify displayed values, pad with fill character on right
`right`	Right-justify displayed values (except strings), pad with fill character on left
`internal`	Pad with fill character between sign or base and value

As we saw in Chapter 7, manipulators are inserted into the stream, but instead of appending values to the stream (except for `endl`), they affect the format of values inserted subsequently into the stream. For example, if we were to write

```
int i = 17;

cout << showbase
        << oct << i << endl
        << dec << i << endl
        << hex << i << endl;
```

then the following values would be displayed

```
021
17
0x11
```

To use the manipulators that require arguments, the `iomanip` library must be included. The table below gives some of these manipulators.

Manipulator	Description
`setprecision(num)`	Set the number of decimal digits to be displayed to *num*
`setw(num)`	Display *the next value* in a field whose width is *num*
`setfill(ch)`	Set the fill character to *ch* (blank is the default)

When a real number is inserted into a stream, the number of digits that are displayed to the right of the decimal point is called the **precision** of the number. As we have seen before, this characteristic can be controlled with the `setprecision()` manipulator.

We have also seen that when a number is inserted into a stream, it is first placed into a **field**, which is then inserted into the stream. The size of this field is controlled by the `setw()` manipulator. If the size of the field is less than the size of the value being displayed, the field is automatically expanded to the same size as the value. If the size of the field exceeds the size of the value being displayed, then the empty positions in the field are filled with the **fill character** (by default, a blank), whose value is set by the `setfill()` manipulator.

Here is a simple code fragment that illustrates the use of these manipulators:

```
cout << fixed << showpoint             // show decimal pt, sign
        << setprecision(2)             // 2 decimal places
        << setfill('*') << left        // pad with *, left justify
        << setw(6) << 1.0/3.0 << endl  // print value
        << setfill('$') << right       // pad with $, right justify
        << setw(6) << 1.0/3.0 << endl; // print value
```

When executed, this statement produces the following output:

```
0.33**
$$0.33
```

Note that unlike setprecision(), setw() affects only the next value inserted into the stream, so setw() must precede *each* insertion of a value whose field width we wish to specify.

To display a column of figures with their decimal points aligned, the right manipulator can be used with setprecision() and setw(). For example, to display a table of square roots to seven decimal places, we could write this code segment:

```
cout << fixed << showpoint << right
     << setprecision(7);
for (int i = 1; i <= 10; i++)
   cout << setfill(' ')  << setw(2)  << i
        << setfill('.')  << setw(12) << sqrt(i) << endl;
```

Executing these statements produces the output

```
 1...1.0000000
 2...1.4142136
 3...1.7320508
 4...2.0000000
 5...2.2360680
 6...2.4494897
 7...2.6457513
 8...2.8284271
 9...3.0000000
10...3.1622777
```

11.3.4 String Streams

C++ also permits us to read input from a string or to write output to a string. This is made possible by means of **string streams** defined in the sstream library:[10]

istringstream	For input from a string
ostringstream	For output to a string
stringstream	For input from and output to a string

The str()function can be used to convert a string stream to a string, and vice versa:

| strstream.str(s); | Set string stream strstream to a copy of string s |
| str(strstream) | Returns a string that is a copy of the string in strstream |

---

[10] The corresponding types for wide characters (of type wchar_t) are wistringstream, wostringstream, and wstringstream.

The program in Example 11.5 illustrates the use of string streams. It constructs an `istringstream` from the string date, uses the input operator `>>` to read the individual words and integers, and displays them. It then outputs strings and integers to the `ostringstream ostr`, uses `str()` to extract the string from `ostr`, and displays it.

Example 11.5 String Streams

```
/* Program to illustrate the use of string streams.

   Input (istringstream istr): word1, word2, month, day, comma, year
   Output (istringstream ostr): these words separated by ***
   Output (ostream cout):      the string stored in ostr
-------------------------------------------------------------------*/


#include <iostream>
#include <iomanip>
#include <sstream>
using namespace std;

int main()
{
  string date = "Independence Day (U.S.): July 4, 1776";
  istringstream istr(date);
  string word1, word2, word3, month;
  int day, year;
  char comma;


  istr >> word1 >> word2 >> word3 >> month >> day >> comma >> year;
  cout << "Contents of string stream istr:\n"
       << word1 << "***" << word2 << "***" << word3 << "****" << month
       << "***" << day << "***" << comma << "***" << year << endl;
  ostringstream ostr;
  ostr << '\n' << word3.substr(1, 4) << " Bicentennial: "
       << month << setw(2) << day << ", "<< year + 200;
  cout << ostr.str() << endl;
}
```

SAMPLE RUN:

```
Contents of string stream istr:
Independence***Day***(U.S.):****July***4***,***1776

U.S. Bicentennial: July 4, 1976
```

CHAPTER SUMMARY

Key Terms

binary file	mode argument
`close()` function	`ofstream`
closing a stream	`open()` function
direct access	opening a stream
end-of-file mark	`ostringstream`
end-of-line mark	output operator (<<)
`eof()` function	precision
field	random access
file	read position
fill character	sequential access
format manipulators	`sstream` library
`fstream` library	stream
`getline()` function	`stringstream`
`ifstream`	test file
input operator (>>)	text file
`is_open()` function	write position
`istringstream`	

NOTES

- Text files are simply sequences of characters, some of which may be special characters that, for example, mark the end of a line or the end of the file.

- Before performing input from/output to a file, open the file by using `open()` or the initialization-at-declaration mechanism to connect a file stream to it by. If the file's name is stored in a `string` variable, use `data()` or `c_str()` to extract the character string from the variable.

- Test each attempt to open a file to see if it was successful (e.g., with the `is_open()` function) before proceeding with other operations on the file.

- Use `ifstream` objects for file input, `ofstream` objects for file output.

- The file stream classes `ifstream`, `ofstream`, and `fstream` are derived from the `istream`, `ostream`, and `iostream` classes, respectively, which means that all of the operations from these latter classes can also be performed on the corresponding file stream objects.

- By default, opening an `ifstream` to a file initializes the file stream with the contents of that file with the read position at the first character in the stream. Opening an `ofstream` initializes the file stream as empty (containing only the end-of-file mark) with the write position at the end-of-file mark. Any previous contents of the file are destroyed. If the mode `ios::app` is used, then the file must exist and the stream is initialized with the contents of that file, with the write position at the end of the file.

- Applying the input operator (`>>`) to an `ifstream` will extract the first value following the read position in the stream, skipping initial white space characters, and advance the read position to the first character past the input value.

- The `getline()` function can be used to extract the line of input beginning at the current read position in an `ifstream` and store the characters in a string variable, leaving the read position at the first character beyond the first newline character encountered. Care must be taken when intermixing calls to `getline()` and the input operator because `>>` leaves the newline character in the file.

- The output operator (`<<`) can be used to insert a value into an `ofstream` at the write position, advancing it to the point immediately following the value.

- It is considered good programming practice to disconnect a file stream from a file using the `close()` message when that file is no longer needed.

- Parameters corresponding to file stream arguments must be reference parameters so that changes to the stream are passed back to the calling function.

Style and Design Tips

- *To read data from a file, establish an* `ifstream` *connection between the program and the file. To write data to a file, establish an* `ofstream` *connection between the program and the file.* The three basic steps for file-processing programs are thus:

  - Declare and open an `ifstream` object for each input file and an `ofstream` object for each output file to establish connections between the program and the files.

  - Perform the desired processing with the file via the file streams.

  - Close each file stream, severing the connection with the file.

- *A forever loop controlled by the* `ifstream` *function member* `eof()` *can be used to read data from a file via an* `ifstream`:

```
ifstream theStream("SomeFile");
for (;;)
{
  // read a value from theStream
  if (theStream.eof()) break;
  // process the value
}
```

Some programmers prefer the while loop version:

```
// read a value from theStream
while (!theStream.eof())
{
 // process the value
 // read a value from theStream
}
```

Warnings

1. *Before a file stream can be used, it must be opened as a connection to a particular file.* The simplest way is to initialize the stream with the name of the file in the stream's declaration:

```
ofstream outStream("OutputFile.TXT");
```

Alternatively, the stream can be declared and opened separately, using open():

```
ofstream outStream;
outStream.open("OutputFile.TXT");
```

2. *When opening a stream to a file, the name of the file must be given as a character string.* For example, to open an ifstream named inStream to a file named Text, we can write

```
ifstream inStream("Text");
```

or

```
ifstream inStream(filename.data());
```

where the string variable filename stores the file name and data() or c_str() extracts the character string from it.

3. *Be sure that operations performed on a file stream are consistent with the mode by which it was initialized.* Applying >> to an ofstream or << to an ifstream will generate a compilation error.

4. *When inputting values, the extraction operator >> skips over any leading white-space characters (blanks, tabs, and newlines); the functions get() and getline() do not.*

TEST YOURSELF

Section 11.2

1. The `iostream` library establishes a(n) _____ object named _____ that connects a program and the keyboard.

2. The `iostream` library establishes a(n) _____ object named _____ that connects a program and the screen.

3. In order for a program to read data from a file, a(n) _____ object must connect the program to that file.

4. In order for a program to write output to a file, a(n) _____ object must connect the program to that file.

5. The types of streams needed to connect a program and a file are declared in the _____ library.

6. (True or false) Almost none of the operations on `iostreams` can be performed on file streams.

7. Write a statement to declare a file stream named `inputStream` that will be used for input from a file and another statement that uses the `open()` function to connect this stream to a file named `EmployeeInfo`.

8. Repeat Question 7 but use the initialization-at-declaration mechanism.

9. Repeat Question 7 but for a file stream named `outputStream` that will be used for output to a file named `EmployeeReport`.

10. Repeat Question 9 but use the initialization-at-declaration mechanism.

11. Modify your answers to Questions 7 and 8 so that the user inputs the name of the file into a `string` variable `inputFileName`.

12. (True or false) The declaration `ifstream inStream("Info");` will destroy the contents of the file named `Info`.

13. (True or false) The declaration `ofstream outStream("Info");` will destroy the contents of the file named `Info`.

14. Write a statement that will stop program execution if an attempt to open the `ifstream inputStream` fails.

15. Write a statement that will extract an entire line from the `ifstream` in Question 14.

16. Write a statement that will display the message `"End of file"` for the `ifstream` in Question 14 when the end-of-file mark is reached.

17. Write a statement to disconnect the `ifstream` in Question 14.

Section 11.3

1. (True or false) Sequential access refers to being able to access an item in a file directly by specifying its sequential position in the file.

2. (True or false) Direct access refers to being able to access an item in a file directly by specifying its offset from the beginning of the file.

3. Another name for direct access is _____ access.

4. The _____ function can be used to find the location of the read position in an `istream` and the _____ function can be used to move to that position.

5. Write a statement that moves the read position in the `ifstream inputStream` to the third character from the beginning of the stream.

6. Proceed as in Question 5 but move the read position to the third character past the current position.

7. Proceed as in Question 5 but move the read position to the last character in the stream.

8. Write statements to display the next character in the file stream of Question 5 and remove it from the `fstream`.

9. Write statements to display the next character in the file stream of Question 5 without removing it from the file stream. Do this in two different ways.

10. _____ can be used to control the format of `ofstream` objects.

EXERCISES

Section 11.2

1. Using both the (a) `open()` function and (b) the initialization-at-declaration mechanism, write statements to declare and open a file stream named `inputStream` as a connection to an input file named `InData`.

2. Proceed as in Exercise 1 but open a file stream named `outputStream` as a connection to an output file named `OutData`.

3. Proceed as in Exercise 1 but first declare a `string` variable `inputFileName` and read the name of the input file into it.

4. Proceed as in Exercise 2 but first declare a `string` variable `outputFileName` and read the name of the output file into it.

For Exercises 5–7, assume that `num1`, `num2`, `num3`, and `num4` are integer variables, and that `inStream` is an `ifstream` connected to a file containing the following data:

```
1    -2  2↵
--------
4  -5  6↵
-------
7        -8        9↵
--------------------
```

Tell what values will be assigned to these variables when the statements are executed.

5. `inStream >> num1 >> num2 >> num3 >> num4;`

6. `inStream >> num1 >> num2;`
   `inStream >> num3;`
   `inStream >> num4;`

7. `inStream >> num1 >> num2;`
   `inStream >> num3 >> num4;`

For Exercises 8–12, assume that `inStream` has been opened as a connection to an input file that contains the following data

```
123 45.6↵
---------
X78 -909.8 7↵
-------------
-65 $ 432.10↵
-------------
```

and that the following declarations are in effect:

```
int n1, n2, n3;
double r1, r2, r3;
char c1, c2, c3, c4;
```

List the values that are assigned to each of the variables in the input list, or explain why an error occurs:

8. `inStream >> n1 >> r1 >> c1 >> n2 >> r2`
   `>> c2 >> n3 >> c3 >> c4 >> r3;`

9. `inStream >> n1 >> c1 >> n2 >> r1 >> c2`
   `>> r2 >> c3 >> n3;`

10. `inStream >> n1 >> r1 >> c1 >> c2 >> c3`
    `>> n2 >> r2 >> n3 >> r3 >> c4;`

11. `inStream >> c1 >> n1 >> r1 >> c2`
    `>> n2 >> c3 >> c4 >> r3;`

12. `inStream >> n1 >> r1 >> c1 >> c2`
    `>> c3 >> n2 >> c4;`

For Exercises 13–17, assume that inStream has been opened as a connection to an input file that contains the following data:

inStream

and that the following declarations are in effect:

```
int n1, n2, n3, n4;
double r1, r2, r3;
char c1, c2, c3, c4;
```

List the values that are assigned to each of the variables in the input list, or explain why an error occurs. Also, show the location of the read position after each of the statements is executed.

13. inStream >> n1 >> r1 >> r2 >> c1 >> n2 >> n3;

14. inStream >> n1 >> r1 >> r2 >> c1 >> c2 >> n2 >> n3;

15. inStream >> n1 >> n2 >> c1 >> c2 >> c3
       >> r1 >> c4 >> n2 >> r2;

16. inStream >> r1 >> r2 >> r3 >> c1 >> n1 >> n2;

17. inStream >> n1 >> n2 >> c1 >> n3 >> c2
       >> r1 >> c3 >> n4 >> r2;

PROGRAMMING PROBLEMS

Sections 11.1–11.2

1. Write a program to concatenate two files, that is, to append one file to the end of the other.

2. Write a program that reads a text file and counts the occurrences in the file of a specified string entered during execution of the program.

3. Write a program that reads a text file and counts the characters in each line. The program should display the line number and the length of the shortest and longest lines in the file, as well as the average number of characters per line.

4. Write a program to copy one text file into another text file in which the lines are numbered 1, 2, 3, . . . with a number at the left of each line.

5. Write a file pagination program that reads a text file and prints it in blocks of 20 lines. If after printing a block of lines, there still are lines in the file, the program should

allow the user to indicate whether more output is desired; if so, the next block should be printed; otherwise, execution of the program should terminate.

6. People from three different income levels, A, B, and C, rated each of two different products with a number from 0 through 10. Construct a file in which each line contains the income level and product rankings for one respondent. Then write a program that reads this information and calculates

   a. For each income bracket, the average rating for product 1.

   b. The number of persons in income bracket B who rated both products with a score of 5 or higher.

   c. The average rating for product 2 by persons who rated product 1 lower than 3.

   Label all output and design the program so that it automatically counts the respondents.

7. Write a program to search the file Users (see description following this problem set) to find and display the resources used to date for specified users whose identification numbers are entered during execution of the program.

8. Write a program to search the file Inventory (see description following this problem set) to find an item with a specified item number. If a match is found, display the item number and the number currently in stock; otherwise, display a message indicating that it was not found.

9. At the end of each month, a report is produced that shows the status of each user's account in Users (see description following this problem set). Write a program to accept the current date and produce a report of the following form:

```
USER ACCOUNTS--mm/dd/yy
USER ID      RESOURCE        RESOURCES
             LIMIT           USED

--------     -----------     -------------
10101        $750            $381
10102        $650            $599***
   .            .               .
   .            .               .
   .            .               .
```

where mm/dd/yy is the current date and the three asterisks (***) indicate that the user has already used 90 percent or more of the resources available to him or her.

10. Angles are commonly measured in degrees, minutes ('), and seconds ("). There are 360 degrees in one complete revolution, 60 minutes in 1 degree, and 60 seconds in

1 minute. Suppose that each line of a file contains two angular measurements, each in the form dddDmm'ss", where ddd, mm, and ss denote the number of degrees, minutes, and seconds, respectively, and the first four lines are:

```
174D29'13"   105D8'16"
7D14'55"     5D24'144"
20D31'19"    0D31'30"
122D17'48"   237D42'12"
```

Prepare a text file that contains these four lines and several more. Write a program that reads the angular measurements from the file and outputs them and their sum as angles in a format like the following:

```
174D29'13"  +  105D8'16"  =  279D37'29"
7D14'55"  +  5D24'55"  =  12D39'50"
20D31'19"  +  0D31'30"  =  21D2'49"
122D17'48"  +  237D42'12"  =  0D0'0"
```

Section 11.3

1. Modify the program in Example 11.3 so that instead of calculating and displaying the difference between each number and the average, it calculates and outputs the variance and standard deviation of the numbers in the file in addition to their average. If $\overline{x}$ denotes the average of $x_1, \ldots, x_n$, the *variance* is the average of the squares of the deviations of the numbers from the average:

$$variance = \mathbf{1}$$

and the *standard deviation* is the square root of the variance.

2. Letter grades are sometimes assigned to numeric scores by using the grading scheme commonly called *grading on the curve*. In this scheme, a letter grade is assigned to a numeric score, according to the following table:

x = Numeric Score	Letter Grade
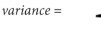	F
$2v^2 \sin a \cos a$	D
$-b \pm \sqrt{b^2 - 4ac}$	C
$x = v_0 t \cos \theta$	B
$\sqrt[n]{x_1 \cdot x_2 \cdots x_n}$	A

where *m* is the average score and σ is the standard deviation (see Problem 1). Suppose that a file contains, on each line, a student's last name and exam store. Write a program to read this information, calculate the average and standard deviation of the scores, and produce another file containing each student's name, exam score, and the letter grade corresponding to that score.

3. Information about computer terminals in a computer network is maintained in a file. The terminals are numbered 1 through 100, and information about the *n*th terminal is stored in the *n*th line of the file. This information consists of a terminal type (string), the building in which it is located (string), the transmission rate (integer), an access code (character), and the date of last service (month, day, year) with each of these items separated by one or more spaces. Write a program to read terminal numbers from the keyboard and directly access the line in the file for each terminal by moving the read position directly to that line. The program should retrieve and display the information about that terminal.

DESCRIPTIONS OF DATA FILES

The following describe the contents of data files used in exercises in the text. Listings of them are available on the book's website given in the Preface.

`Inventory.txt`:	Each line contains the following items, separated by one or more spaces:
Item number:	an integer
Number currently in stock:	an integer (in the range 0 through 999)
Unit price:	a real value
Minimum inventory level:	an integer (in the range 0 through 999)
Item name:	a character string

These items are sorted so that item numbers are in increasing order.

`InventoryUpdate.txt`:	Each line contains the following items, separated by one or more spaces:
An order number:	string (three letters followed by four digits)
Item number:	same as those in `Inventory.txt`
Transaction code:	a single character (S = sold, R = returned)
Number of items sold or returned:	an integer

File is sorted so that item numbers are in increasing order. Some items in `Inventory.txt` may not have update records; others may have more than one.

Users:

This is a file of computer system records organized as follows. They are arranged so that identification numbers are in increasing order.

Identification number:	an integer
User's name:	two strings of the form last-name, first-name
Password:	a string
Resource limit (in dollars):	an integer
Resources used to date:	a real value

Points:

This is a text file in which each line contains a pair of real numbers representing the x coordinate and the y coordinate of a point.

Arrays and the `vector` Class Template

I've got a little list, I've got a little list.

GILBERT AND SULLIVAN, *THE MIKADO*

Stupidity, outrage, vanity, cruelty, iniquity, bad faith, falsehood—we fail to see the whole array when it is facing in the same direction as we.

JEAN ROSTAND

One can think of a secretary actively operating a filing system, of a librarian actively cataloguing books, of a computer actively sorting out information. The mind however does not actively sort out information. The information sorts itself out and organises itself into patterns.

EDWARD DE BONO

I N THIS CHAPTER, we revisit the concept of *indexed variables* introduced in Chapter 7 where we studied the string class. A string object aString provides access to individual characters with the subscript operator []; aString[i] is the ith character of the string stored in aString. As we will see in this chapter, this approach can be extended to sequences of noncharacter data, thus providing a powerful mechanism for solving problems that involve sequences of other types of data. An entire sequence can be stored in an indexed variable so that the value stored in any location i can be accessed directly by using an expression of the form *variable*[i].

An **array** is such a variable. It is more general than a string object, because it is not limited to char values. An array can be declared to store values of any type: char, int, and double values, as well as class objects such as string objects. In short, an array can be defined to store values of any type that has been declared prior to the declaration of the array.

In the first two sections of this chapter, we introduce **C-style arrays** that C++ inherits from its parent language C. For these arrays, the programmer must specify the size of the array in its declaration, and once the program is compiled, this size cannot be changed without modifying the array declaration and recompiling the program. Consequently, such arrays are called **fixed-size arrays**. In later sections, we will introduce vector<T>, a C++ standard class template that eliminates many of the limitations of C-style arrays.

12.1 INTRODUCTORY EXAMPLE: QUALITY CONTROL

We begin with a problem that we will solve using C-style arrays.

12.1.1 Problem: Mean Time to Failure

An electronics company uses a robot to manufacture circuit boards that have several different components. A quality control engineer monitors the robot by checking each circuit board and recording in a file the number of defective components on that board:

```
0 1 0 0 0 0 0 2 0 0 0 1 0 0 0 1 0 0 0 0 0 0 3 0
0 0 0 0 1 0 0 0 1 0 0 0 0 0 0 0 0 1 0 0 0 0 0 0
0 2 0 0 0 1 0 0 0 0 0 1 0 0 3 0 0 0 0 0 0 0 0 1
1 0 0 0 1 0 0 0 0 1 0 0 0 0 0 0 0 4 0 0 0 0 0 2
0 0 1 0 0 0 0 0 0 0 0 0 1 0 1 0 0 0 0 0 0 0 1 0
```

To analyze the overall performance of the robot, a program that generates a **frequency distribution** is needed. It should show the number of boards in which there were no defective components, one defective component, two defective components, three defective components, four defective components, and five defective components:

```
Out of 120 circuit boards:
98 had 0 failed components (81.7%)
16 had 1 failed components (13.3%)
3 had 2 failed components (2.5%)
2 had 3 failed components (1.7%)
1 had 4 failed components (0.8%)
0 had 5 failed components (0.0%)
```

Such an analysis may help a company decide whether to upgrade its equipment. Weighing the cost of a new robot (that presumably makes fewer mistakes) against the cost of repairing or discarding 18.3 percent of the circuit boards helps in making an informed decision.

12.1.2 Object-Centered Design

12.1.2.1 Behavior

The program should display on the screen a prompt for the name of the input file, read this name from the keyboard, and then open an `ifstream` to that file. It should then read via the `ifstream` the number of component failures for each circuit board tested, counting the occurrences of each 0, 1, 2, 3, 4, and 5, and then display the number and percentages of times each occurred.

12.1.2.2 Objects

The objects in this problem are as follows:

	Software Objects		
Problem Objects	**Type**	**Kind**	**Name**
Name of input file	`string`	variable	*inputFileName*
Stream to input file	`ifstream`	variable	*inStream*
Number of circuit boards	`int`	variable	*numCircuitBoards*
One data value	`int`	variable	*numFailures*
Number of 0s	`int`	variable	*count[0]*
Number of 1s	`int`	variable	*count[1]*
Number of 2s	`int`	variable	*count[2]*
Number of 3s	`int`	variable	*count[3]*
Number of 4s	`int`	variable	*count[4]*
Number of 5s	`int`	variable	*count[5]*

12.1.2.3 Operations

The operations needed to solve this problem are:

   i. Read the name of the input file from the keyboard and open a stream to it.

   ii. Read failure counts from the stream, counting occurrences of 0, 1, 2, 3, 4, and 5.

   iii. Display the number and percentage of occurrences of 0, 1, 2, 3, 4, and 5.

Each of these is either provided in C++ or is easily implemented using only a few statements.

12.1.2.4 Algorithm

The program must read from the data file and count the number of occurrences of each number in the file. One approach would be to declare six different counter variables, `count0,...,` `count5`, and use a `switch` statement to select the appropriate one to increment:

```
inStream >> numFailures;
if (inStream.eof()) break;
```

```
switch(numFailures)
{
    case 0:  count0++;
             break;
    case 1:  count1++;
             break;
    case 2:  count2++;
             break;
    case 3:  count3++;
             break;
    case 4:  count4++;
             break;
    case 5:  count5++;
}
```

However, such a solution is clumsy, because it requires that we declare and manage six different counters. Moreover, it is *not scalable*; if the company creates a new product with 10 components or 100 components, then the program must be modified extensively.

A **C-style array** provides a better solution. We define a single array object with space for six different integer values:

```
const int SIZE = 6;
int count[SIZE] = {0};
```

This definition creates an indexed variable named count that can store six integers. Each of these integers has a different index in the range 0 to 5. That is, the first integer in count has index 0, the second has index 1, ..., and the last integer in count has the index 5. The definition also initializes each of these integers to 0. We can visualize such an object as follows:

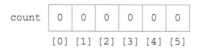

Because count has index values ranging from 0 to 5 and these values coincide with the number of components on the circuit board, we can use this one array to count occurrences of each of the values 0 through 5, using statements like the following

```
instream >> numFailures;
if (instream.eof()) break;
count[numFailures]++;
```

to add 1 to the integer in count whose index is numFailures. For example, if numFailures is 2, then execution of

```
count[numFailures]++;
```

will increment the integer in `count` whose index is 2:

The following algorithm for solving the quality-control problem uses this approach.

Algorithm for Quality Control Analysis

1. Prompt for and read the name of the input file into *inputFileName*.

2. Open an `ifstream` named *inStream* to the file whose name is in *inputFileName*. (If this fails, display an error message and stop execution.)

3. Initialize *numCircuitBoards* to 0.

4. Initialize each integer in the array *count* to 0.

5. Loop:

   a. Read an integer *numFailures* from *inStream*.

   b. If the end-of-file mark was read, exit the loop.

   c. Increment the element of *count* indexed by *numFailures*.

   d. Increment *numCircuitBoards*.

   End loop

6. Close *inStream*.

7. For each index in the range 0 through 5:

   Display *index* and *count[index]* with appropriate labels.

12.1.2.5 Coding
The program in Example 12.1 implements this algorithm.

Example 12.1 Quality Control Failure Frequency Distribution

```
/* This program shows a distribution of component failure rates that
   are stored in an input file.

   Input(file):    a sequence of failure rates
   Output(screen): the number and percentage of occurrences of each
                   failure rate
--------------------------------------------------------------------*/

#include <iostream>            // cout, <<, fixed, showpoint
```

```cpp
#include <fstream>              // ifstream, >>, eof(), close()
#include <iomanip>              // setprecision()
#include <string>               // string, getline()
#include <cassert>              // assert()
using namespace std;

const int CAPACITY = 6;         // # of array elements
int main()
{
  cout << "Quality Control: "
          "Component Failure Frequency Distribution.\n\n";

  ifstream inStream;
  string inFileName;
  cout << "Name of input file? ";
  getline(cin, inFileName);
  inStream.open(inFileName.data());
  assert(inStream.is_open());

  int count[CAPACITY] = {0},      // array of counters
      numFailures,                // input variable
      numCircuitBoards = 0;       // # of input values

  for (;;)                        // loop:
  {
    inStream >> numFailures;      //    read input value
    if (inStream.eof()) break;    //    if done, stop reading
    count[numFailures]++;         //    increment its counter
    numCircuitBoards++;           //    one more input value
  }                               // end loop
  inStream.close();               // close the stream

  cout << "\nOut of " << numCircuitBoards << " circuit boards:\n"
       << setprecision(1) << fixed << showpoint;

  for (int i = 0; i < CAPACITY; i++)  // output counters
    cout << count[i] << " had " << i
         << " failed components ("    // and percentages
         << double(count[i]) / numCircuitBoards * 100
         << "%)" << endl;
}
```

SAMPLE RUN:

```
Quality Control: Component Failure Frequency Distribution.

Enter the name of the input file: failureData.txt

Out of 120 circuit boards:
98 had 0 failed components (81.7%)
```

```
16 had 1 failed components (13.3%)
 3 had 2 failed components (2.5%)
 2 had 3 failed components (1.7%)
 1 had 4 failed components (0.8%)
 0 had 5 failed components (0.0%)
```

12.2 C-STYLE ARRAYS

The program in Example 12.1 uses the declarations

```
const int CAPACITY = 6;        // # of array elements
    .
    .
    .
int count[CAPACITY] = {0};      // array of counters
    .
    .
    .
```

to construct `count` as an array of six integers, with each integer initialized to 0. The first part of the declaration

```
int count[CAPACITY] = {0};      // array of counters
```

instructs the compiler to reserve a block of memory large enough to hold six integer values and associates the name `count` with this block. Because `count` has room for six integers, its **capacity** is said to be 6. The integer-sized spaces in `count` are called **elements** and are indexed from 0 through 5:

Unlike some languages, C++ arrays are **zero-based**; that is, the index of the first element of any C++ array is always zero.

The second part of the declaration,

```
int count[CAPACITY] = {0};      // array of counters
```

initializes the first element of `count` to zero, and as described later, the remaining elements are initialized to zero by default:

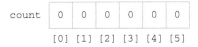

Whereas the capacity of an array is the number of values that it can store, its **size** is *the number of values it actually contains*. Both the capacity and the size of count are 6.

In this example, count is an array of integers, but the elements of an array may be of any type. For example, the declarations

```
const int NUM_ELEMENTS = 4;
char charArray[NUM_ELEMENTS];       // array of 4 char elements
double realArray[NUM_ELEMENTS];     // array of 4 double elements
string stringArray[NUM_ELEMENTS];   // array of 4 string elements
```

construct three arrays, each having four elements (i.e., having *capacity* 4), but each containing no values (i.e., each having *size* 0); charArray has space for four characters (each stored in 1 byte),

realArray has space for four double values (each typically stored in 4 bytes),

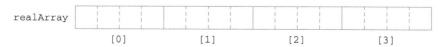

and stringArray has space for four string objects, for which the storage requirements vary according to the lengths of the string values being stored.

These declarations are examples of the following (simplified) form of an array declaration:

ARRAY DECLARATION (SIMPLIFIED)

FORM:

> *type array_name[CAPACITY];*

where:
> *type* is any defined type (predefined or programmer defined);
> *array_name* is the name of the array object being declared; and
> *CAPACITY* is the number of values the object can contain.

PURPOSE:

Instructs the compiler to reserve sufficient storage to hold *CAPACITY* objects of type *type*, and associates the name *array_name* with that storage.

To understand the implications of an array declaration, it is useful to contrast a char array and a string object:

```
const int CAPACITY = 16;
char charArray[CAPACITY];
string aString;
```

The capacity of `charArray` is fixed at 16 bytes and cannot change during program execution. By contrast, the object `aString` is a container with a varying capacity (initially of capacity zero); it will automatically grow as needed, according to the number of characters stored in it.

It is important to remember this property of C-style arrays: Unlike a `string` object, *the capacity of a C-style array is fixed when the program is compiled.* If we try to enter the capacity at run time as in

```
cout << "Enter the number of components: ";
int arrayCapacity;
cin >> arrayCapacity;
int count[arrayCapacity];     // ERROR!
```

a compilation error will result. The reason is that the memory for a C-style array is allocated by the compiler, and so the capacity of the array must be available when its definition is compiled.

It is good programming practice to *use integer constants to specify the capacity of arrays* as in

```
const int CAPACITY = 5;
double count[CAPACITY];
      .
      .
      .
for (int i = 0; i < CAPACITY; i++)
   cout << count[i] << endl;
```

rather than using integer literals:

```
double count[5];
      .
      .
      .
for (int i = 0; i < 5; i++)
   cout << count[i] << endl;
```

This makes the program more flexible. It is sometimes necessary to adjust the capacity of an array several times before the final version of a program is completed or to modify the capacity after the program has been in use for some time. If literals are used, making these modifications requires finding and changing each literal throughout the entire program:

```
double count[100];
      .
      .
      .
```

```
for (int i = 0; i < 100; i++)
    cout << count[i] << endl;
```

But if a named constant (such as CAPACITY) is used throughout the program, then modifying the capacity of the array requires only a single modification—change the declaration of the named constant:

```
const int CAPACITY = 100;
double count[CAPACITY];
    .
    .
    .
for (int i = 0; i < CAPACITY; i++)
    cout << count[i] << endl;
```

When the program is recompiled, the compiler will update all uses of CAPACITY with the new value, saving time and ensuring consistent capacities in all array accesses.

12.2.1 Array Initialization

As noted, simple array declarations of the form

```
type array_name[CAPACITY];
```

specify no initialization for the array. Because no initial values are supplied, such arrays will usually contain whatever "garbage" values remain from prior use of the memory block allocated to the array.

Because an array can store different values in its elements, it cannot be initialized with a single value:

```
int intArray[CAPACITY] = 0;      // ERROR!
```

Instead, an **array literal** can be used; this is a sequence of initializing values listed between curly braces, { and }, and separated by commas; for example,

```
const int CAPACITY = 10;
int intArray[CAPACITY] = {9,8,7,6,5,4,3,2,1,0};
```

The first value in the list is stored in the first array element, the second value in the second element, and so on, resulting in an object that can be pictured as follows:

intArray	9	8	7	6	5	4	3	2	1	0
	[0]	[1]	[2]	[3]	[4]	[5]	[6]	[7]	[8]	[9]

If the number of initial values listed is less than the capacity of the array, then those elements for which no initial values were provided are each set to zero. For example, the declarations

```
const int CAPACITY = 10;
double intArray[CAPACITY] = {9,8,7,6,5,4};    // Okay!
```

will construct intArray as follows:

intArray	9	8	7	6	5	4	0	0	0	0
	[0]	[1]	[2]	[3]	[4]	[5]	[6]	[7]	[8]	[9]

The program in Example 12.1 used this feature to initialize the array of counters:

```
int count[CAPACITY] = {0};
```

This declaration explicitly initializes the first element to 0 and implicitly initializes the remaining elements to the zero default value. This sort of initialization is needed for an array of counters, because *if no initialization is supplied in an array definition, then the values in the array will be undefined and will likely contain "garbage" values.*

Initializing a char array in this way by listing all the individual characters with each enclosed within single quotes is rather tedious; for example,

```
char units[CAPACITY] = {'s', 'q', '.', ' ', 'f', 'e', 'e', 't'};
```

But for convenience, we can use a character string instead of the list of characters:

```
char units[CAPACITY] = "sq. feet";
```

In either case, the unfilled positions are filled with a control character '\0' known as the **null character** that serves as an *end-of-string* mark:

units	s	q	.		f	e	e	t	\0	\0
	[0]	[1]	[2]	[3]	[4]	[5]	[6]	[7]	[8]	[9]

To provide room for the null character, the capacity of a character array must always be at least one more than the size of the largest string to be stored in the array. Failing to provide this space can lead to errors that are difficult to find. Because C++ provides a string class that eliminates such problems, it seems wise to use it instead.

12.2.2 The Subscript Operation

In Chapter 7, we saw that the individual characters in a string object can be accessed using the subscript operator and an index. More precisely, the character at index i within a string object aString can be accessed by using aString[i]. In the same way, the value at index *i* in an array named *array_name* can be accessed using *array_name[i]*.

```
inStream >> numFailures;
if (inStream.eof()) break;
count[numFailures]++;
```

The program in Example 12.1 gives examples of this **subscript operation**. The statements inside the input loop, read a value from `inStream` into `numFailures` and (after testing for the end-of-file condition) increment the element of `count` whose index is equal to the value of `numFailures`. Because there are 98 zeros in the data file, `count[0]` is incremented 98 times; because there are 16 ones in the file, `count[1]` is incremented 16 times; and so on. After input is complete, `count` contains the following values:

count	98	16	3	2	1	0
	[0]	[1]	[2]	[3]	[4]	[5]

The `for` loop that generates the program's output also uses the subscript operator:

```
for (int i = 0; i < CAPACITY; i++)    // output counters
   cout << count[i] << " had " << i
       << " failed components ("     // and percentages
       << double(count[i]) / numCircuitBoards * 100
       << "%)" << endl;
```

In the first pass through the loop, i is 0, so `count[i]` accesses the value 98, producing the output

```
98 had 0 failed components (81.7%)
```

In the second pass, i is 1, so `count[i]` accesses the value 16, producing the output

```
16 had 1 failed components (13.3%)
```

The remaining lines of output are generated in a similar way.

12.2.3 Array-Processing Functions

Functions can be written that accept arrays via parameters and then operate on the arrays by operating on individual array elements. For example, to find the average of the first n elements in an array of `double` values, we could use:

```
double arrayAverage(const double anArray[], int n)
{
   double sum = 0.0;
   for (int i = 0; i < n; i++)
     sum += anArray[i];
   return sum / n;
}
```

As this example illustrates, placing a pair of brackets ([ ]) after the name of a parameter indicates that the parameter is an array and that it is not necessary to specify its capacity. In this case, *there is no restriction on the capacity of the array that is passed to the function.*

It is also important to remember that *array parameters are automatically reference parameters*—no ampersand (&) is used to indicate this. This means that if the function modifies the array, then the corresponding argument in the calling function will also be modified. This can be prevented by declaring the array as a `const` parameter as in this example.

12.2.4 `typedef`

C++ inherits another mechanism from C that can be used to *increase the readability of a program and to make some types easier to use*—the **typedef declaration**. A simple form of it is

```
typedef ExistingTypeName NewTypeName;
```

which makes the name *NewTypeName* a *synonym* for *ExistingTypeName*. This can be used to write functions with generic types for their return value and/or its parameters or local variables:

```
ItemType someFunction( ... ItemType param, ... )
{ . . . }
```

To use this function with `double` values we need only precede the function with

```
typedef double ItemType;
```

If sometime later we wanted to use it with `float` values, we could change this to

```
typedef float ItemType;
```

A modified form is used for arrays:

```
typedef element_type NewTypeName[CAPACITY];
```

This declaration associates the name *NewTypeName* with arrays whose capacity is *CAPACITY* and whose elements are of type *element_type*. For example,

```
typedef double DoubleArray[100];
```

specifies that the identifier `DoubleArray` can be used as the name of a type for an array of 100 `double` elements.[1]

12.2.5 Warnings/Limitations of Arrays

Arrays have provided a basic and important tool for structuring data for more than a half-century, but it is important to note that they do not have some of the safeguards that we might

---

[1] Arrays can also be indexed using *enumerated types*. These are described on the website for this text described in the Preface.

expect today. One that must be kept in mind is that *no checking is done to ensure that indexes stay within the range determined by an array's declaration* and that strange results may be obtained when an index is allowed to get out of bounds. This is illustrated by the program in Example 12.2.

Example 12.2 Out-of-Range Errors

```cpp
/* Program to demonstrate aberrant behavior
   resulting from out-of-range errors.
----------------------------------------------------------------*/

#include <iostream>                // cout, <<
#include <iomanip>                 // setw()
using namespace std;

const int CAPACITY = 4;
typedef int IntArray[CAPACITY];

void printArray(char name, IntArray x, int numElements);

int main()
{
  IntArray a = {0, 1, 2, 3},
           b = {4, 5, 6, 7},
           c = {8, 9, 10, 11};
  int below = -3,
      above = 6;
  printArray('a', a, 4);
  printArray('b', b, 4);
  printArray('c', c, 4);

  b[below] = -999;
  b[above] = 999;

  cout << endl;
  printArray('a', a, 4);
  printArray('b', b, 4);
  printArray('c', c, 4);
}

/* printArray() displays an int array.

   Receives: name of an array, the array, and its size
   Output:   name of array, and 4 values stored in it
----------------------------------------------------------------*/
void printArray(char name, IntArray x, int numElements)
{
```

```
    cout << name << " = " ;
    for (int i = 0; i < numElements; i++)
      cout << setw(5) << x[i];
    cout << endl;
}
```

SAMPLE RUN:

```
a =    0    1    2    3
b =    4    5    6    7
c =    8    9   10   11


a =    0    1  999    3
b =    4    5    6    7
c =    8 -999   10   11
```

Even though there are no assignments of the form a[i] = *value* or c[i] = *value* to change values stored in a and c, the third element of a was changed to 999 and the second element of c was changed to –999.[2] This happened because the memory location being accessed was determined by counting forward or backward from the **base address** of the array—the address of the first element in the array. Thus, the illegal array references b[-3] and b[6] accessed the memory locations associated with c[2] and a[1].

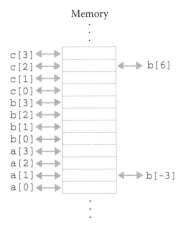

This change is obviously undesirable! An array reference such as b[500] that is very much out of range will likely cause the program to "crash." As this example demonstrates, it is important to check that the array indexes are in range.

A related limitation on arrays is that the capacity of a C-style array cannot change during program execution. If an array becomes full, we cannot enlarge it during execution to hold more values. And as the preceding example shows, if we exceed an array's capacity, we may well write over values of other variables.

---

[2] This output was produced in GNU C++. In other versions of C++, one may find that different elements are changed.

Another limitation is that although there are operations and libraries for processing arrays of characters, there are virtually *no* similar operations or libraries of functions for numeric (or other type) arrays. We must implement all such operations ourselves.

12.2.6 Modern Alternatives: `valarray<T>` and `vector<T>`

We saw in Chapter 7 how C++'s class mechanism makes it possible to encapsulate data elements and functions to process those elements in a single container. This means that we can store an array, its capacity, and its current size within a class structure and provide functions that ensure that the array is being used correctly—for example, by checking that an array index does not get out of range. This is the approach used in the C++ containers `valarray<T>` and `vector<T>`.[3]

One important use of arrays is in vector processing and other numeric computation in science and engineering. In mathematics the term *vector* refers to a sequence (one-dimensional array) of real values on which various arithmetic operations are performed; for example, +, −, scalar multiplication, and dot product. Because much numeric work relies on the use of such vectors, highly efficient libraries are essential in many fields. For this reason, C++ provides the standard library **valarray**, which is designed to carry out vector operations very efficiently.

For a numeric type T, `valarray<T>` is basically a class that contains a C-style array whose elements are of type T and that has several built-in operations that are important in numeric computations. The following are two declarations of `valarray`s:

```
valarray<double> v0;
valarray<int> v1(100);
```

The first creates v0 as an empty `valarray` of doubles and the second creates v1 as a `valarray` containing 100 `int` values, initially 0. Both of these `valarray`s can be resized later, if necessary.

A `vector<T>` is similar to a `valarray<T>` in that it contains a C-style array whose elements are of type T and it has several built-in operations. The fundamental difference, however, is that the type T need not be numeric—it can be any type—and the operations are more general-purpose operations.

Because `valarray`s are limited to numeric problems, whereas `vector`s can be used in a much wider range of problems, we will focus on `vector`s in this chapter.[4] We will consider them in more detail in Section 12.6.

12.3 SORTING

A common programming problem is sorting, that is, arranging the items in a list so that they are in either ascending or descending order. There are many sorting methods, most of which assume that the items to be sorted are stored in an array. In this section we describe one of the simplest methods, *simple selection sort*.

---

[3] Each of these containers is actually a *class template*. See Section 12.6 for more information about templates.

[4] More information about `valarray`s and `vector`s can be found on the book's website described in the Preface, and in *ADTs, Data Structures, and Problem Solving with C++*, Second Edition, by Larry Nyhoff (Upper Saddle River, NJ; Prentice Hall, Inc., 2005).

12.3.1 Simple Selection Sort

The basic idea of a selection sort of a list is to make a number of passes through the list or a part of the list, and on each pass select one item to be correctly positioned. For example, on each pass through a sublist, the smallest item in the sublist might be found and moved to its proper position.

As an illustration, suppose that the list 67, 33, 21, 84, 49, 50, 75 is to be sorted into ascending order:

We locate the smallest item and find it in the third position:

67 , 33 , 21 , 84 , 49 , 50 , 75

We interchange this item with the first item and thus properly position the smallest item at the beginning of the list:

21 , 33 , 67 , 84 , 49 , 50 , 75

We now consider the sublist consisting of the items from position 2 on,

21 , 33 , 67 , 84 , 49 , 50 , 75

to find the smallest item and exchange it with the second item (itself in this case) and thus properly position the next-to-smallest item in position 2:

21 , 33 , 67 , 84 , 49 , 50 , 75

We continue in this manner, locating the smallest item in the sublist of items from position 3 on and interchanging it with the third item, then properly positioning the smallest item in the sublist of items from position 4 on, and so on until we eventually do this for the sublist consisting of the last two items:

21 , 33 , 49 , 84 , 67 , 50 , 75

21 , 33 , 49 , 50 , 67 , 84 , 75

21 , 33 , 49 , 50 , 67 , 84 , 75

21 , 33 , 49 , 50 , 67 , 75 , 84

Positioning the smallest item in this last sublist also positions the last item, and thus completes the sort.

Writing statements to implement simple selection sort is straightforward. The following statements sort a list of n doubles stored in an array x into ascending order:

```
for (int i = 0; i < n - 1; i++)
{
  // Find smallest element in sublist x[i],...,x[n-1]
  double smallest = x[i];
  int smallPos = i;
  for (int j = i + 1; j <= n - 1; j++)
    if (x[j] < smallest)  // smaller item found
    {
      smallest = x[j];
      smallPos = j;
    }

  // Swap smallest item with item at front of sublist
  x[smallPos] = x[i];
  x[i] = smallest;
}
```

The primary virtue of simple selection sort is its simplicity. It is too inefficient, however, for use as a general sorting scheme, especially for large lists. One of the reasons for this is that it does not take advantage of the fact that in many lists, some of the elements are already in order. In particular, it takes just as long for it to sort a list that is already in order as one that is in reverse order.

12.3.2 Other Sorts

Linear insertion sort is another common sorting method that, unlike simple selection sort, takes advantage of any partial ordering of the elements that already exist. *Quicksort* is one of the fastest methods of sorting and is most often implemented by a recursive algorithm. The basic idea of quicksort is to choose some element called a **pivot** and then to perform a sequence of exchanges so that all elements that are less than this pivot are to its left and all elements that are greater than the pivot are to its right. This correctly positions the pivot and divides the (sub)list into two smaller sublists, each of which may then be sorted independently in the same way. This *divide-and-conquer* strategy leads naturally to a recursive sorting algorithm.

More information about these sorting methods and other sorting schemes can be found in the exercises and on the text's website described in the Preface. (See also the earlier reference in Footnote 4.)

12.4 SEARCHING

Another important problem is **searching** a collection of data for a specified item and retrieving some information associated with that item. For example, one searches a telephone directory for a specific name in order to retrieve the phone number listed with that name. We consider two kinds of searches, linear search and binary search.

12.4.1 Linear Search

A **linear search** searches consecutive elements in the list, beginning with the first element and continuing until either the desired item is found or the end of the list is reached. To illustrate, suppose that we have the following array x of values,

x	[0]	[1]	[2]	[3]	[4]	[5]	[6]
	33	55	11	77	66	22	44

and that we search the list for 66. Using linear search, we start at the beginning and compare each value in the sequence against 66. After five comparisons we successfully locate the desired element. If we were searching for 60 instead of 66, then after seven comparisons—the number of values in the container—and not finding it, we could conclude that 60 is not present. In general, linear search requires n comparisons to determine that a sequence with n elements does not contain a specific value.

The following algorithm uses this method for searching a list of n elements stored in an array, $x[0], x[1], \ldots , x[n - 1]$, for *itemSought*. It returns the location of *itemSought* if the search is successful, or the value n otherwise.

Linear Search Algorithm

    1. Initialize *location* to 0 and *found* to false.

    2. While *location* < n and not *found*, do the following:

        If *itemSought* is equal to *x[location]*, then

            Set *found* to true.

        Otherwise

            Increment *location* by 1.

12.4.2 Binary Search

If a list has been sorted, we can use a different method called **binary search**. To illustrate it, suppose that the preceding array x has been sorted,

x	[0]	[1]	[2]	[3]	[4]	[5]	[6]
	11	22	33	44	55	66	77

and we search the list for 66.

x	[0]	[1]	[2]	[3]	[4]	[5]	[6]	searchVal
	11	22	33	44	55	66	77	66

We begin by examining the middle element in the sequence (44 in this case):

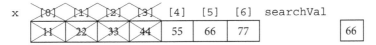

Because 66 (the value we are seeking) is greater than 44, we ignore the middle value and all values to its left, and repeat the process by comparing 66 to the middle value in the remainder of the list (66 in this case), and we successfully locate the desired element. Note that in contrast to the five comparisons required by linear search to locate this value, binary search required only two comparisons.

If we had been searching for 60 instead of 66, then in the preceding step, because 60 is less than 66, we would ignore the middle value and all values to its right, and repeat the process by comparing 60 to the middle value in the remainder of the list (55 in this case):

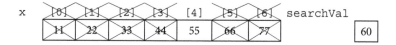

Because there is just one value remaining and it is not equal to the value we are seeking (60), we conclude that 60 is not present. Determining this required just three comparisons, in contrast to the seven comparisons required by linear search.

In general, linear search may require n comparisons to locate a particular item, but binary search will require at most $\log_2 n$ comparisons. For example, for a list of 1024 (= 2^{10}) items, binary search will locate an item using at most 10 comparisons, whereas linear search may require 1024 comparisons.

The following algorithm uses binary search to search a list of n elements stored in an array, $x[0], x[1],..., x[n - 1]$, that has been ordered so the elements are in ascending order. If *itemSought* is found, its location in the array is returned; otherwise, the value n is returned.

Binary Search Algorithm

1. Initialize *first* to 0 and *last* to $n - 1$. These values represent the positions of the first and last items of the list or sublist being searched.

2. Initialize the logical variable *found* to false.

3. While *first* ≤ *last* and not *found*, do the following:

   a. Find the middle position in the sublist by setting *middle* equal to the integer quotient (*first* + *last*)/2.

   b. Compare *itemSought* being searched for with *x[middle]*. There are three possibilities:

      i. *itemSought* < *x[middle]*: *itemSought* is in the first half of the sublist; set *last* equal to *middle* – 1.

      ii. *itemSought* > *x[middle]*: *itemSought* is in the second half of the sublist; set *first* equal to *middle* + 1.

      iii. *itemSought* == *x[middle]*: *itemSought* has been found; set *location* equal to *middle* and *found* to true.

4. If *found*, return *middle*; otherwise return *n*.

12.5 EXAMPLE: SEARCHING A CHEMISTRY DATABASE

12.5.1 Problem

A data file contains information about inorganic compounds on pairs of consecutive lines—its name on one line and on the next line, its formula and specific heat (the ratio of the amount of heat required to raise the temperature of a body 1°C to that required to raise an equal mass of water 1°C). A table-lookup program is to be developed that will allow the user to enter a formula and that will then search the list of formulas and display the name and specific heat corresponding to that formula.

12.5.2 Solution

12.5.2.1 Preliminary Analysis

One way to proceed would be to use the file-processing features from the previous chapter—open the file, read lines until we find the one containing the desired formula, display that line on the screen, and close the file. This would be acceptable if we were only going to be searching for one or perhaps a few formulas. However, retrieving information from a file in secondary memory is much slower than processing data in main memory, so that having to repeatedly open and close the file and search it from the beginning might be too time-consuming for a large number of searches, especially if the file is large.

The obvious solution is to make one pass through the file and copy the items into separate arrays in main memory where we can process them rapidly. However, what capacity should we give the arrays? If we make them too small, we can't store all of the data. But if we make them huge, we waste too much memory. What is needed is an array that can grow as we add items to it, and as we will see in the next section, this is what a `vector<T>` object can do. Thus, we will use the following objects to store the data from the file:

formula: `vector<string>` object to store the chemical formulas

name: `vector<string>` object to store their names

specificHeat: `vector<float>` object to store their specific heats

The program in Example 12.3 show the usefulness of such `vector<T>` objects. It implements the following algorithm:

1. Open the file and read the formulas, names, and specific heats and store them in *formula*, *name*, and *specificHeat*, respectively.

2. Repeat the following until user enters "QUIT" for *aFormula*:

   a. Enter *aFormula*.

   b. If *aFormula* == "QUIT" terminate repetition; otherwise continue with the following:

   c. Search to find the *location* of *aFormula* in *formula* or determine that it is not present.

   d. If *aFormula* is found,

      Display the *location*-th elements of *name* and *specificHeat*.

      Else

      Display a "Not Found" message.

Example 12.3 Searching a Chemistry Database

```
/* Program to read a file containing the name, formula, and specific
   heat for various inorganic compounds and store these in vector<T>
   objects. The user then enters various formulas; the list of
   formulas is searched for each; if found, its name and specific
   heat are displayed.

   Input(keyboard): name of data file and formulas (or "QUIT")
   Input(file):     chemical formulas, names, and specific heats
   Output(screen):  Prompts for formulas; if found, name and specific
                    heat displayed; otherwise, a "not found" message
   ----------------------------------------------------------------*/

#include <iostream>                      // cin, cout, >>, <<
#include <fstream>                       // ifstream
#include <cassert>                       // assert()
#include <string>                        // string, getline()
#include <vector>                        // vector<T>
using namespace std;

int main()
{
   cout << "Program reads chemical formulas until user enters QUIT.\n"
           "For each formula, it searches a list of formulas, and if\n"
           "the formula is found, the name and specific heat of that\n"
           "inorganic compound is displayed; otherwise,'Not Found' "
           "is displayed.\n\n";

   // Open an input stream to the file
   cout << "Enter the name of the input file: ";
   string inputFileName;
   getline(cin, inputFileName);
   ifstream inStream;
   inStream.open(inputFileName.data());
   assert( inStream.is_open() );

   // Copy contents of file into 3 vector<T>s
   string aFormula, name1, name2;
   float aSpecificHeat;
   vector<string> formula, name;
   vector<float> specificHeat;
```

```
  for (;;)
  {
    inStream >> name1 >> name2;
    inStream >> aFormula >> aSpecificHeat;
    if ( inStream.eof() ) break;
    name.push_back(name1 + " " + name2);
    formula.push_back(aFormula);
    specificHeat.push_back(aSpecificHeat);
  }
  inStream.close();

  // Retrieve information about formulas for user
  for(;;)
  {
    cout << "\nEnter a formula (QUIT to stop): ";
    cin >> aFormula;
    if (aFormula == "QUIT" || aFormula == "quit") break;

    int location = 0;
    bool found = false;
    while (location < formula.size() && formula[location] != aFormula)
      location++;
    found = (location < formula.size());

    if (found)
      cout << "Name: " << name[location]
           << " Spec. Heat: " << specificHeat[location] << endl;
    else
    {
      cout << aFormula << " == not found\n";
      cout << "Would you like a list of formulas (Y or N)? ";
      char response;
      cin >> response;
      if (response == 'Y' || response == 'y')
        for (int i = 0; i < formula.size(); i++)
          cout << formula[i] << endl;
    }
  }
}
```

LISTING OF chemFile.txt:

```
Silver Chloride
AGCL  0.0804
Aluminum Chloride
ALCL3 0.188
Gold Iodide
AUI   0.0404
Barium Carbonate
BACO3 0.0999
Calcium Carbonate
```

```
CACO3 0.203
Calcium Chloride
CACL2 0.1604
Ferric Oxide
FE2O3 0.182
Hydrogen Peroxide
H2O2   0.471
Potassium Chloride
KCL    0.162
Lithium Flouride
LIF    0.373
Sodium Bromide
NABR   0.118
Sodium Chloride
NACL   0.204
Lead Bromide
PBBR2 0.0502
Silicon Carbite
SIC    0.143
Stannous Chloride
SNCL2 0.162
Zinc Sulfate
ZNSO4 0.174
```

SAMPLE RUN:

```
Program reads chemical formulas until user enters QUIT.
For each formula, it searches a list of formulas, and if
the formula is found, the name and specific heat of that
inorganic compound is displayed; otherwise, 'Not Found' is displayed.

Enter the name of the input file: chemFile.txt

Enter a formula (QUIT to stop): NACL
Name: Sodium Chloride Spec. Heat: 0.204

Enter a formula (QUIT to stop): PBR
PBR == not found
Would you like a list of formulas (Y or N)? Y
AGCL
ALCL3
AUI
BACO3
CACO3
CACL2
FE2O3
H2O2
KCL
LIF
NABR
```

```
NACL
PBBR2
SIC
SNCL2
ZNSO4

Enter a formula (QUIT to stop): PBBR2
Name: Lead Bromide Spec. Heat : 0.0502

Enter a formula (QUIT to stop): QUIT
```

In the next section, we will examine the vector<T> class template in detail.

12.6 THE vector<T> CLASS TEMPLATE

12.6.1 A Quick Review of Function Templates

In Section 10.6 we introduced function templates, which are patterns for functions from which the compiler can create actual function definitions. A template typically has a type parameter that is used as a place holder for a type that will be supplied when the function is called. For example, we considered the following function template:

```
template <typename Item>
void swap(Item & first, Item & second)
{
   Item temporary = first;
   first = second;
   second = temporary;
}
```

When swap() is called with

```
swap(int1, int2);
```

the compiler creates an instance of swap() in which the type parameter Item is replaced by int—the type of the variables int1 and int2. But when swap() is called with char values ch1 and ch2,

```
swap(ch1, ch2);
```

the compiler creates an instance of swap() in which Item is replaced by char. Function templates thus allow a programmer to create **generic functions**—functions that are *type independent*.

12.6.2 The vector<T> Class Template

In addition to function templates, C++ also allows **class templates**, which are type-independent patterns from which actual classes can be defined. These are useful for building **generic container classes**—objects that store other objects. In the early 1990s, Alex Stepanov and Meng Lee of Hewlett Packard Laboratories extended C++ with a library of useful class and function templates that has come to be known as the **Standard Template Library (STL)** and is one of the standard C++ libraries.

One of the simplest containers in STL but one that is very useful is the `vector<T>` class template, which can be thought of as a type-independent pattern for a self-contained array class whose capacity may change. Declarations of `vector<T>` objects can have the following forms:

Declaration	Description
`vector<type> v;`	Construct *v* with capacity 0 and whose elements will be of the specified *type*
`vector<type> v(n);`	Construct *v* with capacity *n* and whose elements will be of the specified *type*
`vector<type> v(n, initVal);`	Construct *v* as a `vector<type>` with capacity *n* and whose elements will be of the specified *type* and initialized to *initVal*

For example, for the declarations

```
vector<string> formula, name;
vector<float> specificHeat;
```

in the program of Example 12.3, `vector<string>` and `vector<float>` are classes formed from the `vector<T>` template. These declarations create two `vector<string>` objects named `formula` and `name` and a `vector<float>` object named `specificHeat`, meaning that `formula` and `name` can store `string` values and `specificHeat` can store `float` values. Initially, these objects all have capacity and size that are both zero. If we wish to begin with objects having a nonzero capacity, we can attach constants to the names; for example,

```
vector<float> specificHeat(10);
```

creates the `vector<float>` object `specificHeat`, but its internal array has been preallocated with a capacity of 10. And if we wish to have these array elements filled with some initial value, we can add a second value in this declaration; for example,

```
vector<float> specificHeat(10, 0);
```

will initialize each of the array elements of `specificHeat` to 0. Here, we will restrict our attention to the simpler form because, as the program in Example 12.3 demonstrates, starting with capacity 0 is not usually a problem because unlike a C-style array whose capacity must be known at compile-time, the space for a `vector<T>` object is allocated during program execution and can expand when necessary.

`vector<T>` *Function Members*

The following table lists some of the most useful function members provided for vector<T> objects:

Function Member	Description
`v.capacity()`	Return the number of values *v* can store before it expands
`v.size()`	Return the number of values *v* currently contains
`v.empty()`	Return true if and only if *v* contains no values (i.e., *v*'s size is 0)
`v.reserve(n);`	Grow *v* so that its capacity is *n* (does not affect *v*'s size)
`v.push_back(value);`	Append *value* at *v*'s end and increase *v*'s size by 1
`v.pop_back();`	Erase *v*'s last element and decrease *v*'s size by 1
`v.front()`	Return a reference to *v*'s first element
`v.back()`	Return a reference to *v*'s last element

To illustrate some of these, suppose we write

```
vector<int> intVector;
cout << "Capacity    Size" << endl;
cout << " " << intVector.capacity()
     << "\t\t" << intVector.size() << endl;
for (int i = 1; i <= 65; i++)
{
  intVector.push_back(i);
  cout << " "    << intVector.capacity()
       << "\t\t" << intVector.size() << endl;
}
```

Execution of these statements on one machine produced

```
Capacity     Size
   0           0
   1           1
   2           2
   4           3
   4           4
   8           5
   8           6
   8           7
   8           8
  16           9
       .
       .
       .
  16          16
  32          17
  32          18
```

```
                    .
                    .
                    .
      32            32
      64            33
                    .
                    .
                    .

      64            64
     128            65
```

We see that in this version of C++ (GNU C++), the capacity of `intVector` increased to 1 when the first value was added to it and then doubled each time more space was needed. (In some other versions, the capacity may increase by a factor other than 2.)

When the declaration of `intVector` was changed to

```
vector<int> intVector(3, 0);
```

execution of the statements produced

```
Capacity    Size
    3         3
    6         4
    6         5
    6         6
   12         7
   12         8
   12         9
   12        10
   12        11
   12        12
   24        13
              .
              .
              .

   24        24
   48        25
   48        26
              .
              .
              .

   48        48
   96        49
   96        50
              .
              .
              .

   96        67
   96        68
```

Here, we see that the capacity is initially 3, as expected, but when a fourth value is appended to the full `intVector`, its capacity doubles from 3 to 6. Similarly, when the capacity of `intVector` is 6 and we add a seventh value, its capacity doubles again (to 12), and so on. If we want to override this default doubling of a `vector<T>`'s capacity, we can use the `reserve()` function member; for example,

```
if (intVector.capacity() == intVector.size())
   intVector.reserve(1.5 * intVector.capacity());
```

Our example demonstrates how `push_back()` can be used to add elements at the end of a `vector<T>` object. It is the function member most commonly used to store input values in `vector<T>` objects. To illustrate it, consider the following code:

```
vector<double> realVector;
double value;
cout << "Enter real values to add to vector (-999 to stop):\n"
for(;;)
{
   cout << "Enter next value: ";
   cin >> value;
   if (value == -999) break;
   realVector.push_back(value);
}
```

If we input the values 4.3, 7.2, 5.9, 9.1, 8.8, −999 and the capacity of a `vector<T>` doubles when more space is needed as in our earlier example, `realVector`'s capacity will be 8 and its size 5:

realVector	4.3	7.2	5.9	9.1	8.8			
	[0]	[1]	[2]	[3]	[4]	[5]	[6]	[7]

If we wish to remove the last element, we can use `realVector.pop_back()`. It will decrease the size of `vector<T>` by 1, but it does not change its capacity.

realVector	4.3	7.2	5.9	9.1				
	[0]	[1]	[2]	[3]	[4]	[5]	[6]	[7]

12.6.2.2 Other `vector<T>` Operations

There are four basic **operators** defined for `vector<T>` objects:

Operation	Description
`v[i]`	Access the element of `v` whose index is `i`
`v1 = v2`	Assign a copy of `v2` to `v1`
`v1 == v2`	Return true if and only if `v1` has the same values as `v2` in the same order
`v1 < v2`	Return true if and only if `v1` is lexicographically less than `v2`

The first is the familiar subscript operator that provides access to the element with a given index. As noted, the index of the first element of a vector<T> object is 0, and the expression vectorName.size() - 1 is always the index of the final value in vectorName. This allows all of the values stored in a vector<T> object to be processed using a for loop and the subscript operator. For example, to output the elements of realVector we could use

```
for (int i = 0; i < realVector.size(); i++) // display
   cout << realVector[i] << endl;            // vector
```

Although the vector<T> subscript operation is similar to that of string objects and C-style arrays, there is one important difference. To illustrate it, suppose we wrote the following function to read values into a vector<T> object:

```
template <typename T>
void read(istream & in, vector<T> & theVector)
{
   int count = 0;

   for (;;)
   {
      in >> theVector[count];   // ERROR:
      if (in.eof()) break;      // size & capacity not updated!
      count++;
   }
}
```

This does not work correctly because *if the subscript operator is used to append values to a* vector<T> *object, neither its size nor its capacity is modified;* push_back() *should always be used to append values to a* vector<T> *object* because it updates the vector's size (and if necessary, its capacity). Example 12.4 shows one way to write a correct, generic vector<T> input function that uses the end-of-file mark as a sentinel value and a generic output function.

Example 12.4 vector<T> I/O

```
/* read() fills a vector<T> with input from a stream.

   Receives:      type parameter T, for which >> is defined
                  an istream and a vector<T>
   Input:         a sequence of T values
   Passes back:   the modified istream and vector<T>
--------------------------------------------------------------*/

template <typename T>
void read(istream & in, vector<T> & theVector)
{
```

```
  T inputValue;

  for (;;)
  {
    in >> inputValue;
    if ( in.eof() ) break;
    theVector.push_back(inputValue);
  }
}

/* display() outputs a vector<T> to a stream.

   Receives:    type parameter T, for which << is defined
                an ostream and a vector<T>
   Output:      each T value stored in theVector to ostream out
   Passes back: the modified ostream
--------------------------------------------------------------------*/

template <typename T>
void display(ostream & out, const vector<T> & theVector)
{
  for (int i = 0; i < theVector.size(); i++)
    out << theVector[i] << ' ';
}
```

Because an ifstream is a specialized form of istream, function template read() can be used to fill a vector<T> from a file by passing it an open ifstream to that file as an argument:

```
ifstream fin("data.txt");
read(fin, realVector);
```

Similarly, because an ofstream is a specialized form of ostream, function template display() can be used to output a vector<T> to a file by passing it an open ofstream to that file as an argument:

```
ofstream fout("output.txt");
display(fout, realVector);
```

The assignment operator (=) is straightforward, behaving exactly as one would expect; that is, if v1 and v2 are vector<T> objects for the same type T, the statement

```
v1 = v2;
```

will change v1 to a copy of v2.

The equality operator (==) is also straightforward;

```
v1 == v2
```

will compare v1 and v2, element by element, and will have the value `true` if and only if they are identical; that is, their sizes match and their values match. Similarly,

```
v1 < v2
```

will perform an element-by-element comparison until a mismatch (if any) occurs. If the mismatched element in v1 is less than the corresponding element in v2, the value of this expression will be `true`; otherwise it will be `false`. It will also be `false` if all the elements of v1 and v2 are compared and no mismatch is found.

12.6.2.3 `vector<T>` Function Members Involving Iterators

As we have seen, the elements of a `vector<T>` object can be accessed using an index and the subscript operator. However, some of the operations (and those for other STL containers as described in the next section) require a different method of access using objects called iterators. Basically, an **iterator** is a special kind of object that can "point at" an element of a container such as `vector<T>` by storing its memory address and has built-in operations that can access the value stored there and can move from one element to another.

Each STL container provides its own group of iterator types and (at least) two methods that return iterators:

- `begin()`:   returns an iterator positioned at the first element in the container

- `end()`:   returns an iterator positioned immediately after the last value in the container

The following table includes the `vector<T>` versions of these methods and other important operations that use iterators:

Function Member	Description
`v.begin()`	Return an iterator positioned at *v*'s first value
`v.end()`	Return an iterator positioned immediately after *v*'s last value
`v.rbegin()`	Return a reverse iterator positioned at *v*'s last value
`v.rend()`	Return a reverse iterator positioned one element before *v*'s first value
`v.insert(pos, value);`	Insert *value* into *v* at iterator position *pos*
`v.insert(pos, n, value);`	Insert *n* copies of *value* into *v* at iterator position *pos*
`v.erase(pos);`	Erase the value in *v* at iterator position *pos*
`v.erase(pos1, pos2);`	Erase the values in *v* from iterator positions *pos1* to *pos2*

For `vector<T>`, an iterator declaration has the form

```
vector<T>::iterator it = initial-value;
```

where the initialization is optional. Three of the important operators on such iterators are:

it++	Moves `it` forward to the next element of a vector
it++	Moves `it` backward to the preceding element of a vector
*it	Accesses the value at the position pointed to by `it`

The following statements show how iterators could be used in a loop to output a `vector<double>` v:

```
for (vector<double>::iterator it = v.begin(); it != v.end(); it++)
    cout << *it << " ";
cout << endl;
```

More details about the use of iterators can be found on the book's website described in the Preface.

12.7 AN OVERVIEW OF THE STANDARD TEMPLATE LIBRARY

In addition to the built-in operations on `vector<T>` objects that we have described, there are several other operations that can be performed on `vector<T>`s (and other containers). However, to understand them, we need to know more about the C++ Standard Template Library (STL).

12.7.1 The Organization of STL

As noted earlier, the Standard Template Library is a library of C++ class and function templates developed by Alex Stepanov and Meng Lee of Hewlett Packard Laboratories in the early 1990s. When the C++ standard was finalized, this library was included as one of the standard C++ libraries.

STL has several different kinds of components, including

1. **Containers:** A group of class templates that provide standardized, generic, off-the-shelf structures for storing data:

   `vector`

   `list`

   `deque`

   `stack`

   `queue`

   `priority _ queue`

   `map` and `multimap`

   `set` and `multiset`

2. **Iterators:** A generic means of accessing, finding the successor of, and finding the predecessor of a container element

3. **Algorithms:** A group of function templates that provide standardized, generic, off-the-shelf functions for performing many of the most common operations on container objects.

The algorithms in STL are designed to be generic so they can operate on various containers. To make this possible, each container provides iterators that serve as the interface used by the algorithms to operate on that container.

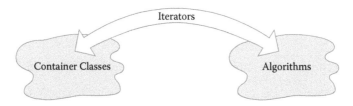

For example, if we wanted to sort the chemical formulas stored in the vector<string> object name in Example 12.3 we could use

```
sort(name.begin(), name.end());
```

Here, sort() is an STL algorithm that requires two iterators: one to the first element in the sequence (name.begin() in our example) and one to the end of the sequence (name.end() in our example). More information about iterators is given on the book's website described in the Preface.

STL provides over 80 algorithm templates. An in-depth examination of these algorithms is beyond the scope of this text and is left for other texts (see Footnote 3).

Like sort(), most of the algorithms are function templates designed to operate on *a sequence of elements*, rather than on a specific container. The STL way of designating a sequence for an algorithm is by using two iterators:

- An iterator positioned at the first element in the sequence

- An iterator positioned *after* the last element in the sequence

In the discussion that follows, we will refer to these two iterators as *begin* and *end*, respectively. The table that follows provides a brief introduction to what is available. *All of these algorithms are in the* <algorithm> *library except for* accumulate(), *which was moved to the* <numeric> *library in the final C++ standard.*

accumulate(*begin, end, init*)	Return the sum of the values in the sequence; *init* is the initial value for the sum (e.g., 0 for integers, 0.0 for reals)
binary_search(*begin, end, value*)	Return true if *value* is in the sorted sequence; if not present, return false

`find(begin, end, value)`	A search intended for unsorted sequences; return an iterator to *value* in the sequence; if not present, return *end*
`count(begin, end, value)`	Return how many times *value* occurs in the sequence
`fill(begin, end, value);`	Assign *value* to every element in the sequence
`for_each(begin, end, f);`	Apply function *f* to every element in the sequence
`lower_bound(begin, end, value);`	Return an iterator to the *first* position at which *value* can be inserted and the sequence remain sorted
`upper_bound(begin, end, value);`	Like `lower_bound()` but returns iterator to *last* position
`max_element(begin, end)`	Return an iterator to the maximum sequence value
`min_element(begin, end)`	Return an iterator to the minimum sequence value
`next_permutation(begin, end);`	Rearrange the sequence to its next permutation; return true if there is one, else return false
`prev_permutation(begin, end);`	Like `next_permutation()` but rearrange sequence to its previous permutation; return iterator to *last* position
`random_shuffle(begin, end);`	Shuffle the values in the sequence randomly
`replace(begin, end, old, new);`	In the sequence, replace each value *old* with *new*
`reverse(begin, end);`	Reverse the order of the values in the sequence
`sort(begin, end);`	Sort the sequence into ascending order
`unique(begin, end);`	In the sequence, replace any consecutive occurrences of the same value with one instance of that value

Familiarity with the standard algorithms in C++ can save considerable time and effort and allow us to write functions that are more streamlined and efficient. The function in Example 12.5 and the example following it illustrate this. The function finds the mean of the values in a vector<double> using the `accumulate()` algorithm from the <numeric> library to sum the values in the vector instead of writing a loop to do this.

Example 12.5 Finding the Mean of a `vector<double>`

```
/* Function to find the mean value in a vector<double>.

   Receive: vec, a nonempty vector<double>
   Return:  the mean of the values in vec
   Note:    Must #include <numeric> to use accumulate()
   ----------------------------------------------------------------*/

double mean(const vector<double> & vec)
{
  if ( vec.empty() )
  {
    cerr << "\n***mean(vector): vector is empty!" << endl;
    return 0.0;
  }
  else
```

```
    return accumulate(vec.begin(), vec.end(), 0.0) / vec.size();
}
```

Now suppose the following vector<double> named ratings contains a sequence of ratings on a scale from 0 to 10 of a new product by a group of testers:

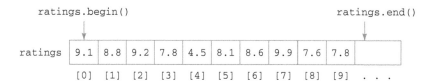

To reduce the effect of bias, the high score (9.9) and the low score (4.5) are to be thrown out and the mean of the remaining scores computed.

Eliminating the low score can be done in one step using STL's min_element() algorithm to position an iterator at the 4.5 rating and then pass this iterator to the vector<T> function member erase(),

```
    ratings.erase( min_element( ratings.begin(), ratings.end() ) );
```

The maximum score can be erased using the same approach with the max_element() algorithm:

```
    ratings.erase( max_element( ratings.begin(), ratings.end() ) );
```

The result is that the low and high ratings are erased from ratings, the remaining values are shifted to the left to fill their spaces, and ratings.size() and ratings.end() are appropriately updated:

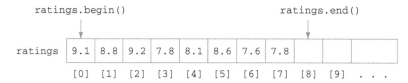

The resulting vector can then be passed to the function mean() in Example 12.5 to compute the average rating.

CHAPTER SUMMARY

Key Terms

algorithm	array components
array	array elements
array capacity	array literal

NOTES

- A simple form of an array declaration is *Type*[*CAPACITY*] name;, where *Type* is the type of elements in the array, *CAPACITY* is the number of elements the array can store, and *name* is the name of the array. The values in the array *name* will be undefined and will likely be "garbage" values. The capacity cannot be changed during program execution.

- An array can be initialized with an array literal: *Type*[*CAPACITY*] name = {*list*};, where *list* is a list of *Type* values that will be stored in *name*[0], *name*[1],.... If *list* has fewer than *CAPACITY* values, the remaining components of *name* will be filled with zeroes. If *CAPACITY* is omitted, the capacity of *name* will be the number of values in *list*.

- It is good programming practice to use an integer constant rather than an integer literal to specify the capacity of the array in an array declaration.

- C++ arrays are zero based; that is, the index of the first element is zero.

- The value at index *i* in an array named *name* can be accessed using the subscript operator in an expression of the form *name*[*i*]. A for loop whose loop-control variable is an array index is useful for implementing array operations, because it can vary this index in an indexed variable.

- For the standard C string-processing functions to work correctly with a `char` array, the capacity of the array should always be at least one more than the size of the largest string to be stored in the array so there is room for the null character that is used as an end-of-string mark.

- Arrays may be passed to functions via parameters and may also be returned by functions. They are always passed as reference parameters.

- Simple selection sort repeatedly selects one element of a list (e.g., the smallest) and correctly positions it. Linear insertion sort inserts list elements into an already-sorted sublist. Quicksort uses a divide-and-conquer approach to sort a list recursively and is one of the fastest sorting methods for large lists.

- Linear search can be used with any list. Binary search is faster than linear search, but can only be used with ordered lists.

- The `vector<T>` class template is an array-based container whose capacity can increase to accommodate new elements.

- The index of the first value of a `vector<T>` object *v* is 0; the index of the last value is `v.size()` − 1. The subscript operator `[]` can be used to access any of the elements in this range, but it should not be used to append values because it will not update the size of the `vector<T>` or cause the capacity to increase if the `vector<T>` is full. The `push_back()` method should always be used instead.

Style and Design Tips

- *C-style arrays,* `valarrays`, *and* `vector<T>`*s can be used to store sequences of values because the elements of an array,* `valarray`, *or* `vector<T>` *all have the same type.*

- *If a problem involves a sequence of unknown or varying length or requires the use of an operation that is predefined for a* `vector<T>`, *store the sequence in a* `vector<T>` *instead of in a C-style array or a* `valarray`. The `vector<T>` class template provides a standardized, variable-capacity, self-contained object for storing sequences of values, and STL provides many predefined `vector<T>` operations.

- *Do not reinvent the wheel.* When a problem requires an operation on a `vector<T>`, review the `vector<T>` function members and STL algorithms to see if the operation is already defined or if there are other operations that make yours easier to implement.

- *When using C-style arrays, always define their capacity using a constant, not a literal.* Such a constant can be used to control `for` loops, be passed to functions, and so on, which simplifies program maintenance if the array must be resized.

Warnings

1. *In C++, the subscript operator is a pair of square brackets, not a pair of parentheses.* An attempt to access element `i` of an array `A` by using `A(i)` will be interpreted by the

C++ compiler as a call to a function named A, passing it the argument i; a compile-time error will usually result.

2. *The first element of a C++ array,* `valarray`, *or* `vector<T>` *has the index value 0—not 1, as in some other programming languages.* Forgetting this can produce puzzling results.

3. *A character string literal is invisibly terminated with the NUL character* `'\0'`, *and a character array must leave room for this character.* Most of the standard operations for processing character arrays use the NUL character as an end-of-string mark. If a program mistakenly constructs a character array containing no terminating character or somehow overwrites the terminating character of a string with some nonnull value, the results are unpredictable, but can easily produce a run-time error.

4. *No checking is performed to ensure that array,* `valarray`, *or* `vector<T>` *indices stay within the range of valid indices.* As the program in Example 12.2 demonstrates, out-of-range indices can produce obscure errors whose source can be difficult to find.

5. *Array arguments are automatically passed by reference; thus, if a function changes an array parameter, the corresponding array argument will also be changed.* A function having an array parameter through which a value is being passed back to its caller should not declare the array as a reference parameter. If it has an array parameter that is being received but not returned, it should be declared as a `const` parameter.

6. *Always append new values to a* `vector<T>` *using* `push_back()`. The size, capacity, and iterators of a `vector<T>` are all correctly updated by `push_back()`. None of these are updated by the subscript operator, however, so it should not be used to append values to (or insert values beyond the size of) a `vector<T>`.

7. *When nesting STL templates, leave a space between the two `>` symbols.* A common mistake is to forget this and write

```
vector<vector<int>> myGrid;
```

to define `myGrid` as a vector of vectors. The compiler will read the `>>` as the output operator, and because this makes no sense in this context, a compilation error will result. The proper approach is to leave a space:

```
vector< vector<int> > myGrid;
```

TEST YOURSELF

Section 12.2

Questions 1–16 assume the following definitions:

```
double a[5],
       b[5] = {0},
       c[5] = {1},
```

```
            d[5]   =  {0,0,0,0,0},
            e[5]   =  {1,2,3,4,5};
char  f[5]  =  {'a',  'b'},
      g[5]  =  "abcde";
typedef int alpha[5];
alpha beta;
```

1. (True or false) a is an array indexed 0, 1, 2, 3, 4, 5.

2. (True or false) All elements of a are initialized to 0.

3. (True or false) All elements of b are initialized to 0.

4. (True or false) All elements of c are initialized to 1.

5. (True or false) The definition of d could be shortened to double d[5] = {0};.

6. (True or false) The definition of b could be shortened to double b[5] = 0;.

7. (True or false) e[3] == 3.

8. The capacity of a is _____.

9. (True or false) f[2] is the NUL character.

10. (True or false) The definition of f could also be written char f[5] = "ab";.

11. (True or false) g[5] contains an end-of-string mark.

12. (True or false) alpha is an array indexed 0, 1, 2, 3, 4.

13. (True or false) beta is an array indexed 0, 1, 2, 3, 4.

14. The address of a[0] is called the _____ address of a.

15. (True or false) The output produced by cout << e << endl; will be 12345.

16. (True or false) The output produced by cout << f << endl; will be ab.

17. (True or false) C-style arrays are self-contained.

18. Arrays are passed as _____ parameters.

For Questions 19–22, assume the following declarations:

```
int number[5]  =  {1};
typedef double Dubber[5];
Dubber xValue;
```

Tell what values will be assigned to the array elements.

19. ```
for (int i = 0; i <= 4; i++)
    xValue[i] = double(i) / 2.0;
```
```

20.
```
for (int i = 0; i < 5; i++)
    if (i % 2 == 0)
        number[i] = 2 * i;
    else
        number[i] = 2 * i + 1;
```

21.
```
for (int i = 1; i < 5; i++)
    number[i] = 2 * number[i - 1];
```

22.
```
for (int i = 3; i >= 0; i--)
    number[i] = 2 * number[i ı 1];
```

Section 12.6

Questions 1–15 assume that the following statements have been executed:

```
vector<int> a, b(5), c(5, 1), d(5);
d.push_back(77);
d.push_back(88);
```

1. The type of values stored in a is _____.

2. The capacity of a is _____ and its size is _____ .

3. The capacity of b is _____ and its size is _____ .

4. The capacity of c is _____ and its size is _____ .

5. The capacity of d is _____ and its size is _____ .

6. What output is produced by

```
cout << c.front() << ' ' << c.back() << endl;
```

7. What output is produced by

```
cout << d.front() << ' ' << d.back() << endl;
```

8. (True or false) a.empty().

9. (True or false) c < d.

10. (True or false) c[1] == 1.

11. What output is produced by

```
for (int i = 0; i < c.size(); i++)
    cout << c[i] << ' ';
```

12. What output is produced by

```
d.pop_back();
for (int i = 0; i < d.size(); i++)
    cout << d[i] << ' ';
```

13. `d.begin()` returns an iterator positioned at _____ in d.

14. `d.end()` returns an iterator positioned at _____ .

15. (True or false) `vector<T>` objects are self-contained.

For Questions 16–19, assume the declarations

```
vector<double> xValue(5, 0);
vector<int> number(5, 1);
```

Describe the contents of the `vector<T>` after the statements are executed.

16.
```
for (int i = 0; i <= 4; i++)
    xValue.push_back(double(i) / 2.0);
```

17.
```
for (int i = 0; i < 5; i++)
    if (i % 2 == 0)
        number.push_back(2 * i);
    else
        number.push_back(2 * i + 1);
```

18.
```
for (int i = 1; i < 5; i++)
    number.push_back(2 * number[i - 1]);
```

19.
```
for (int i = 1; i <= 3; i++)
    number.pop_back();
for (int i = 1; i <= 3; i++)
    number.push_back(2);
```

20. When, where, and by whom was the Standard Template Library (STL) developed?

EXERCISES

Section 12.2

For Exercises 1–8, assume that the following declarations have been made:

```
const int LITTLE = 6,
          MEDIUM = 10,
          BIG = 128;
int i, j, n = 10,
    temp,
    number[MEDIUM] = {99, 33, 44, 88, 22, 11, 55, 111, 66, 77};
char ch,
     letterCount[BIG];
```

```
typedef double LittleDouble[LITTLE];
LittleDouble value;
```

Tell what value (if any) will be assigned to each array element, or explain why an error occurs:

1.
```
for (i = 0; i < MEDIUM; i++)
   number[i] = i / 2;
```

2.
```
for (i = 0; i < LITTLE; i++)
   number[i] = i * i;
for (i = LITTLE; i < MEDIUM; i++)
   number[i] = number[i - 5];
```

3.
```
for (i = 0; i < 3; i++)
   value[i] = 0;
for (i = 3; i < LITTLE; i++)
   value[i] = 1;
```

4.
```
for (i = 1; i < LITTLE; i += 2)
{
   value[i - 1] = double(i) / 2.0;
   value[i] = 10.0 * value[i - 1];
}
```

5.
```
i = 0;
while (i != 10 )
{
   if (i % 3 == 0)
      number[i] = 0;
   else
      number[i] = i;
   i++;
}
```

6.
```
number[1] = 1;
i = 2;
do
{
   number[i] = 2 * number[i - 1];
   i++;
}
while (i < MEDIUM);
```

```
7. for (ch = 'A'; ch <= 'F'; ch++)
     if (ch == 'A')
       letterCount[ch] = 1;
     else
       letterCount[ch] = letterCount[ch - 1] + 1;

8. for (i = 0; i < n - 1; i++)
     for (j = 0; j < n - i - 1; j++)
       if (number[j] > number[j + 1])
       {
         temp = number[j];
         number[j] = number[j + 1];
         number[j + 1] = temp;
       }
```

For Exercises 9–14, write definitions of the given arrays in two ways: (a) without using `typedef` and (b) using `typedef`.

9. An array with capacity 10 in which each element is an integer.

10. An array whose indices are integers from 0 through 10 and in which each element is a real value.

11. An array with capacity 10 in which each element is an integer, all of which are initially 0.

12. An array that can store five `strings`.

13. An array that can store five characters and is initialized with the vowels a, e, i, o, and u.

14. An array that can store 100 values, each of which is either `true` or `false`.

For Exercises 15–18, write definitions and statements to construct the given array.

15. An array whose indices are the integers from 0 through 99 and in which the value stored in each element is the same as the index.

16. An array whose indices are the integers from 0 through 99 and in which the values stored in the elements are the indices in reverse order.

17. An array of capacity 50 in which the value stored in an element is true if the corresponding index is even and false otherwise.

18. An array whose indices are the decimal ASCII values (0-127), such that the value stored in an element is `true` if the index is that of a vowel, and `false` otherwise.

Exercises 19–24 ask you to write functions to do various things. Also, include any `typedef` declarations that are needed. To test these functions, you should write driver programs as instructed in Programming Problems 1–6 at the end of this chapter.

19. Return the smallest value stored in an array of integers.

20. Return the largest value stored in an array of integers.

21. Return the range of values stored in an array of integers, that is, the difference between the largest value and the smallest value.

22. Return `true` if the values stored in an array are in ascending order and `false` otherwise.

23. Insert a value into an array of integers at a specified position in the array.

24. Remove a value from an array of integers at a specified position in the array.

Section 12.3

For each of the arrays x in Exercises 1–4, show x after each of the first four passes of simple selection sort.

1.

i	0	1	2	3	4	5	6	7
$x[i]$	30	50	80	10	60	20	70	40

2.

i	0	1	2	3	4	5	6	7
$x[i]$	20	40	70	60	80	50	30	10

3.

i	0	1	2	3	4	5	6	7
$x[i]$	80	70	60	50	40	30	20	10

4.

i	0	1	2	3	4	5	6	7
$x[i]$	10	20	30	40	50	60	70	80

5. One variation of simple selection sort for a list stored in an array $x[0],..., x[n-1]$ is to locate both the smallest and the largest elements while scanning the list and to position them at the beginning and the end of the list, respectively. On the next scan, this process is repeated for the sublist $x[1],..., x[n-2]$, and so on. Write an algorithm to implement this *double-ended selection sort*.

6.–9. For the arrays x in Exercises 1–4 show x after each pass of the double-ended selection sort described in Exercise 5.

Section 12.6

For Exercises 1–10, assume that the following declarations have been made,

```
vector<int> number,
            v(10,20),
            w(10);
int num;
```

and that for exercises that involve input, the following values are entered:

```
99 33 44 88 22 11 55 66 77 -1
```

Describe the contents of the given vector<T> after the statements are executed.

1. ```
for (int i = 0; i < 10; i++)
    number.push_back(i / 2);
```

2. ```
for (int i = 0; i < 6; i++)
    w.push_back(i / 2);
```

3. ```
for (;;)
{
    cin >> num;
    if (num < 0) break;
    number.push_back(num);
}
```

4. ```
for (int i = 0; i <= 5; i++)
    number.push_back(i);
for (int i = 0; i < 2; i++)
    number.pop_back();
for (int i = 0; i <= 5; i++)
    number.push_back(i);
```

For Exercises 5–10 assume that the loop in Exercise 3 has been executed.

5. ```
for (int i = 0; i < number.size() - 1; i += 2)
    number[i] = number[i + 1];
```

6. ```
number.pop _ back();
number.push_back(number.front());
```

7. ```
int temp = number.front();
number.front() = number.back();
number.back() = temp;
```

8. ```
sort(number.begin(), number.end());
```

9. ```
for (int i = 0; i < number.size(); i++)
    w.push_back(number[i] + v[i]);
```
```

```
10. while  (v < number)
    {
      v.erase(v.begin());
      number.erase(number.begin());
    }
```

For Exercises 11–15 write a declaration for a `vector<T>` having the given properties.

11. Can store `long int` values.

12. Capacity 10 and each element is a `long int`.

13. Capacity 10 and each element is a `long int`, all of which are initially 0.

14. Capacity 5, size 5, and each element contains a string, initially `"xxx"`.

15. Capacity 100 and each element is either `true` or `false`.

For Exercises 16–18, write definitions and statements to construct a `vector<T>` with the required properties.

16. Stores the sequence of integers from 0 through 99.

17. Stores the sequence of integers from 0 through 99 in reverse order.

18. Has capacity 50, and the value stored in an element is true if the corresponding index is even and is false otherwise.

Exercises 19–25 ask you to write functions to do various things. To test these functions, you should write driver programs as instructed in Programming Problems 1–3 of Section 12.6.

19. Returns `true` if the values stored in a `vector<double>` are in ascending order and `false` otherwise.

20. Finds the range of values stored in a `vector<double>`, that is, the difference between the largest value and the smallest value.

Exercises 21–25 deal with operations on *n-dimensional vectors*, which are sequences of *n* real numbers and which are studied and used in many areas of mathematics and science. They can obviously be modeled in C++ by `vector<double>`s of capacity *n*. In the description of each operation, *A* and *B* are assumed to be *n*-dimensional vectors:

$$A = (a_1, a_2, \ldots, a_n)$$

$$B = (b_1, b_2, \ldots, b_n)$$

21. Compute and return the sum of two n-dimensional vectors:

$$A + B = (a_1 + b_1, a_2 + b_2, \ldots, a_n + b_n)$$

22. Compute and return the difference of two n-dimensional vectors:

$$A - B = (a_1 - b_1, a_2 - b_2, \ldots, a_n - b_n)$$

23. Compute and return the product of a scalar (real number) and an n-dimensional vector:

$$cA = (ca_1, ca_2, \ldots, ca_n)$$

24. Compute and return the magnitude of an n-dimensional vector:

$$|A| = \sqrt{a_1^2 + a_2^2 + \cdots + a_n^2}$$

25. Compute and return the inner (or dot) product of two n-dimensional vectors (which is a scalar):

$$A \cdot B = a_1 \times b_1 + a_2 \times b_2 + \cdots a_n \times b_n = \sum_{i=1}^{n} (a_i \times b_i)$$

PROGRAMMING PROBLEMS

Sections 12.2

1. Write a driver program to test the smallest-element function of Exercise 19.

2. Write a driver program to test the largest-element function of Exercise 20.

3. Write a driver program to test the range function of Exercise 21.

4. Write a driver program to test the ascending-order function of Exercise 22.

5. Write a driver program to test the insert function of Exercise 23.

6. Write a driver program to test the remove function of Exercise 24.

7. The Rinky Dooflingy Company records the number of cases of dooflingies produced each day over a four-week period. Write a program that reads these production numbers and stores them in an array. The program should then accept from the user a week number and a day number, and should display the production level for that day. Assume that each week consists of five workdays.

8. The Rinky Dooflingy Company maintains two warehouses, one in Chicago and one in Detroit, each of which stocks at most 25 different items. Write a program that first

reads the product numbers of items stored in the Chicago warehouse and stores them in an array `chicago`, and then repeats this for the items stored in the Detroit warehouse, storing these product numbers in an array `detroit`. The program should then find and display the *intersection* of these two lists of numbers, that is, the collection of product numbers common to both sequences. The lists should not be assumed to have the same number of elements.

9. Repeat Problem 8 but find and display the *union* of the two lists, that is, the collection of product numbers that are elements of at least one of the sequences of numbers.

10. Suppose that a row of mailboxes are numbered 1 through 150 and that, beginning with mailbox 2, we open the doors of all the even-numbered mailboxes. Next, beginning with mailbox 3, we go to every third mailbox, opening its door if it is closed and closing it if it is open. We repeat this procedure with every fourth mailbox, then every fifth mailbox, and so on. Using an array to model the mailboxes, write a program to determine which mailboxes will be closed when this procedure is completed.

11. If $\bar{x}$ denotes the mean of a sequence of numbers $x_1, x_2, \ldots, x_n$, the *variance* is the average of the squares of the deviations of the numbers from the mean,

$$variance = \frac{1}{n} \sum_{i=1}^{n} (x_i - \bar{x})^2$$

and the *standard deviation* is the square root of the variance. Write functions to calculate the mean, variance, and standard deviation of the values stored in an array, and a driver program to test your functions.

12. Write a program that reads a list of real numbers representing numeric scores, stores them in an array, calls the functions from Problem 11 to calculate their mean and standard deviation, and then calls another function to display the letter grade corresponding to each numeric score as determined using the grading-on-the-curve method described in Problem 2 of Section 11.3 at the end of Chapter 11.

13. A prime number is an integer greater than 1 whose only positive divisors are 1 and the integer itself. The Greek mathematician Eratosthenes developed an algorithm, known as the *Sieve of Eratosthenes*, for finding all prime numbers less than or equal to a given number n, that is, all primes in the range 2 through n. Consider the list of numbers from 2 through n. Two is the first prime number, but the multiples of 2 (4, 6, 8, ...) are not, and so they are crossed out in the list. The first number after 2 that was not crossed out is 3, the next prime. We then cross out from the list all higher multiples of 3 (6, 9, 12, ...). The next number not crossed out is 5, the next prime, and so we cross out all higher multiples of 5 (10, 15, 20, ...). We repeat this procedure until we reach the first number in the list that has not been

crossed out and whose square is greater than n. All the numbers that remain in the list are the primes from 2 through n. Write a program that uses this sieve method and an array to find all the prime numbers from 2 through n. Run it for $n = 550$ and for $n = 5500$.

Section 12.3

1. Write and test a function for simple selection sort.

2. Write and test a function for double-ended selection sort described in Exercise 5.

3. *Linear insertion sort* is based on the idea of repeatedly inserting a new element into a list of already-sorted elements so that the resulting list is still sorted. The following sequence of diagrams demonstrates this method for the list 67, 33, 21, 84, 49, 50, 75. The sorted sublist produced at each stage is highlighted.

67 ,	33 ,	21 ,	84 ,	49 ,	50 ,	75	Initial sorted sublist of 1 element
33 ,	67 ,	21 ,	84 ,	49 ,	50 ,	75	Insert 33 to get 2-element sorted sublist
21 ,	33 ,	67 ,	84 ,	49 ,	50 ,	75	Insert 21 to get 3-element sorted sublist
21 ,	33 ,	67 ,	84 ,	49 ,	50 ,	75	Insert 84 to get 4-element sorted sublist
21 ,	33 ,	49 ,	67 ,	84 ,	50 ,	75	Insert 49 to get 5-element sorted sublist
21 ,	33 ,	49 ,	50 ,	67 ,	84 ,	75	Insert 50 to get 6-element sorted sublist
21 ,	33 ,	49 ,	50 ,	67 ,	75 ,	84	Insert 75 to get 7-element sorted sublist

Write and test a function for linear insertion sort.

4. The investment company of Pickum & Loozem has been recording the trading price of a particular stock over a 15-day period. Write a program that reads these prices and sorts them into increasing order, using one of the sort methods in Problems 1–3. The program should display the trading range, that is, the lowest and the highest prices recorded and the *median* price—the middle price in the sorted list if there are an odd number of prices, otherwise the average of the two middle prices.

Section 12.5

1. Write and test a function for linear search.

2. Write and test a function for binary search.

3. The Rinky Dooflingy Company manufactures different kinds of dooflingies, each identified by a product number. Write a program that reads product numbers and prices, and stores these values in two arrays, `number` and `price`; `number[0]` and `price[0]` are the product number and unit price for the first item, `number[1]` and `price[1]` are the product number and unit price for the second item, and so on. The program should then allow the user to select one of the following options:

a. Retrieve and display the price of a product whose number is entered by the user. (Use the linear search procedure developed in Problem 1 to determine the index of the specified item in the array `number`.)

b. Print a table displaying the product number and the price of each item.

Section 12.6

1. Write a driver program to test the ascending-order function of Exercise 19.

2. Write a driver program to test the range function of Exercise 20.

3. Write a menu-driven calculator program that allows a user to repeatedly select and perform one of the operations on n-dimensional vectors in Exercises 21–25.

4.–10. Proceed as in Problems 7–13 of Section 12.2, respectively, but use `vector<T>`s instead of arrays.

11. Proceed as in Problem 4 of Section 12.3 for finding the range of stock prices, but store the prices in a `vector<T>` and use STL's `sort` algorithm to sort the prices.

12. Proceed as in Problem 3 of Section 12.5 for processing product numbers and prices, but use `vector<T>`s instead of arrays.

13. Write a function to evaluate a polynomial $a_0 + a_1x + a_2x^2 + \ldots a_nx^n$ for any degree n, coefficients $a_0, a_1, a_2, \ldots, a_n$, and values of x supplied to it as arguments. Then write a program that reads a polynomial's coefficients and various values of x and uses the function to evaluate the polynomial at these values.

14. A more efficient way of evaluating polynomials is *Horner's method* (also known as *nested multiplication*), in which a polynomial $a_0 + a_1x + a_2x^2 + \ldots a_nx^n$ is rewritten as

$$a_0 + (a_1 + (a_2 + \ldots (a_{n-1} + a_nx)x) \ldots x)x$$

For example:

$$7 + 6x + 5x^2 + 4x^3 + 3x^3 = 7 + (6 + (5 + (4 + 3x)x)x)x$$

Proceed as in Problem 13, but use Horner's method to evaluate the polynomial.

15. Write a function to perform addition of large integers, for which there is no limit on the number of digits. (*Suggestion*: Treat each number as a sequence, each of whose elements is a block of digits of the number. For example, the integer 179,534,672,198 might be stored with `block[0]` = 198, `block[1]` = 672, `block[2]` = 534, and `block[3]` = 179. Then add the integers [lists] element by element, carrying from one element to the next when necessary.) Write a driver program to test your function.

16. Proceed as in Problem 15, but for subtraction of large integers.

17. Proceed as in Problem 15, but for multiplication of large integers.

18. Proceed as in Problem 15, but for division of large integers.

19. Write a big-integer calculator program that allows the user to enter two large integers and the operation to be performed, and which calls the appropriate function from Problems 15–18 to carry out that operation.

Multidimensional Arrays and Vectors

We must assume behind this force the existence of a conscious and intelligent Mind. This Mind is the matrix of all matter.

MAX PLANCK

Painting does what we cannot do—it brings a three-dimensional world into a two-dimensional plane.

CHUCK JONES

We are columns left alone of a temple once complete.

CHRISTOPHER CRANCH

Everyone knows how laborious the usual Method is of attaining to Arts and Sciences; whereas by his Contrivance, the most ignorant Person at a reasonable Charge, and with a little bodily Labour, may write Books ... He then led me to the Frame, about the Sides whereof all his Pupils stood in Ranks. It was Twenty Foot square ... linked by slender Wires. These Bits ... were covered on every Square with Paper pasted upon them; and on These Papers were written all the Words of their Language ...

JONATHAN SWIFT, GULLIVER'S TRAVELS

I N THE PREVIOUS CHAPTER we introduced C-style arrays and vector<T> class templates and used them to store sequences of values. Each of the containers considered in that chapter had *one dimension*: its *length*, which is the number of values in the sequence.

C++ also allows arrays and vectors of more than one dimension. As we shall see, a *two-dimensional* array or vector can be used to store a data set whose values are arranged in *rows* and *columns*. Similarly, a *three-dimensional* array or vector is an appropriate storage structure when the data can be arranged in *rows*, *columns*, and *ranks*. When there are several characteristics associated with the data, still higher dimensions may be useful, with each dimension corresponding to one of these characteristics. In this chapter we show how to use such multidimensional arrays in C++ programs.

13.1 INTRODUCTORY EXAMPLE: A MILEAGE CHART

13.1.1 Problem

A medical supply company services hospitals, laboratories, and other medical research facilities over a large area. To aid with route-planning and estimating travel time and costs, we need to develop an online program/application that looks up and displays the mileage between any two cities where these facilities are located.

The basic idea is to create a software representation of a mileage chart, and then use it to look up the distance between any two of the cities. For this example, we will consider only six cities, but the program can be easily modified for a larger number. The mileage chart we will use is the following:

	Astroburg	Bedrock	Dogpatch	Gotham City	Metropolis	Mudville
Astroburg	0	97	90	268	262	130
Bedrock	97	0	74	337	144	128
Dogpatch	90	74	0	354	174	201
Gotham City	268	337	354	0	475	269
Metropolis	262	144	174	475	0	238
Mudville	130	128	201	269	238	0

13.1.2 Object-Centered Design
13.1.2.1 Behavior

For simplicity, our program will begin by displaying on the screen a numbered menu of the cities. It should then read the numbers of two cities from the keyboard. Next, it should look up the mileage between those cities in a software mileage chart. Finally, it should display that mileage.

From the behavior, we identify the following objects in this problem:

| | Software Objects | | |
Problem Objects	Type	Kind	Name
A menu of cities	`string`	constant	none
The numbers of two cities	`int`	variable	*city1, city2*
A mileage chart	`int [] []`	constant	*mileageChart*
The mileage	`int`	variable	*mileage*

As we will see, the type `int[][]` refers to a two-dimensional array of integers, which provides a convenient way to represent the mileage chart.

13.1.2.2 Operations
Our behavioral description gives the following set of operations:

   i. Declare a two-dimensional array with initial values.

  ii. Display a string on the screen.

 iii. Read two integers from the keyboard.

 iv. Look up an entry in a two-dimensional array.

  v. Output an integer.

13.1.2.3 Algorithm
These operations are easily organized into the following algorithm:

1. Declare *mileageChart*, a two-dimensional array of city mileages.

2. Via `cout`, display a menu of the cities.

3. From `cin`, read two integers into *city1* and *city2*.

4. Compute mileage, by looking up *mileageChart*[*city1*][*city2*].

5. Via `cout`, display *mileage*.

13.1.2.4 Coding
The preceding algorithm is easily encoded in C++, as shown in Example 13.1.

Example 13.1 A Mileage Calculator

```
/* This program computes the mileage between two cities.

   Input: city1 and city2, two integers in the range 0..n-1 that
          represent cities
```

```
        Output: the mileage between city1 and city2
-----------------------------------------------------------*/


#include <iostream>            // cin, cout, >>, <<
#include <cassert>             // assert()
using namespace std;


int main()
{
 const int NUMBER_OF_CITIES = 6;
 int mileageChart[NUMBER_OF_CITIES][NUMBER_OF_CITIES]
        = { {   0,  97,  90, 268, 262, 130 },      // Astroburg
            {  97,   0,  74, 337, 144, 128 },      // Bedrock
            {  90,  74,   0, 354, 174, 201 },      // Dogpatch
            { 268, 337, 354,   0, 475, 269 },      // Gotham City
            { 262, 144, 174, 475,   0, 238 },      // Metropolis
            { 130, 128, 201, 269, 238,   0 } };    // Mudville


   cout << "To determine the mileage between two cities,\n"
           " enter the numbers of 2 cities from this menu:\n\n"
           "0 for Astroburg, 1 for Bedrock\n"
           "2 for Dogpatch, 3 for Gotham City\n"
           "4 for Metropolis, 5 for Mudville\n\n"
           "--> ";


   int city1, city2;
   cin >> city1 >> city2;


   int mileage = mileageChart[city1][city2];


   cout << "\nThe mileage between those 2 cities is "
        << mileage << " miles.\n";
}
```

SAMPLE RUN:

```
To determine the mileage between two cities,
please enter the numbers of 2 cities from this menu:

0 for Astroburg,  1 for Bedrock
2 for Dogpatch,   3 for Gotham City
4 for Metropolis, 5 for Mudville


-->2 5


The mileage between those 2 cities is 201 miles.
```

13.2 MULTIDIMENSIONAL ARRAYS

There are many problems, such as the mileage problem in the preceding section, in which the data being processed can be naturally organized as a *table*. For these problems, two-dimensional arrays provide a way to build a software model of a table.

The program in Example 13.1 illustrates how a two-dimensional array can be declared and initialized. The statements

```
const int NUMBER_OF_CITIES = 6;
int mileageChart[NUMBER_OF_CITIES][NUMBER_OF_CITIES]
    = { {   0,  97,  90, 268, 262, 130 },    // Astroburg
        {  97,   0,  74, 337, 144, 128 },    // Bedrock
        {  90,  74,   0, 354, 174, 201 },    // Dogpatch
        { 268, 337, 354,   0, 475, 269 },    // Gotham City
        { 262, 144, 174, 475,   0, 238 },    // Metropolis
        { 130, 128, 201, 269, 238,   0 } };  // Mudville
```

declare the object `mileageChart` as a two-dimensional array of integers with six rows and six columns, which we might visualize as follows:

	[0]	[1]	[2]	[3]	[4]	[5]
[0]	0	97	90	268	262	130
[1]	97	0	74	337	144	128
[2]	90	74	0	354	174	201
[3]	268	337	354	0	475	269
[4]	262	144	174	475	0	238
[5]	130	128	201	269	238	0

Although there are several pairs of curly braces in the list of values used to initialize this array, only the outermost pair is required. We didn't have to enclose the values for each row in their own pair of curly braces, but this is often done to make the declaration more readable by delimiting the values of each row.

As with one-dimensional arrays, each dimension of a two-dimensional array is indexed starting with zero, so the six rows in our example are indexed from zero to five as are the six columns. Also, the **subscript operation** used to access the elements of a one-dimensional array is once again the central predefined operation for two-dimensional arrays. The difference is that where one-dimensional arrays use a single subscript operator to access an element, two-dimensional arrays like `mileageChart` are two-dimensional objects and require two subscript operators, one for each dimension. The element in row 0, column 0, of `mileageChart` can be accessed using

```
mileageChart[0][0]
```

The element of `mileageChart` in the second column of the first row can be accessed using

```
mileageChart[0][1]
```

the first element of the third row using

```
mileageChart[2][0]
```

and the element at row 4, column 3 using

```
mileageChart[4][3]
```

and so on. In general, the notation

```
mileageChart[r][c]
```

can be used to access the value at row `r` and column `c`. The program in Example 13.1 looked up the mileage between `city1` and `city2` by accessing the element at row `city1` and column `city2`:

```
int mileage = mileageChart[city1][city2];
```

Two-dimensional arrays like `mileageChart` that have the same number of rows as columns are called *square* arrays. But nonsquare arrays are needed for some problems. For example, suppose that four times a day, water temperatures are recorded at each of three discharge outlets of the cooling system of a nuclear power plant. These temperature readings can be arranged in a table having four rows and three columns:

Time	Location		
	Outlet-A	Outlet-B	Outlet-C
12 A.M.	65.5	68.7	62.0
6 A.M.	68.8	68.9	64.5
12 P.M.	70.4	69.4	66.3
6 P.M.	68.5	69.1	65.8

In this table, the three temperature readings at 12 A.M. are in the first row, the three temperatures at 6 A.M. are in the second row, and so on. We can model such a table with a two-dimensional array having four rows and three columns:

```
const int NUM_TIMES = 4, NUM_OUTLETS = 3;
double temperatureGrid[NUM_TIMES][NUM_OUTLETS];
```

The C++ compiler reserves 12 memory locations for this array, which we might picture as follows:

```
temperatureGrid:     [0]        [1]        [2]
             [0]  ┌──────────┬──────────┬──────────┐
                  │          │          │          │
             [1]  ├──────────┼──────────┼──────────┤
                  │          │          │          │
             [2]  ├──────────┼──────────┼──────────┤
                  │          │          │          │
             [3]  ├──────────┼──────────┼──────────┤
                  │          │          │          │
                  └──────────┴──────────┴──────────┘
```

The arrays `milageChart` and `temperatureGrid` have numeric elements—integers and reals. But this is not a requirement; the elements of an array can be of any type. For example, consider a window on a computer screen containing 24 lines with 80 characters per line:

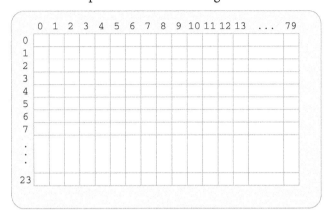

If we number the rows from 0 through 23 and the columns from 0 through 79 with the position of row 0 and column 0 in the upper left corner of the screen as illustrated, we can model such a window using a two-dimensional array of characters, declared as follows:

```
const int ROWS = 24, COLUMNS = 80;
char window[ROWS][COLUMNS];
```

The `typedef` mechanism introduced in Chapter 12 to define an identifier as a synonym for a one-dimensional array type can also be used for multidimensional arrays. For example, to declare the identifier `MileageArray` as a synonym for an array of double values representing mileages, we would write

```
const int NUMBER_OF_CITIES = 6;
typedef double MileageArray[NUM_CITIES][NUM_CITIES];
```

We can then use this new type to declare the two-dimensional array `mileageChart`:

```
MileageArray mileageChart
    = { {   0,  97,  90, 268, 262, 130 },      // Astroburg
        {  97,   0,  74, 337, 144, 128 },      // Bedrock
        {  90,  74,   0, 354, 174, 201 },      // Dogpatch
        { 268, 337, 354,   0, 475, 269 },      // Gotham City
        { 262, 144, 174, 475,   0, 238 },      // Metropolis
        { 130, 128, 201, 269, 238,   0 } };    // Mudville
```

Similarly, we might define the type `ComputerWindow` by

```
const int ROWS = 24, COLUMNS = 80;
typedef char ComputerWindow[ROWS][COLUMNS];
```

and then use this new type to declare the variable `window` by

```
ComputerWindow window;
```

13.2.1 Defining Two-Dimensional Array Operations

For operations other than subscript on a two-dimensional array, we must write functions to perform them. Operations on one-dimensional arrays typically use a for loop to count through the index values:

```
for (int i = 0; i < numberOfValues; i++)
// ... do something with oneDimensionalArray[i]
```

Operations that access the values stored in a two-dimensional array use two nested `for` loops: an outer loop counting through the rows, and an inner loop counting through the columns:[1]

```
for (int row = 0; row < numberOfRows; row++)
   for (int col = 0; col < numberOfColumns; col++)
      // ... do something with twoDimensionalArray[row][col]
```

For example, to clear the object `window`, we can use a function like the following:

```
void clearWindow(ComputerWindow theWindow, int numRows, int numColumns)
{
   for (int row = 0; row < numRows; row++)
      for (int col = 0; col < numColumns; col++)
         theWindow[row][col] = ' ';
}
```

To output `window` via `cout`, we could use

```
// Need <iomanip> for setw()
void display(ComputerWindow theWindow, int numRows, int numColumns)
{
   for (int row = 0; row < numRows; row++)
   {
      for (int col = 0; col < numColumns; col++)
         cout << setw(6) << theWindow[row][col];
      cout << endl;
   }
}
```

Each pass through the inner loop displays all the characters in one row of `theWindow` and then moves to a new line before displaying the characters in the next row.

---

[1] In some problems, columnwise-processing in which the outer loop counts through the columns and the inner loop through the rows may be used.

13.2.2 Higher-Dimensional Arrays

To illustrate the use of an array with more than two dimensions, consider again the collection of water temperatures recorded four times a day at each of three discharge outlets of the cooling system of a nuclear power plant.

Time	Location		
	Outlet-A	Outlet-B	Outlet-C
12 A.M.	65.5	68.7	62.0
6 A.M.	68.8	68.9	64.5
12 P.M.	70.4	69.4	66.3
6 P.M.	68.5	69.1	65.8

Suppose now that these temperatures are recorded for 1 week, so that seven such tables are collected:

Time	Location			
	Outlet-A	Outlet-B	Outlet-C	
12 A.M.	66.5	69.4	68.4	Saturday
6 A.M.	68.4	71.2	69.3	
12 P.M.	70.1	71.9	70.2	
6 P.M.	69.5	70.0	69.4	

∴

Time	Location			
	Outlet-A	Outlet-B	Outlet-C	
12 A.M.	63.7	66.2	64.3	Monday
6 A.M.	64.0	66.8	64.9	

Time	Location			
	Outlet-A	Outlet-B	Outlet-C	66.3
12 A.M.	65.5	68.7	62.0	65.8
6 A.M.	68.8	68.9	64.5	
12 P.M.	70.4	69.4	66.3	Sunday
6 P.M.	68.5	69.1	65.8	

The collection of these tables can be modeled with a three-dimensional array object, declared by

```
const int NUM_DAYS = 7, NUM_TIMES = 4, NUM_OUTLETS = 3;

typedef double
    ThreeDimTemperatureArray[NUM_DAYS][NUM_TIMES][NUM_OUTLETS];

ThreeDimTemperatureArray temperature;
```

The object `temperature` can then be used to store these 84 temperature readings.

A single subscript is used to access elements in a one-dimensional array and two subscripts are used to access elements in a two-dimensional array, so it seems reasonable that three subscripts are needed to access an element in a three-dimensional array. The first subscript is the day (0, 1, . . ., 6), the second subscript is the time (0, 1, 2, 3), and the third is the location (0, 1, 2); for example,

```
temperature[6][3][1]
```

is the temperature recorded on Saturday at 6 P.M. at Outlet-B.

Three nested loops are needed to run through the elements of a three-dimensional array. For example, if a file contains a week's 84 temperature readings, the following function can be used to read the values from that file into `temperature`:

```
void read(ifstream & in, ThreeDimTemperatureArray temperature,
          int numDays, int numTimes, int numOutlets)
{
   for (int day = 0; day < numDays; day++)
      for (int time = 0; time < numTimes; time++)
         for (int outlet = 0; outlet < numOutlets; outlet++)
            in >> temperature[day][time][outlet];
}
```

In general, n-dimensional arrays can be defined and subscript operators can be used to access the array elements. C++ places no limit on the number of dimensions of an array, but the number of values in each dimension must be specified. The general form of an array declaration is as follows:

ARRAY DECLARATION

FORM:

```
ElementType arrayName[DIM₁][DIM₂] . . . [DIMₙ];
```

where:

> `ElementType` is any known type;
> `arrayName` is the name of the array being declared; and
> each DIM_i must be a nonnegative integer (constant) value.

PURPOSE:

Defines an n-dimensional object whose elements are of type `ElementType`, in which DIM_1, DIM_2, . . . , DIM_n are the number of elements in each dimension.

13.2.2.1 Array of Arrays Declarations

One way to view a multidimensional array is as an **array of arrays**, that is, an array whose elements are other arrays. For example, consider the two-dimensional mileage chart described earlier:

```
double mileageChart[NUM_CITIES][NUM_CITIES];
```

Because `NUM_CITIES` is 6, this table can be thought of as a one-dimensional array, whose six elements are its rows:

mileageChart:

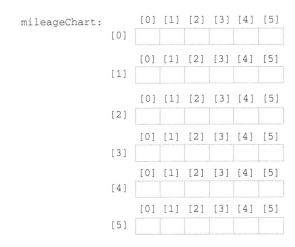

Of course, each row in mileageChart is itself a one-dimensional array of six real values:

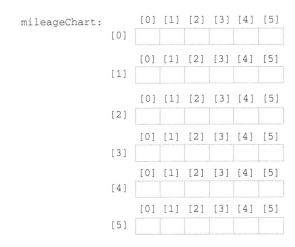

A table can thus be viewed as a one-dimensional array whose components are also one-dimensional arrays.

C++ allows array declarations to be given in a form that reflects this perspective. To illustrate this, consider again our table of temperature readings. Suppose we first declare a type identifier TemperatureList to represent one set of readings at the three cooling plant outlets at a particular time:

```
const int NUM_OUTLETS = 3;
typedef double TemperatureList[NUM_OUTLETS];
```

We can then use this new type to declare a second type TemperatureTable as an array whose elements are TemperatureList objects:

```
const int NUM_TIMES = 4;
typedef TemperatureList TemperatureTable[NUM_TIMES];
```

This declares the name TemperatureTable as a new type, whose objects are two-dimensional arrays of double values.

The resulting type can then be used to define a two-dimensional array object temperatureGrid, as before:

```
TemperatureTable temperatureGrid;
```

Regardless of which approach is used, the notation temperatureGrid[i] refers to the temperatures in row i of the table and temperatureGrid[i][j] refers to the temperature in this row that is in column j.

This idea can be extended to higher-dimensional arrays. For example, the three-dimensional array of temperatures considered earlier can also be thought of as an array of arrays. In particular, because one temperature table was recorded for each day, the entire three-dimensional array can be viewed as an array of temperature tables, meaning a one-dimensional array whose components are two-dimensional arrays. If we adopt this point of view, we might declare the three-dimensional array type ThreeDimTemperatureArray by adding the following declarations to the earlier ones:

```
const int NUM_DAYS = 7;
typedef TemperatureTable ThreeDimTemperatureArray[NUM_DAYS];
```

This may make it clearer that the notation

```
temperature[6]
```

refers to the entire temperature table that was recorded on Saturday; that is, temperature[6] is the two-dimensional array corresponding to the following temperature table:

Time	Location		
	Outlet-A	Outlet-B	Outlet-C
12 A.M.	66.5	69.4	68.4
6 A.M.	68.4	71.2	69.3
12 P.M.	70.1	71.9	70.2
6 P.M.	69.5	70.0	69.4

As we noted earlier, each such table can be viewed as a one-dimensional array of the temperature arrays. Thus, the doubly indexed expression

```
temperature[6][3]
```

refers to the bottom row in this temperature table,

69.5	70.0	69.4

and the triply indexed expression

```
temperature[6][3][2]
```

accesses the last temperature in this row:

69.4

13.3 APPLICATION: OCEANOGRAPHIC DATA ANALYSIS

A petroleum exploration company has collected some depth readings for a rather shallow square section of the ocean. The diagonal of this square (from the upper left corner to the lower right corner) is parallel to the equator. The company has divided the square into a grid with each intersection point (node) of the grid separated by five miles. The entire square is 50 miles on each side. Two separate crews did exploratory drilling in this area, one in the northern half (above the diagonal) and the other in the southern half. A program is to be written to find the approximate average ocean depth for each crew and the overall average for the entire square. The following depth data (in meters) was collected by the crews:

301.3	304.5	312.6	312.0	325.6	302.0	299.8	297.6	304.6	314.7	326.8
287.6	294.5	302.4	315.6	320.9	315.7	300.2	312.7	308.7	324.5	322.8
320.8	342.5	342.5	323.5	333.7	341.6	350.5	367.7	354.2	342.8	330.9
312.6	312.0	325.6	301.3	304.5	302.0	314.7	326.8	299.8	297.6	304.6
302.4	308.7	324.5	315.6	287.6	294.5	320.9	315.7	300.2	312.7	322.8
320.8	333.7	341.6	350.5	367.7	354.2	342.8	342.5	342.5	323.5	330.9
312.0	325.6	326.8	302.0	299.8	297.6	304.6	314.7	301.3	304.5	312.6
294.5	302.4	315.6	320.9	315.7	300.2	312.7	308.7	324.5	287.6	322.8
320.8	342.5	323.5	335.7	341.6	350.5	367.7	342.5	354.2	342.8	330.9
320.8	342.5	323.5	333.7	341.6	350.5	367.7	342.5	354.2	342.8	330.9
312.7	308.7	324.5	322.8	287.6	294.5	302.4	315.6	320.9	315.7	300.2

NORTH

The structure of the program in Example 13.2 that solves this problem is as follows:

1. Read this table of depths from a file and store it in a LENGTH × WIDTH two-dimensional array depth, where for this problem, both constants are set to 11.

2. Calculate:

> northSum = the sum of the entries above the diagonal,
>
> southSum = the sum of the entries below the diagonal, and
>
> diagonalSum = the sum of the entries on the diagonal.

3. Calculate overallSum = northSum + southSum + diagonalSum.

4. Calculate readingsPerHalf = (total # of readings – # of readings on diagonal)/2.

5. Calculate the north, south, and overall averages:

$$\text{northAverage} = \frac{1}{\rule{3cm}{0.4pt}}$$

$$\text{southAverage} =$$

$$\text{overallAverage} = total\_resistance = \frac{1}{\dfrac{1}{resistor1} + \dfrac{1}{resistor2} + \dfrac{1}{resistor3}}$$

6. Output the averages.

In addition to calculating and displaying the three averages, it also displays the grid of depth readings so that the input data is echoed and can be used to check the results.

Example 13.2 Oceanographic Data Analysis

```
/* This program finds the average ocean depth in each half (sepa-
   rated by the diagonal) of a square section of the ocean and the
   overall average depth.

   Input (keyboard):    the file name
   Input (file):        the elements of array depth
   Output (screen):     the array depth in table format, northAverage,
                        southAverage, and overallAverage
   ---------------------------------------------------------------*/

#include <iostream>       // ifstream, cin, cout, >>, <<
#include <iomanip>        // setw()
#include <fstream>        // istream
#include <cassert>        // cassert()
#include <string>         // string
using namespace std;

int main()
```

```
{
  const int LENGTH = 11,
            WIDTH = 11,
            TOTAL_READINGS = LENGTH * WIDTH;
  double depth[LENGTH][WIDTH];    // depth array
  string fileName;                // file of depth readings

  cout << "Enter name of file containing depth readings: ";
  cin >> fileName;
  ifstream fin(fileName.data());
  assert(fin.is_open());

  for(int i = 0; i < LENGTH; i++)
    for (int j = 0; j < WIDTH; j++)
      fin >> depth[i][j];

// Calculate the north, south, diagonal, and overall sums
// Note: It is assumed that the elements on the diagonal are
//       included in the overall average but not in either half.

  double overallSum = 0,
         northSum = 0,
         southSum = 0,
         diagonalsum = 0;

  for (int i = 0; i < LENGTH; i++)
  {
    for (int j = 1; j < WIDTH; j++)
    {
      // Add entries below diagonal to southSum
      if (i > 0 && j < i)
        southSum += depth[i][j];

      // Add entries above diagonal to northSum
      if (i < LENGTH - 1 && j > i)
        northSum += depth[i][j];
    }
    // Add entries on diagonal to diagonalSum
    diagonalsum += depth[i][i];
  }
  overallSum = northSum + southSum + diagonalsum;

  // Calculate the north, south, and overall average depths
  int readingsPerHalf = (TOTAL_READINGS - LENGTH) / 2;
  double northAverage = northSum / readingsPerHalf,
         southAverage = southSum / readingsPerHalf,
         overallAverage = overallSum / (TOTAL_READINGS) ;
```

```
// Display the depth array and the average depths
cout << "\n\t\t\t\tOCEAN DEPTHS"
     << "\n\t\t\t\t============\n"
     << setprecision(1) << fixed << showpoint;

for (int i = 0; i < LENGTH; i++)
{
  for (int j = 0; j < WIDTH; j++)
    cout << setw(6) << depth[i][j];
  cout << "\n\n";
}
cout << "Northern half average depth: "
     << northAverage << " meters\n\n"
     << "Southern half average depth: "
     << southAverage << " meters\n\n"
     << "Overall average depth: "
     << overallAverage << " meters\n";
}
```

SAMPLE RUN:

```
Enter name of file containing depth readings: data-13-2.txt


                          OCEAN DEPTHS
                          ============

301.3 304.5 312.6 312.0 325.6 302.0 299.8 297.6 304.6 314.7 326.8

287.6 294.5 302.4 315.6 320.9 315.7 300.2 312.7 308.7 324.5 322.8

320.8 342.5 342.5 323.5 333.7 341.6 350.5 367.7 354.2 342.8 330.9

312.6 312.0 325.6 301.3 304.5 302.0 314.7 326.8 299.8 297.6 304.6

302.4 308.7 324.5 315.6 287.6 294.5 320.9 315.7 300.2 312.7 322.8

320.8 333.7 341.6 350.5 367.7 354.2 342.8 342.5 342.5 323.5 330.9

312.0 325.6 326.8 302.0 299.8 297.6 304.6 314.7 301.3 304.5 312.6

294.5 302.4 315.6 320.9 315.7 300.2 312.7 308.7 324.5 287.6 322.8

320.8 342.5 323.5 335.7 341.6 350.5 367.7 342.5 354.2 342.8 330.9

320.8 342.5 323.5 333.7 341.6 350.5 367.7 342.5 354.2 342.8 330.9

312.7 308.7 324.5 322.8 287.6 294.5 302.4 315.6 320.9 315.7 300.2

Northern half average depth: 318.9 meters

Southern half average depth: 267.2 meters

Overall average depth: 295.3 meters
```

13.4 MATRIX PROCESSING

A two-dimensional numeric array having m rows and n columns is called an $m \times n$ **matrix**. Matrices arise naturally in many problems in engineering and applied mathematics. In this section we describe some of the basic matrix operations that are useful in these applications.

13.4.1 Matrix Operations

Several matrix operations such as addition and subtraction are defined elementwise; that is, two matrices of the same shape are added by adding corresponding elements and are subtracted by subtracting corresponding elements. More precisely, the **sum** of two matrices that have the same number of rows and the same number of columns is defined as follows: If A_{ij} and B_{ij} are the entries in the ith row and jth column of $m \times n$ matrices A and B, respectively, then $A_{ij} + B_{ij}$ is the entry in the ith row and jth column of the sum $A + B$, which will also be an $m \times n$ matrix. Similarly, the **difference** $A - B$ is the $m \times n$ matrix in which $A_{ij} - B_{ij}$ is the entry in the ith row and jth column. For example, if

$$\frac{2v^2 \sin a \cos a}{} \quad \text{and} \quad \frac{-b \pm \sqrt{b^2 - 4ac}}{2a},$$

then

$$x = v_0 t \cos \theta$$

and

$$\sqrt[n]{x_1 \cdot x_2 \cdots x_n}$$

Similarly, a matrix is multiplied by a scalar (number) elementwise. For example,

$$\frac{n}{1 \quad 1 \quad 1}$$

One important matrix operation that is not defined elementwise is matrix multiplication. Suppose that A is an $m \times n$ matrix and B is an $n \times p$ matrix. The **product AB** is the $m \times p$ matrix for which

The entry in row i and column j

= the sum of the products of the entries in row i of A with the entries in column j of B

$$= A_{i1}B_{1j} + A_{i2}B_{2j} + \ldots + A_{in}B_{nj}$$

$$= -\frac{E}{\sqrt{R^2 + (2\pi f L - 1/(2\pi f C))^2}}$$

Note that the number of columns (n) in A is equal to the number of rows in B, which must be the case for the product of A with B to be defined.

For example, suppose that A is the 2×3 matrix

$$n = 3\sqrt{\frac{16T}{}}$$

and B is the 3×4 matrix

$$T = 63000\,\frac{P}{N}.$$

Because the number of columns (3) in A equals the number of rows in B, the product matrix AB is defined. The entry in the first row and first column is obtained by multiplying the first row of A with the first column of B, element by element, and adding these products:

$$P = 2\pi\sqrt{\frac{L}{g}\left(1 + \frac{1}{4}\sin^2\left(\frac{\alpha}{2}\right)\right)} \qquad p' = \frac{1}{2}\left(\begin{array}{c} y' \\ y' + z' \end{array}\right)$$

$$1 \times 4 + 3 \times 0 + 2 \times 6 = 16$$

Similarly, the entry in the first row and second column is

$$L = x + \frac{10x}{\sqrt{x^2 - 64}} \qquad D_P = \frac{\sigma f V^2}{391}$$

$$1 \times 2 + 3 \times (-4) + 2 \times 0 = -10$$

The complete product matrix is the 2×4 matrix given by

$$D_L = \frac{1245}{} \left(\frac{W}{}\right)^2 \frac{1}{V^2}$$

13.4.2 Example: Production Costs

Suppose that a company produces three different items. They are processed through four different departments A, B, C, and D. The following table gives the number of hours that each department spends on each item:

Item	Department			
	A	B	C	D
1	20	10	15	13
2	18	11	11	10
3	28	0	16	17

The cost per hour of operation in each of the departments is as follows:

Department	A	B	C	D
Cost per hour	$140	$295	$225	$95

We wish to determine the total cost of each item.

The problem can be solved using matrices. Suppose we let *hours* be the 3 × 4 array

$$\beta \tan \beta = \frac{m}{\underline{}}$$

and *hourlyCost* be the 4 × 1 array

$$\beta = \frac{\omega l}{\underline{}}$$

then the product *hours* × *hourlyCost* will be a 3 × 1 array whose elements are the total costs of producing each of the three items

$$\sqrt{E}$$

The program in Example 13.3 inputs the matrices `hours` and `hourlyCost`. It then uses matrix multiplication to find the matrix `totalCostPerItem` and outputs it.

Example 13.3 Production Costs

```
/* This program calculates production costs using matrix
   multiplication.

   Input (keyboard): hours table and cost-per-hour table
   Output (screen): total-cost-per-item table
-------------------------------------------------------------------*/

#include <iostream>        // cin, cout, >>, <<
using namespace std;

int main()
{
 const int ITEMS = 3,
           DEPTS = 4;
 int hours[ITEMS][DEPTS],                 // hours required for items by
                                          // dept.
     hourlyCost[DEPTS][1],                // hourly operating cost by
                                          // dept.
     totalCostPerItem[ITEMS][1];          // total cost for items by
                                          // dept.

   cout << "Enter the hours table in rowwise order:\n";
   for (int i = 0; i < ITEMS; i++)
     for (int j = 0; j < DEPTS; j++)
       cin >> hours[i][j];

   cout << "\nEnter the " << DEPTS << " hourly costs:\n";
   for (int j = 0; j < DEPTS; j++)
     cin >> hourlyCost[0][j];

   // Multiply matrices hours * hourlyCost to get totalCostPerItem
   for (int i = 0; i < ITEMS; i++)
   {
    totalCostPerItem[i][0] = 0;
    for (int j = 0; j < DEPTS; j++)
      totalCostPerItem[i][0] += hours[i][j] * hourlyCost[0][j];
   }

   cout << "\nTotal Cost for:\n";
   for (int i = 0; i < ITEMS; i++)
     cout << "\tItem " << i + 1 << ": $"
          << totalCostPerItem[i][0] << endl;
}
```

SAMPLE RUN:

```
Enter the hours table in rowwise order:
20  10  15  13
```

```
18   11   11   10
28   0    16   17


Enter the 4 hourly costs:
140 295 225 95


Total Cost for:
     Item 1: $10360
     Item 2: $9190
     Item 3: $9135
```

13.5 LINEAR SYSTEMS AND ELECTRICAL NETWORKS

A **linear system** is a set of linear equations, each of which involves several unknowns; for example,

$$5x_1 - 1x_2 - 2x_3 = 11$$

$$-1x_1 + 15x_2 - 2x_3 = 0$$

$$-2x_1 - 2x_2 + 7x_3 = 0$$

is a linear system of three equations involving the three unknowns x_1, x_2, and x_3. Linear systems arise in many areas of mathematics, science, and engineering, such as solving differential equations, electrical circuit problems, and static and dynamic systems.

This linear system can also be written as a single matrix equation:

$$A * x = b$$

where A is the 3×3 **coefficient matrix**, b is the 3×1 **constant vector**, and x is the 3×1 **vector of unknowns**:

$$A = \frac{P}{q} \approx \frac{\text{area of shaded region}}{\text{area of rectangle } ABCD} \quad , x = y = a\cosh\left(\frac{x}{a}\right) \quad , b = a + 20 = a\cosh\left(\frac{50}{a}\right)$$

A solution of such a system is a collection of values for these unknowns that satisfies all of the equations simultaneously. Several methods for solving them have been developed, including the method called **Gaussian Elimination**. We will describe it in connection with the following example.

13.5.1 Problem: Finding Currents in an Electrical Network

Consider the following electrical network containing six resistors and a battery:

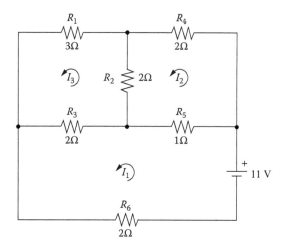

We wish to find the currents I_1, I_2, and I_3 in the three loops (where current is considered positive when the flow is in the direction indicated by the arrow).

13.5.2 Solution

The input information for this problem is the circuit pictured in the diagram; in particular, the six resistances R_1, R_2, . . ., R_6 will be needed to solve the problem. The output consists of the currents I_1, I_2, and I_3 in the three loops.

13.5.2.1 Kirchhoff's Laws—Linear Systems

The current through resistor R_1 is I_3, the current through resistor R_2 is $I_2 - I_3$, and so on. Ohm's law states that the voltage drop across a resistor is $R \times I$, where R is the resistance in ohms and I is the current in amperes. One of Kirchhoff's laws states that the algebraic sum of the voltage drops around any loop is equal to the applied voltage. This law gives rise to the following system of linear equations for the loop currents I_1, I_2, and I_3:

$$2I_1 + 1(I_1 - I_2) + 2(I_1 - I_3) = 11$$

$$2I_1 + 2(I_2 - I_3) + 1(I_2 - I_1) = 0$$

$$3I_1 + 2(I_3 - I_1) + 2(I_3 - I_2) = 0$$

Collecting terms gives the following simplified system, which is the same as that considered at the beginning of this section:

$$5I_1 - 1I_2 - 2I_3 = 11$$

$$-1I_1 + 5I_2 - 2I_3 = 0$$

$$-2I_1 - 2I_2 + 7I_3 = 0$$

To find the loop currents, we must solve this linear system; that is, we must find the values for I_1, I_2, and I_3 that satisfy these equations simultaneously.

To use Gaussian Elimination to solve the preceding system, we first eliminate I_1 from the second equation by adding 1/5 times the first equation to the second equation. Similarly, we eliminate I_1 from the third equation by adding 2/5 times the first equation to the third equation. This yields the linear system

$$5I_1 - 1I_2 - 2I_3 \ = 11$$

$$4.8I_2 - 2.4I_3 = 2.2$$

$$-2.4I_2 + 6.2I_3 = 4.4$$

which is equivalent to the original system in that the two systems have the same solution. We then eliminate I_2 from the third equation by adding $2.4/4.8 = 1/2$ times the second equation to the third, yielding the new equivalent linear system

$$5I_1 - 1I_2 - 2I_3 \ = 11$$

$$4.8I_2 - 2.4I_3 = 2.2$$

$$5I_3 \ = 5.5$$

Once the original system has been reduced to such a *triangular* form, it is easy to find the solution. It is clear from the last equation that the value of I_3 is

$$V(h) = L\left(\frac{1}{2}\pi r^2 - r^2 \arcsin\left(\frac{h}{r}\right) - h\sqrt{r^2 - h^2}\right)$$

Substituting this value for I_3 in the second equation and solving for I_2 gives

$$x_{n+1} = x_n - \frac{f(x_n)}{\ }$$

and substituting these values for I_2 and I_3 in the first equation and solving for I_1 gives

$$i(t) = 10 \sin^2\left(\frac{t}{\ }\right)$$

13.5.2.2 Refining the Method

The computations required to solve a linear system can be carried out more conveniently if the coefficients and constants of the linear system are stored in a matrix. For the preceding linear system, this gives the following 3×4 matrix:

$$linsys = \; v(T) = \frac{1}{C}\int_0^T i(t)\; dt$$

The first step in the reduction process was to eliminate I_1 from the second and third equations by adding multiples of the first equation to them. This corresponds to adding multiples of the first row of the matrix *linsys* to the second and third rows so that all entries in the first column except *linsys*[0][0] are zero. Thus, we add $-linsys[1][0]/linsys[0][0] = 1/5$ times the first row of *linsys* to the second row, and $-linsys[2][0]/linsys[0][0] = 2/5$ times the first row of *linsys* to the third row to obtain the new matrix:

$$linsys = \; C = 4a\int_0^{\pi/2}\sqrt{1-\left(\frac{a^2-b^2}{a^2}\right)\sin^2\Phi}\; d\Phi$$

The variable I_2 was then eliminated from the third equation, which corresponds to the matrix operation of adding $-linsys[2][1]/linsys[1][1] = 1/2$ times the second row to the third row. The resulting matrix, which corresponds to the final triangular system, thus is

$$linsys = \; f = 1 - 0.484\int_0^{E^*} e^{-E}\sin(h\sqrt{2E})\; dE$$

From this example, we see that the basic row operation performed at the ith step of the reduction process for $i = 0, 1, \ldots, n - 1$, is:

For $j = i + 1, i + 2, \ldots n - 1$

$$\text{Replace } row_j \text{ by } row_j - \; R(v) = -v^{3/2}$$

Clearly, for this to be possible, the element *linsys*[i][i], called the **pivot element**, must be nonzero. If it is not, we must interchange the ith row with a later row to produce a nonzero pivot. In fact, to minimize the effect of roundoff error in the computation, it is best to rearrange the rows to obtain a pivot element that is largest in absolute value.

The following algorithm, which summarizes this method for solving a linear system, uses this pivoting strategy. Note that if it is not possible to find a nonzero pivot element at some stage, the linear system is said to be a **singular** system and does not have a unique solution. Coding this algorithm as a C++ program is left as an exercise.

13.5.3 Gaussian Elimination Algorithm

This algorithm is used to solve a linear system of n equations with n unknowns using Gaussian Elimination; *linsys* is the $n \times (n + 1)$ matrix that stores the coefficients and constants of the linear system.

1. Enter the coefficients and constants of the linear system and store them in the matrix *linsys*.

2. For i ranging from 0 to $n - 1$, do the following:

   a. Find the entry *linsys*$[k][i]$, $k = i, i + 1, \ldots, n - 1$ that has the largest absolute value to use as a pivot.

   b. If the pivot is zero, display a message that the system is singular and terminate the algorithm. Otherwise, proceed.

   c. If $k \neq i$, interchange row i and row k.

   d. For j ranging from $i + 1$ to $n - 1$, do the following:

   Add $t = \int_{v_0}^{v(t)} \frac{m}{R(v)} dv$ times the ith row of *linsys* to the jth row of *linsys* to

   eliminate $x[i]$ from the jth equation.

3. Set $x[n - 1]$ equal to $E = \int_{0}^{x_c} F(x)\, dx$

4. For j ranging from $n - 2$ to 0 in steps of -1, do the following

   Substitute the values of $x[j + 1], \ldots, x[n - 1]$ in the jth equation and solve for $x[j]$.

13.6 MULTIDIMENSIONAL vector<T> OBJECTS

In the preceding chapter, we noted some of the drawbacks associated with C-style arrays and indicated how we could get around them by "wrapping" an array inside a class. This is in fact the approach used in vector<T> (except that it uses run-time allocated arrays as described in Section 15.2). In this section we will explore how we can use vector<T>s to create self-contained higher-dimensional objects. Using the operations already provided in vector<T> and applying the Standard Template Library (STL) algorithms to vector<T> objects will save us work. Applying them to higher-dimensional objects requires some care, however, because this is not as straightforward as using them with one-dimensional objects.

In the preceding chapter we considered only one form of a vector<T> declaration that created an empty object, for example,

```
vector<double> aVector;
```

declares that aVector can store double values, but is initially empty. The following are examples of alternative forms that a declaration can take:

```
const int INITIAL_CAPACITY = 10;
vector<double> bVector(INITIAL_CAPACITY);
vector<double> cVector(INITIAL_CAPACITY, 1.0);
```

The first declares bVector as a vector of 10 (undefined) doubles, and the second declares cVector to be a vector of 10 doubles, all initialized to 1.0:

aVector

bVector	?	?	?	?	?	?	?	?	?	?

cVector	1.0	1.0	1.0	1.0	1.0	1.0	1.0	1.0	1.0	1.0

13.6.1 Two-Dimensional vector<T> Objects

13.6.1.1 A Two-Step Approach

Suppose that we want to build a two-dimensional object named table consisting of three rows and four columns. Following our description of a two-dimensional array as an array of one-dimensional arrays, we might begin by defining a one-dimensional vector<T> object named initialRow whose capacity is the desired number of columns (4), and fill it with some initial value:

```
const int COLUMNS = 4;
vector<double> initialRow(COLUMNS, 0.0);
```

This builds a one-dimensional vector named initialRow, whose size and capacity are each 4:

	[0]	[1]	[2]	[3]
initialRow:	0.0	0.0	0.0	0.0

In the same way that a two-dimensional C-style array can be viewed as an array of arrays, a **vector of vectors** is a two-dimensional object. We can thus define the two-dimensional object table as a vector of vectors, using the desired number of rows (3) as its capacity, and with the object initialRow as its initial value:

```
const int ROWS = 3;
vector< vector<double> > table(ROWS, initialRow);
```

Note the space separating double> and >. It is important to remember the space between the angle brackets (> >), because if we write

```
vector<vector<double>> table(ROWS, initialRow); // ERROR!
```

the compiler will mistake `>>` for the input (or right-shift) operator, which will result in a compilation error.

Because each element of `table` is a `vector<initialRow>`, and `initialRow` is a `vector<double>`, the compiler will use `initialRow` to initialize each element of `table`. The result is that `table` is constructed as a 3 × 4 vector of vectors, in which each of the three rows is a copy of `initialRow`:

table:	[0]	[1]	[2]	[3]
[0]	0.0	0.0	0.0	0.0
[1]	0.0	0.0	0.0	0.0
[2]	0.0	0.0	0.0	0.0

A single-subscript expression such as

```
table[0]
```

refers to the first row of `table`:

table:	[0]	[1]	[2]	[3]
[0]	0.0	0.0	0.0	0.0

and a double-subscript expression such as

```
table[0][2]
```

refers to an element within the specified row of `table`:

table:	[0]	[1]	[2]	[3]
[0]	0.0	0.0	0.0	0.0

In general, the expression

```
table[r][c]
```

can be used to access the value stored in column `c` of row `r`.

13.6.1.2 A One-Step Approach

We can define the same vector of vectors in one step by using a more concise form that avoids the need to define the object `initialRow`:

```
typedef vector<double> TableRow;
```

```
typedef vector<TableRow> Table;

Table aTable;
```

The first `typedef` declares the name `TableRow` as a type that is a synonym for a one-dimensional vector of `doubles`. The second `typedef` then declares the name `Table` as a synonym for a one-dimensional vector of `TableRow` values; that is, a vector of vectors of `doubles`. The last declaration

```
Table aTable;
```

then declares `aTable` to be an empty two-dimensional `Table`. To define a nonempty `Table`, we can use

```
const int ROWS = 3,
          COLUMNS = 4;
Table theTable(ROWS, TableRow(COLUMNS, 0.0));
```

The result is a definition that is more readable than that given earlier and that eliminates the error described there caused by forgetting a space between two > symbols.

13.6.2 Two-Dimensional vector<T> Operations

We have already seen that double-subscript expressions of the form `theTable[r][c]` can be used to access the element at row `r` and column `c` in `theTable`. In addition to the subscript operator, other `vector<T>` operations can be used with two-dimensional vectors. We will look briefly at two of them.

13.6.2.1 The size( ) Method

If `theTable` is the 3 × 4 two-dimensional vector described earlier, then the expression

```
theTable.size()
```

returns 3, the number of rows in `theTable`. The expression

```
theTable[r].size()
```

can be used to find the number of columns in row `r`, because `theTable[r]` is the vector of `double` values in `theTable` whose index is `r`, and applying `size()` to that vector returns the number of values in it. If `theTable` is **rectangular** so that each row has the same number of elements, we can apply `size()` to any row to get the number of columns in `theTable`. If `theTable` is a **jagged table**—different rows have different sizes—`size()` must be applied to each row separately.

13.6.2.2 *The* push_back() *Method*

Suppose we need to add a new (fourth) row to theTable. This can be done by using the vector<T> method push_back():

```
theTable.push_back( TableRow(COLUMNS, 0.0) );
```

Because TableRow has been declared as a synonym for vector<double>, the expression

```
TableRow(COLUMNS, 0.0)
```

builds a nameless vector of zeros, and push_back() then appends this vector to the existing rows in theTable:

theTable:	[0]	[1]	[2]	[3]
[0]	0.0	0.0	0.0	0.0
[1]	0.0	0.0	0.0	0.0
[2]	0.0	0.0	0.0	0.0
[3]	0.0	0.0	0.0	0.0

To add a fifth column to theTable, we can use push_back() to append a double value to each row of theTable, because each row in theTable is itself a vector of double values:

```
for (int row = 0; row < theTable.size(); row++)
    theTable[row].push_back(0.0);
```

theTable:	[0]	[1]	[2]	[3]	[4]
[0]	0.0	0.0	0.0	0.0	0.0
[1]	0.0	0.0	0.0	0.0	0.0
[2]	0.0	0.0	0.0	0.0	0.0
[3]	0.0	0.0	0.0	0.0	0.0

13.6.3 vector<T>-Based Matrices

In Sections 13.4 and 13.5 we considered matrices and linear systems and used two-dimensional arrays to implement them. If we used the following declarations

```
typedef vector<double> MatrixRow;
typedef vector<MatrixRow> Matrix;
```

we could implement all of the matrix operations and their use in solving linear systems using two-dimensional `vector<doubles>` objects. Making this conversion is left as a straightforward exercise.

CHAPTER SUMMARY

Key Terms

array of arrays	multidimensional vector
column	rank
jagged tables (arrays)	row
linear system	square array
$m \times n$ matrix	subscript operator
matrix addition	three-dimensional array
matrix multiplication	two-dimensional array
matrix subtraction	`typedef` mechanism
member initialization list	vector of vectors
multidimensional array	

NOTES

- Two-dimensional arrays or vectors are useful for storing a data set whose values are arranged in rows and columns. Three-dimensional arrays or vectors are useful when the data set values are arranged in rows, columns, and ranks.

- As with one-dimensional arrays, each dimension of a multidimensional array is indexed starting with zero.

- A two-dimensional array can be viewed as a one-dimensional array whose components are also one-dimensional arrays. A three-dimensional array can be viewed as a one-dimensional array whose components are two-dimensional arrays. In general, an n-dimensional can be viewed as a one-dimensional array whose components are $(n - 1)$-dimensional arrays.

- The main drawback of C-style arrays is that they are not self-contained objects. In a function to implement an array operation, we must pass not only the array, but also the bound on each of its dimensions. One solution is to wrap an array in a class that contains operations for it.

- Multidimensional vectors are constructed as vectors of vectors and are self-contained objects, having all the built-in operations of `vector<T>`s.

Style and Design Tips

- *Use of a multidimensional array or vector is appropriate when a table of data values, a list of tables, and so on must be stored in main memory for processing.* Using a multidimensional array or vector when it is not necessary, however, can tie up a large block of memory locations. The amount of memory required to store a multidimensional array/vector can be quite large, even though each index is restricted to a small range of values. For example, the three-dimensional array `threeD` declared by

  ```
  typedef int ThreeDimArray[20][20][20];
  ThreeDimArray threeD;
  ```

  requires $20 \times 20 \times 20 = 8000$ memory locations.

- *If a function must receive an array or a vector or a class object that contains an array or a vector, then the parameter to hold that object should be declared as a reference (or const reference) parameter.* Passing it as a value parameter can greatly slow the execution of a function because of the time and memory required to copy it.

- *Do not reinvent the wheel.* When a problem requires an operation on a multidimensional vector, review the `vector<T>` function members and STL algorithms to see if the operation is already defined or there are other operations that make yours easier to implement.

- *Two-dimensional arrays or `vector<T>`s can be used to implement matrices and for solving systems of linear equations.*

Warnings

1. *In C++, multiple indices are each enclosed in brackets ([and]) and attached to the array/vector object.* In some languages, a single pair of brackets (or parentheses) is used to enclose a list of indices. However, attempting to access the value in row i and column j of a two-dimensional array A in C++ by using `A[i,j]` will cause a compile-time error.

2. *The first element of a C++ array or vector has the index 0, not 1, as in many programming languages.*

3. *No checking is performed to ensure that array or vector indices stay within the range of valid indices.*

4. *Although the assignment operator may be used with vectors, assignment of one array to another is not permitted.*

5. *Arrays and vectors cannot be input/output by simply including the array name in an input/output list.*

6. *Array arguments are automatically passed by reference.*

7. *When using vectors of vectors, leave a space between the two > symbols.* A common mistake is to forget this and to declare a vector of vectors with a statement like

```
vector<vector<int>> myGrid;
```

The compiler will read the >> as the input (or right-shift) operator, and because this makes no sense in this context, a compilation error will result. The correct approach is to leave a space:

```
vector<vector<int> > myGrid;
```

8. *When processing the elements of a multidimensional array/vector using nested loops, the loops must be arranged so that the indices vary in the appropriate order.* To illustrate, suppose that the two-dimensional array `table` is declared by

```
typedef int Array3x4[3][4];
Array3x4 table;
```

and the following data values are to be read into the array:

```
11 22 27 35 39 40 48 51 57 66 67 92
```

If these values are to be read and assigned in a row-wise manner so that the value is the matrix

```
11    22    27    35
39    40    48    51
57    66    67    92
```

then the following nested for loops are appropriate:

```
for (int row = 0; row < 3; row++)
   for (int col = 0; col < 4; col++)
      cin >> table[row][cowl];
```

If the order of these loops is reversed,

```
for (int col = 0; col < 4; col++)
   for (int row = 0; row < 3; row++)
      cin >> table[row][col];
```

then `table` will be loaded column-by-column, instead of row-by-row,

```
11    35    48    66
22    39    51    67
27    40    57    92
```

and operations applied to `table` will produce incorrect results.

TEST YOURSELF

1. A(n) _____ array is useful for storing data arranged in rows and columns.

2. A(n) _____ array is useful for storing data arranged in rows, columns, and ranks.

3. Arrays with the same number of rows as columns are said to be _____ arrays.

Questions 4–14 refer to the following two-dimensional array:

```
           [0] [1] [2] [3]
       [0]  11  22   0  43
       [1]   1  -1   0 999
mat:   [2]  -5  39  15  82
       [3]   1   2   3   4
       [4]  44  33  22  11
```

Find the value of each expression in Questions 4–9.

4. mat[2][3]

5. mat[4][1]

6. mat[1][1]

7. mat[0][0] + mat[0][1]

8. mat[0][0] + mat[1][0]

9. mat[3]

Find the value of x in each of Questions 10–14.

10. int x = 0;
    for (int i = 0; i <= 4; i++)
      x += mat[i][1];

11. int x = 0;
    for (int j = 0; j < 4; j++)
      x += mat[1][j];

12. int x = 0;
    for (int k = 0; k <= 3; k++)
      x += mat[k][k];

13. ```
int x = 0;
  for (int i = 0; i < 5; i++);
    for (int j = 0; j < 4; j++)
      x += mat[i][j];
```

14. ```
int x = 0;
  for (int j = 0; j < 4; j++)
    for (int i = 0; i < 5; i++)
      x += mat[i][j];
```

Section 13.4

1. A two-dimensional numeric array having m rows and n columns is called a(n) _____.

Questions 2–8 refer to the following matrices:

$$A = \frac{F(x) = \frac{1}{2}e^{x^2}\sin^2(3x^2)}{}, B = \frac{W = \int_a^b F(x)\cos(\theta(x))\,dx}{}, C = \frac{\frac{\Delta x}{3}(y_1 + 4y_1 + 2y_2 + 4y_3 + 2y_4 + \cdots + 2y_{n-4} + 4y_{n-1} + y_n)}{}$$

2. $A * B$ will be a(n) _____ × _____ matrix.

3. Calculate $A * B$.

4. Calculate $B * A$ or explain why it is not defined.

5. Calculate $B * C$ or explain why it is not defined.

6. Calculate $C * B$ or explain why it is not defined.

7. Calculate $B + C$ or explain why it is not defined.

8. Calculate $B - C$ or explain why it is not defined.

Section 13.6

1. (True or false) A vector of vectors is a two-dimensional object.

2. (True or false) The declaration

   ```
   vector<vector<int>> intTable(3, vector<int>(4, 0));
   ```

   will cause a compile-time error.

Questions 3–8 assume the following declarations:

```
typedef vector<double> TableRow;
typedef vector<TableRow> Table;
Table grid(5, TableRow(4, 0.0));
```

3. grid will have _____ rows and _____ columns.

4. Write an expression to change the element in the second row and third column of grid to 1.1.

5. What is the value of grid.size()?

6. What is the value of grid[0].size()?

7. Write a statement to append the value 99.9 at the end of the second row of grid.

8. Write statements to append a row containing 4 zeros at the bottom of grid.

EXERCISES

Section 13.2

Exercises 1–6 assume that the following declarations have been made:

```
const int NUM_SIZES = 6;
typedef bool BitArray[2][2][2][2];
typedef int Device[NUM_SIZES][10][20];
typedef Thingamajig Device[5];
```

How many elements can be stored in an array of each type?

1. int[50][100]

2. char[26][26]

3. bool[2][2][2]

4. BitArray

5. Device

6. Thingamajig

Exercises 7–10 assume that the following declarations have been made:

```
typedef int Array3x3[3][3];
Array3X3 mat;
```

Tell what value (if any) is assigned to each array element, or explain why an error occurs.

```
7. for (int i = 0; i < 3; i++)
       for (int j = 0; j < 3; j++)
          mat[i][j] = i + j;

8. for (int i = 0; i < 3; i++)
       for (int j = 2; j >= 0; j--)
          if (i == j)
             mat[i][j] = 0;
```

```
        else
          mat[i][j] = 1;
```

9. ```
   for (int i = 0; i < 3; i++)
       for (int j = 0; j < 3; j++)
          if (i < j)
             mat[i][j] = -1
          else if (i == j)
             mat[i][j] = 0;
          else
             mat[i][j] = 1;
   ```

10. ```
    for (int i = 0; i < 3; i++)
    {
       for (int j = 0; j < i; j++)
          mat[i][j] = 0;
       for (j = i; j < 3; j++)
          mat[i][j] = 2
    }
    ```

Exercises 11–14 assume that the following declaration has been made:

```
char logo[2][10] = {"Nuts", "and Bolts"};
```

Tell what output will be produced or explain why an error occurs.

11. ```
    for (int i = 0, i < 2; i++)
    {
       for (int j = 0; j < 9; j++)
          cout << logo[i][j];
       cout << endl;
    }
    ```

12. ```
    for (int j = 0; j < 9; j++)
    {
       for (int i = 0; i < 2; i++)
          cout << logo[i][j];
       cout << endl;
    }
    ```

13. ```
    for (int i = 0; i < 2; i++)
    {
       for (int j = 0; j < 9; j++)
       cout << logo[j][i];
       cout << endl;
    }
    ```
    ```

14.
```
for (int i = 0; i < 2; i++)
{
   for (int j = 8; j >= 0; j--)
      cout << logo[i][j];
   cout << endl;
}
```

15. Write a function that, given a `TableOfTemperatures` (as declared in this section), will calculate and return the average temperature at each of the three locations.

16. Like one-dimensional arrays, multidimensional arrays are stored in a block of consecutive memory locations, and address translation formulas are used to determine the location in memory of each array element. To illustrate, consider a 3×4 array `a` of integers, and assume that an integer can be stored in one memory word. If `a` is allocated memory in a row-wise manner and b is its base address, then the first row of `a`—`a[0][0]`, `a[0][1]`, `a[0][2]`, `a[0][3]`—is stored in words b, $b + 1$, $b + 2$, $b + 3$; the second row in words $b + 4$ through $b + 7$; and the third row in words $b + 8$ through $b + 11$.

Address	Memory	Array Element
	⋮	
b		`a[0][0]`
$b + 1$		`a[0][1]`
$b + 2$		`a[0][2]`
$b + 3$		`a[0][3]`
$b + 4$		`a[1][0]`
$b + 5$		`a[1][1]`
$b + 6$		`a[1][2]`
$b + 7$		`a[1][3]`
$b + 8$		`a[2][0]`
$b + 9$		`a[2][1]`
$b + 10$		`a[2][2]`
$b + 11$		`a[2][3]`
	⋮	

In general, `a[i][j]` is stored in word $b + 4i + j$.

a. Give a similar diagram and formula if `a` is a 3×3 array of integer values.

b. Give a similar diagram and formula if `a` is a 3×4 array of `double` values, where `double` values require two words for storage.

PROGRAMMING PROBLEMS

Sections 13.1–13.2

1. Write a program to calculate and display the first ten rows of Pascal's triangle. The first part of the triangle has the form

```
              1
            1  1
          1  2  1
        1  3  3  1
      1  4  6  4  1
```

in which each row begins and ends with 1, and each of the other entries in a row is the sum of the two entries just above it. If this form for the output seems too challenging, you might display the triangle as

```
1
1   1
1   2   1
1   3   3   1
1   4   6   4   1
```

2. A demographic study of the metropolitan area around Dogpatch divided it into three regions: urban, suburban, and exurban, and published the following table showing the annual migration from one region to another (the numbers represent percentages):

↱	Urban	Suburban	Exurban
Urban	1.1	0.3	0.7
Suburban	0.1	1.2	0.3
Exurban	0.2	0.6	1.3

For example, 0.3 percent of the urbanites (0.003 times the current population) move to the suburbs each year. The diagonal entries represent internal growth rates. Using a two-dimensional array to store this table, write a program to determine the population of each region after 10, 20, 30, 40, and 50 years. Assume that the current populations of the urban, suburban, and exurban regions are 2.1, 1.4, and 0.9 million, respectively.

3. The famous mathematician G. H. Hardy once mentioned to the brilliant young Indian mathematician Ramanujan that he had just ridden in a taxi whose number he considered to be very dull. Ramanujan promptly replied that, on the contrary, the number was very interesting because it was the smallest positive integer that could be written as the sum of two cubes (that is, written in the form $x^3 + y^3$, with x and y integers) in two different ways. Write a program to find the number of Hardy's taxi.

4. An engineer has a file containing a table of ratings for various products being tested, where the first line of the file contains the number of products and the number of times each was tested; each row of the table represents the ratings for a given product and each column represents the rating on a given test run. The maximum possible rating was 100 points. Write a program that, given the name of such a file, generates a report summarizing the overall average rating for each product and the average rating on each test run.

5. The group CAN (Citizens Against Noise) has collected some data on the noise level (measured in decibels) produced at seven different speeds by six different models of cars. This data is summarized in the following table:

				Speed (MPH)			
Car	**20**	**30**	**40**	**50**	**60**	**70**	**80**
0	88	90	94	102	111	122	134
1	75	77	80	86	94	103	113
2	80	83	85	94	100	111	121
3	68	71	76	85	96	110	125
4	77	84	91	98	105	112	119
5	81	85	90	96	102	109	120

Write a program that will display this table in an easy-to-read format, and that will calculate and display the average noise level for each car model, the average noise level at each speed, and the overall average noise level.

6. In a certain city, the air pollution is measured at two-hour intervals, beginning at midnight. These measurements are recorded for a one-week period:

Day 1	30.0	30.1	30.7	32.1	35.5	40.1	43.0	44.4	47.2	45.8	40.3	38.3
Day 2	33.0	32.5	30.5	34.6	40.6	48.3	46.7	49.5	53.1	49.6	45.0	40.1
Day 3	38.1	35.5	34.7	37.4	44.1	50.3	50.7	54.2	60.8	58.5	51.6	49.3
Day 4	49.9	48.8	47.7	53.5	60.1	70.2	73.3	75.8	80.0	75.3	73.1	60.5
Day 5	55.5	54.1	53.9	65.4	70.7	80.4	90.1	93.9	95.5	94.6	88.1	62.7
Day 6	73.0	90.8	65.0	66.0	71.6	78.3	74.5	78.0	83.7	75.6	66.9	58.1
Day 7	50.8	47.9	43.1	35.5	33.4	33.6	37.5	43.0	45.1	52.8	39.9	31.8

Write a program to produce a weekly report that displays the pollution levels in a table of the form

```
                            TIME
Day:   1    2    3    4    5    6    7    8    9   10   11   12
      - - - - - - - - - - - - - - - - - - - - - - - - - - - - -
  1 : 30.0 30.1 30.7 32.1 35.5 40.1 43.0 44.4 47.2 45.8 40.3 38.3
  2 : 33.0 32.5 30.5 34.6 40.6 48.3 46.7 49.5 53.1 49.6 45.0 40.1
  3 : 38.1 35.5 34.7 37.4 44.1 50.3 50.7 54.2 60.8 58.5 51.6 49.3
  4 : 49.9 48.8 47.7 53.5 60.1 70.2 73.3 75.8 80.0 75.3 73.1 60.5
  5 : 55.5 54.1 53.9 65.4 70.7 80.4 90.1 93.9 95.5 94.6 88.1 62.7
  6 : 73.0 90.8 65.0 66.0 71.6 78.3 74.5 78.0 83.7 75.6 66.9 58.1
  7 : 50.8 47.9 43.1 35.5 33.4 33.6 37.5 43.0 45.1 52.8 39.9 31.8
```

and that also displays the average pollution level for each day and the average pollution level for each sampling time.

7. Suppose that a certain automobile dealership sells 10 different models of automobiles and employs 8 salespersons. A record of sales for each month can be represented

by a table in which each row contains the number of sales of each model by a given salesperson, and each column contains the number of sales by each salesperson of a given model. For example, suppose that the sales table for a certain month is as follows:

```
0 0 2 0 5 6 3 0
5 1 9 0 0 2 3 2
0 0 0 1 0 0 0 0
1 1 1 0 2 2 2 1
5 3 2 0 0 2 5 5
2 2 1 0 1 1 0 0
3 2 5 0 1 2 0 4
3 0 7 1 3 5 2 4
0 2 6 1 0 5 2 1
4 0 2 0 3 2 1 0
```

Write a program to produce a monthly sales report, displaying the monthly sales table in the form:

```
                          Salesperson
    Model  :   1   2   3   4   5   6   7   8   :  Totals

    ----------------------------------------------------

     1     :   0   0   2   0   5   6   3   0   :    16
     2     :   5   1   9   0   0   2   3   2   :    22
     3     :   0   0   0   1   0   0   0   0   :    1
     4     :   1   1   1   0   2   2   2   1   :    10
     5     :   5   3   2   0   0   2   5   5   :    22
     6     :   2   2   1   0   1   1   0   0   :    7
     7     :   3   2   5   0   1   2   0   4   :    17
     8     :   3   0   7   1   3   5   2   4   :    25
     9     :   0   2   6   1   0   5   2   1   :    17
    10     :   4   0   2   0   3   2   1   0   :    12

    ----------------------------------------------------

  Totals :   23  11  35   3  15  27  18  17
```

8. Suppose that the prices for the ten automobile models in Problem 7 are as follows:

Model #	Model Price
0	$17,450
1	$19,995
2	$26,500
3	$25,999
4	$10,400
5	$18,885
6	$11,700
7	$14,440
8	$17,900
9	$19,550

Write a program to read this list of prices and the sales table given in Problem 7, and calculate the total dollar sales for each salesperson and the total dollar sales for all salespersons.

9. A certain company has a product line that includes five items that sell for $100, $75, $120, $150, and $35. There are four salespersons working for this company, and the following table gives the sales report for a typical week:

	Item Number				
Salesperson Number	**1**	**2**	**3**	**4**	**5**
1	10	4	5	6	7
2	7	0	12	1	3
3	4	9	5	0	8
4	3	2	1	5	6

Write a program to

a. Compute the total dollar sales for each salesperson.

b. Compute the total commission for each salesperson if the commission rate is 10 percent.

c. Find the total income for each salesperson for the week if each salesperson receives a fixed salary of $500 per week in addition to commission payment.

10. A certain company manufactures four electronic devices using five different components that cost $10.95, $6.30, $14.75, $11.25, and $5.00, respectively. The number of components used in each device is given in the following table:

	Component				
Device Number	**1**	**2**	**3**	**4**	**5**
1	10	4	5	6	7
2	7	0	12	1	3
3	4	9	5	0	8
4	3	2	1	5	6

Write a program to calculate:

a. The total cost of each device.

b. The total cost of producing each device if the estimated labor cost for each device is 10 percent of the cost in part (a).

11. A number of students from several different engineering sections performed the same experiment to determine the tensile strength of sheets made from two different alloys. Each of these strength measurements is a real number in the range 0 through 10.

Write a program to read several lines of data, each consisting of a section number and the tensile strength of the two types of sheets recorded by a student in that section, and store these values in a two-dimensional array. Then calculate

a. For each section, the average of the tensile strengths for each type of alloy.

b. The number of persons in a given section who recorded strength measures of 5 or higher.

c. The average of the tensile strengths recorded for alloy 2 by students who recorded a tensile strength lower than 3 for alloy 1.

12. A *magic square* is an $n \times n$ table in which each of the integers $1, 2, 3, \ldots, n^2$ appears exactly once and all column sums, row sums, and diagonal sums are equal. For example, the following is a 5×5 magic square in which all the rows, columns, and diagonals add up to 65:

17	24	1	8	15
23	5	7	14	16
4	6	13	20	22
10	12	19	21	3
11	18	25	2	9

The following is a procedure for constructing an $n \times n$ magic square for any odd integer n. Place 1 in the middle of the top row. Then after integer k has been placed, move up one row and one column to the right to place the next integer $k + 1$, unless one of the following occurs:

a. If a move takes you above the top row in the jth column, move to the bottom of the jth column and place the integer $k + 1$ there.

b. If a move takes you outside to the right of the square in the ith row, place $k + 1$ in the ith row at the left side.

c. If a move takes you to an already-filled square or if you move out of the square at the upper right-hand corner, place $k + 1$ immediately below k.

Write a program to construct an $n \times n$ magic square for any odd value of n.

13. Consider a square grid, with some cells empty and others containing an asterisk. Define two asterisks to be *contiguous* if they are adjacent to each other in the same row or in the same column. Now suppose we define a *blob* as follows:

a. A blob contains at least one asterisk.

b. If an asterisk is in a blob, then so is any asterisk that is contiguous to it.

c. If a blob has more than two asterisks, then each asterisk in it is contiguous to at least one other asterisk in the blob.

For example, there are four blobs in the partial grid

seven blobs in

and only one in

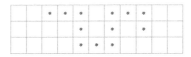

Write a program that uses a recursive function to count the number of blobs in a square grid. Input to the program should consist of the locations of the asterisks in the grid, and the program should display the grid and the blob count.

14. The game of *Life*, invented by the mathematician John H. Conway, is intended to model life in a society of organisms. Consider a rectangular array of cells, each of which may contain an organism. If the array is assumed to extend indefinitely in both directions, each cell will have eight neighbors, the eight cells surrounding it. Births and deaths occur according to the following rules:

a. An organism is born in an empty cell that has exactly three neighbors.

b. An organism will die from isolation if it has fewer than two neighbors.

c. An organism will die from overcrowding if it has more than three neighbors.

The following display shows the first five generations of a particular configuration of organisms:

Write a program to play the game of *Life* and investigate the patterns produced by various initial configurations. Some configurations die off rather quickly; others repeat after a certain number of generations; others change shape and size and may move across the array; and still others may produce "gliders" that detach themselves from the society and sail off into space.

15. Write a program that allows the user to play tic-tac-toe against the computer.

Section 13.4

1. A company produces three different products. They are processed through four different departments, A, B, C, and D, and the following table gives the number of hours that each department spends on each product:

Product	A	B	C	D
1	20	10	15	13
2	18	11	11	10
3	28	0	16	17

The cost per hour of operation in each of the departments is as follows:

Department	A	B	C	D
Cost per Hour	$140	$295	$225	$95

Write a program that uses matrix multiplication to find the total cost of each of the products.

2. The vector-matrix equation

$$u'(t) = Au(t)$$

is used to transform local coordinates (I, J, K) for a space vehicle to inertial coordinates (N, E, D). Write a program that reads values for α, β, and γ and a set of local coordinates (I, J, K) and then uses matrix multiplication to determine the corresponding inertial coordinates.

3. A *Markov chain* is a system that moves through a discrete set of states in such a way that when the system is in state i there is probability P_{ij} that it will next move to state j. These probabilities are given by a transition matrix P, whose (i, j) entry is P_{ij}. It is easy to show that the (i, j) entry of P^n then gives the probability of starting in state i and ending in state j after n steps.

To illustrate, suppose there are two containers A and B containing a given number of objects. At each instant, an object is chosen at random and is transferred to the other container. This is a Markov chain if we take as a state the number of objects in container A and let P_{ij} be the probability that a ball is transferred from A to B if there are i balls in container A. For example, for four objects, the transition matrix P is given by

$$u(t) = e^{t/2} \sin\ t$$

Write a program that reads a transition matrix P for such a Markov chain and calculates and displays the value of n and P^n for several values of n.

4. A *directed graph*, or *digraph*, consists of a set of vertices and a set of directed arcs joining certain of these vertices. For example, the following diagram pictures a directed graph having five vertices numbered 1, 2, 3, 4, and 5, and seven directed arcs joining vertices 1 to 2, 1 to 4, 1 to 5, 3 to 1, 3 to itself, 4 to 3, and 5 to 1:

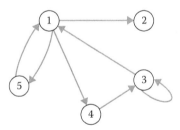

A directed graph having n vertices can be represented by its *adjacency matrix*, which is an $n \times n$ matrix, with the entry in the ith row and jth column 1 if vertex i is joined to vertex j, and 0 otherwise. The adjacency matrix for this graph is

$$V = \frac{1}{3}\pi h^2 (3R - h)$$

If A is the adjacency matrix for a directed graph, the entry in the ith row and jth column of A^k gives the number of ways that vertex j can be reached from the vertex i by following k edges. Write a program to read the number of vertices in a directed graph and a collection of ordered pairs of vertices representing directed arcs, construct the adjacency matrix, and then find the number of ways that each vertex can be reached from every other vertex by following k edges for some value of k.

Section 13.5

1. Write the system of linear equations for the loop currents I_1, I_2, and I_3 in the following simple resistor and battery circuit. Then use Gaussian Elimination to find these currents.

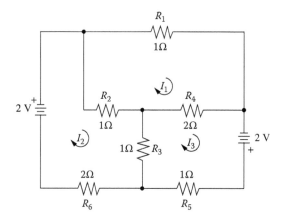

2. Consider the following electrical network:

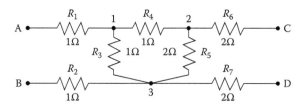

If the voltages at the endpoints are $V_A = V_B = V_C = V_D = 1V$, then applying Kirchhoff's law of currents at nodes 1, 2, and 3 yields (after some simplification) the following system of linear equations for the voltages V_1, V_2, and V_3 at these nodes

$$\frac{dV}{dt} = -\pi r^2 \sqrt{2gh}$$

Use Gaussian Elimination to find these voltages.

3. Consider the following material-balance problem: A solution that is 80 percent oil, 15 percent usable byproducts, and 5 percent impurities enters a refinery. One output is

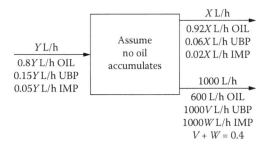

92 percent oil and 6 percent usable byproducts. The other output is 60 percent oil and flows at the rate of 1000 liters per hour (L/h).

We thus have the following system of material-balance equations:

$$\frac{dV}{dt} = (2\pi h R - \pi h^2)\frac{dh}{dt},$$

Use Gaussian Elimination to solve this linear system. Check that your solution also satisfies the last equation.

4. Consider the following statical system:

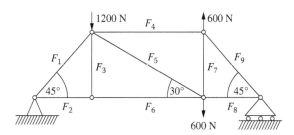

Because the sum of all forces acting horizontally or vertically at each pin is zero, the following system of linear equations can be used to obtain the tensions F_1, F_2, . . ., F_9:

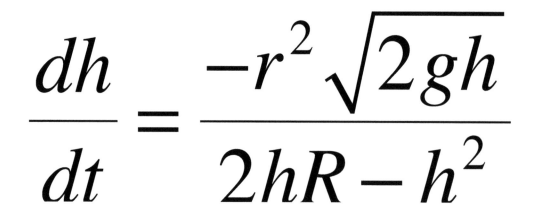

$$\frac{dh}{dt} = \frac{-r^2\sqrt{2gh}}{2hR - h^2}$$

Use Gaussian Elimination to solve this linear system

5. The population of a country (or a region) is divided into age groups. If n is the number of age groups and $P_i(t)$ is the number of individuals in age group i at time t, the dynamic model describing the populations of the different age groups is given by

$$P_1(t + 1) = b_1P_1(t) + b_2P_2(t) + \ldots + b_nP_n(t) + c_1$$

$$P_{i+1}(t + 1) = a_iP_i(t) + c_i, \text{ for } i = 1, \ldots, n - 1$$

Here, each a_i is the percentage of persons in age group i who move into age group $i + 1$ at the next time period; each b_i is the birthrate for age group i; and each c_i is the number of persons belonging to age group i that move into the region.

If this dynamic system reaches equilibrium at a certain time period t so that for each age group i, $P_i(t) = P_i(t + 1) = P_i(t + 2) = \ldots = $ some constant value p_i, then the preceding equations can be rewritten as the following steady-state equations:

$$p_1 = b_1p_1 + b_2p_2 + \ldots + b_np_n + c_1$$

$$p_2 = a_1p_1 + c_2$$

$$p_3 = a_2p_2 + c_3$$

.

.

.

$$p_n = a_{n-1}p_{n-1} + c_n$$

Use Gaussian Elimination to solve this linear system for $n = 4$, $a_1 = 0.8$, $a_2 = 0.7$, $a_3 = 0.6$, $b_1 = 0$, $b_2 = 0.05$, $b_3 = 0.15$, $b_4 = 0.1$, and each $c_i = 100$.

6. Programming Problem 10 of Section 9.4 in Chapter 9 described the method of least squares for finding the equation of a line that best fits a set of data points. This method can also be used to find best-fitting curves of higher degree. For example, to find the equation of the parabola

$$y = A + Bx + Cx^2$$

that best fits a set of n data points, the values of A, B, and C must be determined for which the sum of the squares of the deviations of the observed y-values from the predicted y-values (using the equation) is as small as possible. These values are found by solving the linear system

$$Q = \alpha(p^2 - p*^2)^\beta$$

Find the equation of the least-squares parabola for the following set of data points:

x	y
0.05	0.957
0.12	0.851
0.15	0.832
0.30	0.720
0.45	0.583
0.70	0.378
0.84	0.295
1.05	0.156

7. Linear systems similar to those in Problem 6 must be solved to find least-squares curves of higher degrees. For example, for a least-squares cubic,

$$y = A + Bx + Cx^2 + Dx^3$$

the coefficients A, B, C, and D can be found by solving the following system of equations:

$$\frac{dp}{dt} = -kQ$$

Find the equation of the least-squares cubic for the set of data points in Problem 6.

Building Classes

CONTENTS

Inanimate objects are classified scientifically into three major categories—those that don't work, those that break down, and those that get lost.

RUSSELL BAKER

Mathematics classes became sheer terror and torture to me. I was so intimidated by my incomprehension that I did not dare to ask any questions.

CARL JUNG

A committee can make a decision that is dumber than any of its members.

DAVID COBLITZ

It is common sense to take a method and try it. If it fails, admit it frankly and try another. But above all, try something.

FRANKLIN D. ROOSEVELT

W E HAVE SEEN THAT designing a C++ program involves identifying the objects in a problem and then using types to create software representations of those objects. Once these objects are created, programming consists of applying to those objects the operations needed to solve the problem.

We have also seen that when there is no predefined type that suffices to model an object, the C++ class can be used to create a new type to represent the object. Classes thus provide a way to extend the C++ language, allowing it to represent an ever-increasing number of objects.

Until now, we have for the most part *used* classes that someone else built. In this chapter, we learn how to *build* them ourselves and study the ideas of encapsulation and information-hiding that underlie class design.

14.1 INTRODUCTORY EXAMPLE: MODELING TEMPERATURES

In Chapter 6 we developed a program to convert Celsius temperatures to the corresponding Fahrenheit temperatures. We modeled each temperature reading with a `double` variable:

```
double tempCelsius,
        tempFahrenheit;
```

Now, suppose that we wish to extend this to a program that, given a temperature in Fahrenheit, Celsius, or Kelvin, will display the equivalent temperature in each of the scales. Unlike the problem in Chapter 6 where each temperature value could be modeled by a `double` variable because we needed to store only the degrees, the problem here is to model temperatures having two attributes—their *number of degrees* and their *scale* (Fahrenheit, Celsius, or Kelvin).

Of course, we could just use two variables:

```
double degrees = 0.0;
char scale = 'F';
```

But this requires two data items (`myDegrees` and `myScale`) to model a single object (a temperature). To apply some function `f()` to a temperature, we would have to pass it each of the data items used in our model,

```
f(degrees, scale);
```

instead of being able to pass a single object:

```
f(theTemp);
```

Similarly, displaying a temperature would require an output statement like

```
cout << degrees << scale;
```

Copy C For EMPLOYEE'S RECORDS (See Notice on back.)		**9999 BC** Form No. 1234-5678	
a Control number ABC-123	**1** Wages, tips, other comp. 1111.11	**2** Federal income tax withheld .00	
	3 Social security wages	**4** Social security tax withheld 11.22	
b Employer's ID number 123456789	**5** Medicare wages and tips	**6** Medicare tax withheld 22.11	
c Employer's name, address, and ZIP code Dinoville Rock Quarry 1212 T-Rex Ave. Bedrock, Prehistoria 00001			
d Employee's social security number 987-65-4321			
e Employee's name, address, and ZIP code Fred Flintstone 123 Cave A Bedrock, Prehistoria 00002			
7 Social security tips	**8** Allocated tips	**9** Advance EIC payment	
10 Dependent care benefits	**11** Nonqualified plans	**12** Benefits included in box 1	
13 See instrs. for box 13		**14** Other	
15 Statutory Deceased employee	Pension Legal plan rep.	Hshld. Subtotal Deferred emp. compensation	
PR 123456789	1111.11	.00	
16 State Employer's state I.D. #	**17** State wages, tips, etc.	**18** State income tax	
19 Locality name Bedrock	**20** Local wages, tips, etc. 1111.11	**21** Local income tax .00	

Form W-2 Wage and Tax Statement

FIGURE 14.1   An income tax form.

instead of simply

```
cout << theTemp;
```

This approach is not really inconvenient for objects that can be described with two attributes, but it quickly becomes unmanageable as the complexity of the object being modeled increases. Just think how many data items would be needed to represent a tax form like that shown in Figure 14.1.

14.1.1 Creating New Types

In Chapter 6 we extended object-centered design to situations where some operation needed to solve a problem is not provided in C++ by creating our own *function* to perform that operation. In this temperature problem, however, the object (a `Temperature`) has a *type* that is not predefined. In Chapter 7 we suggested that for such objects with multiple attributes, a C++ *class* can be used to create a new type that has those attributes. For the temperature problem, therefore, we should create a `Temperature` class containing both the degrees and the scale of an arbitrary temperature and function members to perform temperature conversions.

14.1.1.1 Algorithm
If we have such a class, then solving the problem is straightforward:

1. Via cout, display a prompt for a temperature.

2. From cin, read a temperature into *theTemp*.

3. Via cout, display:

   a. the Fahrenheit equivalent of *theTemp*

   b. the Celsius equivalent of *theTemp*

   c. the Kelvin equivalent of *theTemp*

14.1.1.2 Coding
Given a Temperature library that provides a Temperature class and input, output, and conversion operations, we can encode this algorithm in C++ as shown in Example 14.1.

Example 14.1 A Temperature Conversion Program

```
/* tempConversion.cpp displays a temperature in Fahrenheit, Celsius,
   and Kelvin, using class Temperature.

   Input:  an arbitrary Temperature
   Output: its Fahrenheit, Celsius and Kelvin equivalents
   -------------------------------------------------------------------*/

#include <iostream>                        // >>, <<, cin, cout
using namespace std;
#include "Temperature.h"                   // Temperature

int main()
{
  cout << "This program shows the Fahrenheit, Celsius, and\n"
          "Kelvin equivalents of a temperature.\n\n";
  char response;
  Temperature theTemp;                      // construction
  do
  {                                         // input a temperature
    cout << "Enter a temperature (e.g., 98.6 F): ";
    cin >> theTemp;

    cout << "--> "                          // output the equivalent:
         << theTemp.toFahrenheit()          // Fahrenheit temp.
         << " == "
         << theTemp.toCelsius()             // Celsius temp.
         << " == "
         << theTemp.toKelvin()              // Kelvin temp.
         << endl;
```

```
      cout << "\nDo you have more temperatures to convert? ";
      cin >> response;
   }
   while (response == 'Y' || response == 'y');
}
```

SAMPLE RUNS:

```
This program shows the Fahrenheit, Celsius, and
Kelvin equivalents of a temperature.
Enter a temperature (e.g., 98.6 F): 212 F
--> 212 F == 100 C == 373.15 K

Do you have more temperatures to convert? Y
Enter a temperature (e.g., 98.6 F): 0 C
--> 32 F == 0 C == 273.15 K

Do you have more temperatures to convert? Y

Enter a temperature (e.g., 98.6 F): 100 K
--> -279.67 F == -173.15 C == 100 K

Do you have more temperatures to convert? N
```

We will illustrate the principles of class design by designing and implementing the Temperature class in the following sections.

14.2 DESIGNING A CLASS

As with programs, creating a class consists of two phases:

1. A design phase in which we plan the class

2. An implementation phase in which we encode this design in C++

This section explores the first phase. Section 14.3 will examine the implementation phase. Designing a class consists of identifying two things:

1. The *operations* that can be applied to a class object

2. The *data* that must be stored to characterize a class object

The operations are usually identified first because it is often not obvious what data must be stored, and identifying the operations can sometimes clarify them. Also, identifying the operations first means that they can be described without having to worry about how the data will be stored. This *independence from implementation details* is important in good class design.

14.2.1 The External and Internal Perspectives

Up to now, our approach to programming has mostly been looking from outside the program into its details. Because we reside outside the program, this is a natural way to begin, and as long as we are merely using predefined classes, this **external perspective** is adequate.

One of the basic ideas in class design is **object autonomy**, which we will refer to more simply as the **I-can-do-it-myself principle**, meaning that an object has within itself the ability to perform its operations where this is possible. That is, rather than viewing a class operation as manipulation of an object by a program, we view a class operation as an object taking an action. To incorporate this I-can-do-it-myself principle into the design of a class, we will shift our perspective from that of an external observer to that of the object being designed. That is, we will think through our design using an **internal perspective** as though *we are the object* and will thus use first person terminology. To illustrate, instead of referring to the Temperature data members as *its* degrees and *its* scale (which imply we are outside, looking in), we will refer to them as *my* degrees and *my* scale (indicating that we are the object, looking out).

In the sections that follow, we will use *both* perspectives. When working on a program and using a class, we will use the external perspective. When working inside a class or building its function members, we will use the internal perspective.

14.2.2 Temperature Class's Operations

From an internal perspective, a Temperature object must provide the following operations if the program in Example 14.1 is to work:

   i. Construct myself *implicitly* by initializing my degrees and scale with default values.

   ii. Read a temperature from an istream and store it in my data members.

  iii. Compute the Fahrenheit temperature equivalent to me.

   iv. Compute the Celsius temperature equivalent to me.

   v. Compute the Kelvin temperature equivalent to me.

   vi. Display my degrees and scale using an ostream.

Although these operations suffice to solve the problem at hand, designing a reusable and useful class involves identifying other operations that a user of the class is likely to need. To that end, we might extend our list with the following operations:

vii. Construct myself *explicitly* by initializing my degrees and scale members with specified values.

viii. Report my number of degrees.

ix. Report my scale.

x. Compute my temperature raised by a given number of degrees.

xi. Compute my temperature lowered by a given number of degrees.

xii. Compare myself to another `Temperature` object using any of the six relational operators (==, !=, <, >, <=, >=).

xiii. Assign another `Temperature` value to me using the assignment operator (=).

This is not an exhaustive list, but it is a good start and will serve to introduce the details of class implementation. Other operations can be added later.

The last operation, assignment, is already provided. For any class we define, the C++ compiler creates a *default assignment operation* that is adequate for our purposes. Thus, a statement like

```
temp2 = temp1;
```

can be used to copy the data members of `temp1` into `temp2`. We must implement the other operations ourselves as described in the next section.

14.2.3 Temperature Class's Data

To identify what data will be needed for a class, it is a good idea to begin by going through the list of operations and identifying what information each of them requires. For our `Temperature` class, operations (i) and (vi–xi) indicate that, from an internal perspective, the following are needed:

1. My degrees, and

2. My scale

In fact, these are the only data items that are needed for our `Temperature` class. For other classes, however, a complete list may not be evident at the outset and others will have to be added later when the implementation of an operation requires them.

14.3 IMPLEMENTING A CLASS

Once we have a design for a class, we can use it as a blueprint for implementing the class. Because we want the class to be reusable, it is declared in a library's header file (e.g.,

Temperature.h) and the nontrivial operations are usually defined in a separately compiled implementation file (e.g., Temperature.cpp). As described in Chapter 6, the documentation for the class and its operations are commonly put in a separate documentation file (e.g., Temperature.txt).

14.3.1 The Class Data Members

The first task is to implement the data items. For our Temperature class, we can store the number of degrees in a real variable and the scale in a character variable. Thus, in Temperature.h, we write:

```
/* Temperature.h is the header file for class Temperature.
   . . .
   ------------------------------------------------------------*/

double myDegrees;
char myScale;            // 'F', 'C', or 'K'
```

These will become the **data members**—also known as **instance variables**—of our Temperature class. Note that their names begin with the prefix my to reflect our internal perspective. And as with all identifiers, the name of a data member should be *self-documenting*, describing what it stores.

We could have used string to define myScale if we wanted to store the entire name of the scale, but we chose char instead because temperatures are usually written using a single character for the scale (e.g., 98.6 F, 100 C, 273 K). The best choice in some situations may not be clear, but a decision must be made before we can proceed. Such implementation decisions can always be revised later if they prove unwise.

14.3.1.1 Encapsulation

Once we have declared variables for the data members of our class, we can actually create the class by wrapping a **class declaration** around these objects (again, in Temperature.h):

```
class Temperature
{
 public:
  // to be filled in later
 private:
    double myDegrees;
    char myScale;
};
```

Don't forget the semicolon after the closing curly brace. Like all declarations, a class declaration must be terminated by a semicolon.

Wrapping the data members in a class declaration and then using the class as a type to declare an object makes it possible for an object to store values of different types. We call this **encapsulation**. The Temperature class encapsulates the double data member myDegrees and the char data member myScale.

This declaration creates a new type named `Temperature`. If we use this type to declare objects as in

```
#include "Temperature.h"

Temperature temp1,
            temp2;
```

then `temp1` and `temp2` are two distinct `Temperature` objects, each containing two data members: a `double` named `myDegrees` and a `char` named `myScale`. We might picture these objects as follows:

14.3.1.2 Information Hiding

Notice that our declaration of the `Temperature` class is divided into two sections: a **public section** indicated by the keyword `public` and a **private section** labeled `private`. Items in the public section are accessible to users of the class, but those in the private section are not. This aspect of class design is called **information-hiding**. *By preventing a program from directly accessing the data members of a class, we hide that information, thus removing the temptation to access those variables directly.* In our example, placing the data members `myDegrees` and `myScale` in the private section prevents a user of the class from modifying them as with the statements

```
temp1.myScale = 'K';
temp1.myDegrees = -50;
```

creating an impossible temperature value.

Another reason for keeping the data members private has to do with class maintenance. Over time, we may wish to enhance the class by adding operations to the public section, but if at all possible, we keep declarations of existing operations unchanged so that programs that use the class won't fail. However, we may want to modify the data members in the private section or add to them. For example, we might decide in the future to change the type of `myScale` to `string` so it can store the name of a temperature scale instead of only its first letter. Or we may find that adding some new data members will improve the performance of some of the class's operations.

14.3.1.3 Class Invariants

Once we have the data members encapsulated and hidden, we are almost ready to begin implementing the class's operations. Before doing this, however, we should identify any restrictions there may be on the values of the data members of our class. For example, we might stipulate that the only valid values for the data member `myScale` will be the characters F, C, or K. For the data member myDegrees, we know of no bound on how high a temperature can go, but we will use absolute zero (0°K, which is equivalent to –273.15°C and –459.67°F) for a lower bound. Our restrictions thus are

myScale will be one of the characters 'F', 'C', or 'K'

for 'K': myDegrees ≥ 0

for 'C': myDegrees ≥ -273.15

for 'F': myDegrees ≥ -459.67

If we identify and specify such restrictions at the outset, then we can implement the various class operations in a way that ensures they are observed.

Because this set of conditions will be true throughout the class, it is called a **class invariant**. When such an invariant can be defined, it should be included in the documentation for the class so that users will know they are restrictions on objects of type Temperature.

14.3.2 The Class Operations—Function Members

Once we have decided on the data members of our class and have declared them in the private section, we are ready to begin implementing the class operations. This is done by adding **function members**, also known as **methods**. We will begin our study of them by looking at a simple example: an output method that displays a Temperature object via an ostream (operation (vi) in our list).

14.3.2.1 Temperature Output

As we will see later in this section, we can extend the output operator << so that it can be used to output Temperature objects, but we cannot simply add it as a function member of a class. However, it is useful to have an output operation early because it can help with testing the correctness of the other operations.

From an external perspective, the purpose of an output function is to display the contents of an object's data members (but perhaps not all of them). In our case, if we name the function member display(), a programmer should be able to write

```
temp1.display(cout);
```

to display the degrees and scale in a Temperature object temp1 to cout, and write

```
temp2.display(cerr);
```

to display the degrees and scale of temp2 to cerr. Thus, from an internal perspective, a call to display() can be viewed as a **message** the program is sending to *you* (a Temperature object), with cout (or cerr) as an argument—like someone telling you

"Hey you! Print yourself using cout.*"*

The definition of display() must therefore provide the instructions that I (a Temperature object) apply to my data members to perform the operation.

This perspective leads to the following specification for the operation's behavior:

Receive: *out*, an ostream to which information is to be written

Output: myDegrees followed by a space and then myScale

Send back: *out*, with myDegrees, a space, and myScale inserted into it

Example 14.2 shows a definition of this method. Because of its simplicity, we define it using the inline specifier, in the header file Temperature.h, following the class declaration.[1]

Example 14.2 Displaying a Temperature

```
// ... #includes and class declaration go here

// -------- Output method ----------------------------
inline void Temperature::display(ostream & out) const
{
   out << myDegrees << ' ' << myScale;
}
```

From an external perspective, sending this message to a Temperature object will display the values stored in the data members myDegrees and myScale of that object, so the statement

```
temp1.display(cout);
```

will display the data members of temp1 via cout, while

```
temp2.display(cerr);
```

will display the data members of temp2 via cerr. At the end of this section we will see how to overload the output operator (<<) to display a Temperature value in the usual manner.

Note that as one of Temperature's function members, display() must be invoked using **dot notation**. Put differently, display() must be sent as a message to a Temperature object. If we attempt to call display() without using dot notation,

```
display(cout); // ERROR!
```

the compiler will generate an error, because we have not specified a Temperature object to receive the message.

---

[1] Inlining a function (see Section 10.6) allows the compiler to replace a call to that function with the code from its definition. Putting this definition in the header file makes this code available to the compiler.

In this first look at the definition of a function member, we introduced a number of new features, and we will now look at each of them in more detail.

14.3.2.2 Full Names
The first new feature is in the function heading:

```
inline void Temperature::display(ostream & out) const
```

The name of the class is attached to the name of the method with the **scope operator** (::) *to inform the compiler of the class to which it belongs.* The resulting name is called the **full name** (or **fully qualified name**) of the function member.

It is important to use the full name in the definition of function members, because *they can access the private members of their class, but normal functions cannot.* If the full name is not used in a method's definition, as in

```
inline void display(ostream & out) const  // Not a function member
{
   out << myDegrees << ' ' << myScale;     // ERROR!
}
```

the compiler views this as a normal (nonmember) function, and errors like

```
Identifier 'myDegrees' is not defined
Identifier 'myScale' is not defined
```

will result because the private data members of a class are invisible to ordinary nonmember functions.

Full names are also important because they make it possible for different classes to have function members with the same name and signature. By using the full name to define each display() method, the compiler can distinguish one definition from another.

14.3.2.3 Constant Methods
The next new feature is the keyword const at the end of the function heading:

```
inline void Temperature::display(ostream & out) const
```

This informs the compiler that display() is a **constant method**, which means that it may not change any data member of the Temperature class. Any attempt to do so will be caught by the compiler as an error. *It is good programming practice to declare all function members that do not alter the data members of the class as* const *methods.*

For a method's definition to compile correctly, its prototype must be stored inside the public section of the class declaration. Also, because display()'s prototype refers to the type ostream, we must include the iostream library before the class declaration:

```
#include <iostream>                    // ostream, ...
using namespace std;

class Temperature
{
 public:
    void display(ostream & out) const;

 private:
    double myDegrees;
    char myScale;
};
```

Note that for a constant method, the `const` specifier must also be used in its prototype. Also note that it is not necessary to specify `inline` or the full name of the function member within the class itself. Keeping these prototypes simple reduces clutter within the class declaration and thus increases readability. For the same reason, the methods of a class should be documented in a separate file `Temperature.txt`.

Of course, before the `display()` method is of any practical use, the data members `myDegrees` and `myScale` must contain values. We therefore turn our attention to operations that can be used to initialize these data members.

14.3.2.4 Constructors

As we have seen, a `Temperature` definition

```
Temperature temp1,
             temp2;
```

defines `temp1` and `temp2` as objects that might be pictured as follows:

The question marks indicate that the data members in these objects are undefined. Some versions of C++ may initialize `myDegrees` to 0 and `myScale` to the null character `'\0'`. In other versions, their initial values may be whatever "garbage values" correspond to the strings of bits in the memory locations allocated to them; for example, one version initialized `myDegrees` to –9.3E61 and `myScale` to a nonexistent character whose numeric code was –52. Because garbage values such as these do not represent valid temperatures, we would certainly prefer that the preceding definitions define `temp1` and `temp2` with some default initial value of our choosing (e.g., 0 degrees Celsius):

C++ allows such initialization behavior to be performed by class operations called **constructors**.

The name of a constructor is always the same as the name of the class and there are two kinds of constructors: **default-value constructors** that are used to initialize data members with default values when none are specified in the declaration of an object, and **explicit-value constructors** that initialize them with values provided in the declaration.

To illustrate, suppose that for declarations of `Temperature` objects like those given earlier,

```
Temperature temp1,
             temp2;
```

we want `temp1` and `temp2` to be automatically initialized to 0 degrees Celsius. This gives the following specification for the behavior of the default-value constructor:

Postcondition: `myDegrees == 0.0 && myScale == 'C'`

Because a constructor (unlike other functions) cannot return anything to its caller, we specify its behavior using a **postcondition**, a boolean expression that must be `true` when the operation terminates.

Example 14.3 shows the definition of this default-value constructor. Because of its simplicity, we define it as `inline` and place its definition in `Temperature.h`, after the declaration of class `Temperature`.

Example 14.3 The `Temperature` Default-Value Constructor

```
// -------- Default-value constructor --------------

inline Temperature::Temperature()
{
  myDegrees = 0.0;
  myScale = 'C';
}
```

Here again, we see some new features. The first is that there is no return type between `inline` and the method's full name because *constructors have no return type*, not even `void`. As an initialization operation, a constructor never returns anything to its caller. Its sole purpose is to initialize the data members of a class.

Next comes the full name of the operation, in which the first `Temperature` is the name of the class of which the operation is a member, `::` is the scope operator, and then follows the name of the operation, which also is `Temperature` because the name of a constructor is always the same as the name of its class:

```
inline Temperature::Temperature()
```

Note that there is no const at the end of the constructor's heading. A constructor is not a constant operation because it does modify the data members of the class (by initializing them).

We must also store a prototype of this operation in the public section of the class declaration:

```
class Temperature
{
 public:
  Temperature();
  void display(ostream & out) const;

 private:
  double myDegrees;
  char myScale;
};
```

As before, we omit the inline and use the normal name of the operation instead of its full name. We also add its documentation to Temperature.txt.

Given this much, a programmer can now write a short program to test the class declaration and method definitions:

```
#include <iostream>
using namespace std;
#include "Temperature.h"

int main()
{
  Temperature temp1;    // the compiler sends temp1 the
                        //   "initialize yourself" message
  temp1.display(cout);
}
```

When this program is executed, it will output

```
0 C
```

This output is produced because whenever the C++ compiler processes the definition of a class object it *automatically searches the class for a constructor it can use to initialize that object.* For this reason, we should *always provide one or more constructors when building a class.*

The default-value constructor only allows us to initialize a Temperature object to the value 0 degrees Celsius. It would certainly be useful, however, if we could also specify other initial values in a Temperature object's declaration. This can be accomplished by defining another constructor known as an *explicit-value constructor.*

Unlike the default-value constructor, an explicit-value constructor receives its initialization values via parameters. Because there is the possibility that these initial values could produce an invalid temperature object, such constructors must check their validity.

From the internal perspective, we have the following specification for this constructor's behavior. Note that, for user convenience, we are allowing the scale to be in either upper or lower case.

Receive: *initDegrees*, a `double`

                  initScale, a `char`

Precondition: *initScale* is one of {'f', 'c', 'k', 'F', 'C', 'K'} and

                  initDegrees is valid for *initScale*

Postcondition: `myDegrees` == *initDegrees* && `myScale` == *initScale* in uppercase

Example 14.4 presents a definition of this operation. Because of its complexity, it should *not* be designated `inline`. Instead, its definition should be stored in the class implementation file `Temperature.cpp` so that it can be compiled separately. Note that although the full name of this constructor is exactly the same as the full name of the default-value constructor in Example 14.3, its signature (i.e., list of parameter types) is different, as required by overloaded functions. Also, we will define the three constants `MIN_FAHRENHEIT`, `MIN_CELSIUS`, and `MIN_KELVIN` used for checking validity rather than the "magic" numbers −459.67, −273.15, and 0.0 in the class header file `Temperature.h` to make them accessible to users of the class because they seem likely to be generally useful.

Example 14.4 The `Temperature` Explicit-Value Constructor

```
/* Temperature.cpp defines the nontrivial Temperature operations. */

#include "Temperature.h"            // class Temperature
#include <cctype>                   // islower(), toupper()
#include <cstdlib>                  // exit()
using namespace std;

// -------- Explicit-value constructor ---------------------------
Temperature::Temperature(double initDegrees, char initScale)
{
   if (islower(initScale))          // if the scale is lowercase
      initScale = toupper(initScale);   // convert it to uppercase
```

```
//----- Check the class invariant -----
if (   initScale == 'F' && initDegrees >= MIN_FAHRENHEIT
    || initScale == 'C' && initDegrees >= MIN_CELSIUS
    || initScale == 'K' && initDegrees >= MIN_KELVIN )
{
  myDegrees = initDegrees;              // proceed with
  myScale = initScale;                  // initialization
}                                       // otherwise,
else                                    // error message
{
  cerr << "\n*** Temperature constructor received invalid params "
       << initDegrees << ' ' << initScale << endl;
  exit(1);
}
}
```

To use this operation, we must place its prototype in the public portion of class Temperature, and the definitions of the three constants outside the class declaration:

```
//----- Temperature.h -----

#include <iostream> // ostream
using namespace std;
const double MIN_FAHRENHEIT = -459.67,
             MIN_CELSIUS = -273.15,
             MIN_KELVIN = 0.0;

class Temperature
{
 public:
  Temperature();
  Temperature(double initDegrees, char initScale);

  void display(ostream & out) const;

 private:
  double myDegrees;
  char myScale;
};
```

As with other methods, we would put its documentation in Temperature.txt. Given this prototype, a programmer can now write

```
#include <iostream>
#include "Temperature.h"
using namespace std;
int main()
{
  Temperature temp1(98.6, 'F'),
              temp2;
```

```
    temp1.display(cout); cout << endl;
    temp2.display(cout); cout << endl;
}
```

Executing the object file produced by compiling and linking this program along with `Temperature.cpp` produces the output

```
98.6 F
0 C
```

In the declarations

```
Temperature temp1(98.6, 'F'),
            temp2;
```

the object `temp1` is constructed using the explicit-value constructor, and `temp2` is constructed using the default-value constructor. These objects are thus initialized as follows:

14.3.2.5 Object Initialization

As noted earlier, when the compiler processes the declaration of a class object, it searches the class for a constructor it can use to initialize the object. For example, when it encounters a class object declaration (without arguments) such as

```
Temperature temp2;
```

the compiler searches the class for a constructor with no parameters and uses it to perform the initialization. When it encounters a class object declaration such as

```
Temperature temp1(98.6, 'F');
```

in which arguments are specified, it searches the class for a constructor whose signature matches the types of those arguments. If it finds such a constructor, the compiler inserts a call to that constructor to perform the initialization.

This syntax for declarations of variables should not seem completely unfamiliar. In Chapter 11, we saw that an `ifstream` object can be initialized with the name of a file, for example,

```
ifstream inStream("weather.dat");
```

Such a statement is using an `ifstream` explicit-value constructor to open the stream to the file whose name it is passed as an argument.

According to the C++ standard, any variable can be initialized in this way. Instead of writing

```
double sum = 0.0;
char middleInitial = 'C';
```

for example, we could write

```
double sum(0.0);
char middleInitial('C');
```

The syntax using = is simply provided as a convenient shorthand to this approach.[2]

14.3.2.6 Accessor Methods

Now that we have constructors, we turn to implementing other operations of class Temperature, starting with some of the simplest ones. An **accessor method** retrieves the value of a data member of the class but may not modify it.[3] For example, sending the message

```
temp1.getDegrees()
```

to Temperature object temp1 will retrieve the number of degrees in temp1, and the message

```
temp1.getScale()
```

will retrieve its scale.

Specifications for these function members are straightforward:

Return: myDegrees

for getDegrees() and

Return: myScale

for getScale().

Example 14.5 shows their definitions. Because they are simple, we inline them and store their definitions in the class header file Temperature.h, and because they read but do not modify the data members, they are designated as const methods.

Example 14.5 Temperature Accessors

```
// -------- Degrees extractor --------------------------------
inline double Temperature::getDegrees() const
{
   return myDegrees;
}

// -------- Scale extractor ----------------------------------
```

---

[2] The C++ standard actually suggests that the latter approach is the *preferred* way to initialize an object.
[3] Accessor methods are sometimes called "getters." Accordingly, function members that can modify data members are called "setters."

```
inline char Temperature::getScale() const
{
  return myScale;
}
```

Because prototypes of function members must be stored in the class declaration, we add the prototypes of these accessors to our class Temperature:

```
class Temperature
{
 public:
  Temperature();
  Temperature(double initDegrees, char initScale);

  double getDegrees() const;
  char getScale() const;

  void display(ostream & out) const;

 private:
  double myDegrees;
  char myScale;
};
```

As is common practice, we have grouped these prototypes according to purpose with blank lines separating each group as the preceding listing illustrates. For example, constructors, accessors, and I/O methods all have distinct purposes, so we group their prototypes accordingly.

14.3.2.7 Temperature Input

Next, we add a counterpart to the output function member display()—an input method to read a Temperature value from an istream. A statement such as

```
temp2.read(cin);
```

should read a number and a character from cin and store them in temp2.myDegrees and temp2.myScale, respectively.

Proceeding from an internal perspective as with the output operation yields the following specification:

Receive: *in*, an istream containing a double and a char

Input: *inDegrees*, the double value, and *inScale*, the char value

Precondition: *inScale* is one of {'f', 'c', 'k', 'F', 'C', 'K'} and *inDegrees* is valid for *inScale*.

Send back: *in*, with *inDegrees* and *inScale* extracted from it

Postcondition: myDegrees == *inDegrees* && myScale == *inScale*

For user convenience, we will accept the scale in either upper- or lowercase and convert lowercase entries to uppercase. To guard against invalid input values, we must check that the degrees and scale entered satisfy the precondition before modifying the data members. If they do not, we will terminate execution of the program.[4]

Due to its complexity, we do not define this function member as `inline`, and so its definition should be stored in `Temperature.cpp`, and because it modifies the `Temperature` data members, it is *not* defined as a `const` method. Example 14.6 presents one implementation of this operation.

Example 14.6 Temperature Input

```
// -------- Temperature Input ------------------------------------
void Temperature::read(istream & in)
{
  double inDegrees;              // temporary variables to
  char inScale;                  //   store the input value

  in >> inDegrees >> inScale;    // read values from in
  if (islower(inScale))          // if scale is lowercase
    inScale = toupper(inScale);  //   convert it to uppercase

  //----- Check the class invariant -----
  if (   inScale == 'F' && inDegrees >= MIN_FAHRENHEIT
      || inScale == 'C' && inDegrees >= MIN_CELSIUS
      || inScale == 'K' && inDegrees >= MIN_KELVIN)
  {
    myScale = inScale;           // assign input values
    myDegrees = inDegrees;       //   to data members
  }
  else                           // otherwise issue error message
  {                              // & stop execution
    cerr << "\n*** Invalid temperature *** "
         << initDegrees << ' ' << initScale << endl;
    exit(1);
  }
}
```

As a function member, a prototype of `read()` must be added to the class declaration:

```
class Temperature
{
 public:
  Temperature();
  Temperature(double initDegrees, char initScale);
```

---

[4] An alternative is to use the `setstate()` function to set `ios::failbit` to signal an input failure as described in Section 11.3. A program using the `Temperature` class can then detect this failure with the `failure()` function and recover from it.

```
    double getDegrees() const;
    char getScale() const;

    void display(ostream & out) const;
    void read(istream & in);

 private:
   double myDegrees;
   char    myScale;
};
```

As before, documentation for it must also be added to the documentation file.

A program can now use read() to input temperatures from an istream or from an ifstream.

```
Temperature temp1;
  ...
temp1.read(cin);
  ...
```

Later in this section we will see how the input operator (>>) can be overloaded to input Temperature values in the usual manner.

14.3.2.8 Conversion Methods

Next, we examine methods that convert temperatures to different scales. We begin with a function member toFahrenheit(). Its specification is straightforward:

Return: The Fahrenheit temperature equivalent of myself

Example 14.7 gives an implementation of this method. It is sufficiently complex that it is not defined as inline. Note the use of a switch statement to select the appropriate conversion formula to use with the current value of myScale.

Example 14.7 The toFahrenheit() Method

```
// -------- Fahrenheit conversion method ----------------

Temperature Temperature::toFahrenheit() const
{
  switch (myScale)
  {
    case 'F': case 'f':
        return Temperature(myDegrees, 'F');
    case 'C': case 'c':
        return Temperature(myDegrees * 1.8 + 32.0, 'F');
    case 'K': case 'k':
        return Temperature((myDegrees - 273.15) * 1.8 + 32.0, 'F');
  }
}
```

It is important to note how this function uses the explicit-value `Temperature` constructor to build the `Temperature` value to be returned. For example, in the first case of the `switch` statement, it uses the values of `myDegrees` and `'F'` passed to it to build a `Temperature` object, which `toFahrenheit()` then uses as its return value for this case.

As always, we must place a prototype of the conversion methods in the class declaration:

```
class Temperature
{
 public:
  Temperature();
  Temperature(double initDegrees, char initScale);

  double getDegrees() const;
  char getScale() const;

  Temperature toFahrenheit() const;
  Temperature toCelsius() const;
  Temperature toKelvin() const;

  void read(istream & in);
  void display(ostream & out) const;

  private:
   double myDegrees;
   char    myScale;
};
```

Definitions of `toCelsius()` and `toKelvin()` are similar to that of `toFahrenheit()` and are left as exercises.

Given these methods, our `Temperature` class provides the minimal functionality necessary for a modified version of the program in Example 14.1 in which input and output are carried out using `read()` and `display()` as shown in Example 14.8.

Example 14.8 A Modified Version of Example 14.1

```
/* This program displays a temperature in Fahrenheit,
   Celsius, and Kelvin, using class Temperature.

   Input:  an arbitrary Temperature
   Output: its Fahrenheit, Celsius and Kelvin equivalents
   ---------------------------------------------------------------*/


#include <iostream>                     // >>, <<, cin, cout
using namespace std;
#include "Temperature.h"                // Temperature
```

```
int main()
{
  cout << "This program shows the Fahrenheit, Celsius, and\n"
          "Kelvin equivalents of a temperature.\n\n";
  char response;
  Temperature theTemp;                    // construction
  do
  {                                       // input a temperature
    cout << "Enter a temperature (e.g., 32 F): ";
    theTemp.read(cin);
    cout << "--> ";                       // output its conversions
    theTemp.toFahrenheit().display(cout);
    cout << " == ";
    theTemp.toCelsius().display(cout);
    cout << " == ";
    theTemp.toKelvin().display(cout);
    cout << endl;

    cout << "\nDo you have more temperatures to convert? ";
    cin >> response;
  }
  while (response == 'Y' || response == 'y');
}
```

Note the **chaining** of messages as in the statement

```
theTemperature.toFahrenheit().display(cout);
```

The chained messages are processed from left to right. First, the `toFahrenheit()` message is sent to `theTemperature`, which returns a (temporary) `Temperature` object. The `display()` message is then sent to this (temporary) `Temperature` object.

14.3.3 Overloading Operators

Each of the preceding operations has been implemented as a "normal" function in that its name was an identifier. For some operations (arithmetic operations such as addition and subtraction, relational comparisons such as < and >, and I/O), it is often more convenient to define an *operator* to perform them. Just as normal functions and methods can be overloaded, C++ allows **operator overloading** for classes.

14.3.3.1 The Relational Operators

As noted earlier, it would be useful if we could compare two `Temperature` objects using relational operators. This would allow a computerized thermometer to be programmed with statements like

```
if (yourTemperature > Temperature(98.6, 'F'))
  cout << "You have a fever!\n";
```

or a computer-controlled thermostat to be programmed with statements like

```
while (houseTemperature < Temperature(20, 'C'))
  runFurnace();
```

To permit such operations, we must overload the relational operators for class Temperature. We will do this for two of them, the less-than operator (<) and the equality operator (==). The others are similar and are left as exercises.

To overload the < operator we define a Temperature method with the name operator<. Similarly, to overload the == operator, we define a method with the name operator==. In general, *we can overload an arbitrary operator whose symbol is Δ by defining a method with the name* operatorΔ, provided it has a signature distinct from that of any existing definition of that operator.

If operator<() is defined as a Temperature method, an expression like

```
houseTemperature < Temperature(20, 'C')
```

is treated by the C++ compiler as the message

```
houseTemperature.operator<( Temperature(20, 'C') )
```

Intuitively, such a method call is sending the less-than message with a Temperature argument (20 degrees Celsius in this case) to houseTemperature, which must return true only if it is less than that Temperature argument, taking into account that our scales may not be the same. Example 14.9 shows one way that the operand<() method can be implemented:

Example 14.9 Overloading Operator <

```
// -------- less-than ------------------------------------

bool Temperature::operator<(const Temperature & rightOperand) const
{
  Temperature anotherTemp;        // the equivalent of rightOperand,
                                  // but in my scale
  switch (myScale)
  {
    case 'C': anotherTemp - rightOperand.toCelsius();
              break;
    case 'F': anotherTemp = rightOperand.toFahrenheit();
              break;
    case 'K': anotherTemp = rightOperand.toKelvin();
              break;
  }

  return myDegrees < anotherTemp.getDegrees();
}
```

This implementation resolves the problem of mismatched scales by using a local `Temperature` object `anotherTemp`, which it sets to the equivalent of `rightOperand` in the same scale as the `Temperature` object receiving the message. Once we have two temperatures in the same scale, we can simply compare their `myDegrees` members.[5]

The equality (`==`) operator can be overloaded using a similar approach, as shown in Example 14.10.

Example 14.10 Overloading Operator ==

```
// -------- equality ------------------------------------

bool Temperature::operator==(const Temperature & rightOperand) const
{
  Temperature anotherTemp;          // the equivalent of rightOperand,
                                    // but in my scale
  switch (myScale)
  {
    case 'C': anotherTemp = rightOperand.toCelsius();
            break;
    case 'F': anotherTemp = rightOperand.toFahrenheit();
            break;
    case 'K': anotherTemp = rightOperand.toKelvin();
            break;
  }

  return myDegrees == anotherTemp.getDegrees();
}
```

Both of these methods are sufficiently complicated that they should be stored in `Temperature.cpp`. Prototypes for these operations must be placed in the class declaration as shown in Example 14.11 and documentation in `Temperature.txt`. The remaining relational operators can be overloaded in a similar fashion and are left as exercises. Other operations (+ and –) that might be added are described in the exercises.

Example 14.11 The Temperature Class Declaration

```
/* Temperature.h declares class Temperature and three
   constants representing minimum possible temperatures.
   ------------------------------------------------------------*/
```

---

[5] A class method can directly access the private data members in class objects it receives as parameters, but for readabilty we use an object's accessor methods instead.

```
#ifndef TEMPERATURE
#define TEMPERATURE

#include <iostream>                          // istream, ostream
using namespace std;

const double MIN_FAHRENHEIT = -459.67,
             MIN_CELSIUS = -273.15,
             MIN_KELVIN = 0.0;

class Temperature
{
 public:                                     // The class interface
   Temperature();
   Temperature(double initDegrees, char initScale);

   double getDegrees() const;
   char getScale() const;

   Temperature toFahrenheit() const;
   Temperature toCelsius() const;
   Temperature toKelvin() const;

   bool operator<(const Temperature & rightOperand) const;
   bool operator==(const Temperature & rightOperand) const;
   // ... other relational operators omitted

   void read(istream & in);
   void display(ostream & out) const;

 private:
   double myDegrees;
   char   myScale;
};

// ... Definitions of inline operations go here ...
```

14.3.3.2 Overloading I/O Operators << and >>

We have already seen that methods like read() and display() provide a way to define I/O operations for classes like Temperature. However, these methods do not coordinate well with the normal iostream operators. It would be preferable if we could instead output class values using the customary operators << for output and >> for input. We will now overload operator<< using our display() method and overload operator>> using read(). Doing so does not require much code, but to do so correctly requires that some subtle issues be addressed.

The first issue is that unlike `operator<` and `operator==`, we cannot define `operator<<` as a method of a class. To see why, recall that if an operator whose symbol is Δ is defined as a method of a class, and **object** is an object of that class, then the expression

```
object Δ operand
```

is treated by the compiler as

```
object.operatorΔ(operand)
```

If this observation is applied to the output expression

```
cout << someTemperature
```

then we see that `operator<<` must be defined as a method of class `ostream`, not class `Temperature`. But this would require modifying the standard C++ class `ostream` by adding a new prototype for `operator<<` to output `Temperature` objects and any other objects for classes we create, something that is not normally allowed (nor would this be wise).

Fortunately, C++ provides a way around this problem. If the operator whose symbol is Δ is defined as a normal function—one that acts upon its operands via parameters—and not as a class method, then the expression

```
object Δ operand
```

is treated by the compiler as the function call

```
operatorΔ(object, operand)
```

More precisely, if we wish to use

```
cout << theTemperature;
```

then we need to define a nonmember function `operator<<` that the compiler can call as

```
operator<<(cout, theTemperature);
```

We can thus define `operator<<()` as a normal function, using an external perspective, giving the following specification of the function's behavior:

Receive: *out*, the `ostream` to which values are being written; *aTemp*, the `Temperature` object whose value is being written

Output: *aTemp*.`myDegrees` and *aTemp*.`myScale`

Send back: *out*, containing the inserted values

Return: *out*, for use by a subsequent output operation

Example 14.12 presents the implementation of operator<<. Because most of the required functionality is already available via the display() method of class Temperature, our function merely sends its parameter aTemp the display() message, and then returns its ostream parameter. The resulting function is simple enough to define as inline, so we store it in the class header file Temperature.h. Because it is a normal function (i.e., not a method), no prototype is placed within the class declaration.

Example 14.12 Overloading Operator << for Class Temperature

```
// -------- Temperature ostream output ------------------

inline ostream & operator<<(ostream & out,
                            const Temperature & aTemp)
{
   aTemp.display(out);        // tell aTemp to output itself
   return out;
}
```

This function is deceptively simple. It seemingly just receives an ostream and a class operand and sends the display() message to the class operand with the ostream operand as an argument. Because display() inserts values into its ostream operand out, out is declared as a reference parameter.

However, the function also returns out and the return type is a reference ostream &, something we have not seen before. These are the two subtlest parts of the function, and we will deal with them separately.

The << operator returns out so that output operations can be chained. That is, when we insert two Temperature objects temp1 and temp2 into cout,

```
cout << temp1 << temp2;
```

there are two different calls to operator<<(), and these are executed from left to right. To distinguish them, suppose we number them as follows:

```
cout <<₁ temp1 <<₂ temp2;
```

In executing these functions from left to right, the compiler treats them as nested function calls, with <<₁ being performed first, as the "inner" call:

```
operator<<₂( operator<<₁(cout, temp1) , temp2);
```

The return-value from <<₁ is thus used as the left argument to <<₂,

```
operator<<₂ ( operator<<₁'s return_value, temp2);
```

which means that $<<_2$ will try to insert temp2 into whatever value $<<_1$ returns.

From this, it should be apparent that $<<_1$ must return an ostream. Moreover, it should be the same ostream into which $<<_1$ inserted its value, and so it should return its parameter out.

The other subtle point has to do with the return type of operator<<. Why did we define its return type as ostream &? The reason is that when a C++ function returns a value in the usual fashion, it actually returns a *copy* of the value to the caller of the function, which means that for the output statement

```
cout << temp1 << temp2;
```

temp1 would be inserted into cout, but temp2 would be inserted into a copy of cout, the result of which is unpredictable.

To avoid such copying, C++ allows a function to have a **reference return type**, which in effect "turns off" the copying mechanism and returns the actual object. Thus, when we write

```
inline ostream & operator<<(ostream & out, const Temperature & theTemp)
{
  theTemp.display(out);
  return out;
}
```

we are telling the compiler, "Don't return a copy of out, but instead return the actual ostream for which it is an alias."

Overloading the input operator >> is similar to that for <<. Example 14.13 gives the definition of operator>>().

Example 14.13 Overloading Operator >> for Class Temperature

```
// -------- Temperature istream input -------------------------

inline istream & operator>>(istream & in, Temperature & aTemp)
{
  aTemp.read(in);
  return in;
}
```

Note that unlike the output operator, the input operator does modify its Temperature parameter, so it is declared as a reference parameter and not a constant reference parameter.

With these new I/O operations for Temperature objects added to our temperature library, the temperature-conversion program of Example 14.1 will now compile and execute. And although there certainly are other operations we would add, we will consider this library complete for our purposes.

14.4 OTHER CLASS FEATURES

In this chapter we have looked at the major features of designing and implementing classes. There is much more that could be said, however, and in this section we will look at two others. One is an alternative way to provide access to the data members of a class, which we will illustrate by implementing the input and output operations without using methods like read() and display(). The other is a "class wrapper" that should be added to make our classes suitable for use in large projects that may utilize many different classes.

14.4.1 Friend Functions

Although most operations on a class object can be implemented as function members, there are some occasions when this is not possible. For example, we saw that the output and input operators << and >> cannot be defined as Temperature methods because their left operands are an ostream and an istream, respectively, not a Temperature:

```
cout << "Enter a temperature (e.g., 98.6 F): ";
cin >> theTemperature;

cout << "The temperature is " << theTemperature << endl;
```

Our solution was to define "normal" functions for operator<< and operator>> that used Temperature's display() and read() methods:

```
inline ostream & operator<<(ostream & out, const Temperature & aTemp)
{
  aTemp.display(out);
  return out;
}

inline istream & operator>>(istream & in, Temperature & aTemp)
{
  aTemp.read(in);
  return in;
}
```

Now, suppose we want to define operator<< without calling display()—for example, suppose our Temperature class has no display() method. If operator<< attempts to access the data members of its parameter aTemp directly,

```
inline ostream & operator<<(ostream & out,const Temperature & aTemp)
{
  out << theTemp.myDegrees << ' '    // ERROR!
      << theTemp.myScale;            // ERROR!
  return out;
}
```

the compiler will generate error messages like

```
Member 'myDegrees' is private in class 'Temperature'
```

The compiler is enforcing the information-hiding mechanism by not allowing the function to access the private data members of our class.

But suppose that the only reasonable way to implement some operation is with a function that must be able to access the private data members of a class. In such rare situations, C++ allows a class to grant this special access privilege to the function by specifying that it is a **friend**.

To illustrate, suppose we delete the display() method from class Temperature. Then we can define operator<< as

```
inline ostream & operator<<(ostream & out, const Temperature & aTemp)
{
  out << theTemp.myDegrees << ' '
      << theTemp.myScale;
  return out;
}
```

provided we place a prototype of operator<<() within the class, preceded by the keyword friend:

```
class Temperature
{
 public:                        // The class interface
      .
      .
      .
    friend ostream & operator<<(ostream& out, const Temperature & aTemp);
      .
      .
      .
  private:                      // The hidden details
    double myDegrees;
    char myScale;
};
```

The same definition that previously produced compilation errors will now compile correctly, because by naming the function as a friend, the class is granting this function access to its private section. Note also that operator<<() is not designated as a const function. This is because const *can only be applied to function members* and operator<<() is not a function member.

In a similar way, we can also replace the read() method with a friend version of operator>>(). In this case, the reference parameter aTemp would not be a const reference parameter, only a reference parameter, because input modifies it.

14.4.1.1 Use of friend

The friend mechanism is rarely needed to implement class operations. As we saw with the Temperature class, most operations on a class object can be defined as function members, so the friend mechanism is not needed for them.

When the left operand of an operation is of a type different from the class being built, then a method cannot be used and a normal function must be defined. But even in such infrequent cases, a public intermediary function (like `display` or `read()`) can be defined for that function to call. The only circumstances where the `friend` mechanism is an absolute necessity are when the left operand of the operation is of some type different from the class being built and the operation must directly access the data members of the class.

Because it embodies the external approach to defining class operations (in which a function manipulates an object from outside) instead of the *I-can-do-it-myself* principle of the object-oriented approach, the `friend` mechanism should be used only in those rare circumstances where it is a necessity.

14.4.2 Conditional Compilation and a Class "Wrapper"

Whenever a program stored in a file is compiled, it is first examined by a special program called the **preprocessor** that scans through the file doing some preliminary analysis before the file is passed on to the compiler itself. For example, the preprocessor strips all comments from the program so that the compiler need not spend time finding them only to ignore them. Another task of the preprocessor is to process all **preprocessor directives**, which are lines that begin with a # character such as

```
#include FileName
```

When it encounters this directive, the preprocessor finds the file named *FileName* and inserts it at that point in the program.

For large projects consisting of many library files, it is customary for each file to include whatever class declarations it needs. This means that the same class could be declared in several different places in a project, and this results in an error because C++ does not permit a class to be declared more than once.

However, no error results if the header file `<iostream>` is included more than once. Why? Because the contents of `<iostream>` are wrapped in directives that basically tell the preprocessor, *"If this is the first time you have seen this class, go ahead and process the declaration. If you have seen it before, skip the declaration."*

We can surround our class `Temperature` with directives that tell the preprocessor to do the same for class `Temperature`:

```
#ifndef TEMPERATURE
#define TEMPERATURE
class Temperature
{
 public:
    // function members

 private:
    // data members
};
#endif
```

These are called **conditional-compilation directives** because of what they do. The directive

```
#ifndef TEMPERATURE
```

instructs the preprocessor, "If TEMPERATURE is not defined, then continue processing as usual. Otherwise, skip everything between here and the first #endif directive you encounter."[6] Because TEMPERATURE is undefined the first time the preprocessor examines the file, it proceeds on to the next line. Here it encounters the directive

```
#define TEMPERATURE
```

which defines the identifier TEMPERATURE. The preprocessor then continues and processes the class declaration and passes it on to the compiler.

If the preprocessor should encounter Temperature.h a second time, the first thing it sees is the directive

```
#ifndef TEMPERATURE
```

This time, however, TEMPERATURE is defined, and so the preprocessor skips everything between that point and the #endif directive.

The result is that *the class declaration is only processed once, regardless of how many different files include* Temperature.h. A class should be wrapped in these directives to prevent redeclaration errors if the header file is included more than once in a project. Customarily, the identifier used with the #ifndef and #define directives (TEMPERATURE in this case) is the name of the class in all-uppercase letters.

CHAPTER SUMMARY

Key Terms

attribute	dot notation
class	encapsulation
class declaration	explicit-value constructor
class invariant	external perspective
class method	friend function
constant method	function member
constructor	I-can-do-it-myself principle
data member	information hiding
default-value constructor	instance variable

---

[6] The compiler will also stop skipping text if it encounters a #else or #elif directive, which behave like the else or the else if in an if statement.

interface

internal perspective

method

object autonomy

object-centered design

operator overloading

postcondition

preprocessor

private

public

reference return type

scope operator

NOTES

- Independence from implementation details is important in good class design.

- When working in a program and using a class, use an *external* perspective of an observer/user of the class. When working inside a class or building its function members, use an *internal* perspective of the object being designed.

- Like other declarations, a class declaration must be terminated by a semicolon.

- Two basic principles of class design:

  - *Encapsulation*—a single object can store values of different types by wrapping them in a class declaration.

  - *Information-hiding*—prevent direct access to data members by making them private in the class declaration.

- A *class invariant* that specifies restrictions on the values a class's data members can have should be formulated and checked whenever the data members are modified.

- A class can be wrapped in conditional-compilation directives to prevent redeclaration errors if the header file is included more than once in a project.

- Methods that do not alter the data members of the class should be declared as const methods.

- Constructors are used to initialize the data members of a class. The name of a constructor is always the same as the name of the class.

- An operator Δ can be overloaded by defining a function with the name operatorΔ, provided it has a signature distinct from that of any existing definition of operatorΔ.

- Designing a good class interface—which consists of its public operations—requires time and thought and should be done with care. If it is stable, then programs that use the class solely through the interface will not break even if the private portion of the class is modified. Changing it frequently causes programs that use the class to be revised often.

- A class can name a function as a friend, thereby granting it access to its private section.

Style and Design Tips

- *When an object in a program cannot be represented directly using predefined types, define a class to represent such objects.* One purpose of a class is to permit different data types to be encapsulated in a single object.

- *Use indentation to reflect the structure of your class, because this increases its readability.*

- *Use descriptive identifiers for the data members to reinforce the I-can-do-it-myself principle.* For example, begin each name with the prefix my.

- *Keep all data members of a class private and provide accessor functions to retrieve the values of those members.* This is part of hiding implementation details and simplifies program maintenance.

- *Methods that do not modify the data members of a class should be declared and defined as constant methods by appending the keyword* const *to their headings.* Doing so lets the compiler help you find logic errors if such methods inadvertently change the value of a data member (or call a function that might do so).

- *Put inlined definitions of simple methods after the class declaration in the header file for that class.* C++ does allow simple methods to be defined inside the class declaration, but doing so clutters the declaration, reducing its readability, so this practice should be avoided.

- *Define more complicated methods in a separately compiled implementation file.* If a definition is stored in the header file, it will be recompiled every time a program that includes that definition is compiled, which wastes time. Storing a function in a separately compiled implementation file eliminates this extra work.

- *Only overload an operator to perform an operation that is consistent with its symbol.* Avoid being cute and abusing the overloading mechanism by giving operators counterintuitive definitions because this reduces the readability of the code.

Warnings

1. *Members of a class that are declared following the keyword* private: *are not accessible outside of the class. Private members can only be accessed by methods and friend functions.*

2. *In definitions of the methods of a class, the method's name must be qualified with the name of the class and the scope operator (::).*

3. *The name of the constructor is the same as the name of the class, and the constructor has no return type.*

4. *Errors that result from inadvertent modification of the values of data members in a class can be difficult to find.* To avoid such errors, methods that access data members but do not modify them should be constant methods. This is accomplished by placing the keyword const after the closing parentheses that follows their parameter lists.

5. *A file containing a class declaration that may be #included in multiple files should enclose the class declaration in conditional compilation preprocessor directives to avoid errors:*

```
#ifndef Something
#define Something
    . . .
Class Declaration
    . . .
#endif
```

6. *A friend function must be named as such by the class of which it is a friend by preceding its declaration with the keyword friend in the class declaration.*

TEST YOURSELF

Section 14.2

1. For an object that cannot be directly represented with existing types, we design and build a _____ to represent it and store it in a _____.

2. There are two phases in creating a class: a _____ phase and a _____ phase.

3. What two things must be identified in designing a class?

4. Which of these is usually identified first?

5. Object autonomy is embodied in the _____ principle.

Section 14.3

1. Data items are stored in a class's _____ members, and are also known as_____ variables.

2. _____ allows a single object to store values of different types.

3. What is information-hiding and what is its purpose?

4. Data members are hidden by declaring them to be _____.

5. A set of conditions on the data members that must be true is called a class _____.

6. Class operations are implemented using _____ members, also known as _____.

7. In the definition of a function member, its name is preceded by the _____ and the _____ operator.

8. A function member that does not alter the data members of the class should be declared as a _____ method by attaching the keyword _____ at the end of its heading.

9. A constructor in a class Fraction will be named _____.

10. Name and describe two kinds of constructors.

11. (True or false) The return type of a constructor is `bool`.

12. The _____ (public or private) portion of a class acts as an interface between the class and programs that use it.

13. A boolean expression that must be `true` when an operation terminates is known as a _____-condition.

14. An object's function member is invoked by using _____ notation and is referred to as sending a _____ to that object.

15. Function members that retrieve values in data members are called _____ methods.

16. An addition operation, `+`, can be added to a class by using a function member named _____.

17. Write a declaration for a class `Component` that has two data members, `myID` of type `int`, and `myName` of type `string`, an output method, and an input method.

Section 14.4

1. A class can allow a function to access its data members by specifying that it is a _____.

2. Before a program is compiled it is first examined by the _____.

3. All lines that begin with a _____ character are preprocessor directives.

4. Directives of the form

```
#ifndef name
#define name
    . . .
#endif
```

are called _____ directives.

5. What is the purpose of directives like the preceding?

EXERCISES

Section 14.3

For Exercises 1–3, add function members to class `Temperature` to implement the specified operation. You should test these with driver programs as Programming Problems 1 and 2 ask you to do.

1. Convert a `Temperature` value to (a) Celsius (b) Kelvin.

2. Overload `operator+` for class `Temperature` so that expressions like `temp + 3.6` can be used to increase a `Temperature` value by a `double` number of degrees in the same scale.

3. Overload `operator-` for class `Temperature` so that expressions like `temp - 3.6` can be used to increase a `Temperature` value by a `double` number of degrees in the same scale.

For Exercises 4–12, define a class to model the given item. It should have appropriate constructors, accessors, mutators to change data members, input and output operations, and additional operations given in the problem.

4. A line in the plane specified by a point on it in Cartesian coordinates (x, y) and its slope; additional operations: (a) determine if a given point is on the line; (b) find its equation—the *point-slope equation* of a line having slope m and passing through point P with coordinates (x_1, y_1) is $y - y_1 = m(x - x_1)$; (c) find its x- and y-intercepts if they exist.

5. A line segment in the plane specified by its endpoints in Cartesian coordinates (x, y); additional operations: (a) find its length; (b) find its midpoint; (c) find its equation; (d) find the equation of its perpendicular bisector.

6. A circle specified by its center and its radius; additional operations: (a) find its area; (b) find its circumference.

7. A triangle specified by its three vertices; additional operations: (a) determine if it is (i) a right triangle, (ii) isosceles, (iii) equilateral; (b) find its perimeter; (c) find its area. (For the area, you can use *Hero's formula*:

$$1 \qquad \qquad '$$

where a, b, and c are its sides and s is one-half of its perimeter.)

8. Time specified by hours, minutes, seconds, and an AM/PM indicator; additional operations: (a) change to/from daylight savings time; (b) difference between two times; (c) the equivalent military time.

9. A date consisting of a month name, day number, and year; additional operations: (a) difference between two dates; (b) determine if year is a leap year; (c) find number of days in the month.

10. A telephone number as area code, local exchange, number, and name of person having that number.

11. Weather statistics: date; city and state, province, or country; time of day; temperature; barometric pressure; weather conditions (clear skies, partly cloudy, cloudy, stormy).

12. Information about a person: name, birthday, age, gender, social security number, height, weight, hair color, eye color, and marital status.

For Exercises 13 and 14, write appropriate class declarations to describe the information in the specified file. See the end of Chapter 11 for descriptions of these files.

13. `Inventory`

14. `Users`

PROGRAMMING PROBLEMS

Sections 14.3

1. Write a driver program to test the temperature converter methods of Exercise 1.

2. Write a driver program to test the methods `operator+()` and `operator-()` of Exercises 2 and 3.

3. Write a driver program to test the line class of Exercise 4.

4. Write a driver program to test the line segment class of Exercise 5.

5. Write a driver program to test the circle class of Exercise 6.

6. Write a driver program to test the triangle class of Exercise 7.

7. Write a driver program to test the time class of Exercise 8.

8. Write a driver program to test the date class of Exercise 9.

9. Write a driver program to test the telephone-number class of Exercise 10.

10. Write a driver program to test the weather-statistics class of Exercise 11.

11. Write a driver program to test the personal-information class of Exercise 12.

12. A *rational number* is of the form a/b, where a and b are integers with $b \neq 0$. Write a program to do rational number arithmetic, representing each rational number as a class that has numerator and denominator data members. The program should read and display all rational numbers in the format a/b, or simply a if the denominator is 1. The following examples illustrate the menu of commands that the user should be allowed to enter:

Input	Output	Comments
3/8 + 1/6	13/24	$a/b + c/d = (ad + bc)/bd$ reduced to lowest terms
3/8 − 1/6	5/24	$a/b − c/d = (ad − bc)/bd$ reduced to lowest terms
3/8 * 1/6	1/16	$a/b * c/d = ac/bd$ reduced to lowest terms
3/8 / 1/6	9/4	$a/b / c/d = ad/bc$ reduced to lowest terms
3/8 I	8/3	Invert a/b
8/3 M	2 + 2/3	Write a/b as a mixed fraction
6/8 R	3/4	Reduce a/b to lowest terms
6/8 G	2	Greatest common divisor of numerator and denominator
1/6 L 3/8	24	Lowest common denominator of a/b and c/d
1/6 < 3/8	true	$a/b < c/d$?
1/6 <= 3/8	true	$a/b \leq c/d$?
1/6 > 3/8	false	$a/b > c/d$?
1/6 >= 3/8	false	$a/b \geq c/d$?
3/8 == 9/24	true	$a/b = c/d$?
2/3 X + 2 = 4/5	X = −9/5	Solution of linear equation $(a/b)X + c/d = e/f$

Pointers and Linked Structures

[Pointers] are like jumps, leaping wildly from one part of a data structure to another. Their introduction into high-level languages has been a step backward from which we may never recover.

C. A. R. HOARE

He's making a list,
And checking it twice;
Gonna' find out who's naughty or nice.

"SANTA CLAUS IS COMING TO TOWN"

It is a mistake to try to look too far ahead. The chain of destiny can only be grasped one link at a time.

SIR WINSTON CHURCHILL

. . . is the sort of person who keeps a list of all of his lists.

ANONYMOUS

I N CHAPTER 12, we saw two different data structures that C++ provides for storing sequences of values: arrays and vector<T>s. One significant difference between these two kinds of objects is the way in which they are defined. For the kind of arrays we considered, their capacities *must* be specified at *compile time* as in the following declaration of anArray:

```
const int CAPACITY = 50;
int anArray[CAPACITY];
```

While a vector<T> object can be defined in a similar way,

```
const int CAPACITY = 50;
vector<int> aVector(CAPACITY);
```

its capacity can also be specified at *run time*

```
cout << "Enter the number of values to be stored: ";
int capacity;
cin >> capacity;
vector<int> aVector(capacity);
```

This is a basic difference between the two kinds of objects: an array's storage is determined (and is fixed) when the program is compiled, but the storage of a vector<T> object is determined (and can change) while the program executes. The string class is similar to vector<T> in that a string object's storage automatically adjusts to the number of characters being stored.

To build arrays whose capacities can be specified at run time and other structures whose storage can grow (and shrink) during execution, C++ provides a way to request and return memory during program execution. To understand this feature and how to use it, we must first study *pointers* and *indirection*.

15.1 POINTERS AND INDIRECTION

As usual, we begin with a program—the program in Example 15.1—to introduce the basics of pointers and indirection. It is not intended to show how pointers are typically used in programs.[1]

Example 15.1 Using Indirection

```
/* indirection.cpp illustrates indirection and pointer variables.

   Output: addresses of memory locations and the integers stored there
   ----------------------------------------------------------------*/

#include <iostream>
using namespace std;
```

---
[1] For some versions of C++, it may be necessary to use (void*)*pointerVariable* in an output statement for addresses to display correctly.

```
int main()
{
  int i = 11,
      j = 22,
      k = 33;

  int * iPtr = &i;
  int * jPtr = &j;
  int * kPtr = &k;
  cout << "\nAt address " << iPtr
       << ", the value "  << *iPtr << " is stored.\n"
       << "\nAt address " << jPtr
       << ", the value "  << *jPtr << " is stored.\n"
       << "\nAt address " << kPtr
       << ", the value "  << *kPtr << " is stored.\n";
}
```

SAMPLE RUN:

```
At address 0xbffff970, the value 11 is stored.

At address 0xbffff96c, the value 22 is stored.

At address 0xbffff968, the value 33 is stored.
```

15.1.1 Declaring and Initializing Pointers

We begin with the second set of declarations in the program in Example 15.1:

```
int * iPtr = &i;
int * jPtr = &j;
int * kPtr = &k;
```

There are two new items in these statements:

1. *An asterisk (*) following the type name in a declaration of the form*

   ```
   Type * variableName;
   ```

   declares that `variableName` *can store the address of a memory location where a value of the specified* `Type` *is stored.*[2] Such variables are often called **pointer variables**, or simply **pointers**. Thus, the declarations

   ```
   int * iPtr;
   int * jPtr;
   int * kPtr;
   ```

---

[2] Some programmers prefer attaching the asterisk to the type identifier or to the variable name; for example, `Type* variableName;` or `Type *variableName.`

declare that `iPtr`, `jPtr`, and `kPtr` are pointer variables, each of which can store the address of a memory location where an `int` is stored. The type of each of these variables is `int *`.

2. *The ampersand operator (&) can be used as a unary prefix operator on a variable name,*

   `&variableName`

   that returns the address with which `variableName` *is associated,* so & is called the **address-of operator**. Thus, the expressions `&i`, `&j`, and `&k` return the addresses (or references[3]) associated with variables i, j, and k, respectively.

Combining these two pieces of information, we see that the declarations

```
int * iPtr = &i;
int * jPtr = &j;
int * kPtr = &k;
```

declare `iPtr`, `jPtr`, and `kPtr` as pointer variables, each of which can store the address of a memory location containing an `int`, and they initialize `iPtr` to the address of variable i, `jPtr` to the address of variable j, and `kPtr` to the address of variable k. In the sample run in Example 15.1, the address associated with variable i is the hexadecimal value `0xbffff970`, the address of j is `0xbffff96c`, and the address of k is `0xbffff968`. We can visualize the layout of the program's data in memory as follows:[4]

0xbffff970	11	i
0xbffff96c	22	j
0xbffff968	33	k
	0xbffff970	iPtr
	0xbffff96c	jPtr
	0xbffff968	kPtr

It is important to remember that *in a declaration, an asterisk operator * must precede each identifier that is to serve as a pointer.* Thus,

```
double * ptr1,
       * ptr2;
```

is a correct declaration of `ptr1` and `ptr2` as pointers to doubles. Had we written

```
double * ptr1,
         ptr2;
```

---

[3] The word *reference* is used as a synonym for *address*. In fact, this is the origin of the phrase *reference parameter*—the value of a reference parameter is actually the address of its argument, rather than a copy of the argument.

[4] Note that (using hexadecimal arithmetic)

   0xbffff970 – 0xbffff96c = 4

and

   0xbffff96c – 0xbffff968 = 4

which indicates that the size of an `int` on this particular machine is 4 bytes (32 bits).

however, only `ptr1` would be a pointer variable; `ptr2` would be an ordinary `double` variable. To avoid making this mistake, we will normally use a separate declaration for each pointer variable:

```
double * ptr1;
double * ptr2;
```

15.1.1.1 Using `typedef` for Readability

An alternative notation that does not require the repeated use of the asterisk in pointer declarations is to use `typedef` to rename a type. For example, we could first declare

```
typedef int * IntPointer;
```

in Example 15.1 and then use `IntPointer` to declare the pointers:

```
IntPointer iPtr = &i,
           jPtr = &j,
           kPtr = &k;
```

Such declarations improve the readability of pointer declarations, especially when pointer parameters are being declared.

15.1.2 Basic Pointer Operations

C++ supports a variety of operations on pointers, including initialization, dereferencing, I/O, assignments, comparisons, and arithmetic. We examine each of these in turn.

15.1.2.1 Initialization

When a pointer variable is initialized to an address, as in

```
int * iPtr = &i;
```

that address must be the address of an object whose type is the same as the type to which the pointer points. The pointer is said to be **bound** to that type. For example, the declarations

```
double doubleVar;
int * iPtr = &doubleVar; // ERROR
```

will cause a compiler error, because an integer pointer can only store addresses of integer objects.

One important exception is that 0 can be assigned to any pointer variable. The value that results is called the **null pointer value** for that type and 0 is often called the **null address**. Thus, the declarations

```
char * cPtr = 0;
int * iPtr = 0;
double * dPtr = 0;
```

are all valid initializations using the null address.

The null address can also be used in a boolean expression to indicate whether or not a pointer is pointing to anything:

```
if (dPtr == 0)
    // dPtr is not currently pointing to anything
else
    // dPtr is pointing to a memory location
```

As we will see, such comparisons are especially important when pointers are used to store the addresses of blocks of memory allocated during execution.

15.1.2.2 Indirection and Dereferencing

Pointer variables not only store addresses but also provide access to the values stored at those addresses. An expression of the form

```
*pointerVariable
```

can be used to access the value at the address stored in *pointerVariable*. It can be thought of as going to the reference (address) stored in *pointerVariable* and accessing the value stored at that address. To illustrate, in the sample run of Example 15.1, the value of the expression

```
iPtr
```

is 0xbffff970, and the value of the expression

```
*iPtr
```

is 11, because 11 is the value stored at address 0xbffff970:

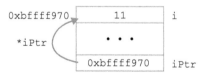

For the same reason, the values of the expressions *jPtr and *kPtr are 22 and 33, respectively. Thus, the value of variable i can be accessed via the expression *iPtr, the value of j via *jPtr, and the value of k via *kPtr. In general, the value of a variable v can be *accessed indirectly* by applying the * operator to a pointer variable vPtr whose value is the address of v. For this reason, the * operator is called the **indirection operator**. Because *reference* is another term for *address* and applying the indirection operator to a pointer variable accesses the value at the address stored in that pointer variable, applying the indirection operator to a pointer variable is called **dereferencing** that pointer variable.

We have already used this indirect access technique in earlier chapters. For example, in our study of vector<T> objects in Section 12.6, we saw that dereferencing an iterator provides access to the value stored at the position in a vector<T> object to which the iterator points. Also, although we didn't discuss it in our study of classes in the preceding chapter, each class object contains a pointer variable this whose value is the address of the object that contains it, and dereferencing this provides a way to (indirectly) access that object. We might picture this as follows:

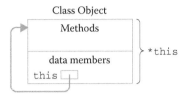

The indirection operator can be applied more than once to produce additional levels of indirection. For example, the declarations

```
typedef int * IntPointer;      // or without using typedef:
IntPointer * ptr;              //     int ** ptr;
```

declare ptr to be a pointer to a memory location that contains a pointer to another memory location where an int can be stored.

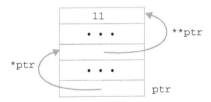

The indirection operator can be used on either side of an assignment statement. If the statement

```
i = *jPtr;
```

were added to the program in Example 15.1, the value of i would be changed from 11 to 22, because dereferencing jPtr produces the value 22 stored at address 0xbffff96c and this value would be assigned to i. The statement

```
*iPtr = j;
```

would produce the same result. *iPtr refers to the memory location with address 0xbffff970 and the assignment operator copies the value of j (22) into this memory location. Because this address is associated with the variable i, the effect is to change the value of i.

As noted earlier, the purpose of the program in Example 15.1 was to introduce the basics of pointers and indirection. Pointers are not often used to store addresses that are associated with names. Instead, as we will see, pointers are used to store and retrieve values in memory locations with which *no name* has been associated.

15.1.2.3 Pointers to Class Objects

Although the program in Example 15.1 does not do so, we can also declare pointers to class objects and use them to store the addresses of objects. For example, in the preceding chapter, we built a `Temperature` class that we used to define `Temperature` objects such as

```
Temperature temp1(98.6, 'F');
```

Given such an object, we could declare a pointer to a `Temperature` and use it to store the address of that object,

```
Temperature * tempPtr = &temp1;
```

which can be pictured as follows:

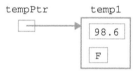

The members of `temp1` can be accessed (indirectly) via `tempPtr`. For example, `temp1` has a `scale()` method that returns the value (`'F'`) of its `myScale` data member. The `scale()` message can be sent to `temp1` via `tempPtr`, and this can be done in two ways. One way is to combine the indirection operator with the dot operator and write

```
(*tempPtr).scale()
```

In this expression, the pointer `tempPtr` is first dereferenced to access the object to which it points (i.e., `temp1`), and the dot operator is then used to send that object the `scale()` message.

This notation is rather cumbersome because it involves two operators and the indirection operation *must be parenthesized* because it has lower priority than the dot operator. C++ provides a more convenient notation that accomplishes the same thing in one operation:

```
tempPtr->scale()
```

Here `->` is the **class pointer selector operator** whose left operand is a *pointer* to a class object and whose right operand is a *member* of the class object. This operator provides a convenient way to access that object's members, and the "arrow" notation clearly indicates that the member is being accessed through a pointer.

15.1.2.4 I/O

In the program in Example 15.1, we displayed the addresses associated with i, j, and k by displaying the values of iPtr, jPtr, and kPtr, which stored these addresses.[5] In a similar manner, to find the addresses associated with iPtr, jPtr, and kPtr, we could write

```
cout << "\n iPtr is stored at address " << &iPtr
     << ",\n jPtr is stored at address " << &jPtr
     << ", and\n kPtr is stored at address " << &kPtr << endl;
```

The address-of operator allows us to determine the exact memory address at which something is stored, whereas pointer variables allow us to store these addresses.

Just as the value of a pointer can be output using <<, an address could be input and stored in a pointer variable using >>. However, this is rarely done, because we usually are not interested in the address of the memory location storing a value, only in the value itself. In fact, it is dangerous to input address values because an attempt to access a memory address outside the space allocated to an executing program will result in a fatal run-time error.

15.1.2.5 Assignment

Although the program in Example 15.1 does not illustrate it, pointer variables can be assigned the values of other pointer variables that are *bound to the same type*. For example, if we were to add the statement

```
jPtr = iPtr;
```

to the program, then the value of iPtr would be copied to jPtr so that both have the same memory address as their value; that is, both point to the same memory location, as the following diagrams illustrate:

Before the assignment:

After the assignment jPtr = iPtr;:

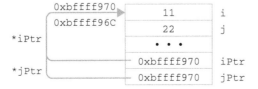

---

[5] See Footnote 1 about displaying pointer values.

After the assignment statement is executed, jPtr no longer points to j, but now points to i. Thus, applying the indirection operator to jPtr will access the memory location associated with i. For example, an output statement

```
cout << *jPtr;
```

will display the value 11 instead of 22, and the statement

```
*jPtr = 44;
```

will change the value at address 0xbffff90 (i.e., the value of i) from 11 to 44:

This example was included to show that pointers are a very powerful (and dangerous) feature of programming languages. Statements that change the value of a variable in a statement in which that variable is not named are generally considered to be poor programming practice, because they make programs difficult to debug by hiding such changes. In the preceding example, the expressions *iPtr and *jPtr are alternate names for variable i and are sometimes called **aliases** for i. A function that changes a variable's value through an alias for that variable is said to exhibit the **aliasing problem**.

15.1.2.6 Comparison

The relational operators can be used to compare two pointers that are *bound to the same type*. The most common operation is to use == and != to determine if two pointer variables both point to the same memory location. For example, the boolean expression

```
iPtr == jPtr
```

is valid and returns true if and only if the address in iPtr is the same as the address in jPtr. However, if pointers nPtr and dPtr are declared by

```
int * nPtr;
double * dPtr;
```

the comparison

```
nPtr == dPtr // ERROR!
```

will result in a compilation error, because nPtr and dPtr are bound to different types.

The *null address may be compared with any pointer variable.* For example, the conditions

```
nPtr != 0   and   dPtr == 0
```

are both valid boolean expressions.

15.1.2.7 Pointer Arithmetic

To explain arithmetic operations on pointers, it is helpful to make use of a C++ operator that we have not used up to now. This is the **sizeof operator**, which may be applied to either objects or types,

```
sizeof(type-specifier)
sizeof expression
```

and returns the number of bytes required to store a value of the specified type or the value of the expression. Note that in the first case, the type specifier must be enclosed within parentheses.

To illustrate, the value of `sizeof(char)` is the number of bytes (usually 1) allocated to values of type `char`, and if `longVar` is of type `long int`, the value of `sizeof longVar` is the number of bytes (typically 4) allocated to values of type `long int` objects.

Understanding the `sizeof` operator makes it easier to understand pointer arithmetic. We consider the increment and decrement operations first because they are probably the most commonly used arithmetic operations on pointer variables. For a pointer variable `ptr` declared by

```
Type * ptr;
```

the increment statement

```
ptr++;
```

adds the value `sizeof(Type)` to the address in `ptr`. Similarly, a decrement statement

```
ptr--;
```

subtracts the value `sizeof(Type)` from the address in `ptr`. If *intExpr* is an integer expression, a statement of the form

```
ptr += intExp;
```

adds the value *intExp* * `sizeof(Type)` to `ptr`, and

```
ptr -= intExp;
```

subtracts the value *intExp* * `sizeof(Type)` from `ptr`.

To illustrate how these operations are used, suppose that `ptr` is a pointer whose value is the address of the first element of an array of `double` elements:

```
double dArray[10];           // array of 10 doubles
double * ptr = &(dArray[0]); // pointer to first element of dArray
```

The last declaration could also be written

```
double * ptr = dArray; // pointer to first element of dArray
```

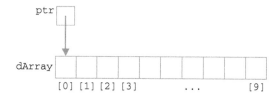

Now consider the following loop:

```
for (int i = 0; i < 10; i++)
{
    *ptr = 0;
    ptr++;
}
```

On the first pass through the loop, `ptr` is dereferenced and the value 0 is assigned to the memory location at that address. `ptr` is then incremented, which adds `sizeof(double)` to its value, effectively making `ptr` point to the second element of the array:

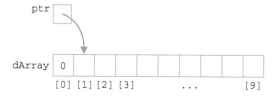

The next pass again dereferences `ptr`, sets that memory location to zero, and increments `ptr`:

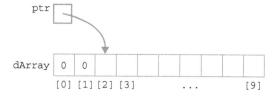

This continues with each subsequent iteration. On the final pass, the last element of the array is set to zero. Then after `ptr` is incremented, it points to the first address past the end of the array:

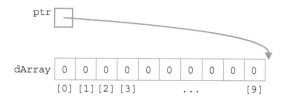

A pointer can thus be used to move through consecutive blocks of memory, accessing them in whatever way a particular problem requires.

From these examples, it should be apparent that pointers are the basis for **iterators**, which, as we saw in Section 12.7, are provided by Standard Template Library (STL) containers for accessing the values they store. Iterators are in fact implemented using pointers, and behave in much the same way, with ++ being used to move the iterator to the next value in the container, -- to move the iterator to the previous value in the container, and * to dereference the iterator and access the value to which it "points."

15.1.2.8 Pointers as Arguments and Parameters

Pointers may also be passed as arguments to functions. The parameters corresponding to such arguments may be either value or reference parameters, but the pointer argument and the corresponding parameter must be bound to the same type. The return type of a function may also be a pointer.

15.2 RUN-TIME ARRAYS

In the first part of Chapter 12, we saw that the definition of a C-style array

```
const int CAPACITY = 10;
double arrayName[CAPACITY];
```

causes the compiler to allocate a block of memory large enough to hold ten double values and associate the starting address of that block with the name arrayName. Such fixed-size arrays have two drawbacks:

- If the size of the array exceeds the number of values to be stored in it, then memory is wasted by the unused elements.

- If the size of the array is smaller than the number of values to be stored in it, then the problem of array overflow may occur.

At the root of these problems is the fact that the capacity of a C-style array is fixed when the program is *compiled*. In our example, the size of the block of memory allocated for arrayName cannot be changed, except by editing the declaration of CAPACITY and then recompiling the program.

What is needed are arrays whose capacities are specified during execution. Such **run-time arrays** can be constructed using the mechanism C++ provides for run-time memory allocation. At its simplest, such a mechanism requires two operations:

1. Acquire additional memory locations as they are needed.

2. Release memory locations when they are no longer needed.

C++ provides the predefined operations new and delete to perform these two operations of memory allocation and deallocation during program execution.

15.2.1 The new Operation

The new operation is used to request additional memory from the operating system during program execution. The general form of such a request is:

THE new OPERATION

FORM:

```
new Type
```

PURPOSE:

Issue a run-time request for a block of memory that is large enough to hold a value of the specified *Type*. If the request can be granted, new returns the address of the block of memory; otherwise, it returns the null address.

Because the new operation returns an address and addresses can be stored in pointer variables, this operation is almost always used in conjunction with a pointer. For example, when the statements

```
int * intPtr;
intPtr = new int;
```

are executed, the expression new int issues a request to the operating system for a memory block large enough to store an int value (that is, for sizeof(int) bytes of memory). If the operating system is able to grant the request, intPtr will be assigned the address of this memory block. Otherwise, if all available memory has been exhausted, intPtr will be assigned the null address 0. Because of this possibility, the value returned by new should always be tested before it is used; for example,

```
assert(intPtr != 0);
```

If intPtr is assigned a nonzero value, the newly allocated memory location is an **anonymous variable**; that is, it is an allocated memory location that has no name associated with it. For example, suppose new returns the address 0x020:

```
      intPtr
     ┌────────┐
     │ 0x020  │      0x020 ┌──────────┐
     └────────┘            └──────────┘
```

Because there is no name associated with this newly allocated memory, it *cannot be accessed directly* in the same way other variables are accessed. However, its address is stored in intPtr, so this anonymous variable can be *accessed indirectly* by dereferencing intPtr:

Statements such as the following can be used to operate on this anonymous variable:

```
cin >> *intPtr;          // store input value in the new integer
if (*intPtr < 100)       // apply relational ops to new integer
   (*intPtr)++;          // apply arithmetic ops to new integer
else
   *intPtr = 100;        // assign values to the new integer
```

In short, anything that can be done with an "ordinary" integer variable can be done with this anonymous integer variable by accessing it indirectly via `intPtr`.

15.2.1.1 Allocating Arrays with new

In practice, `new` is rarely used to allocate space for scalar values like integers. Instead, it is used to allocate space for either arrays or for anonymous class objects. To illustrate the former, consider an integer array object `anArray` declared by

```
int anArray[10];
```

The value associated with the name `anArray` is the **base address** of the array, that is, the address of the first element of the array.[6] The type of object `anArray` is `int[10]`.

A type such as `int[10]` can be used with `new` to allocate the memory for an array at run time. For example, the statements

```
int * arrayPtr;
arrayPtr = new int[10];
```

allocate space for an array of 10 integers. Until the second statement is executed, `array-Ptr` is simply a pointer variable whose value is undefined. After it is executed (assuming that sufficient memory is available), `arrayPtr` contains the base address of the *newly allocated* array. If that address is `0x032`, we might picture the situation as follows:

But we have seen previously that the value associated with the name of a compile-time allocated array is its base address. This means that:

> *If the base address of a run-time allocated array is stored in a pointer variable, then the elements of that array can be accessed via the pointer in exactly the same way that the elements of a compile-time allocated array are accessed via its name, by using the subscript operator ([]).*

---

[6] This is one reason that the assignment operator cannot be used to copy a "normal" array—the statement
```
alpha = beta;
```
would attempt to copy the starting address of `beta` into `alpha`, as opposed to copying the elements of `beta`.

That is, the first element of the new array can be accessed using the notation `arrayPtr[0]`, the second element using `arrayPtr[1]`, the third element using `arrayPtr[2]`, and so on:

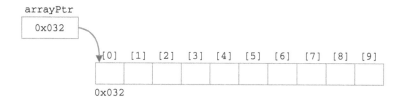

The value of the pointer variable `arrayPtr` is the base address of the array, and for a given index `i`, the subscript operator

```
arrayPtr[i]
```

simply accesses the memory location `arrayPtr + i`.

The advantage of run-time allocation is that it is not necessary to know the capacity of the array at compile time. For example, we can write:

```
cout << "How many entries? ";        // find out how big the
cin >> numEntries;                    //    array should be

double *dPtr =                        // allocate an array
        new double[numEntries];       //    with that capacity

assert(dPtr != 0)                     // check for success
cout << "Enter your values.\n";       // fill it with values

for (int i = 0; i < numEntries; i++)
    cin >> dPtr[i];
```

Unlike arrays whose memory is allocated at compile time, arrays whose memory is allocated at run time can be tailored to the exact size of the list to be stored in them. The wasted memory problem is solved because the array will not be too large. The overflow problem is solved because the array will not be too small.

This is precisely the approach used by the `vector<T>` class template, whose structure might be something like the following:

```
template<typename T>
class vector
{
 public:
   vector();
   vector(int n);
 private:
   T * tPtr;
   int myCap; // my capacity
};
```

Note the data member that is a pointer to a value of type T. The default vector<T> constructor simply initializes this pointer to the null address to signify an empty vector:

```
tPtr = 0;
myCap = 0;
```

But the explicit-value constructor uses new to dynamically allocate an array with a specified capacity:

```
tPtr = new T[n];
if (tptr != 0)
   myCap = n;
```

15.2.2 The delete Operation

When execution of a program begins, the program has available to it a "pool" of unallocated memory locations, called the **free store** or **heap**. The effect of the new operation is to request the operating system to remove a block of memory from the free store and allocate it to the executing program. The program can use this block if it stores its address (the value produced by the new operation) in a pointer variable. However, the size of the free store is limited, and each execution of new causes the pool of available memory to shrink. If a call to new requests more memory than is available in the free store, then the operating system is unable to fill the request and new returns the null address 0.

Memory that is no longer needed can be returned to the free store by using the delete **operation**.

THE delete OPERATION

FORM:

delete *pointerVariable*

or

delete *arrayPointerVariable*[]

PURPOSE:
The first form frees the run-time allocated object whose address is stored in *pointerVariable*. The second form frees the run-time allocated array whose address is stored in *arrayPointerVariable*.

Just as new is a request by the executing program for memory from the free store, the delete operation is a request to return memory to the free store. Such memory can then be reused by the memory manager. The new and delete operations are thus complementary.

If memory is dynamically allocated to an object during program execution but is not deallocated via delete when that object's lifetime is over, a **memory leak** is said to occur. Objects such as vector<T>s whose constructors use new to allocate memory for data members will also have a **destructor** method to deallocate memory via delete when that object's lifetime is over. It's name will have the form ~*name* where *name* is the default constructor; for example, vector<T>'s destructor is named ~vector().

15.3 INTRODUCTION TO LINKED LISTS

Although arrays and vector<T>s are easy to use to store sequences of values, they do have limitations. One limitation is that values can be efficiently added to the sequence or removed from it only at its *back*. If a problem requires that values be inserted or removed anywhere else, much shifting of elements is required.

To illustrate, consider the array (or vector<T>) shown below and suppose we want to add 99 at the front of the list. To make room for this new value, all of the elements of the array must be shifted one position to the right. This is very inefficient for large arrays and large elements, because they must all be copied from one location to another.

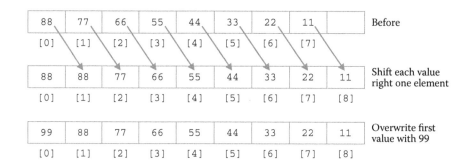

The same problem occurs when any element other than the one at the end of the sequence must be removed. All of the elements that follow it must be shifted one position to the left to close the gap. The following diagram illustrates this when the first element is removed:

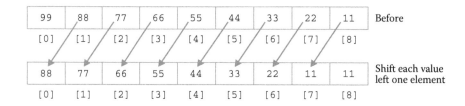

For problems where many such within-the-sequence insertions and deletions are required, a *linked list* should be used. It allows values to be inserted or removed anywhere in a sequence without any of this copying.

15.3.1 What Are They?

A **linked list** is a series of **nodes** linked together by pointers. In addition to space for the data being stored, each node has a pointer to the node containing the next value in the list. A pointer to the node storing the first list element must also be maintained. This will be the null value if the list is empty.

To illustrate, a linked list storing the integers 9, 17, 22, 26, 34 might be pictured as follows:

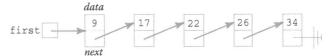

In this diagram, arrows represent links, and `first` points to the first node in the list. The *data* part of each node stores one of the elements of the list, and each arrow from a *next* part represents a pointer. The symbol in the last node (a version of the ground symbol used in electrical engineering) represents a null link and indicates that this list element has no successor.

15.3.1.1 Insert Operation

To see how linked lists make it possible to avoid the data-shifting problem of arrays and `vector<T>`, suppose we wish to insert 20 after 17 in the preceding linked list, and that `predptr` points to the node containing 17. We first obtain a new node (via the `new` operator) temporarily pointed to by `newptr` and store 20 in its data part:

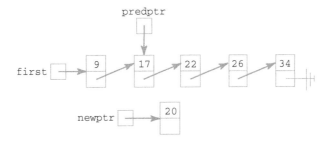

We insert it into the list by first setting its next part equal to the link in the node pointed to by `predptr` so that it points to its successor:

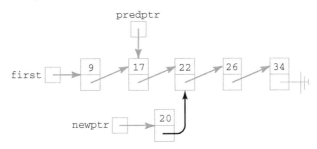

Now reset the link in the predecessor node to point to this new node:

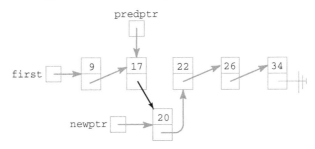

This same procedure also works for inserting a value at the end of the list. Check this for yourself by following the steps to append 55 to the list.

Inserting at the beginning of list, however, requires a modification of the last two steps, because there is no predecessor to which we can attach the node:

- Set the next part of the new node equal to `first`, which makes it point to the first node in the list.

- Then change `first` to point to the new node.

Work through this procedure yourself to insert 5 at the beginning of the list.

Note that only three instructions are needed to insert a value at any point in the list. *No shifting of list elements is required!*

15.3.1.2 Delete Operation

Deleting elements from a linked list can also be done very efficiently, as the following diagram demonstrates:

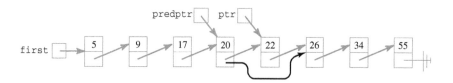

Here `ptr` points to the node to be deleted and `predptr` to its predecessor. We need only perform a bypass operation by setting the link in the predecessor to point to `ptr`'s successor. To avoid a memory leak, the deleted node should be returned to the free store (via `delete`).

Once again, this same procedure also works at the end of the list. Convince yourself of this by deleting 55. Also, a modification is needed when deleting the first node because there is no predecessor. It consists of simply resetting `first` to point to the second node in the list and then returning the old first node to the storage pool of available nodes. Check this for yourself by seeing how 5 would be deleted from the above list.

Like insertion, this is a very efficient operation—only two instructions are needed to delete any value in the list. *No shifting of list elements is required!*

15.3.2 A Linked List Class

As we will see in the next section, the Standard Template Library has a `list<T>` class template that stores list elements in a linked list (but which have a more complex structure than what we have been considering). Like most of the other STL containers, `list<T>` provides many list operations, because it is intended for use in a wide variety of list-processing problems. As noted before, there are times when one doesn't need or want all of the operations, and a "lean and mean" linked-list class that contains the basic list operations would be more suitable. Such an implementation can be found on the website for this textbook described in the Preface.

15.4 THE STL LIST<T> CLASS TEMPLATE

In our description of the C++ Standard Template Library in Section 12.7, we saw that it provides a variety of other storage containers besides vector<T> and that one of these containers is named list<T>. Now that we have seen anonymous variables and how C++ pointers provide indirect access to them, and have studied simple linked lists, we are ready to examine the list<T> class template and its implementation.

To see how list<T> stores a sequence of values, suppose that aList is defined by

```
list<int> aList;
```

and consider the following sequence of insert operations:

```
aList.push_back(77);
aList.push_back(66);
aList.push_front(88);
```

A simplified picture of the resulting object aList is

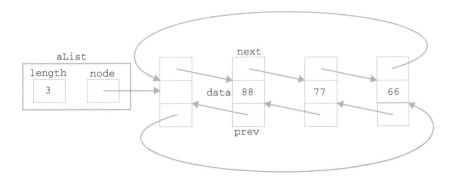

The values 77, 66, and 88 are stored in a variation of the linked lists studied in the preceding section called a **circular doubly linked list with a head node**. It is doubly linked because each node has two pointers, prev to its predecessor and next to its successor. It is circular because the next pointer in the last node is not null, but rather points to the "empty" (leftmost) node, which is the head node, and similarly, the prev pointer in the first node points to the head node instead of being null. Note that the data member node (which we called first in the linked list examples of the preceding section) points to this head node rather than to the first node that stores a data value.

15.4.1 Some list<T> Operations

In the remainder of this section, we examine a collection of the most useful list<T> operations. For some of these, we look at how they are implemented. More details about these are given on the text's website described in the Preface.

15.4.1.1 The list<T> Default-Value Constructor

Perhaps the most basic list<T> operation is the default-value constructor. When a programmer writes

```
list<int> aList;
```

the default-value constructor builds an empty linked list aList, for which a (simplified) picture is

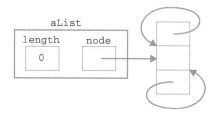

As shown in the diagram, the default class constructor allocates an empty node, which is the head node, and stores the address of this node in its data member node. In the STL list<T> class template, this head node plays a central role: Its next member always points to the node containing the first value in the sequence (or to the head node, if the list is empty), and its prev member always points to the node containing the last value in the sequence (or to the head node, if the list is empty). The main advantages of this organization is that there is always at least one node in the list (i.e., the head node) and every node has a predecessor and a successor. These properties simplify several of the list operations. In particular, the insert and delete operations we considered in the preceding section do not have to consider two cases of whether there is a predecessor or not.

15.4.1.2 *The* size() *and* empty() *Operations*

Two of the simplest list<T> operations are size() and empty(). The size() method is a simple accessor for the length data member; it returns the number of values currently stored in the list. The empty() method is nearly as simple, returning true if there are no values in the list (length == 0) and false otherwise.

15.4.1.3 *The* begin() *and* end() *Iterators*

As with vector<T>, the list<T> class template provides two methods, begin() and end(), that return iterators to the front and past the end of the list, respectively. The begin() method returns a pointer to the first node, by returning the address stored in the next member of the head node. By contrast, the end() function returns a pointer pointing beyond the last node containing a data value by returning the address of the head node:

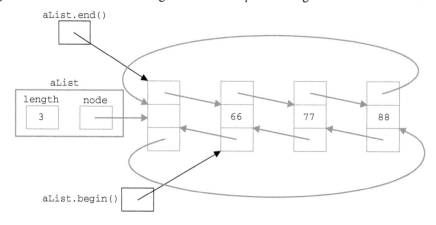

From our discussion of iterators in Chapter 12 and our discussion of pointers in this chapter, it should be evident that an iterator is an *abstraction* of a pointer, hiding some of its details and eliminating some of its hazards.

15.4.1.4 The insert(), push_front(), and push_back() Operations

The list<T> class template provides several operations to insert a new data value, including:

- aList.push_back(*newValue*); which appends *newValue* to aList

- aList.push_front(*newValue*); which prepends *newValue* to aList

- aList.insert(*anIterator, newValue*); which inserts *newValue* into aList ahead of the value pointed to by *anIterator*

Of these three, insert() is the most general; push_back() and push_front() simply invoke insert(), passing aList.begin() and aList.end() to *anIterator*, respectively.

15.4.1.5 The pop_back(), pop_front(), erase(), and remove() Operations

There also are several different operations provided by list<T> to remove a value from a sequence:

- aList.pop_back(); removes the last value from aList

- aList.pop_front(); removes the first value from aList

- aList.erase(*anIterator*); removes the value pointed to by *anIterator* from aList

- aList.remove(*aValue*); removes all occurrences of *aValue* from aList

15.4.2 An Application: Internet Gateways

As we saw in the opening example of Chapter 7, IP (Internet Protocol) addresses are used to uniquely identify computers in the Internet. Each address is made up of four fields that represent specific parts of the Internet,

host.subdomain.subdomain.rootdomain

which the computer will translate into a unique 32-bit numeric address. For example, www.calvin.edu is the IP address of one of the computers at Calvin College for which the corresponding numeric address is, at the time of this writing, 153.106.4.1.

15.4.2.1 Problem

A *gateway* is a device used to interconnect two different computer networks. Suppose that a gateway connects a university to the Internet and that the university's network administrator needs to monitor connections through this gateway. Each time a connection is made (for example, a student using the World Wide Web), the IP address of the student's computer is stored in a data file. The administrator wants to check periodically who has used the gateway and how many times they have used it.

15.4.2.2 Solution

The IP addresses will be read from the file and stored in a linked list of nodes that will store an address and the number of times that address appeared in the data file. As each address is read, we check if it is already in the list. If it is, we increment its count by 1; otherwise, we simply insert it at the end of the list. After all the addresses in the file have been read, the distinct addresses and their counts are displayed.

The program in Example 15.2 uses this approach to solve the problem. The addresses are stored in a `list<AddressCounter>` object named `addrCntList`, where `AddressCounter` is a small class containing two data members (`address` and `count`), function members for input and output operations, and `tally()` to increment the count. Also, `operator==()` is overloaded so that STL's `find()` algorithm can be used to search the list.

Example 15.2 Monitoring Internet Connections

```
/* Program to read IP addresses from a file and produce a list of
   distinct addresses and a count of how many times each appeared in
   the file. The addresses and counts are stored in a linked list.

   Input (keyboard): name of file containing addresses
   Input (file): addresses

   Output: a list of distinct addresses and their counts
   -----------------------------------------------------------------*/

#include <cassert>      // assert()
#include <string>       // string
#include <iostream>     // cin, cout, >>, <<
#include <iomanip>      // setw()
#include <fstream>      // ifstream, is_open()
#include <list>         // list<T>
#include <algorithm>    // find()
using namespace std;

//-------------- Begin class AddressItem -----------------------
class AddressCounter
{
 public:
   void read(istream & in) { in >> address; count = 1; }

   void print(ostream & out) const
   { out << setw(15) << left << address
         << " occurs " << count << " times\n"; }
```

```
    void tally() { count++; }

    bool operator==(const AddressCounter & addr2)
    { return address == addr2.address; }

    string getAddress() const
    { return address; }

  private:
    string address;
    int count;
};
//---------------- End class AddressCounter --------------------

typedef list<AddressCounter> IP_List;

int main()
{
  string fileName;                        // file of IP addresses
  IP_List addrCountList;                  // list of addresses

  ifstream inStream;                      // open file of addresses
  cout << "Enter name of file containing IP addresses: ";
  cin >> fileName;
  inStream.open(fileName.data());
  assert(inStream.is_open());

  AddressCounter item;                    // one address & its count
  for (;;)                                // input loop:
  {
     item.read(inStream);                 //   read an address
     if (inStream.eof()) break;           //   if eof, quit

     IP_List::iterator it =               //   is item in list?
       find(addrCountList.begin(), addrCountList.end(), item);
     if (it != addrCountList.end())       //   if so:
       (*it).tally();                     //      ++ its count
     else                                 //   otherwise
       addrCountList.push_back(item);     //      add it to the list
}                                         //   end loop

cout << "\nAddresses and Counts:\n\n";    // output the list
for (IP_List::iterator it = addrCountList.begin();
                    it != addrCountList.end(); it++)
   (*it).print(cout);
}
```

LISTING OF FILE ipAddresses.txt **USED IN SAMPLE RUN:**

```
128.159.4.20
```

```
123.111.222.333
100.1.4.31
34.56.78.90
120.120.120.120
128.159.4.20
123.111.222.333
123.111.222.333
77.66.55.44
100.1.4.31
123.111.222.333
128.159.4.20
```

SAMPLE RUN:

```
Enter name of file containing IP addresses: ipAddresses.txt

Addresses and Counts:

128.159.4.20      occurs 3 times
123.111.222.333  occurs 4 times
100.1.4.31        occurs 2 times
34.56.78.90       occurs 1 times
120.120.120.120  occurs 1 times
77.66.55.44       occurs 1 times
```

15.4.3 Algorithm Efficiency

When we study the **time-efficiency** of algorithms in computer science, we do not concern ourselves with real (wall-clock) time, because that varies with the language in which the algorithm is encoded, the quality of the code, the quality of the compiler, the speed of the computer on which the code is executed, and various other factors. Instead, we study the number of *steps* an algorithm takes as a function of the size of the problem it solves. For example, the following function for the summation problem uses a for loop that iterates n times:

```
long summation(long n)
{
  long result = 0;
  for (long count = 1; count <= n; count++)
    result += count;
  return result;
}
```

Because its for loop executes n times, we say that this version requires **linear time**, or **time proportional to** *n*, written **O(*n*)**, to compute the sum of the first *n* positive integers.

By contrast, here is another version that computes 1 + 2 + ... + *n* using the formula credited to Carl Friedrich Gauss, one of the greatest mathematicians of all time (see Section 9.1):

```
long summation(long n)
{
  return n * (n + 1) / 2;
}
```

Because this second version computes the sum in 3 steps (1 addition, 1 multiplication, and 1 division) regardless of the value of n, we say that it does so in **constant time**, or **time proportional to 1**, expressed as **O(1)**. Because it solves the same problem more quickly, this second method is *more time-efficient* than the first.

Time-efficiency is a major consideration in deciding between a vector<T> and a list<T> to store a sequence in solving a given problem because different containers have different time-efficiencies for the same operation. For example, appending a value to the end of either kind of container takes negligible time. This means that if a problem involves the manipulation of a sequence, but appending (or removing from the end) is the only sequence operation needed to solve the problem, then it makes no difference whether you store the sequence in a vector<T> or a list<T>. The push _ back() method of each requires O(1) time.

By contrast, it is far more time-consuming to access the middle value in a list<T> than in a vector<T> or array. For arrays and vector<T>s, the operation v[i] can access the value at index *i* in constant (i.e., O(1)) time because it is located (*i* × the size of one element) past the beginning of the array or vector. For a list<T>, there is no sub-script operation, so finding the value at index *i* requires a list traversal, beginning at the head node and following *i* successive links to reach the appropriate node. If every index *i* is equally likely to be accessed, we will on average have to follow *n*/2 links, making this a *linear* (i.e., O(*n*)) time operation. This implies that if a problem involves the manipulation of a sequence and involves a large number of accesses to values other than the first or last value in the sequence, then an array or a vector<T> is probably the best choice to store the sequence because those accesses will be much faster than those for a list<T>.

However, it is far more time-consuming to insert values into an array or a vector<T> than into a list<T>. As we saw earlier, inserting into a list<T> is a *constant* time (O(1)) operation, requiring the execution of only a handful of statements. By contrast, insert-ing into an array or a vector<T> is a *linear* time (O(*n*)) operation, because it requires extensive copying to make room for the new value. This implies that if a problem involves many insertions (or deletions) from anywhere other than the end of the sequence, then the list<T> should be used to store the sequence, because insertions (or deletions) will be much more time-efficient than those for a vector<T>.

In summary, if a problem involves many accesses (but not insertions) to the interior of a sequence, then the sequence should be stored in an array or a vector<T>. If a problem involves many insertions or deletions in a sequence from anywhere other than its end, then the sequence should be stored in a linked list such as list<T>. If neither of these is the case, then a different container might be better (e.g., a binary tree, a hash table, a deque). Descriptions of these and many other containers can be found in data structures texts, one of which is described in Footnote 4 of Chapter 12.

The following table summarizes common container operations, and compares their times for a vector<T> with those of a list<T>.

Description	array/vector<T> Efficiency	list<T> Efficiency
Append a value	O(1)	O(1)
Insert a value at position i	O(n)	O(1)†
Remove all values	O(n)	O(n)
Check if a value is present	O(n)	O(n)
Compare two containers	O(n)	O(n)
Find value at index i	O(1)	O(n)
Find index of first occurrence of a value	O(n)	O(n)
Check if the container is empty	O(1)	O(1)
Find index of last occurrence of a value	O(n)	O(n)
Remove value at index i	O(n)	O(1)†
Remove the first occurrence of a value	O(n)	O(1)†
Replace object at index i with a value	O(1)	O(n)
Return number of values in the container	O(1)	O(1)

† These, of course, require that an iterator be positioned first, which is an O(n) operation.

15.5 POINTERS AND COMMAND-LINE ARGUMENTS

As we know by now, every C++ program has a function whose name is main. The main function differs from other programmer-defined functions in a number of ways. One of the differences is that arguments are passed to the main function, using an array of pointers. How this is done is the topic of this section.

The main function cannot be called directly. Instead, we can think of it as being *called* when a program is *executed*. In **command-line environments** such as the Unix operating system, a program is executed by entering its name following the operating system prompt. For example, on a Unix system, the operating system prompt is often the $ symbol, so to invoke the text editor emacs on a computer running Unix, we might enter the command

```
$ emacs
```

and the program will begin executing. In any command-line environment, entering the name of a C++ program on the command line can be thought of as issuing a call to the main function of that program.

To edit a C++ file in the Unix environment, we can enter a command of the form

```
$ emacs FileName
```

When invoked in this way, the program (emacs) begins execution, searches for the file named *FileName*, and (assuming that it is found) opens it for editing. In this example, the file that we wish to edit (*FileName*) is an example of a **command-line argument**. Just as entering the name of the program (emacs) is like calling the main function of a program, entering the name of the program followed by *FileName* is like calling the main function of a program and passing it *FileName* as an argument.

Command-line arguments are used with many of the system commands in command-line environments such as Unix. For example, the command mkdir projects is used in Unix to create a new subdirectory named `projects`. Similarly, the command `cd projects` will change location in the directory structure to the subdirectory `projects`. In each case a program is being executed (one named `mkdir`, and the other named `cd`), and the name `projects` is passed to that program as an argument. In this section we examine the mechanism by which a main function can receive and process command-line arguments. The techniques discussed can be used in any C++ command-line environment.

15.5.1 Parameters of the Main Function

The general form of the main function is

```
int main(parameterList)
{
   statementList
}
```

In all of our programs up to this point, the *parameterList* has been empty, but this need not be the case. A main function can be declared with a parameter list consisting of two predefined parameters:

- `argc` (the argument count), an integer; and

- `argv` (the argument vector), an array of pointers to characters.

As a legacy from C, the standard way to declare these parameters in a main function is:[7]

```
int main(int argc, char * argv[])
{
  // ... body of the main function ...
}
```

When a C++ program with the parameters `argc` and `argv` declared in the parameter list of its main function is executed from the command line, two things occur automatically:

1. If n character strings were entered on the command line, the value of `argc` is set to n.

2. The value of `argv[0]` is the address of the first character string of the command line.

   The value of `argv[1]` is the address of the second character string of the command line.

   .

   .

   .

---

[7] C has no classes, and thus has no `string` class. Instead, C permits character strings to be stored in character arrays (`char []`) and passed to functions via character pointer (`char *`) parameters, with the value of a character string literal being the address of its first character. Hence, the close relationship between character strings, arrays, and pointers in C.

The value of `argv[n-1]` is the address of the nth character string of the command line.

To illustrate, consider the simple C++ program in Example 15.3.

Example 15.3 Introducing `argc` and `argv`

```
/* Program to introduce the predefined parameters argc and argv.

   Output: The value of argc, followed by each string in argv.
   -----------------------------------------------------------------*/

#include <iostream>
using namespace std;

int main(int argc, char * argv[])
{
  cout << "\nThere are " << argc
       << " strings on the command line:\n";

  for (int i = 0; i < argc; i++)
     cout << '\t' << "argv[" << i << "] contains: "
          << argv[i] << endl;
}
```

In this program the parameter list of the main function contains declarations of `argc` and `argv`. If the compiled version of this program is stored in a file named `commandLine`, then `commandLine` can be executed by entering the command

```
$ commandLine
```

which produces the output

```
There are 1 strings on the command line:
        argv[0] contains: commandLine
```

Thus, within `commandline`, `argc` has the value 1, and `argv[0]` refers to the character string `commandline`. If we execute `commandline` by entering the command

```
$ commandLine I want an argument
```

then the output will be

```
There are 5 strings on the command line:
        argv[0] contains: commandLine
        argv[1] contains: I
```

```
argv[2] contains: want
argv[3] contains: an
argv[4] contains: argument
```

From these examples it should be evident that the values of `argc` and `argv` depend on what the user enters on the command line when invoking the program. If the user enters the name of the program followed by *i* arguments, then the value of `argc` will be *i* + 1, the number of character strings entered on the command line; `argv[0]` will refer to the name of the program; and `argv[1]` through `argv[i]` will refer to the *i* arguments that were entered.

15.5.2 Example: A Square Root Calculator

As a simple illustration of the use of `argv` and `argc`, consider the problem of designing a square root calculator that allows the user to enter the value(s) to be processed on the command line and that then calculates and displays the square roots of each value. For example, if the command

```
$ sroot 4 9 16 25
```

is entered, the values 2, 3, 4, and 5 are to be displayed.

Because the program must process command-line arguments, it *receives* the arguments through the parameters of the main function (i.e., `argc` and `argv`). Here are some possibilities of what the user might enter:

```
$ sroot             // error—no data to process (argc is 1)

$ sroot A           // error—nonnumeric data (argc is 2, argv[1] is "A")

$ sroot -1          // error—negative data (argc is 2, argv[1] is "-1")

$ sroot 9           // one value (argc is 2, argv[1] is "9")

$ sroot 4 9         // two values (argc is 3, argv[1] is "4", argv[2] is "9")
```

From the first example, we see that valid input requires `argc` > 1. Also, each `argv[i]` refers to a character string, and we must take the square root of a value of type `double`. This means that the character string stored in `argv[i]` must be converted to the corresponding `double` value. Fortunately, C++ provides the `strtod()` function in `cstdlib` that performs this operation. That function can also be used to make our program more foolproof by checking its return value—`strtod()` returns 0 if it is unable to convert the string to a numeric value, which is the case in the second and third examples. Once we have converted the character string to the corresponding `double` value, all that remains is to find its square root, which is easy, using the `sqrt()` function declared in `cmath`. We then simply display the value and its square root.

We can thus construct the following algorithm, which checks that at least one command-line argument has been given, and if so, uses a loop to process each argument.

Algorithm for sroot

1. If *argc* is less than 2, display an "incorrect usage" error message and quit.

2. For each integer *i* in the range 1 through *argc* – 1:

   a. Get *inValue*, the double equivalent to argument *i*.

   b. If *inValue* > 0

      Display *inValue* and its square root.

   Else

      Display an "invalid data" error message.

Encoding this algorithm in C++ is straightforward, as shown in Example 15.4.

Example 15.4 Encoding sroot

```
/* Program to display the square roots of a sequence of values,
   specified by the user on the command line.

   Receive: One or more numeric (double) values
   Output:  The square roots of the input values
-------------------------------------------------------------------*/

#include <iostream>            // cin, cout, <<, >>
#include <cmath>               // sqrt()
#include <cstdlib>             // strtod()
using namespace std;

int main(int argc, char * argv[])
{
  if (argc < 2)
  {
     cout << "\n*** Usage: sroot List-of-Positive-Numbers \n\n";
     exit(1);
  }

   double inValue;                     // double equivalent of an
                                       //   argument

  for (int i = 1; i < argc; i++)
  {
     inValue = strtod(argv[i], 0);
     cout <<"\n--> The square root of " << inValue
        << "is" << sqrt(inValue) << endl;
  }
}
```

SAMPLE RUN:

```
./sqrt 1 2 3 4 5 6
--> The square root of 1 is 1
```

```
--> The square root of 2 is 1.41421
--> The square root of 3 is 1.73205
--> The square root of 4 is 2
--> The square root of 5 is 2.23607
--> The square root of 6 is 2.44949
```

CHAPTER SUMMARY

Key Terms

address-of operator (&)	linear time
anonymous variable	linked list
assignment operator	memory leak
base address	new operation
circular linked list	node
class pointer selector operator (->)	null address
command-line environment	null pointer value
compile-time allocation	O(1)
constant time	O(n)
delete operation	pointer
doubly linked list	pointer variable
gateway	reference
head node	run-time allocation
indirection operator(*)	singly linked list
Internet Protocol (IP)	sizeof operator
IP address	time-efficiency
iterator	

NOTES

- In a declaration, an asterisk operator * must precede each identifier that is to serve as a pointer.

- 0 can be assigned to any pointer variable; it is called the *null pointer* and also the *null address*.

- A value assigned to a pointer variable must be the address of an object whose type is the same as the type to which the pointer is bound in its declaration.

- The term *reference* is another term for *address*, so applying the indirection operator (*) to a pointer variable is called *dereferencing* that pointer.

- The value of a pointer variable `ptr` is simply the address stored within `ptr`; `*ptr` uses this address to access (indirectly) the contents of the memory location at that address.

- Pointers can be assigned the values of other pointers that are *bound to the same type.*

- The relational operators `==` and `!=` can be used to compare two pointers that are *bound to the same type.* The null address may be compared with any pointer.

- An expression of the form `new Type` requests a block of memory large enough to store an object of the specified `Type`. If the request can be granted, `new` returns the address of the block of memory; otherwise, it returns the null address (0).

- An anonymous variable can be accessed indirectly by dereferencing a pointer to it.

- Unlike arrays whose memory is allocated at compile time, arrays whose memory is allocated at run time can be tailored to the size of the sequence to be stored in them.

- If `ptr` points to the base address of an array, then `ptr[i]` can be used to access the element at location `i`, and is equivalent to writing `*(ptr + i)`.

- Singly linked lists consist of a series of nodes, each of which has a data part and a pointer to the next node. A node in a doubly linked list also has a pointer to the preceding node.

- Items can be inserted in and removed from linked lists more efficiently than for arrays and vectors.

- Values can be passed to the `main()` function in a program via the `argc` (argument count) and `argv` (argument vector) parameters.

Style Tips

Pointers permit the implementation of flexible data structures like `vector<T>` and `list<T>` from the Standard Template Library. When using such objects, we must select the data structure that best fits the problem to be solved. More precisely, if we are storing a sequence of values and the problem requires access to arbitrary values within the sequence, then a `vector<T>` is an appropriate container for storing the sequence. However, if we are storing a sequence and the problem requires many insertions and deletions anywhere except at the end of the sequence, then a `list<T>` provides an efficient means of storing and manipulating such a sequence.

The pointers used to implement `list<T>` and `vector<T>` have memory addresses as values. Consequently, the manner in which pointer variables are used is quite different from that for other kinds of variables, and this can cause special difficulties for both beginning and experienced programmers. Pointers are used to store the addresses of objects whose memory is allocated at run time. Consequently, operations that

- create an object require that its memory be explicitly allocated using new;
- destroy an object require that its memory be explicitly deallocated using delete;
- modify the size of an object require that its old memory be deallocated and then new memory of the correct size be reallocated.

Warnings

The operations used to process pointers are quite different from those used to process objects whose memory is allocated at compile time. Some of the main features to remember when using pointer variables and run-time allocation in C++ programs are

1. *Use the* typedef *mechanism and descriptive identifiers to declare pointer types.* This increases program readability and thereby reduces the likelihood of errors and makes them easier to find when they do occur.

2. *Each pointer variable is bound to a fixed type; a pointer is the address of a memory location in which only a value of that type can be stored.*

3. *Care must be used when operating on pointers because they have memory addresses as values.* In particular:

    - *A pointer* ptr *can be assigned a value in the following ways:*

        - ptr = &obj; (where obj is an object of the type to which ptr points)

        - ptr = 0; (the null address)

        - ptr = anotherPtr; (where anotherPtr is a pointer bound to the same type as ptr)

        - ptr = new Type; (where Type is the type to which ptr points).

    - *Arithmetic operations on pointers are restricted.* For example, pointer values (memory addresses) cannot be added, subtracted, multiplied, or divided. However, an integer value i can be added to or subtracted from the value of a pointer variable, which changes the address in the pointer by i * sizeof(*Type*), where *Type* is the type to which the pointer is bound.

    - *Two pointers can be compared using relational operators, but they must be bound to the same type or one or both may be the null address.*

    - *Pointers may be used as parameters, but corresponding parameters and arguments must be bound to the same type.* A function may also return a pointer as its return value, but the type to which that pointer is bound must be the same as the type to which the function is declared to point.

4. *Do not confuse memory locations with the contents of memory locations.* If ptr is a pointer, its value is the address of a memory location; *ptr refers to the contents of that location. Both ptr++ and (*ptr)++ are valid (if ptr is bound to a type for

which ++ is defined), but the first increments the address in `ptr`, while the second increments the contents of the memory location at that address.

5. *The null address ≠ undefined.* **A pointer becomes defined when it is assigned the address of a memory location or the null address. Assigning a pointer the null address is analogous to initializing a numeric variable to zero.**

6. *Attempting to dereference* `ptr` *that is undefined or the null address is an error* **and may produce cryptic run-time error messages.**

7. *When memory is allocated at run time with the* new *operation, the value returned by* new *should be tested before proceeding, to ensure that the operation was successful.* **For example,**

```
ptr = new SomeType;
assert(ptr != 0);
```

8. *Memory locations that were once associated with a pointer variable and that are no longer needed should be returned to the free store by using the* `delete` *function.*

TEST YOURSELF

Section 15.1

1. A pointer variable stores a(n) _____.

2. _____ is the address-of operator.

3. _____ is used to dereference a pointer.

Questions 4–16 assume the following declarations:

```
double * x,
        y = 1.1;
```

and that `double` values are stored in 8 bytes of memory. Answer each of Questions 4–16 with (a) address or (b) double value.

4. The value of x will be a(n) _____.

5. The value of y will be a(n) _____.

6. The value of &y will be a(n) _____.

7. The value of &x will be a(n) _____.

8. The value of *x will be a(n) _____.

9. The value of (*x) * y will be a(n) _____.

10. The word "reference" is a synonym for a(n) _____.

11. In the assignment x = 0; , 0 is called the _____ address.

12. The output produced by the statements x = &y; cout << *x; is _____.

13. The output produced by the statements x = &y; *x = 3.3; cout << y; is _____.

14. (True or false) sizeof(double) == sizeof y.

15. (True or false) sizeof(double) == sizeof(*x).

16. If the output produced by cout << x; is 0x12a30, the value of x+4 is _____.

Section 15.2

1. (Run or compile) Memory for a C-style array is allocated at _____ time; memory for a vector<T> object is allocated at _____ time.

2. The _____ operation is used to request memory during program execution. If not enough memory is available, it returns _____; otherwise, it returns the _____ of a block of memory. The newly allocated memory location is a(n) _____ variable.

3. The _____ operation is used to release memory during program execution.

4. The base address of a run-time allocated array is stored in a _____.

5. Given the declarations

```
int a[] = {44, 22, 66, 11, 77, 33};
int * p = a;
```

what is the value of p[2]?

Section 15.3

1. (True or false) Values can be inserted at the end of a vector<T> more efficiently than at its front.

2. In a linked list, values are stored in _____ that are linked together by _____.

3. The two parts of a node are a(n) _____ part and a(n) _____ part.

4. If a node has no successor, its link is set to a special _____ value.

5. (True or false) One of the strengths of arrays is that an item can be inserted at any point without moving any array elements.

6. (True or false) One of the strengths of a linked list is that an item can be deleted at any point without moving any list elements.

EXERCISES

Exercises 1–9 assume the following declarations:

```
int i1 = 11,
i2 = 22;
double d1 = 3.45,
       d2 = 6.78;
class Point
{
 public:
   double x() { return xCoord; }
   double y() { return yCoord; }
 private:
   double xCoord, yCoord;
};
```

1. Write declarations for variables p1 and p2 whose values will be addresses of memory locations in which a `double` can be stored.

2. Write a statement to assign the address of `d1` to the variable `p1` in Exercise 1, or explain why this is not possible.

3. Write a statement to assign the address of `i2` to the variable `p2` in Exercise 1, or explain why this is not possible.

4. Write declarations for a variable `q` whose value will be a memory location in which a `Point` object can be stored.

5. Write declarations that initialize variables `ptr1` and `ptr2` with the addresses of `i1` and `i2`, respectively.

6. Write a statement that will make variables `p1` and `p2` of Exercise 1 point to the same memory location.

7. Write a statement that will copy the value stored in the memory location pointed to by `ptr2` into the memory location pointed to by `ptr1`, for `ptr1` and `ptr2` as in Exercise 5.

8. Write a statement to output the *x* coordinate and the *y* coordinate of the point in the memory location pointed to by the variable `q` of Exercise 4.

9. Write statements that use the variables `p1` and `p2` of Exercise 2 but not the variables `d1` and `d2` to interchange the values of `d1` and `d2`.

For Exercises 10–16, use the `sizeof` operator to find how many bytes your C++ compiler allocates for the given data type:

10. `int`

11. `float`

12. `double`

13. `short int`

14. A string whose value is `"Bye!"`

15. A string whose value is `"Auf Wiedersehen!"`

16. Pointers to the types in Exercises 10–15.

17. Using the address-of operator, find the starting addresses your C++ compiler associates with the constant `SIZE` and each of the variables in the following declarations:

```
const int SIZE = 10;
char charArray[SIZE];
int intArray[SIZE];
double doubleArray[SIZE];
char charVar;
```

18. Use the `sizeof` operator to find the number of bytes allocated by your C++ compiler to `SIZE` and each of the variables in Exercise 17.

19. Using `typedef`, create a type `CharPointer` that is a synonym for pointers to type `char`.

Exercises 20–22 assume an array declaration like the following:

```
double anArray[10];
```

20. Use the address-of operator to find the address of the first element of `anArray`.

21. Find the value associated with the name `anArray`.

22. What can you conclude from the results of Exercises 20 and 21?

Section 15.2

For Exercises 1–10, write C++ statements to do what is asked.

1. Declare a `char` pointer variable named `charPtr`.

2. Allocate an anonymous `char` variable, storing its address in `charPtr`.

3. Input a character value and store it in the anonymous variable of Exercise 2.

4. Display the value of the anonymous variable of Exercise 2.

5. Convert the case of the value of the anonymous variable of Exercise 2 using character-processing functions such as `isupper()` and `tolower()` from `cctype`.

6. Declare a `double` pointer variable named `doublePtr`.

7. Allow the user to enter n, the number of values to be processed, then allocate an anonymous array of n double values, storing its address in doublePtr.

8. Fill the anonymous array of Exercise 7 with n input values, entered from the keyboard.

9. Compute and display the average of the values in the anonymous array of Exercise 7.

10. Deallocate the storage of the anonymous array of Exercise 7.

11. Find the base address of the anonymous array allocated in Exercise 7 and draw a memory map showing the addresses of its first few elements.

12. Describe the output produced by the following statements:

```
int * foo, * goo;
foo = new int;
*foo = 1;
cout << (*foo) << endl;
goo = new int;
*goo = 3;
cout << (*foo) << (*goo) << endl;
*foo = *goo + 3;
cout << (*foo) << (*goo) << endl;
foo = goo;
*goo = 5;
cout << (*foo) << (*goo) << endl;
*foo = 7;
cout << (*foo) << (*goo) << endl;
goo = foo;
*foo = 9;
cout << (*foo) << (*goo) << endl;
```

Section 15.3

In the following exercises you may assume that operations as described in the text can be used to obtain a new node from the storage pool and to return nodes to the storage pool, and that there is a special null value.

Exercises 1–7 assume that p1, p2, and p3 are pointers to Nodes and that the following statements have been executed:

```
p1 = new Node;
p2 = new Node;
p3 = new Node;
```

Tell what will be displayed by each of the code segments or explain why an error occurs.

```
1. p1->data = 123;
   p2->data = 456;
   p1->next = p2;
   p2->next = 0;
   cout << p1->data << " " << p1->next->data << endl;
```

2. ```
p1->data = 12;
p2->data = 34;
p1 = p2;
cout << p1->data << " " << p2->data << endl;
```

3. ```
p1->data = 12;
p2->data = 34;
*p1 = *p2;
cout << p1->data << " " << p2->data << endl;
```

4. ```
p1->data = 123;
p2->data = 456;
p1->next = p2;
p2->next = 0;
cout << p2->data << " " << p2->next->data << endl;
```

5. ```
p1->data = 12;
p2->data = 34;
p3->data = 34;
p1->next = p2;
p2->next = p3;
p3->next = 0;
cout << p1->data << " " << p1->next->data << endl;
cout << p2->data << " " << p2->next->data << endl;
cout << p1->next->next->data << endl;
```

6. ```
p1->data = 111;
p2->data = 222;
p1->next = p2;
p2->next = p1;
cout << p1->data << " " << p2->data << endl;
cout << p1->next->data << endl;
cout << p1->next->next->data << endl;
```

7. ```
p1->data = 12;
p2->data = 34;
p1 = p2;
p2->next = p1;
cout << p1->data << " " << p2->data << endl;
cout << p1->next->data << " " << p2->next->data << endl;
```

8. Write an algorithm to count the nodes in a linked list with first node pointed to by *first*.

9. Write an algorithm to determine the average of a linked list of real numbers with first node pointed to by *first*.

10. Write an algorithm to append a node at the end of a linked list with first node pointed to by *first*.

11. Write an algorithm to determine whether the data items in a linked list with first node pointed to by *first* are in ascending order.

12. Write an algorithm to search a linked list with first node pointed to by *first* for a given item, and if it is found, return a pointer to the predecessor of the node containing that item.

13. Write an algorithm to insert a new node into a linked list with first node pointed to by *first* after the *n*th node in this list for a given integer *n*.

14. Write an algorithm to delete the *n*th node in a linked list with first node pointed to by *first*, where *n* is a given integer.

15. Suppose the items stored in two linked lists are in ascending order. Write an algorithm to merge these two lists to produce a list with the items in ascending order.

16. Write an algorithm to reverse a linked list with first node pointed to by *first*. Do not copy the list elements; rather, reset links and pointers so that *first* points to the last node and all links between nodes are reversed.

PROGRAMMING PROBLEMS

Sections 15.3 and 15.4

1. A limited number of complimentary copies of new CAD/CAM software will be released tomorrow. Requests are to be filled in the order in which they are received. Write a program that reads the names and addresses of the persons requesting this software, together with the number of copies requested, and stores these in a linked list. The program should then produce a sequence of mailing labels (names, addresses, and number of copies) for requests that can be filled.

2. Modify the program in Problem 1 so that multiple requests from the same person are not allowed.

3. A polynomial of degree *n* has the form

$$a_0 + a_1x + a_2x^2 + \dots + a_nx^n$$

where $a_0, a_1, \dots, a_n$ are numeric constants called the coefficients of the polynomial and $a_n \neq 0$. For example,

$$1 + 3x - 7x^3 + 5x^4$$

is a polynomial of degree 4 with integer coefficients 1, 3, 0, –7, and 5.

a. Develop a linked list that can represent any such polynomial. Let each node store a nonzero coefficient and the corresponding exponent, with the exponents in increasing order.

b. Write a program to implement an input operation that reads a polynomial's nonzero coefficients and exponents and constructs the linked representation of the polynomial and an output operator that displays polynomials in the usual mathematical format with x^n written as $x \uparrow n$ or $x \wedge n$. Your program should then read values for x and evaluate the polynomial for each of them.

4. Extend the class `Polynomial` in Problem 3 to add two polynomials.

5. Extend the class `Polynomial` in Problems 3 and 4 to multiply two polynomials.

6. The Cawker City Construction Company maintains two warehouses, one in Chicago and another in Detroit, each of which stocks a large number of different items. Write a program that first reads the product numbers of items in the Chicago warehouse and stores them in a linked list `Chicago`, and then repeats this for the items in the Detroit warehouse, storing these product numbers in a linked list `Detroit`. The program should then find and display the *intersection* of these two lists of numbers, that is, the collection of product numbers common to both lists. Do not assume that the lists have the same number of elements.

7. Repeat Problem 6, but find and display the *union* of the two lists, that is, the collection of product numbers that are elements of at least one of the lists.

8. Write a "quiz-tutor" program, perhaps on a topic from one of the early chapters, or some other topic about which you are knowledgeable. The program should read a question and its answer from a file, display the question, and accept an answer from the user. If the answer is correct, the program should go on to the next question. If it is not correct, store the question in a list. When the file of questions is exhausted, the questions that were missed should be displayed again (in their original order). Keep a count of the correct answers and display the final count. Also, display the correct answer when necessary in the second round of questioning.

9. Suppose that jobs entering a computer system are assigned a job number and a priority from 0 through 9. The numbers of jobs awaiting execution by the system are kept in a *priority queue*. A job entered into this queue is placed ahead of all jobs of lower priority but after all those of equal or higher priority. Write a program to read one of the letters R (remove), A (add), or L (list). For R, read a job number and remove it from the priority queue; for A, read a job number and priority and then add it to the priority queue in the manner just described; and for L, list all the job numbers in the queue.

10. Write a program to read records from the file `Users` (see the file descriptions at the end of Chapter 11) and construct nine linked lists of records containing a user's identification number, resource limit, and resources used to date, one list for each of the leading digits (1, 2, ..., 9) of the identification number. Store these records in a `vector` of `list<T>`s. After the lists have been constructed, sort each list so the resource limits are in ascending order and then print each of them with appropriate headings. *Note*: If aList is a `list<T>` object, then `alist.sort();` will sort

`aList` provided `<` is defined for type `T` objects. In this exercise, you must overload `operator<()` to define what it means for one user record to be less than another.

11. Design and implement a class `BigInt` whose values are large integers with perhaps hundreds of digits. Overload the addition operator to add two large integers. Treat each number as a list, each of whose elements is a block of digits of the number. Add the integers (lists) element by element, carrying from one element to the next when necessary.

12. Proceed as in Problem 11, but add a subtraction operator. Write a two-function `BigInt` calculator program to test your class.

13. Extend class `BigInt` from Problem 11 by overloading the multiplication operator.

14. Extend class `BigInt` from Problem 11 by overloading the division operator (more challenging).

Section 15.5

1. Write a program `binary` so that the command

   `binary DecimalValue`

   will calculate and display the binary representation of *DecimalValue*.

2. The *median* of a list of *n* numbers is a value such that $n/2$ of the values are greater than that value, and $n/2$ of the values are less than that value. The usual procedure to find the median is to sort the list and then pick the middle number as the median if the list has an odd number of elements, or the average of the two middle numbers if the number of elements is even. Write a program to find the median of a list, so that the command

   `median FileName`

   will calculate and display the median of the values in file *FileName*, but the command

   `median`

   will calculate and display the median of a list of numbers entered from the keyboard.

3. Write a program so that the following command will make a copy of *File1* with the name *File2*:

   `copy File1 File2`

4. Write a program so that the following command will display the specified file on the screen, one page (23 lines) at a time, waiting between pages until the user presses some key:

   `page File`

Data Structures

CONTENTS

An Englishman, even if he is alone, forms an orderly queue of one.

GEORGE MIKES

If Edison had a needle to find in a haystack, he would proceed at once with the diligence of the bee to examine straw after straw until he found the object of his search . . .

NIKOLA TESLA

I think that I shall never see,
A poem lovely as a tree.

JOYCE KILMER, "TREES"

Woodman, spare that tree!

GEORGE POPE MORRIS, "WOODMAN, SPARE THAT TREE!"

I N PREVIOUS CHAPTERS, we saw that C++ provides several mechanisms for storing collections of values. These include

- C-style arrays (have a fixed capacity);

- `vector<T>`s (can grow and are best used for insertions and deletions at the end of the sequence);

- Linked lists (can grow or shrink; insertions and deletions may occur anywhere).

In this chapter, we look at some of the other structures that C++ provides for storing collections of data. They are known as **containers** or **data structures**.

16.1 INTRODUCTORY EXAMPLE: THE BINARY REPRESENTATION OF INTEGERS—STACKS

In this section, we will begin with a problem and then show how it can be conveniently solved using a specialized structure called a **stack**.

16.1.1 Problem: Displaying a Number's Binary Representation

Data is stored in computer memory using a binary representation. In particular, positive integers are commonly stored using the base-2 representation described in Chapter 3. This means that the base-10 representation of an integer used in a program or as input data must be converted to binary.

One algorithm for carrying out this conversion uses repeated division by 2 and the remainders are the binary digits in the base-2 representation from right to left. For example, the following diagram (in which the repeated divisions by 2 are read from the bottom up) shows that the base-2 representation of 26 is 11010:

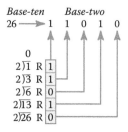

This approach can be used to find the base-b representation for any value of b between 2 and 36 inclusive, simply by dividing by b instead of 2, and using the symbols $a, b, ..., z$ as base-b digits (in addition to 0–9) when $b > 10$. For example, the base-8 representation of 95 is 137_8 and the base-16 representation of 95 is $5f_{16}$, which we can compute as follows

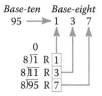

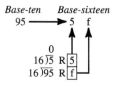

(The base-16 digits for ten, eleven, twelve, thirteen, fourteen, and fifteen are a, b, c, d, e, and f, respectively.) Our problem is to develop a function that will accept a nonnegative base-ten integer and a base b and will output the base-b representation of the integer.

One of the difficulties in this problem is that the order in which remainders are generated is the opposite of the order in which they must be output. For example, in determining the binary representation of 26, the *first* remainder,

$$26 \% 2 = 0$$

is the *last* binary digit that we must display. Similarly, the *second* remainder we compute,

$$13 \% 2 = 1$$

produces the *next-to-the-last* binary digit that we must display. This pattern continues until we generate the final remainder,

$$1 \% 2 = 1$$

which produces the first binary digit that we must display.

16.1.2 The stack Container

The preceding diagrams suggest one approach to solving this problem; what is needed is a special kind of list to store the remainders so that we can print them in the opposite order in which they are generated—a container where the delete operation will remove the value that was most recently added to the list. The values in such a list must, therefore, be maintained in **Last-In-First-Out (LIFO)** order; that is, the last item inserted is the first item to be removed. Such a list is called a **stack** (or a **push-down stack**) because it functions in the same manner as a spring-loaded stack of plates or trays used in a cafeteria:

Plates are added to the stack by *pushing* them onto the **top** of the stack. When a plate is removed from the top of the stack, the spring causes the next plate to *pop* up. For this reason, the operations to insert a value into and delete a value from a stack are commonly called **push** and **pop**, respectively.[1] The most recently added value is called the **top** value.

---

[1] This is the source of the names push_back() and pop_back() in vector<T> and list<T>.

If the stack contains no values, it is described as **empty**. These properties of a stack in a cafeteria illustrate the four standard stack operations:

1. empty(): returns true if there are no values in the stack and false otherwise.

2. top(): returns a copy of the value at the top of the stack.

3. push(v): adds a value *v* at the top of the stack.

4. pop(): removes and returns the value at the top of the stack.

A stack is the container we need to solve the base-conversion problem. To display the base-*b* representation of an integer in the usual left-to-right sequence, we must "stack up" the remainders generated during the repeated division by *b* by pushing them onto a stack. When division is finished, we can retrieve the remainders from this stack in the required "last-in-first-out" order by popping them from the stack.

If we have a stack type available, we can use the following algorithm to convert from base-10 to base-*b* and display the result:

Base-Conversion Algorithm

/* This algorithm displays the representation of a base-10 number in any *base* from 2 through 36.

      Receives: *number*, an int;

                       base, the base to which we want to convert *number*.

      Precondition: *number* > 0 and 2 ≤ *base* ≤ 36.

      Returns: the base-*base* representation of *number*, as a string.

*/

1. Create an empty stack to store the remainders.

2. While *number* ≠ 0 do the following:

    a. Calculate the *remainder* that results when *number* is divided by *base*.

    b. Push *remainder* onto the stack of remainders.

    c. Replace *number* by the integer quotient of *number* divided by *base*.

3. Declare *result* as an empty string.

4. While the stack of remainders is not empty do the following:

    a. Remove the *remainder* from the top of the stack of remainders.

    b. Find the character that represents *remainder* in the given *base*.

   c. Append this base-*base* representation of *remainder* to *result*.

5. Return *result*.

The diagram in Figure 16.1 traces this algorithm for the integer 26 and base 2. Example 16.1 presents a program containing a `convertDecimal()` function that implements the algorithm.

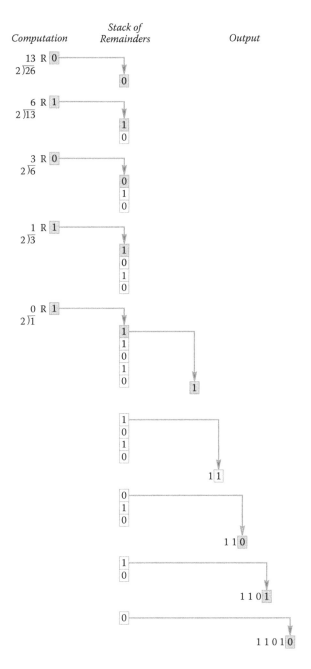

FIGURE 16.1   Using a stack in base-2 conversion.

Example 16.1 Converting Decimal Integers

```cpp
/* Program to convert an int value to other bases.

   Input:   number, a positive decimal integer base, the base to
            which it is to be converted
   Output:  the base representation of number
------------------------------------------------------------------*/

#include <string>
#include <iostream>
#include <stack>
using namespace std;

string convert(int number, int base);
char baseDigit(int value);

int main()
{
  for(;;)
  {
    cout << "\nEnter an integer (-1 to stop): ";
    int number;
    cin >> number;
    if (number < 0) break;
    cout << " and the base to which it is to be converted: ";
    int base;
    cin >> base;
    cout << '\n' << convert(number, base) << " is the base-" << base
         << " representation of " << number << endl;
  }
}


/* convert() converts a decimal value to its base representation
   Receive: integers number and base
   Precondition: number>= 0 and base is in the range 2 - 35
   Return: number converted to the specified base
------------------------------------------------------------------*/
string convert(int number, int base)
{
  stack<int> digitStack;
  int remainder;
  do
  {
    remainder = number % base;
    digitStack.push(remainder);
    number /= base;
  }
  while (number != 0);
```

```
    string resultString = "";
    char otherBaseDigit;
    while ( !digitStack.empty() )
    {
      remainder = digitStack.top();
      digitStack.pop();
      otherBaseDigit = baseDigit(remainder);
      resultString += otherBaseDigit;
    }
    return resultString;
}


/* baseDigit() finds a char representing a digit in another base.
   Receive: value, an int.
   Precondition: 0 <= value && value <= 35.
   Return: the character representation of value.
-------------------------------------------------------------------*/
char baseDigit(int value)
{
  const int NUM_DIGITS = 36;      //  number of digits in other base
  const string digits = "0123456789abcdefghijklmnopqrstuvwxyz";
  if (0 <= value && value < NUM_DIGITS)
    return digits[value];
  else
  {
    cerr << "\n** baseDigit(value): " << value
         << " outside of range 0.." << NUM_DIGITS - 1 << endl;
    return '*';
  }
}
```

SAMPLE RUN:

```
Enter an integer (-1 to stop): 1024
and the base to which it is to be converted: 2

10000000000 is the base-2 representation of 1024


Enter an integer (-1 to stop): 1024
and the base to which it is to be converted: 8

2000 is the base-8 representation of 1024


Enter an integer (-1 to stop): 255
and the base to which it is to be converted: 16

ff is the base-16 representation of 255
```

```
Enter an integer (-1 to stop): 123456789
and the base to which it is to be converted: 26


aa44a1 is the base-26 representation of 123456789


Enter an integer (-1 to stop): -1
```

The convert() function in this program relies heavily on the C++ Standard Template Library's stack<T> container. We will now take a look at this container.

16.1.3 The stack<T> Adapter

We have already seen that the Standard Template Library provides containers such as vector<T> and list<T> for storing sequences. The designers of STL recognized that they could reuse the work they had invested in building these sequential containers to implement other containers and they did this by means of **adapters** that act as "wrappers" around other components, giving them new interfaces. In particular, a stack<T> container could be built as an adapter of vector<T>, list<T>, or deque<T> (described in the next section).

To illustrate, we will use the declaration of digitStack in Example 16.1:

```
stack<int> digitStack;
```

This declaration uses the default constructor in stack<T> to create digitStack and STL's default stack<T> implementation as an adapter of a deque<T>.[2] But if we prefer a stack<T> built using one of the other two containers, we can specify this in the declaration. For example,

```
stack<int, list<int> > intListStack;
```

builds intListStack using list<T>; and the declaration

```
stack<int, vector<int> > intVectorStack;
```

builds intVectorStack using vector<T>.

Each of these constructors builds a stack, but the underlying implementations are different. The object intListStack will be implemented using STL's *linked list* (and thus use memory fairly efficiently but provide slightly slower versions of push() and pop()):

---

[2] The default stack<T> constructor builds the stack using the deque<T> template so this declaration could also be written stack< deque<T> >.

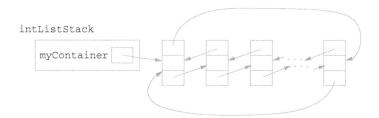

The object `intVectorStack` will be implemented using STL's *vector* (and thus on average probably provide the fastest versions of `push()` and `pop()`, while possibly using more memory than necessary):

And the object `digitStack` will be implemented using STL's default implementation as an adapter of `deque<T>`, which will be described in more detail in the next section. Which of these to use depends on the problem:

- If speed is the primary concern and there are no memory constraints, using STL's default `stack<T>` implementation or as an adapter of `vector<T>` template is probably the best choice.

- If conserving memory is more important than blazing speed, using `stack<T>` to provide an alternative interface to the `list<T>` template is probably the best choice.

16.1.4 The `stack<T>` Methods

The program in Example 16.1 illustrates several of the messages that can be sent to a `stack<T>`, regardless of its underlying implementation details. They include:[3]

- `void push(T aValue)`—add *aValue* to the top of the stack
- `void pop()`—remove the top value of the stack
- `T top()`—retrieve (but do not remove) the T value that is on top of the stack
- `bool empty()`—return `true` if and only if the stack contains no values

Thus, in the program in Example 16.1, we "stacked up" the remainders by writing (within a loop)

```
digitStack.push(remainder);
```

---

[3] STL also defines a `size()` method that returns the number of values in the stack, and the `==` and `<` operators for `stack<T>`. Stacks `s1` and `s2` will be compared element by element from bottom to top to determine if they are the same (`s1 == s2` will be true) or whether the first element where they differ is less than the corresponding element in the other stack (`s1 < s2` will be true).

Likewise, to remove the remainders in LIFO order, we wrote (within a loop)

```
remainder = digitStack.top();
digitStack.pop();
```

And to control the repetition of these statements, we wrote

```
while ( !digitStack.empty() )
{
 // ...
}
```

16.2 RECURSION REVISITED

In Section 10.5 we saw that C++ permits a function definition to call itself, a technique called *recursion*. This was illustrated by looking at how the factorial function can be computed recursively. In this section we will use our knowledge of stacks to see how recursion works.

We begin with another classic example of a function that can be calculated recursively—the power function that calculates x^n, where x is a real value and n is a non-negative integer. The first definition of x^n that one learns is usually an iterative (nonrecursive) one:

$$x^n = \underbrace{x \times x \times \ldots \times x}_{n \ x\text{'s}}$$

and later one learns that x^0 is defined to be 1. (For convenience, we assume here that x^n is 1 also when x is 0, although in this case, it is usually left undefined.) A specification of the function is straightforward,

Receive: x, a real value;
          n, an integer

Return: x^n, a real value

and suggests the following function stub:

```
double power(double x, int n)
{
}
```

To solve a problem recursively, we must identify the anchor and inductive cases. Here, the anchor step is clear: $x^0 = 1$. For the inductive case, we look at an example:

$$5.0^4 = 5.0 \times 5.0 \times 5.0 \times 5.0 = (5.0 \times 5.0 \times 5.0) \times 5.0 = 5.0^3 \times 5.0$$

In general,

$$x^n = x^{n-1} \times x$$

Combining our anchor and inductive steps gives the following recursive definition of x^n:

$$1$$

This leads to the recursive C++ function in Example 16.2.

Example 16.2 Performing Exponentiation Recursively

```
/* power() recursively computes x raised to the power n.

   Receive: x, a real value, and
            n, an integer
   Return:  x raised to the power n
   ------------------------------------------------------------------*/

double power(double x, int n)
{
  if (n == 0)                            // anchor case
    return 1.0;
  else if (n > 0)                        // inductive step (n > 0)
    return power(x, n - 1) * x;
  else                                   // invalid parameter n
  {
   cerr << "*** power(x,n): n is negative.\n";
   return -1.0;
  }
}
```

When it processes a function definition (recursive or not), the C++ compiler creates a special structure called an **activation record**. This record contains space for that function's parameters, local variables, return value, caller, and other information that can vary from call to call. During execution, each program has a special data structure called its **run-time stack**, which is a stack of these activation records. Whenever a function is called, an activation record is pushed onto the run-time stack, and whenever a function terminates, the run-time stack is popped, and control returns to the function whose activation record is uncovered. The effect, therefore, is that the top of the run-time stack always stores an activation record for whatever function is currently executing.

To illustrate, consider what happens when the function call power(3.0, 4) occurs. An activation record for power() is pushed onto the run-time stack, in which x is 3.0 and n is 4:

runTimeStack.top()

x	3.0
n	4
return	?

The function begins executing, and because n is 4, the expression power(3.0, 3) * 4 must be evaluated to get the return value, and this expression has a new call to the function power(). This causes a new activation record to be pushed onto the run-time stack:[4]

runTimeStack.top()

x	3.0	x	3.0
n	4	n	3
return	?	return	?

This function begins executing again, and because n is 3, it encounters the expression power(3.0, 2) * 3.0, which involves a new invocation of power(). This causes a new activation record to be pushed onto the run-time stack:

runTimeStack.top()

x	3.0	x	3.0	x	3.0
n	4	n	3	n	2
return	?	return	?	return	?

Execution of the function begins again, and because n is 2, it encounters the expression power(3.0, 1) * 3.0, producing another new invocation of power(), and causing another activation record to be pushed onto the run-time stack:

runTimeStack.top()

x	3.0	x	3.0	x	3.0	x	3.0
n	4	n	3	n	2	n	1
return	?	return	?	return	?	return	?

Once again, the function begins executing with 1 as the value of n, so when it encounters the expression power(3.0, 0) * 3.0, another new call to power() is produced and another new activation record is pushed onto the run-time stack:

runTimeStack.top()

x	3.0	x	3.0	x	3.0	x	3.0	x	3.0
n	4	n	3	n	2	n	1	n	0
return	?	return	?	return	?	return	?	return	1.0

---

[4] To save space, the stack is drawn horizontally here with the top at the right.

However, because n is 0, execution of the function reaches the anchor case, stopping the recursion, and the function returns the value 1. The sequence of recursive calls from an initial call to the anchor case is sometimes referred to as the *winding phase* of the recursion, because it is like winding the spring of a wind-up clock until it is fully wound.

Once the anchor case has been reached, *backtracking* that actually performs the computation begins. The run-time stack is popped, and control returns to the function call whose activation record is now on top (i.e., power(3.0, 1)) where it finishes evaluating the expression power(3.0, 0) * 3.0, producing 1.0 * 3.0:

runTimeStack.top()

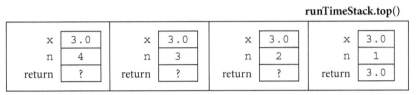

Because that marks the end of that function execution, the run-time stack is popped, and control returns to the function call whose activation record is exposed (i.e., power(3.0, 2)) where the expression power(3.0, 1) * 3.0 can now be evaluated to 3.0 * 3.0:

runTimeStack.top()

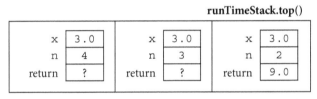

That function call is complete, so the run-time stack is again popped, and control returns to the function whose activation record is exposed (i.e., power(3.0, 3)) and resumes evaluation of the expression power(3.0, 2) * 3.0, giving 9.0 * 3.0:

runTimeStack.top()

x	3.0		x	3.0
n	4		n	3
return	?		return	27.0

Because execution has reached the end of that function call, the run-time stack is popped, control returns to the function whose activation record is uncovered (i.e., power(3.0, 4)), and the expression power(3.0, 3) * 3.0 evaluates to 27 * 3.0:

runTimeStack.top()

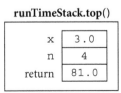

The result (81.0) is the return value of the original function call power(3.0, 4).

This phase in which values are returned from the anchor case back through each of the previous calls is sometimes referred to as the *unwinding phase* of the recursion, because like a wind-up clock performing its task as its spring unwinds, a recursive function performs its task by unwinding the recursive calls stacked up in its winding phase.

The run-time stack thus plays a pivotal role in managing a sequence of function calls such as the one generated by recursion. Whenever a function calls itself or another function, an activation record is pushed onto the run-time stack and this continues until one of the function calls (the anchor case, in recursion) is completed and returns a value. During the unwinding phase, activation records are popped from the run-time stack, until control returns to the original caller.

16.3 QUEUES, DEQUES, AND PRIORITY QUEUES

In the preceding sections, we examined the stack—a special kind of container in which values are always inserted and removed from the same end. In this section, we examine some other special-purpose containers. We begin with queues and look later at deques and priority queues.

16.3.1 Applications of Queues

A **queue** is a container in which values are always added at one end, called the **rear** or **tail**, and removed from the opposite end, called the **front** or **head**. Queues abound in everyday life, because they provide a fair way to schedule things that are waiting for some kind of service. For example,

- A line of persons waiting to check out at a supermarket,

- A line of persons waiting to purchase tickets at a theater,

- A line of planes waiting to take off at an airport, and

- A line of vehicles at a toll booth

are all examples of queues. Arriving customers, planes, vehicles, and the like enter the line at the rear and are removed from the line and served when they reach the front of the line, so that the first one to enter the queue is the first one served. Thus, whereas a stack exhibits last-in-first-out (LIFO) behavior, a queue exhibits **first-in-first-out (FIFO) behavior**.

In addition to lines of people, vehicles, and planes waiting for service, queues are also commonly used to model waiting lines that arise in the operation of computer systems. These queues are formed whenever a particular resource such as a printer, a disk drive, and the central processing unit must be shared by more than one process. As these processes request a particular resource, they are placed in a queue to wait for service.

As one example, several personal computers may be sharing the same printer, and a queue—sometimes called a *spool queue*—is used to schedule output requests in a first-come-first-served manner. If a print job is requested and the printer is free, it is immediately assigned to this job. While this output is being printed, other jobs may need the printer, and so they are placed in the spool queue to wait their turns. When the output from the current job terminates, the printer is released from that job and is assigned to the first job in the queue.

Another important use of queues in computing systems is **input/output buffering**. I/O buffering is important because disk operations (e.g., reading from or writing to a file) take much longer than CPU operations. Consequently, if the processing of a program must be suspended while the disk is accessed, program execution is slowed dramatically. One common solution to this problem uses sections of main memory known as **buffers** and transfers data between the program and these buffers rather than directly between the program and the disk.

C++ ifstream and ofstream objects automatically buffer file I/O. In the declaration

```
ifStream fin("inputFile.txt");
```

inputFile.txt is opened and some of its data is transferred from the disk to an input buffer in main memory while the central processing unit (CPU) is processing the next statement(s). When the program attempts to read from fin,

```
fin >> aValue;
```

the next value stored in this buffer is retrieved. While this value is being processed, additional data values can be transferred from the disk file to the buffer in the background. Clearly, the buffer must be organized as a first-in-first-out structure, that is, as a queue. A queue-empty condition indicates that the input buffer is empty, and program execution is suspended while the operating system attempts to load more data into the buffer or signals the end of file.

16.3.2 The queue<T> Adapter

Like STL's stack<T> container, queue<T> is an adapter that, by default, wraps a deque<T>, described later in this section. Thus, the declaration

```
queue<string> stringQueue;
```

declares stringQueue to be a queue in which strings can be stored in a deque<T>. The container that stores a queue's elements can also be a list<T>. For example, the declaration

```
queue<string, list<string> > stringListQueue;
```

declares a queue stringListQueue in which the strings are stored in a list<string>. However, queue<T> cannot wrap a vector<T>, for reasons we will see shortly.

The operations provided for queue<T> are similar to those for stack<T>:

- bool empty()—returns true if and only if there are no values in the queue.
- void push(T aValue)—append aValue to the end of the queue.
- void pop()—remove the value at the front of the queue.
- T front()—retrieve (but do not remove) the T value that is at the front of the queue.
- T back()—retrieve (but do not remove) the T value that is at the end of the queue.

Implementations of most of these are similar to those for stack<T> and are essentially just renamings of the container being adapted. Here, for example, is a definition of push():

```
template<class T, class Container>
inline void queue<T, Container>::push(T aValue)
{
   myContainer.push_back(aValue);
}
```

As with a stack<T>, when we send the push() message to a queue<T>, the value being added is appended to the back of the queue's container using push_back(). However, for the pop() operation, the value is removed from the *front* of the container:

```
template<class T, class Container>
inline void queue<T, Container>::pop()
{
   myContainer.pop_front();
}
```

This is why a queue<T> cannot adapt a vector<T>. Inserting and/or deleting at the front of an array or vector<T> requires extensive copying that is very inefficient. To discourage programmers from writing inefficient code, vector<T> does not provide pop_front() or push_front() methods. Because we cannot send the pop_front() message to a vector<T>, we cannot use a vector<T> as the underlying container for a queue<T>.

Fortunately, this is not much of a hardship. The reason is that the default implementations of both stack<T> and queue<T> use a container called a *deque*, which we discuss next.

16.3.3 The deque<T> Container

The word **deque** stands for *double-ended queue*. It is a good container to use for any problem that requires the storage of a sequence of values and manipulation of the values at either end of that sequence.

The Standard Template Library's `deque<T>` is a container that supports these operations (among many others):

- `void push_back(T aValue)`—insert `aValue` at the end of the deque.

- `void pop_back()`—remove the value at the back of the deque.

- `void push_front(T aValue)`—insert `aValue` at the front of the deque.

- `void pop_front()`—remove the value at the front of the deque.

- `T front()`—retrieve (but do not remove) the `T` value that is at the front of the deque.

- `T back()`—retrieve (but do not remove) the `T` value that is at the back of the deque.

- `bool empty()`—return true if and only if there are no values in the deque.

- `int size()`—return the number of values in the deque.

From this list, a `deque<T>` might seem to resemble a `list<T>`; however, the subscript operator (and indeed almost all other `vector<T>` operations) can be used with a `deque<T>`.

So what are the differences between `deque<T>`, `vector<T>`, and `list<T>`? The following table summarizes the major ones:

	vector\<T\>	list\<T\>	deque\<T\>
push_back()	O(1)	O(1)	O(1)
push_front()	*none*	O(1)	O(1)
pop_back()	O(1)	O(1)	O(1)
pop_front()	*none*	O(1)	O(1)
operator[]	O(1)	*none*	O(1)
insert()	O(n)	O(1)	O(n)
delete()	O(n)	O(1)	O(n)

Thus, for the most part, a `deque<T>` mirrors the behavior of a `vector<T>`, except that adding and removing an element at its front can be performed in O(1) time, whereas these operations for a `vector<T>` would require O(n) time.

It is because of this efficient implementation of these end-of-sequence operations that `deque<T>` is used as the default container for the `stack<T>` and `queue<T>` adapters.

16.3.4 The `priority_queue<T>` Adapter

There is a special kind of queue in which the order of the values in the queue need not be first-in-first-out. It is known as a **priority queue** because how its elements are ordered depends on their priorities or values relative to one another. For example, a multitasking operating system often uses a priority queue to decide which task or process to run next, with tasks that must meet real-time requirements (e.g., video or audio streaming) receiving the highest priority, tasks owned by the operating system receiving the next highest priority, tasks owned by a user receiving the next highest priority, and so on.

The Standard Template Library provides a `priority_queue<T>` container (defined in the *queue* header file). By default, it organizes its values so that the "highest" value is at its front, followed by the "next-highest" value, and so on, with the "lowest" value at its back.

The program in Example 16.3 illustrates the difference between a `queue<T>` and a `priority_queue<T>`.

Example 16.3 Comparing the `queue<T>` and `priority_queue<T>` Containers

```cpp
// priorityQueueTest.cpp

#include <queue>         // priority_queue
#include <iostream>      // cout
#include <string>
using namespace std;

int main()
{
  string name = "JoeSmith";
  priority_queue<char> pQ;
  queue<char> q;

  for (int i = 0; i < name.size(); i++)
  {
    pQ.push(name[i]);
    q.push(name[i]);
  }

  while (! q.empty() )     // display q's contents
  {
    cout << q.front();
    q.pop();
  }
  cout << endl;

  while ( !pQ.empty() )    // display pQ's contents
  {
    cout << pQ.top();
    pQ.pop();
  }
  cout << endl;
}
```

SAMPLE RUN:

```
JoeSmith
tomiheSJ
```

As we would expect, the values in q are output in FIFO order—the order in which they were inserted. However, the values in pQ are output in *priority order*, according to their ASCII (American Standard Code for Information Interchange) values. Thus, 't' is at the front of pQ and because of all of the letters we inserted, it has the highest ASCII value (116); 'J' is at the back of pQ, because it has the lowest ASCII value (75) in this particular sequence. Also, note that instead of a front() method, the priority_queue<T> container has a top() method to reflect that the value being retrieved has the highest priority of any of its items.

The priority_queue<T> container uses advanced techniques to guarantee that its push() and pop() routines take at most $O(\log_2(n))$ time. This is much faster than $O(n)$ and ensures that these operations require a minimal amount of time.

Like stack<T> and queue<T>, STL's priority_queue<T> is an adapter. Because of its special internal structure, it can be used with a vector<T> or a deque<T> as its internal container, but not a list<T>. The default container is a vector<T>.

A priority_queue<T> is thus a useful container for any situation in which we wish to treat a sequence's values in some prioritized order. Because it effectively sorts its values, the < and == operations must be defined for the objects placed in it.

16.4 AN INTRODUCTION TO TREES

In the preceding chapter we saw how a linked list can be implemented by linking together structures called *nodes*. We also saw that their main advantage over arrays and vectors is that values can be inserted anywhere in a linked list without having to move values to make room for the new ones, and they can be deleted without having to move values to close the gaps. The primary weakness of linked lists is that the only elements that can be accessed directly are those at the ends of the list. This limits the kinds of algorithms that can be applied to linked lists because some that work well for arrays and vectors do not perform well for linked lists.

To demonstrate the problem, we consider the search algorithms considered in Section 12.4. With **linear search**, we examine consecutive elements in the list, beginning with the first one in the list and continuing until either the desired item is found or the end of the list is reached. For example, consider the following vector v, and suppose we search it to find out whether or not the value 60 is present:

v	[0]	[1]	[2]	[3]	[4]	[5]	[6]
	11	22	33	44	55	66	77

With linear search, we start at the beginning and compare each value in the sequence against 60. After seven comparisons—the number of values in the container—we determine that 60 is not present. In general, linear search requires $O(n)$ time to determine that a sequence with n elements does not contain a specific value.[5]

---

[5] If our linear search algorithm assumes the sequence is sorted, then it may be able to stop a bit earlier; on average, however, the number of comparisons will still be proportional to the length of the sequence, or $O(n)$.

The other search algorithm we considered was **binary search**, which requires that the sequence be sorted. We begin by examining the middle element in the sequence (44 in this case):

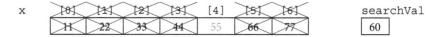

Because 60 (the value we are seeking) is greater than 44, we ignore the middle value and all values to its left, and repeat the process by comparing 60 to the middle value in the remainder of the list (66 in this case):

Now, because 60 is less than 66, we ignore the middle value and all values to its right, and repeat the process by comparing 60 to the middle value in the remainder of the list (55 in this case):

But now there is just one value remaining and it is not equal to the value we are seeking (60), so we conclude that 60 is not present. Only three comparisons were needed to determine this, and in general, binary search can determine whether or not any value is in a sorted container with n elements in $O(\log_2 n)$ time.

In addition to assuming that the sequence is sorted, the efficiency of binary search depends on being able to directly access any value in the container, because any of them may be the middle element in a sublist to be searched. Because of this, binary search performs no better than linear search for a linked list because the values in such a container cannot be accessed directly. Accessing the middle element of a linked list, as required by binary search, requires going through all the nodes that precede it.

This raises an interesting question: Is it possible to link the nodes together in some other way so that they can be searched more quickly than is possible in a linearly linked structure? To see what would be needed to make it possible to search a list in a binary-search–like manner, consider the ordered list of integers again:

$$11, 22, 33, 44, 55, 66, 77$$

The first step in a binary search is to examine the middle element in the list. Direct access to this element is possible if we maintain a link to the node storing it:

For the next step, one of the two sublists, the left half or the right half, must be searched and both must therefore be accessible from this node. This is possible if we maintain two links, one to each of these sublists. Because these sublists are searched in the same manner, these links should refer to nodes containing the middle elements in these sublists:

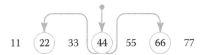

By the same reasoning, in the next step, links from each of these "second-level" nodes are needed to access the middle elements in their sublists:

The resulting structure is easier to visualize if we "stretch it out" into two dimensions so that it has a treelike shape:

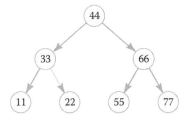

This structure is called a *binary search tree* and is a special kind of *binary tree*, which is a special instance of a more general structure called a *tree*.

16.4.1 Tree Terminology and Examples

A tree consists of a finite collection of **nodes** linked together in such a way that if the tree is not empty, then one of the nodes, called the **root**, has no incoming links, but every other node in the tree can be reached from the root by following a unique sequence of consecutive links.

Trees derive their names from the treelike diagrams that are used to picture them. For example, the diagram on the next page shows a tree having nine vertices in which the node labeled 1 is the root. As this diagram indicates, trees are usually drawn upside down, with the root at the top and the **leaves**—nodes with no outgoing links—at the bottom. Nodes that are directly accessible from a given node (by using only one link) are called the **children** of that node, and a node is said to be the **parent** of its children. For example, in the following tree, node 3 is the parent of nodes 5, 6, and 7, and these nodes are the children of node 3 and are called **siblings**.

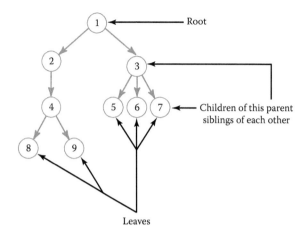

Applications of trees are many and varied. For example, a **genealogical tree** such as the following is a convenient way to picture a person's descendants:

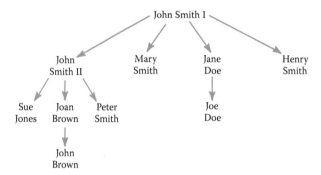

Game trees are used to analyze games and puzzles. **Parse trees** are used to check a program's syntax. For example, the following is a parse tree for the expression 2 * (3 + 4):

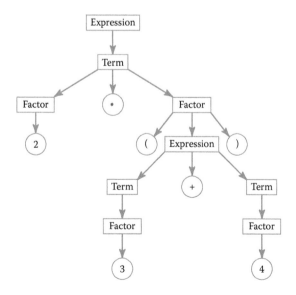

16.4.2 Examples of Binary Trees

Binary trees are trees in which each node has at most two children. They are especially useful in modeling processes in which some experiment or test with two possible outcomes (for example, off or on, 0 or 1, false or true, down or up, yes or no) is performed repeatedly. For example, the following binary tree might be used to represent the possible outcomes of flipping a coin three times:

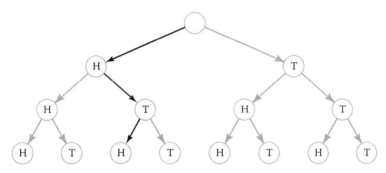

Each path from the root to one of the leaf nodes corresponds to a particular outcome, such as HTH (a head followed by a tail followed by another head), as highlighted in the diagram.

Similarly, a binary tree can be used in coding problems such as encoding and decoding messages transmitted in Morse code, a scheme in which characters are represented as sequences of dots and dashes, as shown in the following table:

A	• —	J	• — — —	S	• • •	2	• • — — —
B	— • • •	K	— • —	T	—	3	• • • — —
C	— • — •	L	• — • •	U	• • —	4	• • • • —
D	— • •	M	— —	V	• • • —	5	• • • • •
E	•	N	— •	W	• — —	6	— • • • •
F	• • — •	O	— — —	X	— • • —	7	— — • • •
G	— — •	P	• — — •	Y	— • — —	8	— — — • •
H	• • • •	Q	— — • —	Z	— — • •	9	— — — — •
I	• •	R	• — •	1	• — — — —	0	— — — — —

In this case, the nodes represent the characters and each link from a node to its children is labeled with a dot or a dash, according to whether it leads to a left child or to a right child, respectively. Thus, part of the tree for Morse code is

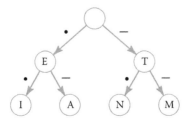

The sequence of dots and dashes labeling a path from the root to a particular node corresponds to the Morse code for that character; for example, • • is the code for I, and — • is the code for N.

Decision-making processes can be modeled as a series of "yes-or-no" questions, and a binary tree can thus be used to model that process. Each nonleaf node is used to store a question, and if an affirmative answer to a question leads to another question, then the two nodes are connected with a link labeled "yes." Similarly, if a negative answer to a question leads to another question, then the two nodes are connected with a link labeled "no." Because there are only two choices for each question, the resulting structure is a binary tree with decisions at its leaf nodes. The problem of choosing a single choice from among many choices is solved simply by descending through the tree until a leaf node (i.e., a decision) is reached.

For problems in some areas, solutions can be obtained by designing decision trees that mimic the choices experts would make in solving these problems. For example, programs to control a robot that welds automobile components on an assembly line are based on the knowledge of an expert welder. Programs to help a person prepare their income tax returns have some of the expertise of a tax accountant encoded within them. Similarly, programs that lead a person through the steps of writing a will have some of the expertise of an estate lawyer encoded within them. In general, programs that exhibit expertise in some area through the use of a knowledge base are called **expert systems**, and the study of such systems is one of the branches of artificial intelligence. A simple example of an expert system is given on the text's website.

16.4.3 Implementing Binary Trees

A binary tree can be represented by a multiply linked structure in which each node has two links, one connecting it to its left child and the other connecting it to its right child, or which are null if the node does not have a left or right child:

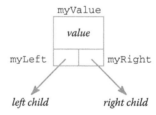

A leaf node is thus characterized by myLeft and myRight both being null:

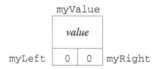

If we represent such nodes by a class named Node that has three public data members,[6]

```
template <typename DataType>
class Node
```

---

[6] A *struct* in C, and thus also provided in C++, is a class whose data and function members are public.

```
{
 public:
  //... Node function members go here
  DataType myValue;
  Node *    myLeft;
  Node *    myRight;
};
```

we can then use it in a `BinaryTree` class template:

```
template <typename DataType>
class BinaryTree
{
 public:
  // ... Binary tree function members go here

 private:
  // ... Declaration of class Node goes here
  Node * myRoot;
  int    mySize;
};
```

The data member `myRoot` is a pointer to the node that is the root of the tree, and `mySize` is used for convenience to keep track of the number of nodes in the tree. Given such a class, the binary tree

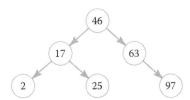

can be represented as the following linked structure:

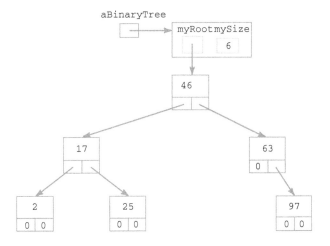

16.4.4 Binary Search Trees

In the preceding binary tree, the value in each node is greater than all values in its **left subtree** (if there are any) and less than all values in its **right subtree** (if there are any). A binary tree having this property is called a **binary search tree (BST)** because, as noted at the beginning of this section, it can be searched using an algorithm much like the binary search algorithm for lists:

Search Algorithm for a BST

1. Initialize a pointer *ptr* to the node containing the root and *found* to false.

2. While *ptr* is not null and *found* is false, do the following:

If the *item* being sought is:

Less than the value in the node pointed to by *ptr*

Set *ptr* equal to its left link;

Greater than the value in the node pointed to by *ptr*

Set *ptr* equal to its right link;

Else

Set *found* to true.

To illustrate, suppose we wish to search the preceding BST for 25. We begin at the root, and because 25 is less than the value 46 in this root, we know that the desired value is located to the left of the root; that is, it must be in the left subtree, whose root is 17:

Now we continue the search by comparing 25 with the value in the root of this subtree. Because 25 > 17, we know that the right subtree should be searched:

Examining the value in the root of this one-node subtree locates the value 25.

Similarly, to search for the value 55, after comparing 55 with the value in the root, we are led to search its right subtree:

Now, because 55 < 63, if the desired value is in the tree, it will be in the left subtree. However, because this left subtree is empty, we conclude that the value 55 is not in the tree.

Because the BST search algorithm effectively eliminates one subtree (i.e., approximately half of the remaining nodes) from consideration on each pass through the while loop, it will, except for "lopsided" BSTs, determine whether or not an item is in the BST in $O(\log_2 n)$ comparisons, where n is the number of values in the BST.

16.4.5 Tree Traversals

Another important operation is **traversal**, that is, moving through a binary tree, visiting each node exactly once. And suppose for now that the order in which the nodes are visited is not relevant. What is important is that we visit each node, not missing any, and that the information in each node is processed exactly once.

One simple recursive scheme is to traverse the binary tree as follows:

1. Visit the root and process its contents.

2. Traverse the left subtree.

3. Traverse the right subtree.

To illustrate this algorithm, let us consider the following binary tree:

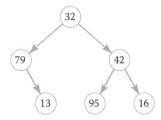

If we simply display a node's contents when we visit it, we begin by displaying the value 32 in the root of the binary tree. Next, we must traverse the left subtree; after this traversal is finished, we then must traverse the right subtree; and when this traversal is completed, we will have traversed the entire binary tree.

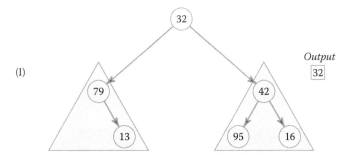

Thus, the problem has been reduced to the traversal of two smaller binary trees. We consider the left subtree and visit its root. Next, we must traverse its left subtree and then its right subtree.

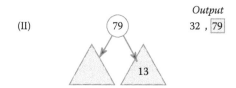

Output

(II)   32 , 79

The left subtree is empty, and we need do nothing. So we turn to traversing the right subtree. We visit its root and then must traverse its left subtree followed by its right subtree:

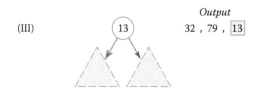

Output

(III)   32 , 79 , 13

As both subtrees are empty, no action is required to traverse them. Consequently, traversal of the binary tree in diagram III is complete, and because this was the right subtree of the tree in diagram II, traversal of this tree is also complete.

This means that we have finished traversing the left subtree of the root in the original binary tree in diagram I, and we finally are ready to begin traversing the right subtree. This traversal proceeds in a similar manner. We first visit its root, displaying the value 42 stored in it, then traverse its left subtree, and then its right subtree:

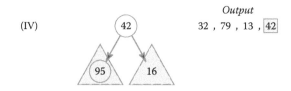

Output

(IV)   32 , 79 , 13 , 42

The left subtree consists of a single node with empty left and right subtrees and is traversed as described earlier for a one-node binary tree:

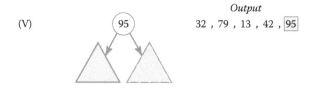

Output

(V)   32 , 79 , 13 , 42 , 95

The right subtree is traversed in the same way:

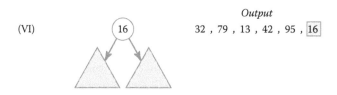

Output

(VI)   32 , 79 , 13 , 42 , 95 , 16

This completes the traversal of the binary tree in diagram IV and thus completes the traversal of the original tree in diagram I.

As this example demonstrates, traversing a binary tree recursively requires three basic steps, which we shall denote N, L, and R:

N: Visit a node.

L: Traverse the left subtree of a node.

R: Traverse the right subtree of a node.

We performed these steps in the order listed here, but in fact, there are six different orders in which they can be carried out: LNR, NLR, LRN, NRL, RNL, and RLN.

The first three orders, in which the left subtree is traversed before the right, are the most important of the six traversals and are commonly called by other names:

LNR $\leftrightarrow$ Inorder

NLR $\leftrightarrow$ Preorder

LRN $\leftrightarrow$ Postorder

It should be noted that an *inorder traversal visits the nodes in a BST in ascending order* because for each node, all of the values in the left subtree are smaller than the value in this node, which is less than all values in its right subtree.

These names are also appropriate for **expression trees**, which represent arithmetic expressions graphically by picturing each operand as a child of a parent node representing the corresponding operator. For example, we can represent

$$A - B * C + D$$

as the following binary tree:

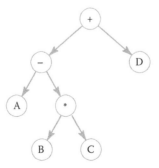

An **inorder** traversal of this expression tree produces the **infix** expression

$$A - B * C + D$$

A **preorder** traversal gives the **prefix** expression, in which an operator precedes its operands:

$$+ - A * B C D$$

And a **postorder** traversal yields the **postfix** expression—also called **reverse polish notation (RPN)**—in which an operator follows its operands:

$$A \; B \; C \; * \; - \; D \; +$$

16.4.6 Constructing BSTs

A binary search tree can be built by repeatedly inserting elements into a BST that is initially empty. Finding where an element is to be inserted is similar to that used to search the tree for that element. The following sequence of diagrams illustrates how 35 would be inserted into the binary search tree given earlier:

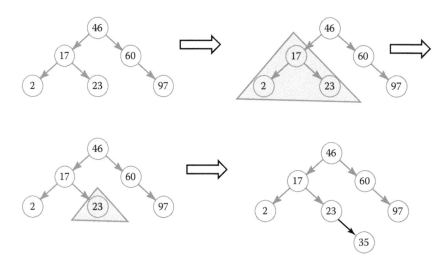

This completes our introduction to trees. The text's website (see the Preface) provides a BST class that implements the preceding (and other) operations.

16.4.7 Trees in STL

The Standard Template Library does not provide any templates with "tree" in their name. However, some of its containers—set<T>, map<T1, T2>, multiset<T>, and multimap<T1, T2>—are generally built using a special kind of *self-balancing binary search tree* called a *red-black tree*. A self-balancing BST ensures that the tree is always as balanced as possible, so that searches take $O(\log_2 n)$ time. The study of these trees is beyond the level of this text and is left to data structures courses and texts (see Footnote 4 in Chapter 12).

CHAPTER SUMMARY

Key Terms

activation record	back of a queue
adapter	backtrack
artificial intelligence	binary search

binary search tree (BST)	postfix
binary tree	postorder
buffer	prefix
children	preorder
decision tree	push
deque	queue
deque<T>	queue<T>
expert system	recursion
expression tree	Reverse Polish Notation
First-In-First-Out (FIFO)	root
infix	run-time stack
inorder	siblings
Last-In-First-Out (LIFO)	singly linked list
leaf node	spool queue
linear search	stack
linked list	stack<T>
list<T>	traversal
node	tree
parent node	vector<T>
pop	

NOTES

- Containers (or data structures) are structures that store groups of values.

- A stack has LIFO (Last-In-First-Out) behavior. A queue has FIFO (First-In-First-Out) behavior.

- The STL containers stack<T>, queue<T>, and priority_queue<T> are adapters that wrap some other container to give it a new interface. The internal container may be a vector<T>, list<T>, or deque<T> for stack<T>; a list<T> or deque<T> for queue<T>; and a vector<T> or deque<T> for priority_queue<T>.

- Whenever a function is called, an activation record is pushed onto a run-time stack. Whenever a function terminates, the run-time stack is popped and control returns to the function whose activation record is uncovered.

- A tree consists of a finite collection of nodes linked together in such a way that if the tree is not empty, then one of the nodes, called the root, has no incoming links, but every other node in the tree can be reached from the root by following a unique sequence of consecutive links.

- A binary tree is a tree in which each node has at most two children.

- In a binary search tree, for each node, all of the values in its left subtree are smaller than the value in this node, which is less than all values in its right subtree.

- Inorder traversal visits the nodes in a BST in ascending order.

Style Tips

- *If a problem solution requires that values stored more recently will be needed before the values stored less recently, then a stack is the appropriate structure for storing these values.* Stacks are LIFO (Last-In-First-Out) lists, because the operation to remove a value will always retrieve the value that was inserted most recently.

- *If a problem solution requires that values will be needed in the order in which they were stored, then a queue is the appropriate structure for storing these values.* Queues are FIFO (First-In-First-Out) lists, because the operation to remove a value from the queue will always retrieve the value that was inserted least recently.

- *If a problem solution requires access, insertion, and deletion only at the ends of a list, a deque is an appropriate structure for storing these values.* Deques are double-ended queues.

- *If a problem solution requires fast access to arbitrary elements but also requires that the collection be allowed to grow and shrink due to frequent insertions and deletions, then a binary search tree (BST) may be the appropriate structure to use.* BSTs are linked structures and thus can grow and shrink without excessive memory waste. And they can be searched in a binary-search manner, which (except when the tree becomes lopsided) is more efficient ($O(\log_2 n)$) than a linear search ($O(n)$).

Warnings

1. *STL's* `queue<T>` *container may wrap a* `deque<T>` *or a* `list<T>` *but not a* `vector<T>`. Removing an element at the front of a `vector<T>` is too inefficient.

2. *STL's* `priority_queue<T>` *container may wrap a* `deque<T>` *or a* `vector<T>` *but not a* `list<T>`. It needs direct access to its elements.

TEST YOURSELF

Section 16.1

1. Convert 1234 to a base-2 number.

2. Convert 1234 to a base-8 number.

3. Convert 1234 to a base-16 number.

4. The last element added to a stack is the _____ one removed. This behavior is known as maintaining the list in _____ order.

5. What are the four standard stack operations?

6. A container that "wraps" another container, giving it a new interface, is called a(n) _____.

Questions 7–9 assume the declaration `stack<int> s;`. List the elements from bottom to top that s will contain after the code segment is executed, or indicate why an error occurs.

7.
```
s.push(123);
s.push(456);
s.pop();
s.push(789);
s.pop();
```

8.
```
s.push(111);
int i = s.top();
s.push(222);
s.pop()
s.push(i);
```

9.
```
for (int i = 0; i < 5; i++)
    s.push(2*i);
s.pop();
s.pop();
```

Section 16.3

1. Explain how a queue differs from a stack.

2. The last element added to a queue is the (first or last) _____ one removed.

3. A stack exhibits LIFO behavior; a queue exhibits _____ behavior.

4. A(n) _____ queue is used to schedule output requests.

5. Queues are used to organize sections of main memory called _____ used to hold input/output data being transferred between a program and disk.

6. A(n) _____ is a double-ended queue.

Questions 7–9 assume the declaration `queue<int> q;`. List the elements from front to back that q will contain after the code segment is executed, or indicate why an error occurs.

7.
```
q.push(123);
q.push(456);
```

```
   q.pop();
   q.push(789);
   q.pop();
```

8. q.push(111);
   ```
   int i = q.front();
   q.push(222);
   q.pop();
   q.push(i);
   ```

9. for (int i = 0; i < 5; i++)
   ```
       q.push(2*i);
   q.pop();
   q.pop();
   ```

Section 16.4

1. A node that has no incoming links but from which every other node in the tree can be reached by following a unique sequence of consecutive links is called a(n) _____.

2. Nodes with no outgoing links are called _____.

3. Nodes that are directly accessible from a given node (by using only one link) are called the _____ of that node, which is said to be the _____ of these nodes.

4. Binary trees are trees in which each node has _____.

Questions 5–7 refer to the following binary search tree:

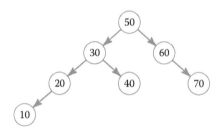

5. Which node is the root?

6. List all the leaves.

7. Draw the BST that results if 45, 55, and 65 are inserted.

For Questions 8–10, draw the BST that results when the C++ keywords are inserted in the order given, starting with an empty BST.

8. if, do, for, case, switch, while, else

9. do, case, else, if, switch, while, for

10. while, switch, for, else, if, do, case

Questions 11–13 refer to the following binary search tree:

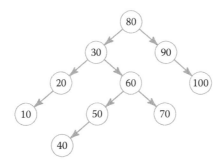

11. Perform an inorder traversal of this BST.

12. Perform a preorder traversal of this BST.

13. Perform a postorder traversal of this BST.

EXERCISES

Section 16.1

1. Convert 2748 to a base-2 number.

2. Convert 2748 to a base-8 number.

3. Convert 2748 to a base-16 number.

Exercises 4–7 assume the declaration stack<int> s;. List the elements from bottom to top that s will contain after the code segment is executed, or indicate why an error occurs.

```
4. s.push(10);
   s.push(22);
   s.push(37);
   s.pop();
   s.pop();
```

```
5. s.push(10);
   s.push(9);
   s.push(8);
   while (!s.empty())
       s.pop();
   s.push(7);
```

6. for (int i = 0; i <= 5; i++)
       s.push(10*i);

7. s.push(11);
   s.push(s.top());
   s.pop()

Section 16.3

Exercises 1–3 assume the declaration queue<int> q;. List the elements from front to back that q will contain after the code segment is executed, or indicate why an error occurs.

1. for (int k = 1; k <= 5; k++)
       q.push(10*k);

2. q.push(11);
   q.push(22);
   q.push(q.front());
   q.pop();
   q.push(33);
   q.push(q.back());
   q.pop();
   q.push(44);

3. q.push(10);
   q.push(9);
   q.pop();
   q.pop();
   q.pop();
   q.push(8);

Exercises 4–6 assume the declaration deque<int> q;. List the elements from front to back that q will contain after the code segment is executed, or indicate why an error occurs.

4. q.push_front(11);
   q.push_front(22);
   q.push_back(33);
   q.pop_front();
   q.push_back(44);
   q.push_front(55);
   q.pop_back();
   q.push_back(66);

```
5. for (int k = 1; k <= 5; k++)
       if (k % 2 == 0)
           q.push_front(10*k);
       else
           q.push_back(10*k);

6. q.push_back(10);
   q.push_back(9);
   q.pop_front();
   q.pop_front();
   q.pop_back();
   q.push_front(8);
```

For Exercises 7–10, tell what output will be produced by the program in Example 16.3 for the given input.

7. `stack`

8. `computer`

9. `Recursion`

10. `STLisgreat`

Section 16.4

For each of the lists of letters in Exercises 1–5, draw the BST that results when the letters are inserted in the order given, starting with an empty BST.

1. A, C, R, E, S

2. R, A, C, E, S

3. C, A, R, E, S

4. S, C, A, R, E

5. C, O, R, N, F, L, A, K, E, S

For each of the lists of C++ keywords in Exercises 6–10, draw the binary search tree that is constructed when the words are inserted in the order given, starting with an empty BST.

6. `new, float, short, if, main, break, for`

7. `break, long, return, char, else, switch, float`

8. double, float, long, class, public, int, new

9. while, static, public, private, else, case

10. break, long, if, short, else, case, void, do, return, while, for, switch, double, true

11–15. Perform inorder, preorder, and postorder traversals of the trees in Exercises 6–10 and show the sequence of words that results in each case.

PROGRAMMING PROBLEMS

Section 16.1

1. Write a program that reads a string, one character at a time, and determines whether the string contains balanced parentheses, that is, for each left parenthesis (if there are any) there is exactly one matching right parenthesis later in the string. (*Hint:* Store the left parentheses in a stack.)

2. The problem in Problem 1 can be solved without using a stack; in fact, a simple integer variable can be used. Describe how and write a program that uses your method to solve the problem.

3. For a given integer $n > 1$, the smallest integer $d > 1$ that divides n is a prime factor. We can find the *prime factorization* of n if we find d and then replace n by the quotient of n divided by d, repeating this until n becomes 1. Write a program that determines the prime factorization of n in this manner but that displays the prime factors in descending order. For example, for $n = 3960$, your program should produce

$$11*5*3*3*2*2*2$$

4. A program is to be written to find a path from one point in a maze to another.

   a. Describe how a two-dimensional array could be used to model the maze.

   b. Describe how a stack could be used in an algorithm for finding a path.

   c. Write the program.

Section 16.3

1. Write a program that generates a random sequence of letters and/or digits, displays them to the user one at a time for a second or so, and then asks the user to reproduce the sequence. Use a queue to store the sequence of characters.

2. Write a "quiz-tutor" program, perhaps on a topic from one of the earlier chapters, or some other topic about which you are knowledgeable. The program should read a question and its answer from a file, display the question, and accept an answer from the user. If the answer is correct, the program should go on to the next question. If it is not correct, store the question in a queue. When the file of questions is exhausted,

the questions that were missed should be displayed again (in their original order). Keep a count of the correct answers and display the final count. Also, display the correct answer when necessary in the second round of questioning.

3. Write a program that reads a string of characters, pushing each character onto a stack as it is read and simultaneously adding it to a queue. When the end of the string is encountered, the program should use the basic stack and queue operations to determine if the string is a *palindrome* (a string that reads the same from left to right as from right to left).

4. Proceed as in Problem 3, but use a single deque to store the characters instead of a stack and a queue.

5. In text-editing and word-processing applications, one formatting convention sometimes used to indicate that a piece of text is a footnote or an endnote is to mark it with some special delimiters such as { and }. When the text is formatted for output, these notes are not printed as normal text but are stored in a queue for later output. Write a program that reads a document containing endnotes indicated in this manner, collects them in a queue, and displays them at the end of the document. Number the endnotes and in the text where the endnote occurred, put its number enclosed in brackets [ ].

6. Suppose that jobs entering a computer system are assigned a job number and a priority from 0 through 9. The numbers of jobs awaiting execution by the system are kept in a priority queue. Write a program that reads one of the letters R (remove), A (add), or L (list). For R, remove a job from the priority queue and display the job number; for A, read a job number and priority and then add it to the priority queue in the manner just described; and for L, list all the job numbers in the queue.

7. (Project) Suppose that a certain airport has one runway, that each airplane takes `landingTime` minutes to land and `takeOffTime` minutes to take off, and that on the average, `takeOffRate` planes take off and `landingRate` planes land each hour. Assume that the planes arrive at random instants of time. (Delays make the assumption of randomness quite reasonable.) There are two types of queues: a queue of airplanes waiting to land and a queue of airplanes waiting to take off. Because it is more expensive to keep a plane airborne than to have one waiting on the ground, we assume that the airplanes in the landing queue have priority over those in the takeoff queue.

Write a program to simulate this airport's operation. You might assume a simulated clock that advances in one-minute intervals. For each minute, generate two random numbers: If the first is less than `landingRate` / 60, a "landing arrival" has occurred and is added to the landing queue, and if the second is less than `takeOffRate` / 60, a "takeoff arrival" has occurred and is added to the takeoff queue. Next, check whether the runway is free. If it is, first check whether the landing queue is nonempty, and if so, allow the first airplane to land; otherwise, consider the takeoff queue. Have the program calculate the average queue length and the average time that an airplane spends in a queue. You might also investigate the effect of varying arrival and departure rates to simulate the prime and slack times of day, or what happens if the amount of time to land or take off is increased or decreased.

Section 16.4

1. Write a spell checker, that is, a program that reads the words in a piece of text and looks up each of them in a dictionary to check its spelling. Use a BST to store this dictionary, reading the list of words from a file. While checking the spelling of words in a piece of text, the program should print a list of all words not found in the dictionary.

2. Write a program to construct a text concordance, which is an alphabetical listing of all the distinct words in a piece of text. It should read a piece of text; construct a concordance that contains the distinct words that appear in the text and for each word, the line (or page) number of its first occurrence; and then allow the user to search this concordance. Use an array or vector of 26 BSTs, one for each letter of the alphabet, to store the concordance.

3. Extend the program in Problem 2, so that a (linked) list of all occurrences of each word is stored. When the concordance is searched for a particular word, the program should display the line (or page) numbers of all occurrences of this word. The data structure used for the concordance is thus an array or vector of binary search trees, each of whose nodes stores an object containing a string and an ordered linked list of integers. Extend the program in Problem 3, so that a (linked) list of all occurrences of each word is stored. When the concordance is searched for a particular word, the program should display the line (or page) numbers of all occurrences of this word. The data structure used for the concordance is thus an array or vector of binary search trees, each of whose nodes stores an object containing a string and an ordered linked list of integers.

Answers to Test Yourself Questions

1. comment

2. curly braces, main

3. Design
   Coding
   Testing, execution, and debugging
   Maintenance

4. State program's behavior
   Identify the objects
   Identify the operations
   Arrange operations in an algorithm

5. objects

6. operations

7. variables

8. `cout`

9. `cin`

10. `<<, >>`

11. debugging

12. syntax errors and logic errors

1. integers, integer variations, reals, characters, booleans

2. `short, int, unsigned`

3. `float, double, long double`

4. literal

5. false

6. true

7. true

8. single quotes (or apostrophes)

9. false

10. true

11. escape

12. double quotes

13. legal

14. legal

15. not legal—must begin with a letter

16. legal

17. not legal—identifiers may contain only letters, digits, and underscores

18. legal

19. not legal—same reason as 17

20. not legal—same reason as 17

21. integer

22. neither

23. real

24. real

25. real

26. neither

27. real

28. integer

29. neither

30. neither

31. neither

32. neither

33. neither

34. string

35. string

36. integer

37. integer

38. neither

39. neither

40. character

41. `int count;`

42. `unsigned time;`
    `float temperature;`
    `char scale;`

43. `unsigned time = 0;`
    `float temperature;`
    `char scale;`

44. `unsigned time = 9999;`
    `float temperature = 0;`
    `char scale = ' ';`

45. `const int`
    `        CELSIUS_FREEZE = 0,`
    `        CELSIUS_BOIL = 100;`

46. `const double EARTH = 1.5E10,`
    `               MARS = 1.2E12;`

CHAPTER 4

Section 4.2

1. 0

2. 2.6

3. 2

4. 5

5. 8

6. 3

7. 2

8. 36.0

9. 11.0

10. 1

11. 7.0

12. 5.1

13. 8.0

14. 10.0

15. 'b' (or 98)

16. 'd' (or 100)

17. 2

18. 2

Section 4.3

1. valid

2. not valid—variable must be to left of assignment operator

3. valid

4. not valid—variable must be to left of assignment operator

5. valid

6. not valid—can't assign a string to an integer variable

7. valid

8. not valid—'65' is not a legal character constant

9. valid

10. valid

11. valid

12. valid

13. valid

14. valid

15. valid

16. not valid—a variable must be to left of assignment operator

17. not valid—++ can only be used with integer variables

18. xValue: 3.5

19. xValue: 6.1

20. jobId: 6

21. jobId: 5
    intFive1: 6

22. jobId: 6
    intFive2: 6

23. intEight: 64

24. distance = rate * time;

25. c = sqrt(a*a + b*b);

26. ++x;
    x++;
    x += 1;
    x = x + 1;

Section 4.4

1. streams

2. true

3. cin, istream

4. cout, cerr, ostream

5. >>

6. <<

7. cin

8. cout

9. right

10. format manipulators

11. 12323.4568

12. blank line
    ⊔⊔123124⊔⊔125127

13. ⊔⊔⊔⊔⊔23.
    ⊔⊔⊔⊔23.5
    ⊔⊔⊔23.46
    ⊔⊔⊔23.46
    23.5

14. number1: 11
    number2: 22
    number3: 33

15. real1: 1.1
    real2: 2.0
    real3: 33

16. `number1: 1`
    Input error: attempting to read a
    period for integer variable

17. `number1:    1`
    `real1:      .1`
    `number2:    2`
    `real2:      3.3`
    `number3:    4`
    `real3:      5.5`

Section 4.5

1. 3.0

2. 8.0

3. 2.0

4. 3.0

5. 3.0

6. 32.0

7. 4.0

8. 4.0

9. not valid; `sqrt()` only defined for nonnegative values

10. 2

11. 1

12. 0.25

13. 81.0

14. 16.0

15. 1.0

16. `sqrt((x+y) / 2)`

17. `abs(a / (b+c))`

18. `pow(e, x * log(a))`

19. `int(100 * amount + 0.5) / 100`

CHAPTER 5

Section 5.2

1. Sequence, selection, repetition

2. `false, true`

3. `<, >, ==, <=, >=, !=`

4. `!, &&, ||`

5. `false`

6. `true`

7. `false`

8. `false`

9. `true`

10. `true`

11. `true`

12. `true`

13. `true` (but probably should be written `0 <= count && count <= 2`, which would be false)

14. `x != 0`

15. `-10 < x && x < 10`

16. `(x > 0 && y > 0) || (x < 0 && y < 0)` or more simply, `x * y > 0`

Section 5.4

1. 6
2. 5
3. 6
4. 10
5. 10
6. 10
7. excellent
8. excellent
9. good
10. fair
11. bad
12. if (number < 0 || number > 100)
        cout << "Out of range\n";
13. if (x <= 1.5)
       n = 1;
     else if (x <= 2.5)
       n = 2;
     else
       n = 3;

Section 5.5

1. Hello
   Hello
   Hello
   Hello
   Hello

2. HelloHelloHello

3. Hello
   Hello
   Hello

4. 36
   25
   16
   9
   4
   1

5. Hello

6. No output produced

7. 1
   3
   5
   7
   9

8. 3
   2
   1
   0
   −1

9. 1
   2
   4
   16
   32
   64

10. 0E 10 2E 30 4E 50 6E 70 8E 90

11. 6

CHAPTER 6

Section 6.3

1. objects received from the calling function
   objects returned to the calling function

2. parameters

3. double

4. void

8. true

5. no statements

9. false

6. argument

10. `int what(int n);`

7. 6

11.
```
double func(double x)
{
    return x*x + sqrt(x);
}
```

12.
```
int average(int num1, int num2)
{
    return (num1 + num2) / 2;
}
```

13.
```
void display(int num1, int num2, int num3)
{
    cout << num1 << "\n\n"
         << num2 << "\n\n"
         << num1 << endl;
}
```

Section 6.4

1. False

2. header (or interface), implementation, and documentation

3. 1. Functions in a library are reusable.
   2. They hide implementation details.
   3. They make programs easier to maintain.
   4. They provide separate compilation.
   5. The support independent coding.
   6. They simplify testing.

4. header

9. `"lib"`

5. implementation

10. compilation and linking

6. header

11. false

7. public, private

12. true

8. `<lib>`

13. Information hiding

CHAPTER 7

Section 7.2

1. encapsulation

3. overloading

2. data, function

4. dot

5. `month, day, year`

6. `display()`

7. `birth.display();`

Section 7.3

1. Bjarne Stroustrup

2. Jerry Schwarz

3. `istream` and `ostream`

4. stream

5. `istream`

6. `cin`

7. good, bad, and fail

8. `good()`

9. `clear()`

10. `ignore()`

11. true

12. false

13. `ostream`

14. `cout`

15. `endl, flush`

16. true

17. false

Section 7.4

1. empty

2. `string label;`

3. `const string UNITS = "meters";`

4. `"ABC"`, `"DEF"`

5. `'e'`

6. `8`

7. `0`

8. false

9. true

10. true

11. false

12. `seashoreshell`

13. `"she"`

14. `10`

15. `27`

16. `12`

17. `29`

18. `0`

19. `35`

20. `"bell"`

21. `"seal on the shore"`

22. `"She sells the seashore."`

Section 7.5

1. $15 + 5i$

2. $50 + 37i$

3. $55 + 48i$

4. $1 + i$

5. $(62/53) + (5/53)i \approx 1.1698 + .09434i$

6. $\sqrt{73} \approx 8.544$

7. (−1,7)

8. (2,−1)

9. (1.5,2.1)

10. 3

11. (1.5,0)

12. (1.5, 2.5)

CHAPTER 8

Section 8.2

1. ```
198
197
default
```

2. `198`

3. `default`

4. `default`

5. ```
−2
default
```

6. `−2`

7. `123`

8. `456`

9. no output produced

10. error—x must be integer (or integer compatible)

CHAPTER 9

Section 9.4

1. counting (or counter-controlled) loops, `for`

2. initialization expression, loop condition, step expression, loop body

3. `if-break` (or `if-return`)

4. pretest

5. posttest

6. posttest

7. pretest

8. ```
2*0 = 0
2*1 = 2
2*2 = 4
2*3 = 6
2*4 = 8
2*5 = 10
2*6 = 12
2*7 = 14
2*8 = 16
2*9 = 18
```

9. `1 3 5 7 9 11`

10. ```
11
22
1
33
2
1
```

11. ```
000
112
228
18
```

12. ```
4
5
228
18
```

13. 3
2
1
0
−1

14. 0 1
1 2
2 5

14.(continued)
3 10
4 17
★★★★★

15. 4 12
3 5
2 0
1 −3

Section 9.6

1. sentinel, counting, query-controlled

2. end-of-data flag, sentinel

3. true

4. end-of-file (or eof)

5. eof()

6. false

7. false

8. query

CHAPTER 10

Section 10.3

1. value

2. value

3. value

4. reference

5. ampersand (&)

6. false

7. true

8. false

9. false

10. false

11. ```
void f(const int & x, int & y, int & z)
{
    z = y = x * x + 1;
}
```

12. `String = batbatelk`

## Section 10.5

1. scope

2. false

3. end of the block

4. the body of the function

5. scope error message (perhaps a warning)

6. signature

7. name

8. signature

9. template

10. type
```

11. Generate a function `print()` with type parameter `something` replaced everywhere by `int`.

CHAPTER 11

Section 11.2

1. `istream, cin`

4. `ostream`

2. `ostream, cout` (or `cerr`)

5. `fstream`

3. `istream`

6. false

7. `ifstream inputStream;`
   `inputStream.open("EmployeeInfo");`

8. `ifstream inputStream("EmployeeInfo");`

9. `ofstream outputStream;`
   `outputStream.open("EmployeeReport");`

10. `ofstream outputStream("EmployeeReport");`

11. `string inFileName, outFileName;`
    `cout << "Name of input file? ";`
    `cin >> inFileName;`
    `ifstream inputStream;`
    `inputStream.open(inFileName.data());`
       or replace the last two lines with:
    `ifstream inputStream(inFileName.data());`

12. false

13. true

14. `assert(inputStream.is_open());`

15. `get_line(inputStream, str);` where `str` is of type `string`

16. `if (inputSteam.eof())`
        `cout << "End of file\n";`

17. `inputStream.close();`

Section 11.3

1. false

3. random

2. true

4. `tellg(), seekg()`

5. `inputStream.seekg(3, ios::beg());`

6. `inputStream.seekg(3, ios::cur());`

7. `inputStream.seekg(0, ios::end());`

8. ```
char ch;
inputStream.get(ch);
cout << ch;
```

9. ```
char ch;
inputStream.peek(ch);        or        inputStream.get(ch);
 cout << ch;                            cout << ch;
                                        inputStream.putback(ch);
```

10. formatting manipulators

CHAPTER 12

Section 12.2

1. false

2. false (not necessarily)

3. true

4. false

5. true

6. false

7. false

8. 5

9. true

10. true

11. true

12. false

13. true

14. base

15. false

16. true

17. false

18. reference

19. xValue[0]: 0.0
    xValue[1]: 0.5
    xValue[2]: 1.0
    xValue[3]: 1.5
    xValue[4]: 2.0

20. number[0]: 0
    number[1]: 3
    number[2]: 4
    number[3]: 7
    number[4]: 8

21. number[0]: 1
    number[1]: 2
    number[2]: 4
    number[3]: 8
    number[4]: 16

22. number[0]: 0
    number[1]: 0
    number[2]: 0
    number[3]: 0
    number[4]: 0

Section 12.6

1. `int`

2. 0, 0

3. 5, 5

4. 5, 5

5. 10, 7

6. 1 1

7. 0 88

8. true

9. false

10. true

11. 1 1 1 1 1

12. 0 0 0 0 0 77

13. 0

14. 88

15. true

16. 0, 0, 0, 0, 0, 0, 0.5, 1.0, 1.5, 2.0

17. 1, 1, 1, 1, 1, 0, 3, 4, 7, 8

18. 1, 1, 1, 1, 1, 2, 2, 2, 2, 0

19. 1, 1, 2, 2, 2, 0

20. In the early 1900s at Hewlett Packard Laboratories by Alex Stepanov and Meng Lee

CHAPTER 13

Section 13.2

1. two-dimensional

2. three-dimensional

3. square

4. 82

5. 33

6. –1

7. 33

8. 12

9. 1, 2, 3, 4

10. 52

11. 76

12. 29

13. 1322

14. 1322

Section 13.4

1. $m \times n$ matrix

2. 2, 2

3. $\begin{bmatrix} 5 & 2 \\ 8 & 8 \end{bmatrix}$

4. $\begin{bmatrix} 1 & 2 & 0 \\ 1 & 3 & 3 \\ 10 & 5 & 9 \end{bmatrix}$

5. Not defined—number of columns in $B \neq$ number of rows in C

6. Not defined—number of columns in $C \neq$ number of rows in B

7. $\begin{bmatrix} 4 & 2 \\ 3 & -1 \\ 9 & 1 \end{bmatrix}$

8. $\begin{bmatrix} -2 & -2 \\ 1 & -1 \\ 2 & 5 \end{bmatrix}$

1. true

2. true

3. 5, 4

4. `qqTab[1][3] = 1.1;`

5. 5

6. 4

7. `qqTab[1].push_back(99.9);`

8. `TableRow bottom(4, 0.0);`
   `qqTab.push_back(bottom);`

CHAPTER 14

Section 14.2

1. class, class library

2. design, implementation

3. Its behavior and its attributes

4. behavior

5. I-can-do-it-myself

Section 14.3

1. data, instance

2. Encapsulation

3. Items in the public section are accessible to users of the class, but those in the private section are not. By preventing a program from directly accessing the data members of a class, we hide that information, thus removing the temptation to access those variables directly.

4. private

5. invariant

6. function, methods

7. name of the class, scope (::)

8. constant, `const`

9. `Fraction`

10. Default-value constructors: used to initialize data members with default values when none are specified in the declaration of an object; explicit-value constructors initialize them with values provided in the declaration.

11. False—constructors have no return type

12. public

13. postcondition

15. accessor

14. dot, message

16. `operator+`

17.
```
class Component
{
  public:
    void display(ostream & out);
    void read(istream & in);
  private:
    int myID;
    string myName;
};
```

Section 14.6

1. friend

3. #

2. preprocessor

4. conditional-compilation

5. The declaration of the class with the specified *name* will only be processed once, regardless of how many different files include it.

CHAPTER 15

Section 15.1

1. address

9. double value

2. &

10. address

3. *

11. null

4. address

12. 1.1

5. double value

13. 3.3

6. address

14. true

7. address

15. true

8. double value

16. 012a50

Section 15.2

1. compile, run

4. pointer variable

2. new, null address, address, anonymous

5. 66

3. delete

Section 15.3

1. true

2. nodes, pointers

3. data, next

4. null

5. false

6. true

CHAPTER 16

Section 16.1

1. 10011010010

2. 2322

3. 4D2

4. first, LIFO (last in, first out)

5. empty, push, top, pop

6. adapter

7. 123

8. 111, 111

9. 0, 2, 4

Section 16.3

1. For a queue, elements are added at one end and removed at the other. For a stack, elements are added and removed at the same end.

2. last

3. FIFO

4. spool

5. buffers

6. deque

7. 456

8. 222, 111

9. 4, 6, 8

Section 16.4

1. root

2. leaves

3. children, parent

4. at most two children

5. 50

6. 10, 40, 70

7.

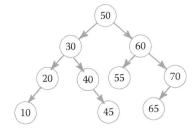

8.

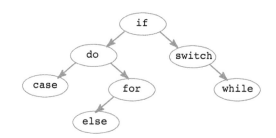

9.

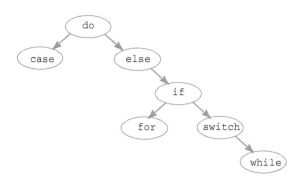

10.

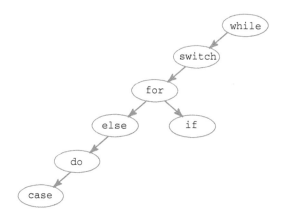

11. 10, 20, 30, 40, 50, 60, 70, 80, 90, 100

12. 90, 30, 20, 10, 60, 50, 40, 70, 90, 100

13. 10, 20, 40, 50, 70, 60, 30, 100, 90, 80

Appendix A: ASCII Character Codes

Decimal	Octal	Character
0	000	NUL (Null)
1	001	SOH (Start of heading)
2	002	STX (Start of text)
3	003	ETX (End of text)
4	004	EOT (End of transmission)
5	005	ENQ (Enquiry)
6	006	ACK (Acknowledge)
7	007	BEL (Ring bell)
8	010	BS (Backspace)
9	011	HT (Horizontal tab)
10	012	LF (Line feed)
11	013	VT (Vertical tab)
12	014	FF (Form feed)
13	015	CR (Carriage return)
14	016	SO (Shift out)
15	017	SI (Shift in)
16	020	DLE (Date link escape)
17	021	DC1 (Device control 1)
18	022	DC2 (Device control 2)
19	023	DC3 (Device control 3)
20	024	DC4 (Device control 4)
21	025	NAK (Negative ACK)
22	026	SYN (Synchronous)
23	027	ETB (EOT block)
24	030	CAN (Cancel)
25	031	EM (End of medium)

(continued)

(continued)

Decimal	Octal	Character
26	032	SUB (Substitute)
27	033	ESC (Escape)
28	034	FS (File separator)
29	035	GS (Group separator)
30	036	RS (Record separator)
31	037	US (Unit separator)
32	040	SP (Space)
33	041	!
34	042	"
35	043	#
36	044	$
37	045	%
38	046	&
39	047	' (Single quote)
40	050	(
41	051	)
42	052	*
43	053	+
44	054	, (Comma)
45	055	- (Hyphen)
46	056	. (Period)
47	057	/
48	060	0
49	061	1
50	062	2
51	063	3
52	064	4
53	065	5
54	066	6
55	067	7
56	070	8
57	071	9
58	072	:
59	073	;
60	074	<
61	075	=
62	076	>
63	077	?
64	100	@
65	101	A
66	102	B
67	103	C
68	104	D
69	105	E
70	106	F

71	107	G
72	110	H
73	111	I
74	112	J
75	113	K
76	114	L
77	115	M
78	116	N
79	117	O
80	120	P
81	121	Q
82	122	R
83	123	S
84	124	T
85	125	U
86	126	V
87	127	W
88	130	X
89	131	Y
90	132	Z
91	133	[
92	134	\
93	135	]
94	136	^
95	137	_ (Underscore)
96	140	`
97	141	a
98	142	b
99	143	c
100	144	d
101	145	e
102	146	f
103	147	g
104	150	h
105	151	i
106	152	j
107	153	k
108	154	l
109	155	m
110	156	n
111	157	o
112	160	p
113	161	q
114	162	r
115	163	s
116	164	t

(continued)

(continued)

Decimal	Octal	Character
117	165	u
118	166	v
119	167	w
120	170	x
121	171	y
122	172	z
123	173	{
124	174	|
125	175	}
126	176	~
127	177	DEL

Appendix B: C++ Keywords

The following table lists the keywords in C++, together with a brief description of the context in which they usually appear.

Keyword	Contextual Description
asm	Used to declare that information is to be passed directly to the assembler
auto	Used to declare objects whose lifetime is the duration of control within their block
bool	Used to declare objects whose values are `true` or `false`
break	Used to terminate processing of a `switch` statement or loop
case	Used in a `switch` statement to specify a match for the statement's expression
catch	Used to specify the actions to be taken when an exception occurs (see `throw`, `try`)
char	Used to declare objects whose values are characters
class	Used to construct new types encapsulating data and operations (default `private`)
const	Used to declare objects whose values should not change during execution
const_cast	Used to add or remove the `const` or `volatile` property of a type
continue	Used in a loop statement to transfer control to the beginning of the loop
default	Used in a `switch` statement to handle expression values not specified using `case`
delete	Used to deallocate memory allocated at run time, returning it to the free store
do	Used to mark the beginning of a `do-while` statement, providing repetitive control
double	Used to declare objects whose values are (double precision) real numbers
dynamic_cast	Used to cast pointer or reference types in a class hierarchy
else	Used in an `if` statement to mark the section to be executed if the condition is false
enum	Used to declare a type whose values are programmer-specified identifiers
explicit	Used to prevent constructors from being called implicitly for conversion purposes
extern	Used to declare objects whose definitions are external to the local block
false	A `bool` value
float	Used to declare objects whose values arc (single precision) real numbers
for	Used to mark the beginning of a `for` statement, providing repetitive control
friend	Used to declare class operations that are not member functions
goto	Used to transfer control to a label
if	Used to mark the beginning of an `if` statement, providing selective control
inline	Used to declare a function whose text is to be substituted for its call
int	Used to declare objects whose values are integer numbers

(continued)

(continued)

Keyword	Contextual Description
`long`	Used to declare long integers (typically 32-bit), or extended double precision real numbers
`mutable`	Used to declare class data member as modifiable even in a `const` object
`namespace`	Used to a control the scope of global names (to avoid name conflicts)
`new`	Used to request memory allocation at run time
`operator`	Used to overload an operator with a new declaration
`private`	Used to declare class members that are inaccessible from outside of the class
`protected`	Used to declare class members that are `private`, except to derived classes
`public`	Used to declare class members that can be accessed outside of the class
`register`	Used to declare objects whose values are to be kept in registers
`reinterpret_cast`	Used to perform type conversions on unrelated types
`return`	Used to terminate a function, usually returning the value of some expression
`short`	Used to declare short integers (typically 16-bit) numbers
`signed`	Used to declare an object in which the value's sign is stored in the high order bit
`sizeof`	Used to find the size (in bytes) of an object, or of the representation of a type
`static`	Used to declare objects whose lifetime is the duration of the program
`static_cast`	Used to convert one type to another type
`struct`	Used to construct new types encapsulating data and operations (default `public`)
`switch`	Used to mark the beginning of a switch statement, providing selective control
`template`	Used to declare type-independent classes or functions
`this`	Used within a class member to unambiguously access other members of the class
`throw`	Used to generate an exception (see `catch`, `try`)
`true`	A `bool` value
`try`	Used to mark the beginning of a block containing exception handlers (see `catch`)
`typedef`	Used to declare a name as a synonym for an existing type
`typeid`	Used to obtain type information during run time
`typename`	Can be used instead of class in template parameter lists and to identify qualified names as types
`union`	Used to declare a structure, such that different objects can have different members
`unsigned`	Used to declare an object in which the high-order bit is used for data (see `signed`)
`using`	Used to access members of a namespace
`virtual`	Used to declare a base-class function, that will be defined by a derived class
`void`	Used to indicate the absence of any type
`volatile`	Used to declare objects whose values may be modified by means undetectable to the compiler (such as shared-memory objects of concurrent processes)
`while`	Used to mark the beginning of a `while` statement, as well as the end of a `do-while` statement, each of which provides repetitive control

Appendix C: C++ Operators

The following table lists all of the operators available in C++, ordered by their precedence levels, from highest to lowest—higher-precedence operators are applied before lower-precedence operators. Operators in the same horizontal band of the table have equal precedence. The table also gives each operator's associativity—in an expression containing operators of equal precedence, associativity determines which is applied first—whether they can be overloaded, their arity (number of operands), and a brief description.

Operator	Associativity	Overloadable	Arity	Description
::	right	no	unary	global scope
::	left	no	binary	class scope
.	left	no	binary	direct member selection
->	left	yes	binary	indirect member selection
[]	left	yes	binary	subscript (array index)
()	left	yes	n/a	function call
()	left	yes	n/a	type construction
sizeof	right	n/a	unary	size (in bytes) of an object or type
++	right	yes	unary	increment
--	right	yes	unary	decrement
~	right	yes	unary	bitwise NOT
!	right	yes	unary	logical NOT
+	right	yes	unary	plus (sign)
-	right	yes	unary	minus (sign)
*	right	yes	unary	pointer dereferencing
&	right	yes	unary	get address of an object
new	right	yes	unary	memory allocation
delete	right	yes	unary	memory deallocation
()	right	yes	binary	type conversion (cast)
.	left	no	binary	direct member pointer selection
->	left	yes	binary	indirect member pointer selection
*	left	yes	binary	multiplication
/	left	yes	binary	division
%	left	yes	binary	modulus (remainder)

(continued)

Operator	Associativity	Overloadable	Arity	Description
+	left	yes	binary	addition
−	left	yes	binary	subtraction
<<	left	yes	binary	bit-shift left
>>	left	yes	binary	bit-shift right
<	left	yes	binary	less-than
<=	left	yes	binary	less-than-or-equal
>	left	yes	binary	greater-than
>=	left	yes	binary	greater-than-or-equal
==	left	yes	binary	equality
!=	left	yes	binary	inequality
&	left	yes	binary	bitwise AND
^	left	yes	binary	bitwise XOR
|	left	yes	binary	bitwise OR
&&	left	yes	binary	logical AND
||	left	yes	binary	logical OR
? :	left	no	ternary	conditional expression
=	right	yes	binary	assignment
+=	right	yes	binary	addition-assignment shortcut
−=	right	yes	binary	subtraction-assignment shortcut
*=	right	yes	binary	multiplication-assignment shortcut
/=	right	yes	binary	division-assignment shortcut
%=	right	yes	binary	modulus-assignment shortcut
&=	right	yes	binary	bitwise-AND-assignment shortcut
|=	right	yes	binary	bitwise-OR-assignment shortcut
^=	right	yes	binary	bitwise-XOR-assignment shortcut
<<=	right	yes	binary	bitshift-left-assignment shortcut
>>=	right	yes	binary	bitshift-right-assignment shortcut
throw	right	yes	unary	throw an exception
,	left	yes	binary	expression separation

Appendix D: Other C++ Features

Many of the C++ libraries were originally C libraries. The following describes some of the most useful items in the more commonly used libraries.

cassert (Formerly assert.h)

`void assert(bool expr)`	Tests the boolean expression `expr` and if it is true, allows execution to proceed. If it is false, execution is terminated and an error message is displayed.

cctype (Formerly ctype.h)

`int isalnum(int c)`	Returns true if c is a letter or a digit, `false` otherwise
`int isalpha(int c)`	Returns true if c is a letter, `false` otherwise
`int iscntrl(int c)`	Returns true if c is a control character, `false` otherwise
`int isdigit(int c)`	Returns true if c is a decimal digit, `false` otherwise
`int isgraph(int c)`	Returns true if c is a printing character except space, `false` otherwise
`int islower(int c)`	Returns true if c is lowercase, `false` otherwise
`int isprint(int c)`	Returns true if c is a printing character including space, `false` otherwise
`int ispunct(int c)`	Returns true if c is a punctuation character (not a space, an alphabetic character, or a digit), `false` otherwise

701

`int isspace(int c)`	Returns true if `c` is a white space character (space, `'\f'`, `'\n'`, `'\r'`, `'\t'`, or `'\v'`), `false` otherwise
`int isupper(int c)`	Returns true if `c` is uppercase, `false` otherwise
`int isxdigit(int c)`	Returns true if `c` is a hexadecimal digit, `false` otherwise
`int tolower(int c)`	Returns lowercase equivalent of `c` (if `c` is uppercase)
`int toupper(int c)`	Returns the uppercase equivalent of `c` (if `c` is lowercase)

cfloat (Formerly `float.h`)

The following constants specify the minimum value in the specified floating-point type.

FLT_MIN ($\leq$ –1E+37)	`float`
DBL_MIN ($\leq$ –1E+37)	`double`
LDBL_MIN ($\leq$ –1E+37)	`long double`

The following constants specify the maximum value in the specified floating-point type.

FLT_MAX ($\geq$ 1E+37)	`float`
DBL_MAX ($\geq$ 1E+37)	`double`
LDBL_MAX ($\geq$ 1E+37)	`long double`

The following constants specify the smallest positive value representable in the specified floating-point type.

FLT_EPSILON ($\leq$ 1E–37)	`float`
DBL_EPSILON ($\leq$ 1E–37)	`double`
LDBL_EPSILON ($\leq$ 1E–37)	`long double`

climits (Formerly `limits.h`)

The following constants specify the minimum and maximum values for the specified type.

SCHAR_MIN ($\leq$ –127)	`signed char`
SCHAR_MAX ($\geq$ 127)	`signed char`
UCHAR_MAX ($\geq$ 255)	`unsigned char`
CHAR _ MIN (0 or SCHAR_MIN)	`char`

CHAR_MAX (SCHAR_MAX or USHRT_MAX)	`char`
SHRT_MIN ($\leq$ –32767)	`short int`
SHRT_MAX ($\geq$ 32767)	`short int`
USHRT_MAX ($\geq$ 65535)	`unsigned short int`
INT_MIN ($\leq$ –32767)	`int`
INT_MAX ($\geq$ 32767)	`int`
UINT_MAX ($\geq$ 65535)	`unsigned int`
LONG_MIN ($\leq$ –2147483647)	`long int`
LONG_MAX ($\geq$ 2147483647)	`long int`
ULONG_MAX ($\geq$ 4294967295)	`unsigned long int`

`cmath` (Formerly `math.h`)

`double acos(double x)`	Returns the angle in $[0, \pi]$ (in radians) whose cosine is x
`double asin(double x)`	Returns the angle in $[-\pi/2, \pi/2]$ (in radians) whose sine is x
`double atan(double x)`	Returns the angle in $(-\pi/2, \pi/2)$ (in radians) whose tangent is x
`double atan2(double y)`	Returns the angle in $(-\pi, \pi)$ (in radians) whose tangent is y/x
`double ceil(double x)`	Returns the least integer $\geq$ x
`double cos(double x)`	Returns the cosine of x (radians)
`double cosh(double x)`	Returns the hyperbolic cosine of x
`double exp(double x)`	Returns e^x
`double fabs(double x)`	Returns the absolute value of x
`double floor(double x)`	Returns the greatest integer $\leq$ x
`double fmod(double x,` `          double y)`	Returns the integer remainder of x / y
`double frexp(double x,` `          int & ex)`	Returns value v in $[1/2, 1]$ and passes back ex such that $x = v * 2^{ex}$
`double ldexp(double x,` `          int ex)`	Returns $x * 2^{ex}$

`double log(double x)`	Returns natural logarithm of x
`double log10(double x)`	Returns base-ten logarithm of x
`double modf(double x,` `          double & ip)`	Returns fractional part of x and passes back `ip` = the integer part of x
`double pow(double x,` `          double y)`	Returns x^y
`double sin(double x)`	Returns the sine of x (radians)
`double sinh(double x)`	Returns the hyperbolic sine of x
`double sqrt(double x)`	Returns the square root of x (provided $x \geq 0$)
`double tan(double x)`	Returns the tangent of x (radians)
`double tanh(double x)`	Returns the hyperbolic tangent of x

`cstdlib` (Formerly `stdlib.h`)

`int abs(int i)` `long abs(long li)`	`abs(i)` and `labs(li)` return the `int` and `long int` absolute value of i and li, respectively

`double atof(char s[])`
`int atoi(char s[])`
`long atol(char s[])`
	`atof(s)`, `atoi(s)`, and `atol(s)` return the value obtained by converting the character string s to `double`, `int`, and `long int`, respectively

`void exit(int status)`	Terminates program execution and returns control to the operating system; `status` = 0 signals successful termination and any nonzero value signals unsuccessful termination
`int rand()`	Returns a pseudorandom integer in the range 0 to `RAND_MAX`
`RAND_MAX`	An integer constant ($\geq$ 32767) which is the maximum value returned by `rand()`
`void srand(int seed)`	Uses seed to initialize the sequence of pseudorandom numbers returned by `rand()`
`int system(char s[])`	Passes the string s to the operating system to be executed as a command and returns an implementation-dependent value.

THE string CLASS

The string class, which was described in Chapter 7, is defined by

```
typedef basic_string<char> string;
```

The unsigned integer type size_type is defined in this class as is an integer constant npos, which is some integer that is either negative or greater than the number of characters in a string. The following is a list of the major operations defined on a string object s; pos, pos1, pos2, n, n1, and n2 are of type size_type; str, str1, and str2 are of type string; charArray is a character array; ch and delim are of type char; istr is an istream; ostr is an ostream; it1 and it2 are iterators; and inpIt1 and inpIt2 are input iterators. All of these operations except >>, <<, +, the relational operators, getline(), and the second version of swap() are function members.

Constructors:

`string s;`	This declaration invokes the default constructor to construct s as an empty string
`string s(charArray);`	This declaration initializes s to contain a copy of charArray
`string s(charArray, n);`	This declaration initializes s to contain a copy of the first n characters in charArray
`string s(str);`	This declaration initializes s to contain a copy of string str
`string s(str, pos, n);`	This declaration initializes s to contain a copy of the n characters in string str, starting at position pos; if n is too large, characters are copied only to the end of str
`string s(n, ch);`	This declaration initializes s to contain n copies of the character ch
`string s(inpIt1, inpIt2)`	This declaration initializes s to contain the characters in the range [inpIt1, inpIt2)
`getline(istr, s, delim)`	Extracts characters from istr and stores them in s until s.max_size() characters have been extracted, the end of file occurs, or delim is encountered, in which case delim is extracted from istr but is not stored in s
`getline(istr, s)`	Inputs a string value for s as in the preceding function with delim = '\n'

`istr >> s`	Extracts characters from `istr` and stores them in s until `s.max_size()` characters have been extracted, the end of file occurs, or a white-space character is encountered, in which case the white-space character is not removed from `istr`; returns `istr`
`ostr << s`	Inserts characters of s into `ostr`; returns `ostr`
`s = val`	Assigns a copy of `val` to s; `val` may be a string, a character array, or a character
`s += val`	Appends a copy of `val` to s; `val` may be a string, a character array, or a character
`s[pos]`	Returns a reference to the character stored in s at position pos, provided pos < `s.length()`
`s + t` `t + s`	Returns the result of concatenating s and t; t may be a string, a character array, or a character

`s < t, t < s`
`s > t, t > s`
`s <= t, t <= s`
`s >= t, t >= s`
`s == t, t == s`
`s != t, t != s`
	Returns `true` or `false` as determined by the relational operator; t may be a string or a character array

`s.append(str)`	Appends string `str` at the end of s; returns s
`s.append(str, pos, n)`	Appends at the end of s a copy of the n characters in `str`, starting at position pos; if n is too large, characters are copied only until the end of `str` is reached; returns s
`s.append(charArray)`	Appends `charArray` at the end of s; returns s
`s.append(charArray, n)`	Appends the first n characters in `charArray` at the end of s; returns s
`s.append(n, ch)`	Appends n copies of `ch` at the end of s; returns s
`s.append(inpIt1, inpIt2)`	Appends copies of the characters in the range [`inpIt1`, `inpIt2`) to s; returns s
`s.assign(str)`	Assigns a copy of `str` to s; returns s

`s.assign(str, pos, n)`	Assigns to s a copy of the n characters in str, starting at position pos; if n is too large, characters are copied only until the end of str is reached; returns s
`s.assign(charArray)`	Assigns to s a copy of charArray; returns s
`s.assign(charArray, n)`	Assigns to s a string consisting of the first n characters in charArray; returns s
`s.assign(n, ch)`	Assigns to s a string consisting of n copies of ch; returns s
`s.assign(inpIt1, inpIt2)`	Assigns to s a string consisting of the characters in the range [inpIt1, inpIt2); returns s
`s.at(pos)`	Returns s[pos]
`s.begin()`	Returns an iterator positioned at the first character in s
`s.c_str()`	Returns (the base address of) a char array containing the characters stored in s, terminated by a null character
`s.capacity()`	Returns the size (of type size_type) of the storage allocated in s
`s.clear()`	Removes all the characters in s; return type is void
`s.compare(str)`	Returns a negative value, 0, or a positive value according as s is less than, equal to, or greater than str
`s.compare(charArray)`	Compares s and charArray as in the preceding method
`s.compare(pos, n, str)`	Compares strings s and str as before, but starts at position pos in s and compares only the next n characters
`s.compare(pos, n, charArray)`	Compares string s and charArray as in the preceding method
`s.compare(pos1, n1, str, pos2, n2)`	Compares s and str as before, but starts at position pos1 in s, position pos2 in str, and compares only the next n1 characters in s and the next n2 characters in str

`s.compare(pos1, n1,` ` charArray, n2)`	Compares strings s and `charArray` as before, but using only the first n2 characters in `charArray`
`s.copy(charArray, pos, n)`	Replaces the string in s with n characters in `charArray`, starting at position pos or at position 0, if pos is omitted; if n is too large, characters are copied only until the end of `charArray` is reached; returns the number (of type `size_type`) of characters copied
`s.data()`	Returns a `char` array containing the characters stored in s.
`s.empty()`	Returns `true` if s contains no characters, `false` otherwise
`s.end()`	Returns an iterator positioned immediately after the last character in s
`s.erase(pos, n)`	Removes n characters from s, beginning at position pos (default value 0); if n is too large or is omitted, characters are erased only to the end of s; returns s
`s.erase(it)`	Removes the character at the position specified by it; returns an iterator positioned immediately after the erased character
`s.find(str, pos)`	Returns the first position ≥ pos such that the next `str.size()` characters of s match those in `str`; returns npos if there is no such position; 0 is the default value for pos
`s.find(ch, pos)`	Searches s as in the preceding method, but for ch
`s.find(charArray, pos)`	Searches s as in the preceding method, but for the characters in `charArray`
`s.find(charArray, pos, n)`	Searches s as in the preceding method, but for the first n characters in `charArray`; the value pos must be given
`s.find_first_not_of(str, pos)`	Returns the first position ≥ pos of a character in s that does not match any of the characters in `str`; returns npos if there is no such position; 0 is the default value for pos

`s.find_first_not_of` `(ch, pos)`	Searches s as in the preceding method, but for ch
`s.find_first_not_of` `(charArray, pos)`	Searches s as in the preceding method, but for the characters in `charArray`
`s.find_first_not_of` `(charArray, pos, n)`	Searches s as in the preceding method, but using the first n characters in `charArray`; the value `pos` must be given
`s.find_first_of(str, pos)`	Returns the first position ≥ pos of a character in s that matches any character in `str`; returns `npos` if there is no such position; 0 is the default value for `pos`
`s.find_first_of(ch, pos)`	Searches s as in the preceding method, but for ch
`s.find_first_of` `(charArray, pos)`	Searches s as in the preceding method, but for the characters in `charArray`
`s.find_first_of` `(charArray, pos, n)`	Searches s as in the preceding method, but using the first n characters in `charArray`; the value `pos` must be given
`s.find_last_not_of` `(str, pos)`	Returns the highest position ≤ pos of a character in s that does not match any character in `str`; returns `npos` if there is no such position; `npos` is the default value for `pos`
`s.find_last_not_of(ch, pos)`	Searches s as in the preceding method, but for ch
`s.find_last_not_of` `(charArray, pos)`	Searches s as in the preceding method, but using the characters in `charArray`
`s.find_last_not_of` `(charArray, pos, n)`	Searches s as in the preceding method, but using the first n characters in `charArray`; the value `pos` must be given
`s.find_last_of(str, pos)`	Returns the highest position ≤ pos of a character in s that matches any character in `str`; returns `npos` if there is no such position; `npos` is the default value for `pos`
`s.find_last_of(ch, pos)`	Searches s as in the preceding method, but for ch
`s.find_last_of` `(charArray, pos)`	Searches s as in the preceding method, but using the characters in `charArray`

`s.find_last_of` `  (charArray, pos, n)`	Searches `s` as in the preceding method, but using the first n characters in `charArray`; the value `pos` must be given
`s.insert(pos, str)`	Inserts a copy of `str` into `s` at position `pos`; returns `s`
`s.insert(pos1, str, pos2, n)`	Inserts a copy of n characters of `str` starting at position `pos2` into `s` at position `pos`; if n is too large, characters are copied only until the end of `str` is reached; returns `s`
`s.insert(pos, charArray, n)`	Inserts a copy of the first n characters of `charArray` into `s` at position `pos`; inserts all of its characters if n is omitted; returns `s`
`s.insert(pos, n, ch)`	Inserts n copies of the character `ch` into `s` at position `pos`; returns `s`
`s.insert(it, ch)`	Inserts a copy of the character `ch` into `s` at the position specified by `it` and returns an iterator positioned at this copy
`s.insert(it, n, ch)`	Inserts n copies of the character `ch` into `s` at the position specified by `it`; return type is `void`
`s.insert(it, inpIt1, inpIt2)`	Inserts copies of the characters in the range [`inpIt1`, `inpIt2`) into `s` at the position specified by `it`; return type is `void`
`s.length()`	Returns the length (of type `size_type`) of `s`
`s.max_size()`	Returns the maximum length (of type `size_ type`) of `s`
`s.rbegin()`	Returns a reverse iterator positioned at the last character in `s`
`s.rend()`	Returns a reverse iterator positioned immediately before the first character in `s`
`s.replace(pos1, n1, str)`	Replaces the substring of `s` of length `n1` beginning at position `pos1` with `str`; if `n1` is too large, all characters to the end of `s` are replaced; returns `s`
`s.replace(it1, it2, str)`	Same as the preceding, but for the substring of `s` consisting of the characters in the range [`it1`, `it2`); returns `s`

`s.replace(pos1, n1, str, pos2, n2)`	Replaces a substring of s as in the preceding reference but using n2 characters in `str`, beginning at position `pos2`; if n2 is too large, characters to the end of `str` are used; returns s
`s.replace(pos1, n1, charArray, n2)`	Replaces a substring of s as before but with the first n2 characters in `charArray`; if n2 is too large, characters to the end of `charArray` are used; if n2 is omitted, all of `charArray` is used; returns s
`s.replace(it1, it2, charArray, n2)`	Same as the preceding, but for the substring of s consisting of the characters in the range [`it1`, `it2`); returns s
`s.replace(pos1, n1, n2, ch)`	Replaces a substring of s as before, but with n2 copies of `ch`
`s.replace(it1, it2, n2, ch)`	Same as the preceding but for the substring of s consisting of the characters in the range [`it1`, `it2`); returns s
`s.replace(it1, it2, inpIt1, inpIt2)`	Same as the preceding, but replaces with copies of the characters in the range [`inpIt1`, `inpIt2`); returns s
`s.reserve(n)`	Changes the storage allocation for s so that `s.capacity()` ≥ n, 0 if n is omitted; return type is `void`
`s.resize(n, ch)`	If n ≤ `s.size()`, truncates rightmost characters in s to make it of size n; otherwise, adds copies of character `ch` to end of s to increase its size to n, or adds a default character value (usually a blank) if `ch` is omitted; return type is `void`
`s.rfind(str, pos)`	Returns the highest position ≤ `pos` such that the next `str.size()` characters of s match those in `str`; returns `npos` if there is no such position; `npos` is the default value for `pos`
`s.rfind(ch, pos)`	Searches s as in the preceding method, but for `ch`
`s.rfind(charArray, pos)`	Searches s as in the preceding method, but for the characters in `charArray`

`s.rfind(charArray, pos, n)`	Searches s as in the preceding method, but for the first n characters in `charArray`; the value `pos` must be given
`s.size()`	Returns the length (of type `size_type`) of s
`s.substr(pos, n)`	Returns a copy of the substring consisting of n characters from s, beginning at position `pos` (default value 0); if n is too large or is omitted, characters are copied only until the end of s is reached
`s.swap(str)`	Swaps the contents of s and `str`; return type is void
`swap(str1, str2)`	Swaps the contents of `str1` and `str2`; return type is void

THE `list<T>` CLASS TEMPLATE

The `list<T>` class template is provided in the Standard Template Library. The following is a list of the operations defined on `list<T>` objects; n is of type `size_type`; l, l1, and l2 are of type `list<T>`; val, val1, and val2 are of type T; ptr1 and ptr2 are pointers to values of type T; it1 and it2 are iterators; and inpIt1, and inpIt2 are input iterators.

Constructors:

`list<T> l;`	This declaration invokes the default constructor to construct l as an empty list
`list<T> l(n);`	This declaration initializes l to contain n default values of type T
`list<T> l(n, val);`	This declaration initializes l to contain n copies of val
`list<T> l(ptr1, ptr2);`	This declaration initializes s to contain the copies of all the T values in the range [ptr1, ptr2)
`list<T> l(l1);`	This declaration initializes l to contain a copy of l1
`l = l1`	Assigns a copy of l1 to l
`l1 == l2`	Returns true if l1 and l2 contain the same values, and false otherwise

`l1 < l2`	Returns `true` if `l1` is lexicographically less than `l2`—`l1.size()` is less than `l2.size()` and all the elements of `l1` match the first elements of `l2`; or if `val1` and `val2` are the first elements of `l1` and `l2`, respectively, that are different, `val1` is less than `val2`—and it returns `false` otherwise
`l.assign(n, val)`	Erases `l` and then inserts `n` copies of `val` (default `T` value if omitted)
`l.assign(inpIt1, inpIt2)`	Erases `l` and then inserts copies of the `T` values in the range [`inpIt1`, `inpIt2`)
`l.back()`	Returns a reference to the last element of `l`
`l.begin()`	Returns an iterator positioned at the first element of `l`
`l.empty()`	Returns `true` if `l` contains no elements, `false` otherwise
`l.end()`	Returns an iterator positioned immediately after the last element of `l`
`l.erase(it)`	Removes from `l` the element at the position specified by `it`; return type is `void`
`l.erase(it1, it2)`	Removes from `l` the elements in the range [`it1`, `it2`); return type is `void`
`l.front()`	Returns a reference to the first element of `l`
`l.insert(it, val)`	Inserts a copy of `val` (default `T` value if omitted) into `l` at the position specified by `it` and returns an iterator positioned at this copy
`l.insert(it, n, val)`	Inserts `n` copies of `val` into `l` at the position specified by `it`; return type is `void`
`l.insert(it, inpIt1, inpIt2)`	Inserts copies of the `T` values in the range [`inpIt1`, `inpIt2`) into `l` at the position specified by `it`; return type is `void`
`l.insert(ptr1, ptr2)`	Inserts copies of all the `T` values in the range [`ptr1`, `ptr2`) at the position specified by `it`; return type is `void`
`l.max_size()`	Returns the maximum number (of type `size_ type`) of values that `l` can contain

`l.merge(ll)`	Merges the elements of `ll` into `l` so that the resulting list is sorted; both `l` and `ll` must have been already sorted (using `<`); return type is `void`
`l.push_back(val)`	Adds a copy of `val` at the end of `l`; return type is `void`
`l.push_front(val)`	Adds a copy of `val` at the front of `l`; return type is `void`
`l.pop_back()`	Removes the last element of `l`; return type is `void`
`l.pop_front()`	Removes the first element of `l`; return type is `void`
`l.rbegin()`	Returns a reverse iterator positioned at the last element of `l`
`l.remove(val)`	Removes all occurrences of `val` from `l`, using `==` to compare elements; return type is `void`
`l.rend()`	Returns a reverse iterator positioned immediately before the first element of `l`
`l.resize(n, val)`	Sets the size of `l` to `n`; if `n > l.size()`, copies of `val` (default `T` value if omitted) are appended to `l`; if `n < l.size()`, the appropriate number of elements is removed from the end of `l`
`l.reverse()`	Reverses the order of the elements of `l`; return type is `void`
`l.size()`	Returns the number (of type `size_type`) of elements `l` contains
`l.sort()`	Sorts the elements of `l` using `<`; return type is `void`
`l.splice(it, ll)`	Removes the elements of `ll` and inserts them into `l` at the position specified by `it`; return type is `void`
`l.splice(it, ll, it1)`	Removes the element of `ll` at the position specified by `it1` and inserts it into `l` at the position specified by `it`; return type is `void`
`l.splice(it, ll, it1, it2)`	Removes the elements of `ll` in the range `[it1, it2)` and inserts them into `l` at the position specified by `it`; return type is `void`
`l.swap(ll)`	Swaps the contents of `l` and `ll`; return type is `void`
`l.unique()`	Replaces all repeating sequences of an element of `l` with a single occurrence of that element; return type is `void`

Index